THE SCIENCE AND PRACTICE OF PIG PRODUCTION

THE SCIENCE AND PRACTICE OF PIG PRODUCTION

COLIN WHITTEMORE NDA BSc PhD DSc FIBiol

Longman
Scientific &
Technical

To family and friends

Longman Scientific & Technical,
Longman Group UK Ltd,
Longman House, Burnt Mill, Harlow,
Essex CM20 2JE, England
and Associated Companies throughout the world.

First published 1993

British Library Cataloguing in Publication Data
A catalogue record for this book is available from the British Library

ISBN 0–582–09220–5

Set in 10 on 12½ pt Plantin Roman

Printed in Singapore

CONTENTS

Preface and
Acknowledgements   xvii

Chapter 1 Introduction   1

Chapter 2 Pig meat and carcass quality   4

Introduction   4
Marketing   5
Consumption patterns   8
Pig meat quality   12
Shape and meatiness   21
Meat from entire male pigs   24
Carcass quality and grading standards   27
Carcass yield: killing-out percentage (ko%)   35
Choice of carcass weight and carcass fatness   36
Problems of variation   40
Manipulation of carcass quality by control of feeding   42
    Protein   44
    Level of feed (energy)   44

Chapter 3 Growth and body composition changes in pigs   48

Introduction   48
General composition of the body   50
Growth curves   53
Lean and protein growth   56

Changing the shape of the potential growth curve 61
Fat and lipid growth 62
The special case of early growth 65
Weight loss by negative fat growth 67
Weight loss by negative protein growth and
compensatory or catch-up growth 73
Growth response to feed supply 75
Hormonal manipulation of growth 80

Chapter 4 Reproduction 84

Introduction 84
Hormones of the oestrous cycle 86
Puberty 91
Ovulation 92
Fertilization 94
Artificial insemination 98
Pregnancy 101
Parturition 106
Lactation 108
Mammary structure 108
Milk synthesis and production 112
Natural suckling and milk ejection 117
*Initiation – Passive withdrawal – Milk ejection –
Termination* 118
Lactation length 121
The interval between weaning and farrowing 124
Productivity summary 127

Chapter 5 Aspects of pig behaviour and welfare 129

Introduction 129
Self-choice feeding 130
Behaviour associated with reproduction 134
Mating 134
Parturition 137
Nursing and sucking 139
Competitive and aggressive behaviour 145
Locomotion, ingestion and elimination 149
Locomotion 150

Ingestion	151
Elimination	153
Injuries and mutilations	154
Behavioural response to stress	157
Welfare	162

Chapter 6 Development and improvement of pigs by genetic selection — 167

The early years	167
The creation and improvement of the twentieth-century breed types	173
Pig improvement strategies	178
First principles and basic tenets	178
Additive genetic variance ($\sigma^2 A$ – Non-additive genetic variance	179
Using hybrid vigour	184
The incorporation of new genes by importation	188
Positive selection within a breeding population	191
Heritability (h^2) – The selection differential (S) – Generation interval (GI) – Pure-breeding (nucleus) herd size – Accuracy of measurement and effectiveness of candidate comparison – Genotype:environment interaction – Index selection – Selection objectives – The progeny test – The performance test – The test regime – Future objectives for within-population selection	192
Breeding plans	224
Use of central testing facility by breeders	224
The independent pedigree (nucleus) breeder of limited herd size	225
The general approach of the breeding company	227
A particular breeding company	231
The commercial breeder/grower	234
The commercial grower	236
New biotechnology techniques	237
Gene transfer	239
Exogenous hormone manipulation	241
Embryo transfer and cloning	241

d+E

C+P

Early life markers   242
Best linear unbiased prediction   243
Sex determination   243

Chapter 7   Disease prevention   245

Introduction   245
Density of stocking   247
Lameness   249
Feed pathogenic contaminants   250
Feed mycotoxins   251
Porcine stress syndrome (PSS)   252
Stomach ulcers, colitis, rectal prolapse, strictures,
atresia ani, hernia and constipation   252
Low numbers of viable sperm   253
Sucking piglet mortality   255
Cannibalism and aggression   256
Infertility in the sow   257
Internal and external parasites   260
Barrier maintenance against infective agents   261
Disease monitoring   264
Diseases associated with infections of the digestive
tract and with diarrhoea   265
  Colibacillosis   265
  Swine dysentery   267
  Transmissible gastro-enteritis (TGE)   267
  Epidemic diarrhoea   268
  Paratyphoid   268
  Porcine intestinal adenomatosis (PIA)   268
Specific diseases of the respiratory tract   269
  Mycoplasma (enzootic) pneumonia   269
  Atrophic rhinitis   270
  Actinobacillus (haemophilus) pleuro pneumonia   270
  Glasser's disease   271
  Swine influenza   272
  General control of respiratory disease   272
Diseases associated with reproductive problems
in the breeding herd   272
  Parvovirus   272
  Mastitis   273
  Aujeszky's disease (pseudorabies)   273

Porcine reproductive and respiratory syndrome
(PRRS)   275
Leptospirosis   275
Streptococcal meningitis   275
Exudative epidermitis (greasy pig disease)   276
Erysipelas   276
African swine fever   277
Mulberry heart disease   277
Practicalities of prevention and avoidance   277

Chapter 8   Energy value of feedstuffs for pigs   280

Introduction   280
Measurement of the energy economy of pigs   282
The possibilities for generalization   287
Energy classification of pig feedstuffs   288
Determination of digestible energy by use of
live animal studies   290
Estimation of digestible energy from chemical
composition   295
The roles of fibre and fat   298
Guide values for energy content of pig feedstuffs   303

**Chapter 9   Nutritional value of proteins and amino
acids in feedstuffs for pigs**   304

Introduction   304
Digestibility of protein   306
Measurement of digestibility and retention of
crude protein (N x 6.25)   310
Factors influencing digestibility of protein   312
    Heat damage   313
    Fur, feather and hide   313
    Abrasion   313
    Rate of passage   314
    Protection and comminution   314
    Anti-nutritional factors   314
    Site of digestion   316
Ileal digestible amino acids   317
Available amino acids   320
Guide values for protein and amino acid
content of pig feedstuffs   322

Chapter 10 Value of fats in pig diets   323

Introduction   323
Value of fats in pig diets   328
Incorporation of diet fats into animal carcass fat   329
Diet compounding using fats and oils   330

Chapter 11 Energy and protein requirements for maintenance, growth and reproduction   333

Introduction   333
Energy   334
   Maintenance   335
   Production   336
   Reproduction   339
Protein   340
   The utilization of digested protein: ideal protein   341
   Maintenance   348
   Production   348
   Reproduction   349
Example calculations for the energy and protein requirements of growing and breeding pigs   349
   Naive calculation of the energy and protein requirements of a young pig   350
   Naive calculation of the energy and protein requirements of a growing pig   351
   Naive calculation of the energy and protein requirements of a finishing pig   351
   Naive calculation of the energy and protein requirements of a pregnant sow   352
   Naive calculation of the energy and protein requirements of a lactating sow   353
   Recommended energy and protein concentrations in diets for different classes of pigs   354
Energy and protein requirements of the breeding boar   355

Chapter 12 Requirements for water, minerals and vitamins   357
Introduction   357
Water   361

Minerals	364
Calcium and phosphorus	364
Sodium, chlorine and potassium	366
Magnesium	366
Trace mineral elements	366
Vitamins	367

Chapter 13 Appetite and voluntary feed intake   **370**

Introduction	370
Appetite as a consequence of nutrient demand	373
Negative feedback: heat losses	374
Positive feedback: relative productivity above maintenance	375
Influence of diet nutrient density	375
Appetite as a consequence of gut capacity	377
Estimation of feed intake in young and growing pigs	378
Estimation of feed intake in sows	382
Appetite as a consequence of feed characteristics	385
Specific dietary ingredients	385
Feed enzymes	385
Environmental gases	388
Addition of dietary organic acids	389
Addition of micro-organisms	389
Addition of antibiotics	390
Flavours, flavour-enhancers and masking agents	391
Form of feed and method of presentation	391

Chapter 14 Diet formulation   **393**

Introduction	393
Energy concentration of the diet	394
Protein:energy ratio (g CP/MJ DE)	395
Protein value (V)	398
Response to dietary concentration of DE	399
Response to protein:energy ratio	400
Response to protein value (V)	404

Relating feed ingredients to pig requirements 404
    Variety and number of feed ingredients 406
    Choice of feedstuff ingredients for particular diets 408
Unit value of nutrients from feedstuffs 413
Number of different diets required 415
Enhancement of the precision of formulation 417
    Better definition of dietary utilizable energy 418
    Better definition of dietary utilizable protein 420
    Conclusion to possibilities for enhancement
    of precision of formulation 423
    Benefits from enhanced precision of formulation 424
Least-cost diet formulation by computer 426
    Description of raw material feedstuff ingredients 427
    Description of the nutrient specification for the diet 428
    Computerized least-cost diet formulation 428
    Changes of ingredients 430

Chapter 15 Optimization of feed supply to growing pigs and breeding sows 432

Introduction 432
Growing pigs 432
    Interactions with level of feed supply 435
    Slaughter weight – Feed conversion efficiency –
    Response to an increase in feed intake –
    Interaction with genotype and sex (potential
    lean-tissue growth rate and predisposition
    to fatness) 435
    Ration scales for growing pigs 440
Breeding sows 444
    General pattern of live weight and fatness
    changes in breeding sows 444
    Feed supply to first mating 448
    Feed supply in pregnancy 449
    Feed supply between weaning and conception 454
    Feeding in lactation 455
    Condition scoring 459
    Rationing scales for breeding sows 461
Feeding the sucking pig 465
Ration control by diet dilution 466
Conclusions 467

Chapter 16 Product marketing 469

Introduction 469
Marketing pig feeds 469
Marketing fully compounded diets 471
Technical support 474
Marketing proprietary protein concentrates 476
Technical support 478
Marketing supplements and pre-mixes 478
Technical support 479
New products 479
Customer complaints 480
Breeding stock supply 482
The importance of the customer in product design 482
Types of breeding stock 483
Selection of breeding stock for sale 485
Description of breeding stock – Health status –
Receipt of breeding stock on to the farm 486
New products 488
Pricing 490
Technical support 491
Supply schedules for replacement stock 492
Marketing slaughter pigs 493
Weight at sale for carcass meat 495
Fatness at point of slaughter 496
Choice of market outlet 497
Marketing opportunities: case calculations 499
Case 1: an alternative contract – Case 2: pigs
which are too fat – Case 3: improving upon
excellence – Case 4: difference between diets –
Sensitivities 499
Conclusion 503

Chapter 17 The environmental requirements of pigs 504

Introduction 504
Temperature 505
Floor surface 506
Behaviour 506
Convection 507
Cooling and heating 508

Metabolic response to cold and heat | 510
Influence of insulation | 511
Ventilation | 511
Space | 514
Summary of housing requirements | 516
Waste | 516
Calculation of housing requirements | 519
Outdoor pig production | 522
Feed delivery systems | 525
Sucking pigs | 525
Weaned pigs | 526
Growers and finishers | 527
Lactating sows | 528
Pregnant sows | 529
Housing layouts | 531

Chapter 18 Production performance monitoring | 550

Introduction | 550
Records – on the pig unit | 550
Naive records | 551
Records – in the office | 553
Records for selection | 555
Commercial records | 555
Targets | 557
Routine control | 563

Chapter 19 Simulation modelling | 567

Introduction | 567
Development of quantitative response prediction simulation models | 568
Diagnosis and target setting | 571
Information transfer | 571
Model-building approaches | 572
Potential growth: limitations to growth | 576
Nutrient (energy and protein) yield from the diet | 577
Energy | 577
Protein | 579
Nutritional requirement for energy | 580

Nutritional requirement for protein 583
Feed intake 585
  On-farm determination 586
  From calculation of perceived nutrient (energy) needs 586
  From knowledge of gut capacity limits 587
  Influence of environmental temperature 588
  Influence of stocking density 588
  Empirical estimates 589
  Breeding sows 589
Composition and growth of the live body of growing pigs 591
Carcass quality 592
Flow diagram 595
Attempting a simulation model to predict the performance of breeding sows 595
Sow weight and body composition 596
Piglet birth weight 597
Survivability 601
Litter size (numbers of piglets) at birth 603
Lactation 604
  Weaning weight 607
Weaning-to-conception interval 608
Maternal weight and fatness 611
Flow diagram 613
Physical and financial peripherals 613

Chapter 20 Conclusion 617

Appendices 620

1. Nutritional guide values of some feed ingredients for pigs 620
2. Guide nutrient specifications for pig diets 630
3. Guide feedstuff ingredient compositions of pig diets 632
4. Description files for raw materials 634

Bibliography 635
Index 637

PREFACE AND ACKNOWLEDGEMENTS

This book is the fifth and by far the most extensive that I have written. It results from a friendly gibe that I should prepare a *proper book*; that is, with the objectives of adequate scale and breadth of coverage to deal comprehensively with the full and international spectrum of the science and practice of pig production.

The temptation to hold a conference and publish the proceedings, or to gather a consortium of authors and edit a many-author title, was resisted. I feel that such books lack cohesion and purpose and fail to demonstrate any integrating philosophy. Here I wish to take the reader through the subject, examining the knowledge offered from a perspective which, although multidimensional, nevertheless springs from one individual source. Such an ambition requires single authorship.

Necessarily, one author and the concept of knowledge in both breadth and depth are to some extent contradictory. The dilemma may be resolved only through learning from others and benefiting from their counsel. This I have done for very many happy years of production practice, university learning, scientific research work and consultancy. I am happy also to be able to recognize formally all those colleagues of international repute who have been so kind as to painstakingly read, referee and, where necessary, amend the chapters which follow.

I owe, therefore, a great debt to all those from whom I have accumulated information, knowledge and wisdom.

However, this precious resource has been subject to filtration, emendation and sometimes rejection, and it has been copiously added to through many years of careful and patient work from the Edinburgh group of scientists.

I acknowledge gratefully that a considerable part of the contents of this book stem originally from the minds and work of others. So whilst thanking from my heart my colleagues in farming practice, in the agriculturally allied trades and in teaching and research in universities and research institutes world-wide, I alone must bear the responsibility for the final form of the book.

Finally, I acknowledge the unstinting moral and practical support of my wife and family and of my secretary throughout the preparation of this text.

COLIN T WHITTEMORE
Edinburgh
Spring 1992

CHAPTER 1

INTRODUCTION

The prime agricultural purpose of the pig is to provide human food (fresh, cured or processed), so pig production is necessarily preoccupied with the quality of the pig meat and the efficiency of its production. One aspect of the efficiency of the pig is its rapid growth rate, another is its high reproductive rate and a third is its omnivorous and accommodating eating habits. Historically, the pig was often seen as a scavenger; a rural scavenger in fields and woodlands, and an urban one using human food by-products and waste. These roles have ensured that the pig has been used as a domesticated agricultural animal widely throughout the world but usually in small population sizes such as would facilitate by-product utilization and encourage simple husbandry practices.

Recent history has seen the demand for high-quality pig meat products rise world-wide due to the increasing world population and improving human nutritional aspirations. This has raised the profile of the growth and reproductive characteristics of the pig. Simultaneously, improvements in other branches of agricultural practice have made available substantial quantities of cereal, legume and oil-seed grains not appropriate – or not needed – for human consumption. The industrialization of human food preparation (such as the manufacture of farinaceous products from cereal flours, and of margarines and other derivatives from vegetable oils) has added to the changing environment by turning the availability of food by-products and waste away from the local and diversified domestic level and towards centralized manufacturing points.

These are but some of the forces that have changed the nature, and indeed the objectives, of pig production over recent decades. Especially there has been created a large-scale integrated agricultural industry, producing substantial quantities of high-quality meat products which may be readily traded within and across international boundaries. The agricultural and biological sciences have been well set to serve pig production over its recent evolution and the result has been a staggering example of the successful application of science into industry to the benefit of humankind. The modern pig industry bears little or no resemblance to that of only a few decades ago; either in its structure, or in its method of trading, or in its definitions of quality in the end-product, or in its breeding, feeding, housing, health and management.

To the observer, the major changes are perhaps perceived in terms of sophistication and scale, and, associated with the latter, intensification. Coping with these changes has demanded of science, and science has provided, new husbandry practices, new understandings of growth, reproduction and health, new appreciations of welfare and the environment, new nutritional approaches, and modern genetic improvement techniques. But above all has been the essential placing of the application of science into the context of latter-day marketing and trading. Biological response must be interpreted not only in terms of physical outcomes but, most importantly, in terms of financial consequences.

While trends in pig production have had some degree of commonality world-wide, there remains tremendous diversity in international production practice and end-product objectives. Thus, although a high proportion of the sciences cognate to pig production are transferable across national barriers, the *way* the scientific information is used may need to vary widely. The most useful aspects of the scientific information base are those that address basic causes of biological and economic response, and not those that lay out the responses themselves in the form of rules and recommendations.

Throughout, the book argues causality and logic before presenting outcomes and tries to avoid the bald presentation of the results of this or that production investigation as if the latter were to embody some form of universal truth; which, of course, it would not. It is fundamental to an international text that along with the description of phenomena must go their understanding. Universality lies in the depth of the understanding of biological and economic processes, not necessarily in the accuracy of their description. Nevertheless, pig production is a quantitative as well as a qualitative activity. Meat of given quantity and definable characteristics requires to be produced from feedstuffs of quantifiable nutritional value when given to specific pig genotypes in describable environmental, managemental and economic circumstances. Where possible, therefore, factors impinging upon the production process are expressed in hard and quantifiable terms. After all, it is only through quantification that the absolute and relative importances of the various aspects of the sciences and practice of pig production can be compared, judged and ultimately used.

The world's pig industry is a success story founded in the application of science to management. The extent of the success thus far is well illustrated in Figures 1.1 and 1.2. But there is no reason for complacency, and a great deal more can be done to improve the level of efficiency of output and the quality of the end-product.

Most pig production enterprises world-wide are under-achieving, and there is a considerable – usually unnecessary – disparity in production efficiency between different nations. Compared to the presently known potential, the world's pig industry falls far short. Application of scientific principles and sound production practices can redress that shortfall.

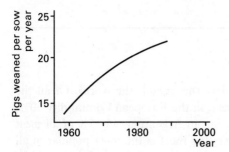

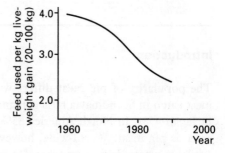

Figure 1.1 Rate of increase in sow productivity (northern European data).

Figure 1.2 Rate of improvement in feed usage per kilogram live-weight gain. The seventies saw the introduction of high-nutrient-density diets in significant quantities (northern European data).

CHAPTER 2

PIG MEAT AND CARCASS QUALITY

Introduction

The popularity of pig meat differs widely throughout the world. Of all the meat eaten in Scandinavia 60% is pigmeat, in the European Community 50%, in Japan 45%, in North America 35%, but in Argentina only 5% of all meat eaten is pig meat. World-wide, however, pig meat is the most popular of all available meats. It may be taken for granted that pig meat product must first be *needed* (ie there must be a market) and next be *safe* to eat. It must also be *efficiently produced* (competitively priced) and of the required level of *quality*. But increasingly meat production must satisfy further criteria in relation to production *method* and production *environment*, which must be both *sustainable* and *ethical*.

Within recent history, the pig has been bred both for supreme fatness (the required commodity being lard) and supreme leanness (the required commodity being muscle). Whether the acceptable pig carcass be fat or lean depends much upon national predilection. As industrialization develops, the desire for lean meat appears to dominate the definition of carcass quality. Eastern European, Japanese and North and South American tastes are, like western Europe, demanding progressively lower levels of meat-fat content. Asian pigs, numerous in China and the Pacific countries, are over-fat by western standards, and it is likely that some desire to reduce fatness will take place there also. Nevertheless, one nation's lean pig remains another nation's fat pig.

 In Europe, the pig meat market used to be for a mass product at a low price. Indeed, general advancement of knowledge in pig production has resulted in the real cost of pig meat now being a mere 30% of what it was 40

4

years ago. Fat reduction, now achieved almost to its optimum level, was initiated originally not by consumer discrimination against fat but by the economics of pig production and pig processing, both of which benefit from a lean carcass. Pig markets still tend to divide into two types: one uses pig meat as a substrate for a huge variety of manufactured meat products and satisfies a continuing need for low-cost meat; while the other, of increasing importance, attends to the consumer supply of pork and bacon meat of high eating quality. This latter market needs carefully prepared cuts, a low level of subcutaneous fat, an adequate level of intramuscular fat and a large mass of lean tissue giving a high perception of eating pleasure. The requirement is meat that is juicy, tender and tasty, avoiding all suggestion of dryness, paleness, taint or toughness.

Producers may be reluctant to accept the restraints asked of them by a meat trade concerned primarily with product quality. Conversely, the trade may be slow to allow producers to profit from the benefits of new techniques and streamlined production practices. Such problems will remain with an industry which continues to be based upon adversarial trading and mistrust at each interface of the production chain. The successful future of pig industries the world over is likely to require the benefits of integration between breeder, producer, feed compounder, meat packer and retailer. The producer needs effective lines of communication with the market. If that communication generates information which is useful, both producer and meat trader will profit.

Marketing

Live pigs are marketed at any time subsequent to weaning (which can be between 6 and 20kg and between 3 and 8 weeks of age). Pigs may be sold to slaughter for meat, or grown on to a greater weight. Pig meat can come from sucking pig at one extreme and, at the other, from 180kg animals destined for the production of large Italian Parma hams. National consumption patterns cause pigs to be finished and consumed at a wide variety of weights between these extremes. As the mature size of many Asian pig types is so much less than European types, slaughter weights are usually less. In many European countries 50–80kg live weight is the traditional *butcher* range for the production of small, fresh pork joints bought and consumed locally, while in Japan, Australia, North and South America and all the countries of Europe, 90–120kg live weight is the traditional range for *industrial processing* at meat packing plants which distribute fresh pork joints, cured bacon products and delicatessen lines. Animals used for the production of hams which are dry-cured by slow and natural processes (such as prosciutto in Italy [140–170kg live weight] and cerrano in Spain), are usually finished to even greater live weights. Cull sows weighing 120–300kg may command a useful price according to the state of the market. Meat from cull sows can be satisfactorily incorporated into a range of products, and it can be cost-effective to sell culled animals in good body condition when the price is satisfactory. Adult male pigs which have been used for breeding are inappropriate for human consumption, the meat being coarse and often tainted. This criticism does not apply to fast-grown entire males of European breed types slaughtered at less than 110kg.

Prime meat is usually found at around 30–60% of mature size (Figure 2.1). At less than 30% there is a tendency towards lack of flavour, whilst at greater than 60% there is a tendency towards lack of tenderness. Pig types of larger mature size merit a higher slaughter weight and it is noticeable that conventional slaughter weights in the European pig industry have been increasing over recent years. Most pigs used for processing and meat packing in industrialized countries have a mature weight of between 250kg and 350kg, while in Asian types mature weight can be as low as 150kg.

The primary pig production industry may segregate into specialized farm types, often breeders or feeders. Specialized breeding units will sell weaned or store pigs at weights of 20–40kg to grow-out units, which will feed them to finished weight. Such a system allows concentration of resources, skills and preferences. Frequently, breeder units are small, being part of a mixed family farming enterprise. Grow-out units, on the other hand, tend to be larger and more industrialized, buying from many sources of supply. Such specialization is now seen as leading to more problems than solutions, consequent primarily upon the ill-effects of transport and trading in young animals, variability in type of pig supplied, and disease incursion. Modern enterprises are best structured as breeder/finisher operations where a single unit both breeds and grows the pigs to final specification. Single-step operations also allow a more effective relationship between ultimate customer and primary producer; thus

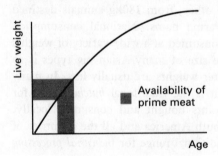

Figure 2.1 Prime meat is found from pigs slaughtered at between 30% and 60% of mature size.

meat quality aspects as perceived by the meat packer and retailer may be effectively relayed back to the breeder. Choice of pig type or hybrid by the breeder is of vital interest to grower, packer and retailer. Any discontinuity of production flow, such as an industry based on separate businesses dealing with the breeding and feeding aspects of the production process, can only serve to obscure an optimum policy and hold back the development of the industry.

Within any one nation, or group of inter-trading nations, the pig industry has potential for expansion or contraction. The extent of this potential is directly linked to the number of producers who have alternative sources of income available to them. When pig prices are good the size of the breeding herd is expanded and new herds are begun. This may be achieved readily by retaining for breeding purposes female stock originally destined for slaughter, or by the purchase of extra hybrid breeding stock replacements. These animals will produce young some 6 months later, and a further 6 months downstream additional slaughter pigs become available for sale. At this point supply is enhanced and prices will ease back. The elasticity of demand for pig products is such that although customers will purchase more pig meat as prices fall and it becomes cheaper, the ability to eat more does not keep pace with the falling price. The process of falling pig prices will continue to take place for approximately a further year or so, at which point breeding herds will begin to be reduced in both size and number. This process may take another year before a shortage of available slaughter pigs becomes evident and prices begin to strengthen again. This simplistic pig cycle (Figure 2.2) is obfuscated by fresh meat butchers equalizing prices of various meats, and also by fluctuations in the world supply of pig-feed ingredients. Grain and soya prices are relatively independent of pig price because grains are used for purposes other than the feeding of pigs. When pig-feed ingredient prices are

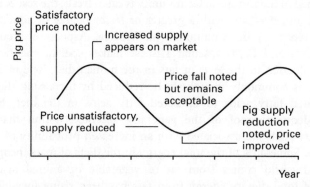

Figure 2.2 The cyclical nature of pig prices. These are further subject to influence from price of competitive meats, prevailing feed price and the availability of capital.

low, producers can tolerate a reduced pig carcass price for a longer time before needing to retract. A combination of favourable international grain prices and favourable pig carcass prices will produce a profit bonanza and an accelerated rate of expansion.

The ebb and flow of the pig cycle as a result of pig and grain prices is more apparent in countries where capital expenditures do not figure largely. For an industrial-based intensive pig industry with high capital investment in its buildings, the availability of money at satisfactory interest rates acts as a regulator on the entrepreneur's ability either to expand when times are good or to contract when they are not.

Consumption patterns

The estimated world consumption of animal product (millions of tonnes) is: milk 400 (steady), pig 63 (rising), beef and veal 48 (steady), poultry 33 (rising), eggs 20 (steady), farmed fish 8 (rising), sheep and goat 6 (steady). Pig meat accounts for more than 40% of the world's meat consumption and is produced from about 850 million pigs. Of these pigs one-third are in China, 10% are in Russia, and 5% are in the USA. Next in line as major pig producers are Brazil, Germany and Japan. The European Community has about 110 million pigs from which come 14 million tonnes of pig meat. This gives the European Community more than 20% of world pig meat production. Within the Community more than 30% of pig meat production is in Germany, about 12.5% is in each of France, The Netherlands, Italy and Spain, while the UK and Denmark have around 10% each. About 25% of pig meat consumed in the European Community is eaten fresh, the rest being cured or processed into products with a greater or lesser degree of added value.

The UK report of the Committee on Medical Aspects of Food Policy relating to Diet and Cardiovascular Disease makes specific mention of the need for a reduction in the level of fats in the human diet. The incidence of heart disease is commonly thought to be reduced by increasing the ratio of polyunsaturated fatty acids to saturated fatty acids in the diet, but there remains inadequate proof of the generally accepted notion that dietary saturated fatty acids are associated with an increased probability of coronary heart disease. Some 40% of the total energy in the diets of many people in the industrialized world comes from fat of vegetable or animal origin. The proportion of total dietary energy from fats has been rising inexorably over recent years; and it is perhaps this simple statistic which is the more relevant.

Meanwhile, there has been a simultaneous dramatic reduction of fat in pig meat down from about 30% of the total carcass in the 1960s to about 15% or less in a 100kg bacon carcass of today. Subcutaneous backfat measurements have been reduced by genetics, nutrition and feeding management by about 0.5mm annually. Pig subcutaneous fatty tissue contains about 70% lipid, and two-thirds or more of all fat on a pig carcass may be found in the subcutaneous depots. For top-grade bacon, backfat depth targets at the P2 site for pigs of 90kg live weight can be as low as 12mm. Given the potential of pig meat to be extremely lean, it may be assumed that on health grounds alone the consumption of pig meat and pig products will continue to further increase.

Medical wisdom advises that fat consumption should fall to around 30% of total diet energy, and that the polyunsaturated:saturated fatty acid ratio in dietary fat should be around 1.0. Presently in many countries of the western world the ratio is about 0.3. The Committee on Medical Aspects of Food Policy has suggested 0.45 as an achievable target. As about half of all diet fat comes from animal products (including milk, milk products, eggs and meat), and as some meat fats have a poor polyunsaturated:saturated fatty acid ratio, there may be consumer benefit in the identification of meat and meat products low in fat, but what fat there is being high in polyunsaturated fatty acids.

Notwithstanding, some fat is necessary in meat to allow optimum meat quality, particularly with regard to succulence, taste, texture and cookability. Linoleic acid (18:2) is an essential dietary nutrient for normal body function, being the precursor – in a somewhat rate limited pathway – for gamma linolenic (18:3) and, in turn, dihomo gamma linolenic (20:3) and arachidonic (20:4) acids; the series being generally involved in membrane function, phospholipid production, hormone manufacture, cardiac function and blood flow. (In '18:2', etc., the number before the colon refers to the number of carbon atoms in the chain, and the number after the colon refers to the number of double bonds. The greater the number of double bonds, the less saturated the fat.) Eicosapentaenoic (20:5) and docosahexaenoic (22:6) acids are found in fish oils and considered particularly helpful in reducing the incidence of heart disease, for blood flow and for brain and associated functions. Because the metabolism of the pig can move fatty acids from dietary sources into the pig's own lipid tissues, the incorporation of health-positive fatty acids from pig-feed fat into pig meat is feasible.

Modern lean strains of pig can produce meat which is itself low in fat; *semitendinosus* muscle contains about 3% fat and *longissimus dorsi* muscle about 1%. The polyunsaturated:saturated fatty acid ratio of the lipid found in lean pig meat is around an excellent 0.8 (that is, close to the ideal medically recommended, and above the target for the diet as a whole), provided the pig

9

diet has not been supplemented with tallow. The presence of unsaturated oils such as soya oil in the diet of the pig is beneficial in improving the proportion of polyunsaturated fatty acids in the pig meat. It is a characteristic of pig fat that it will reflect the fatty acid composition of the diet eaten, thus the C18:2 content of the fat of a pig with soya oil in its diet will be almost three times that of a pig with tallow in its diet. It is also the case that the fat in leaner pigs has a higher polyunsaturated:saturated ratio than the fat in fatter pigs. As leanness improves, so does the fatty acid ratio. It is evident from recent studies at Edinburgh that the meat from modern lean strains of pig not only reaches the medical criteria for low overall fat intake, and for preferred polyunsaturated:saturated fatty acid ratios, but also pig meat can actually exceed the set medical targets and thereby bring about an enhancement of the health value of the diet as a whole.

Most advanced European countries eat about 70–90kg of total meat per person per year; this is about 30% less than in the USA. In the UK much less pig meat is eaten, in both absolute and proportional terms, than in Denmark, Germany or The Netherlands (Table 2.1). In the UK the consumption of imported cured pig meat and pig-meat products accounts for 30–40% of total consumption and comes from a variety of sources, including Denmark, The

Table 2.1 Pig meat consumption world-wide (1990)

	Population (millions)	Pig meat consumption (kg per head)	Notes
Denmark	5	45	1, 2, 4
Germany	80	60	1, 3
The Netherlands	12	40	1, 4
United Kingdom	54	25	1, 4
North America	400	30	2, 4
South America	300	7	
USSR	300	15	
China	1100	20	
Japan	120	15	4

Notes
1. In the last 40 years pig meat consumption in Europe has nearly doubled.
2. Of total meat consumed, pig meat totals 65% in Denmark and 35% in North America.
3. It is possible that in some countries the consumption of pig meat is nearing saturation. The German people eat about as much pig meat as the British people eat all meats combined.
4. The Netherlands and Denmark are more than 200% self-sufficient and are massive exporters. The United Kingdom is 70% self-sufficient and is a significant importer. North America and (especially) Japan are also significant pig meat importers (the latter mostly from Taiwan). The European Community as a whole is 100% self-sufficient.

Netherlands, Poland and Germany.

The annual average meat consumption per person in the European Community is 5kg of sheep meat (falling), 25kg of beef meat (steady), 15kg of poultry (rising) and 35kg of pork and bacon (rising). The European Community is 100% self-sufficient for pig meat overall. The total tonnage of production from the European Community is 12.8 million tonnes, comprising West Germany 3.2, The Netherlands 2.0, France 1.7, Spain 1.3, Denmark 1.2, Italy 1.1, UK 1.0, Belgium and Luxembourg 0.8, Portugal, Greece and Eire 0.5. The European Community has around 12 million sows from which this production comes.

In terms of value for money, pig meat is a better buy than beef or lamb, and rivals chicken and turkey. Pig meat is highly versatile, being used for fresh meat joints, cured joints, bacon, cooked meats (fresh and cured), processed products, pâté, barbecue ribs and many specialized lines of sausages, delicatessen items and so on. The efficiency of transfer of animal

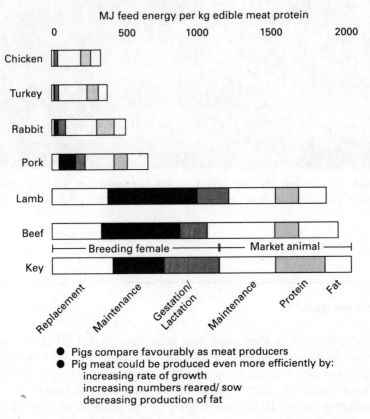

● Pigs compare favourably as meat producers
● Pig meat could be produced even more efficiently by:
    increasing rate of growth
    increasing numbers reared/ sow
    decreasing production of fat

Figure 2.3 Efficiency of transformation of feed to meat (from EAAP long-range study on the future of pig production in Europe).

feed into pig meat is high (Figure 2.3) and continues to improve with new production technology. Increasing the range of pork and bacon products available to consumers (Figure 2.4) is probably the most appropriate way now open for further expansion of the pig meat sector.

Figure 2.4 Sausage smoking with a natural oak wood fire, Spain. Increasing the range of pork products is the best way of increasing consumption in some countries.

Pig meat quality

In comparison with sheep and beef meat there is less effect of anatomical position of the joint on the quality of pig meat. However, back and loin joints (*longissimus dorsi* muscle) are particularly appropriate for using fresh, grilled or fried, while the ham is good for roasting and, of course, for curing. Imaginative cutting techniques (Figure 2.5) do much to increase the product range and the consumption of pig meat. Modern jointing methods trim the subcutaneous fat on the carcass to the level required, use natural muscles as the basis for jointing, produce a range of products suited to current demand and present the meat in a convenient way for handling and cooking.

Fat levels demanded on pork products vary widely amongst pig-eating nations, but there is an almost universal trend toward reduction in desired fat levels unavoidably associated with the meat joint. Often the public demand for lean meat is greater than the indigenous pig type can provide, as is the case in North America. In this event, considerable fat trimming is necessary if the product is to be popular.

Fat reduction has represented the most dramatic aspect of improvement in pig-meat quality over the last 25 years. Pig carcasses have become leaner through an improved understanding of nutritional requirement and through selective breeding for lean types. Pig meat now rivals poultry for leanness, and is considerably less fat than beef or sheep meat. The fat that there is in lean pig meat is also lower in lipid content than is the case with ruminants, and the lipid itself is higher in linoleic acid (polyunsaturates). As the fat

Figure 2.5 Imaginative cutting techniques enhance the product range and broaden consumer choice (courtesy Meat and Livestock Commission, UK).

content of the animal is reduced, so the percentage of water in the fatty tissue that is associated with the meat joints rises. Where the fatness of the pig is gauged through the subcutaneous fat depth at the P2 site, then the percentage of water in dissected fatty tissue may be approximated as: $35 - 1$ (P2mm).

As fat levels reduce, the fatty tissues themselves become softer (due to an increasing proportion of unsaturated fatty acids) and wetter. Soft fat has positive benefit to some curing processes and, of course, in human nutrition contexts. But more usually if the meat is sold with the fat on, either fresh or having been processed ·by modern factory methods, soft fat is seen as disadvantageous. Soft fat is not so pleasant to eat, it makes slicing and packing difficult, and gives the meat a reduced shelf-life. A distinction is required here between lipid depots found within the muscle bundles and the subcutaneous lipid, the latter being the major depot of fat in pigs. A much higher polyunsaturate level can be accepted in intramuscular fat than in subcutaneous fat. If the level of linoleic acid in backfat increases above 150g linoleic acid per kg of fat, the possibility of soft fat problems arises. The ideal ratio of unsaturated:saturated fatty acids in pig fat is around 1.5:1. The use of dietary vitamin E may improve the keeping qualities of the pig meat. Tocopherols are transferred from the diet to the pig fat and help to maintain freshness and delay the onset of rancidity.

Soft subcutaneous fat in pigs is positively associated, in any breed or sex, with leanness itself; although there may also be a tendency for entire males to have slightly softer fat than females, even at the same level of fatness. Unsaturated fatty acids in the diet, especially substitution of stearic (C18:0) with linoleic (C18:2), may influence the type of fat deposited in the body, rendering it softer (Figure 2.6). The content of linoleic acid in pig fat, and its consequent softness, is also related to the rate (in terms of grams per day) at which fat is deposited. At 50g daily fat deposition, tissue fatty acids would comprise about 15% linoleic; while at 200g daily fat deposition the tissue fatty acid would comprise about 10% linoleic.

Leanness, when taken to extremes in pigs, can also lead to the fat becoming lacy and splitting away from the lean (Figure 2.7). Splitting fat can be a problem where the packer wishes to prepare cuts with some fat left on, as is the case with bacon. But fat separation is less of a problem where the joint is sold as lean alone, as would be the case for continental pig loins comprising solely the *longissimus dorsi* muscle.

Backfat depth (usually at the P2 site; Figures 2.8 and 2.9) is often the major criterion of assessment of quality in pig carcasses at the point of payment to the producer. And where the meat is sold with the subcutaneous fat left on, fat depth is absolutely crucial to quality perceptions at the point of purchase and consumption. But subcutaneous fat levels become of minor

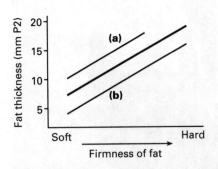

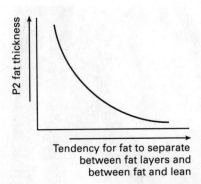

Figure 2.6 Influence of fat thickness upon fat quality. (**a**) relates to pigs fed diets especially high in unsaturated fatty acids such as linoleic, while (**b**) relates to pigs fed diets especially low in unsaturated fatty acids.

Figure 2.7 As fat depth reduces, the tendency for fat to split increases.

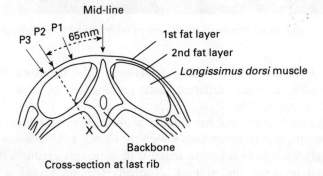

Figure 2.8 Position of intrascope probe measurements P1 (45), P2 (65) and P3 (80), in millimetres from the mid-line taken at the last rib. X indicates the line taken through the *longissimus dorsi* muscle by probes such as the Fat-o-meter and the Hennessy grading probe which measure both fat depth and muscle depth (see also **Figure 2.9**).

importance if the meat is trimmed and prepared before sale.

At the point of purchase, following effective cutting, pig-meat quality is often defined in terms such as the colour, water-holding capacity and firmness of the lean meat itself.

A tendency toward pale, soft and exudative (PSE) muscle (Figure 2.10) is identifiable in about 12% of European carcasses. The severity of the condition is not usually sufficient to reduce the acceptability of the meat, but in some 5% of cases curing and processing potential is reduced and the fresh

Figure 2.9 Probe measurement at the P2 site of both fat and muscle depth (courtesy MLC).

meat sold loses moisture and lacks tenderness. The incidence of PSE is highly variable amongst different meat-packing plants, some showing incidences of up to 25%.

Live pig muscle is red and has a pH of about 7. After slaughter, while the carcass is warm, energy metabolism lowers the muscle pH (adequate amounts of muscle glycogen go to adequate amounts of lactic acid), reduces the colour to pink and increases the wetness. Normally the fall in pH is gradual, ultimately reaching about 5.4 24 hours after slaughter. But if this muscle metabolic activity is prolonged, or excessively rapid, then the pH may drop below 6 within an hour or so of slaughter, instead of the more usual time of 4–6 hours after slaughter. It is the rapidity of the acidification of the muscle which causes the resultant carcass muscle tissues to be unacceptably pale, soft and exudative. PSE is the result of soluble proteins precipitating, the muscle fibre cell membranes leaking as they become depleted of energy and intramuscular fluid rich in lactic acid being exuded from within the fibres. This process is enhanced if the pig is stressed before slaughter and/or if muscle fibre activity continues after death. In general, it may be surmised that there is an increased likelihood of PSE if acidity occurs too rapidly and the meat shows a pH of less than 5.7 measured 45 minutes post-slaughter.

Dark, firm and dry (DFD) muscle is relatively infrequently seen in pigs

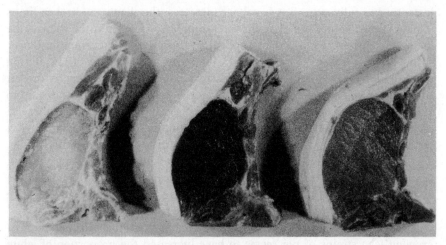

Figure 2.10 Pale, dark and normal pig muscle (courtesy MLC).

and may be attributed to poor pre-slaughter handling and depletion of muscle glycogen through excitement. The pH of such meat tends not to fall and to remain around or above 6.5. It is chronic stress which usually has utilized the glycogen and there will be inadequate levels at the time of slaughter. This leads to inadequate lactic acid and a consequent failure to reach an adequately low pH. In bad cases the meat is inedible. When less acute, the meat lacks flavour, but may be relatively tender.

PSE seems to be related to one or both of two causes: one is mishandling and pig stress, the other is the genetic constitution of the pig. These interact such that susceptible pigs, when stressed, will show a high frequency of PSE meat, whilst if unstressed the rate of PSE can be halved, even in fully susceptible animals. PSE is closely related to the presence in the genotype of the halothane gene. Pigs suffering from porcine stress syndrome (PSS), which is also often associated with the halothane gene, are especially susceptible to PSE. But *all* pigs are susceptible to reduction in meat quality through PSE if they are badly handled and stressed before or during slaughter – especially if mixed into new groups, during the course of transportation to the slaughterhouse, in the slaughterhouse holding pens and at the point of slaughter. Holding pens should be small and, if conditions are good, a dwell-time of 1–4 hours after delivery is usually satisfactory. In hotter weather water sprays can help. Pigs may be rendered unconscious by electrical stunning or by carbon dioxide gas. Although the latter may be more gentle, in some systems there can sometimes be respiratory distress. High voltage stunning (up to 600+ volts) seems to have some benefits over CO_2 and over the use of lower voltages (70–120) which may not instantly and fully stun. But as voltage increases, the chances of broken limbs at slaughter and blood-splash in the carcass increase. The lower the voltage, the longer the current requires to be applied, and application of current for longer than 8 seconds may predispose to PSE. It appears that a voltage of 150, raising some 0.5 amps, for 8 seconds is satisfactory but quicker stunning will result from higher voltages (for example, 210V for 3 seconds), and amperages up to 1.3. Use of hand-held tongs operated amongst a group of pigs held together in a stunning room appears less preferable to stunning at the termination of a conveyor chute along which the pigs quietly walk in an orderly single-file queue. The variable influence of different abattoirs upon strains of pigs of different susceptibility to PSE is shown in Figure 2.11. Susceptibility to PSE meat may be as high as 50% for halothane-positive animals, 10% for halothane-negative animals and 20% for the first cross between the two.

It has been suggested, but remains to be corroborated, that selection for fast-growing lean strains of pig seems to have increased the proportion of white muscle fibres in comparison to red. About 80% of muscle fibres are white in highly improved genotypes, as compared to about 30% in wild boars. White muscle fibres are characterized as being of lower locomotor function, thicker in dimension, having less intramuscular fat and being more able to produce lactic acid but, because of poorer blood supply, less able to remove it. All these characters have recently suggested, but not yet adequately ascertained, to tend toward a susceptibility for lower flavour, reduced tenderness and a higher risk of PSE. Improvement of the quality of lean pig meat, now that the problems of excessive fat are readily resolved by genetic selection, is a major objective of breeding and feeding programmes.

Notwithstanding the negative consequences of the colour of pig meat being too dark, in some countries, such as Japan, a deeper pink colour is preferred to pale white meat. In other nations the reverse is the case, and pork (alluding to chicken) has been referred to as 'the other white meat'. Meat that is particularly lean, and meat from entire male or Duroc pigs, tends to be deeper in colour.

Given that the meat is presented with an acceptably low level of subcutaneous and intermuscular fat (the first requirement of meat quality),

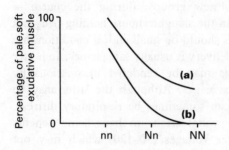

Figure 2.11 Influence of the presence of the 'halothane gene' upon PSE. nn is halothane-positive, NN is halothane-negative, and Nn is the cross-bred. Responses are shown for two processing plants: (a) with poor handling facilities, and (b) with good handling facilities.

subsequent benefit may be achieved from retaining at least 1% of intramuscular fat within the muscle. The eating quality of lean appears to be positively related to its intramuscular fat content. This phenomenon is well known in beef and sheep meat but is also pertinent to the succulence of pork; Figure 2.12 refers to improved white pigs. The improvement of pigs has led to dramatic reductions in fat, which does appear to have caused reductions over the years in tenderness (Table 2.2) but not acceptability, which has risen because, above all, the consumer perceives quality in terms of leanness. With pig meat most of the fat is held in subcutaneous fatty depots, intramuscular fat only increasing from 0.5% in lean pigs to about 3% in fat pigs. In general, the best eating quality is achieved at levels of fatness associated with P2 backfat depths at 100kg live weight of between 8 and 14mm. At below 8mm the quality of the lean falls, while above 14mm the meat is too fatty. Over the range of 0.5–3% intramuscular fat it is possible to show small but important improvements in succulence, tenderness and flavour. Fatness levels of greater than 3% adversely affect eating enjoyment, while at less than 1% fat in the muscle loss of tenderness and succulence may be noted by the consumer. Levels of intramuscular fat of below 1% would be expected to be associated with pigs whose P2 backfat depths were less than 8mm. Recent studies have shown that, even at equal fatness, fast-growing pigs fed *ad libitum* give more tender and more juicy meat than counterparts which have their feed restricted and grow slower. In this work all the pigs were of low fatness, although naturally the pigs fed *ad libitum* are usually fatter than pigs which

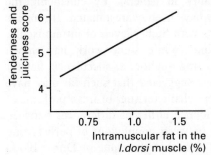

Figure 2.12 There is a small but positive influence of fat in the lean tissue on the eating quality. Note the low levels of fat in these muscle samples from European hybrid pigs.

Table 2.2 Trends in fatness and tenderness scores

Year	Percentage lean meat in carcass	Percentage fat in carcass	Percentage fat in *longissimus dorsi* muscle	Taste and tenderness score of muscle (1–5)
1960	40	40	2.5	3.5
1975	50	30	1.5	3.0
1990	60	20	1.0	2.5

are restricted in their feed allowance because the former have been allowed to eat much more food.

The meat expert, favouring fat and flavour, may define quality differently to the consumer who favours leanness and tenderness. Leanness at the point of purchase may, however, detract from enjoyment at the point of consumption through the predisposition of lean meat to being slightly tougher. Post-slaughter treatment, food preparation and cooking techniques are crucial to the creation of a satisfactory pork product.

There has been some interest in the use of particular breeds, or particular breed mixtures, as a top-crossing sire for the production of high-quality eating pork. For example, the Duroc breed has been cited in both America and Europe as useful to impart superior meat quality to the progeny of Large White × Landrace hybrid mothers. The breed has some reputation for robustness and carries more bone, but is not so prolific as the White hybrid. The Duroc lacks in ham shape and may tend to fatness and slower growth. However, there is an interestingly higher percentage of intramuscular (marbling) fat than with the conventional White breeds (2–4% rather than 1–2%). This extra fatness within the (rather more red-coloured) muscle bundles is said to give extra flavour, tenderness and succulence for fresh (but not cured or processed) meat. As an approximation, the percentage marbling fat in conventional White breeds is approximately equal to $0.1 \times (P2)$, whereas the equivalent relationship for the Duroc employs a multiplier of 0.2. If genuinely required, a similar characteristic could be developed in other breeds.

Possible improvement to eating quality in general, by enhancing the intramuscular fat of any breed type, may have been overestimated. It remains to be proven that fresh meat from breeds with higher levels of intramuscular fat does indeed make for better eating. While some work has shown advantages in intramuscular fat for imparting qualities of succulence, flavour and tenderness, other recent studies have suggested that such fat may have little to offer to the flavour and tenderness characteristics of fresh pork meat. Perhaps these conflicting results are evidence either of differences between populations of the breed, or differences between populations of pork tasting panels. However, there is a developing view that the value of Duroc blood may be in hybrid vigour and hardiness rather than in meat quality characteristics. Maybe the fat content of meat is best manipulated by nutritional means rather than genetics although, while nutrition readily influences overall fatness, the site of deposition of the fat is largely genetically inspired. Further, it is becoming apparent that flavour, tenderness and succulence are unlikely to be best improved in the immediate future by progressive genetic selection; rather these meat characteristics may be enhanced through correct choice of feed, of slaughter weight and slaughter age, pre-, peri- and post-slaughter handling and the cutting, maturing and preparation of the meat prior to consumption.

Supermarkets in industrialized nations now trade the largest market share of pig meat products. They are invoking elements of quality additional to leanness, flavour and tenderness. In particular, perceived needs in terms of quantity and price have been replaced by needs measured in terms of quality and method of production. Supermarket buyers are catering for consumers concerned for the welfare of the pigs, the husbandry methods used, the housing conditions, the breeding and feeding of the pigs, general health and hygiene, absence of specific diseases, freedom from the use of feed additives and growth promoters, animal handling practices, transport and lairage conditions, efficacy of stunning method, carcass handling and carcass treatments and processing conditions. Many of these aspects of pig meat production are likely to be prescribed for producers wishing to enter quality assurance schemes and to provide products to specific supermarket requirements. Meanwhile, given that adequate leanness has now been attained in many populations of pigs, producer attention is being paid to the enhancement of flavour and the increase of the nutritional worth of pig meat through manipulation of the pig diet. This particular discipline is becoming a developing area of pig nutrition research.

Shape and meatiness

Some pig types are more meaty than others and have a characteristic blocky 'meat-line' shape. These types may often but not always carry the 'halothane gene', thus named because the animals react to the anaesthetic halothane. Where this association occurs, the reaction can be used as a marker for muscle shape and meatiness. Halothane reactors are found at various levels in pig populations (Table 2.3) and differ between strains within breeds. Amongst breeds and crosses containing 40–70% of the gene its presence is associated with:

- about 2.5 percentage units more carcass lean;
- about 1% higher killing-out percentage;
- about 10mm reduced carcass length;
- about 25% of the carcasses liable to pale, soft and exudative muscle (5% liability in normal pigs, 25% liability in homozygote halothane-positive pigs);
- one or two pigs less weaned per litter;
- an appetite reduced by about 10–20%;
- growth slowed by about 10–20%;

Table 2.3 Incidence of the 'halothane gene', and associated stress susceptibility, meaty and blocky characteristics

'Breed'[1]	Occurrence (%)
Large White Duroc Hampshire	0
Norwegian Landrace Danish Landrace British Landrace Swedish Landrace	2–25 (variable between strains)
Dutch Landrace Dutch Large White	20–50 (variable between strains)
German Landrace Belgian Landrace Pietrain	70–90 (variable between strains)

[1] Breed definition is no longer an adequate descriptor of pig type or genetic composition. Some breeds termed 'Large White', for example, may show a percentage of halothane-positive individuals in the population. Also many pig 'breeds' available from breeding companies are made from a variety of gene sources and often defy definition other than that provided by the breeder. Such definitions may relate more to what the animal is purported to do, and in what circumstances it should be used, rather than to what breeds or genes are included in the animal's make-up.

- an increase in drip loss;
- bigger *longissimus dorsi* muscle area ($+ 5cm^2$, to attain $45cm^2$ at the P2 site when 90kg live weight);
- up to 12mm more depth of the *longissimus dorsi* muscle at the probe site;
- bigger, more rounded and meatier hams (20% increase in shape score);
- less bone;
- higher mortality in the growing/finishing stage (10% as against 2% in conventional pigs).

Pure-bred Belgian Landrace and Pietrain pig types, and many strains of German Landrace, show benefits over the Large White pig types of up to 4% more lean (61% v. 57% in the carcass sides, Figure 2.13), up to 3% better killing-out (76% v. 73%) and less carcass bone and smaller heads (the lean:bone ratio being about 6:1 as against about 5:1). In consequence of the greater muscle mass and reduced carcass bone, meaty strains of pigs have a higher percentage of lean meat in their carcass at any given backfat depth than do conventional white strains of pigs. Therefore, grading on fat depth alone will erroneously discriminate against blocky meat-type pigs. It is especially where mixed populations of pigs are forwarded to the meat packer that measurements in addition to fat depth should be taken if the correct

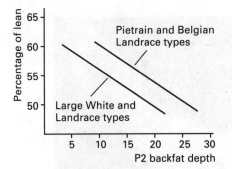

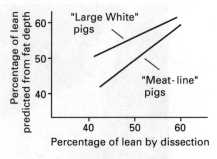

Figure 2.13 At any given fat depth the meat-type breeds of pigs have a higher percentage of lean meat. Their shape is more blocky and muscular and the content of bone is less. Meat pigs which are homozygous for the blocky character will have up to 4% more lean tissue at the same fat depth. Heterozygotes with 50% of the character will have 2% more.

Figure 2.14 Comparative relationships between predicted and actual lean percentage for 'Large White' and 'Meat-line' strains of pigs.

assessment of carcass lean content is to be made. Such additional measurements would be the depth of the eye muscle and a visual assessment score of the shape of the hams themselves. The danger of predicting lean percentages on the basis of fat depth alone is illustrated in Figure 2.14.

Other characteristics of halothane-positive pigs include smaller litters (10–20% less pigs weaned), slower growth rate (10–20%) and a tendency to drop dead when stressed (porcine stress syndrome; PSS), such as when loaded for transportation or held in abattoir lairage.

The characteristic shape of meat strains (Figure 2.15) is short and blocky with well-filled, rounded hams; the rear view being more ω-shaped than n-shaped. The blocky muscling character also shows clearly in size of the loin and in the eye muscle (*longissimus dorsi*) area (Figure 2.16). At 80kg live weight the pure-bred Large White and the Pietrain will have about 35cm^2 and 40cm^2 of eye muscle; at 100kg live weight the two breeds will have respectively 45cm^2 and 55cm^2. In comparison, some American breeds may have only 40cm^2 of eye muscle at 100kg live weight, which would be associated with a lean meat content of only 53% in the carcass. A well-tried breeding strategy is to make a heterozygote top-crossing meat sire line from a halothane negative grandparent female and a halothane positive grandparent male. This parent meat sire is then used on halothane negative parent females to produce the slaughter generation which will exhibit a high level of the benefits from the gene, and a low level of disbenefits. It increasingly appears to be possible to retain characters of meatiness and suppress characters of stress.

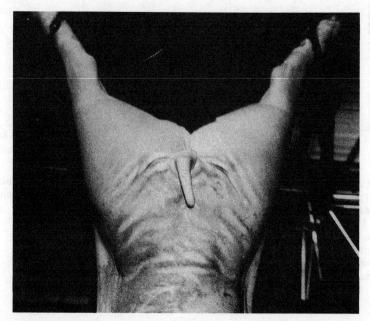

Figure 2.15 Hams of a good quality meat-line cross slaughter pig.

There is great activity and success amongst pig breeders at present to retain all the positive attributes of the halothane gene but to exclude the gene itself, with its associated problems. Genetic selection for rounded ham shape, muscle content and eye muscle area, simultaneous with selection against the halothane gene, in the Large White type of pig is yielding positive results without the need to risk halothane stress susceptibility.

Meat from entire male pigs

Entire males are much less fat than castrates. In the fat, males (being leaner) have a higher proportion of polyunsaturated fatty acids which is a health-positive character. At 90kg an entire male of an improved European strain will have around 10–12% of body lipid, whereas a castrate will have 16–18%. This attribute, together with faster and more efficient growth, makes the entire male an attractive proposition for the production of bacon and pork. Where the production cost for 1kg of lean meat from entire males is taken as 100, equivalent cost for gilts and castrates are 108 and 116.

Figure 2.16 Pig loins. Note that the one on the left is of lesser muscle volume than the one on the right, which is preferred.

In many countries of the world no males are ever castrated and entire pigs are invariably sold for slaughter at all weight ranges. In other countries nearly all the males are obligatorily castrated. In the UK most male pigs used for meat are now sold entire. These go both for fresh meat production and for bacon curing, giving skin-on rashers with large eye muscles and low fat thickness. There remains a view, with some basis in objective evidence, that entire males should in general be slaughtered at less than 105kg if they are to be used for meat. There is a trend world-wide toward the cessation of castration for all meat pigs slaughtered at less than 100kg.

Some people may detect an aroma when bacon from entire males is being cooked; whilst some may find this appetizing, others may dislike the aroma on first encounter. This is not carried through to the table, and in this respect is similar in nature to the aroma of lamb being cooked. However, boar meat may also carry taints which are related to the presence of androstenone and skatole. These are more likely in older animals, above 110kg live weight, which have been grown slowly. It is most unusual for the level of androstenone in pig fat to exceed 1.0μg per gram of fat, but when it does so some consumers may notice an aroma judged as 'less pleasant' than normal cooking odours. Both androstenone and skatole seem to be related in entire males, and both these causes of taints may on occasion also be found at lower concentrations in females. Skatole is formed from tryptophan by gut bacteria, and seems to be taken into the tissue (especially fat) with the help of androstenone. This explains why skatole taint is more common in the male, but any expected relationship between diet levels of tryptophan and skatole taint has as yet eluded the meat scientists. Although skatole levels are higher in males than in females, it is rare for the level to exceed the threshold of human taste perception.

In countless consumer trials over many years, taste panels have been unable to fault bacon and pork products from entire males. In particular, in carcasses of less than 75kg carcass weight the same (very low) level of stronger flavours may be perceived from entire, gilt and castrate meat. There is no clear evidence available showing differences in eating quality or acceptance between the three sexes, and often the greater eye muscle and reduced fat content of the entire male are found by the consumer to be positively beneficial. Although it appears clear that taint may arise from high levels of androstenone and skatole, especially when they occur together, the infrequent expression of this phenomenon can be judged by the many research projects into boar taint which have failed to demonstrate even its existence. It now appears that unacceptable flavours in meat may be as much associated with particular feed ingredients (especially high fat fish meals and various animal and industrial by-products) as with sex. The use of wheat in the feed appears to reduce meat skatole content, while corn gluten can enhance colour. The main effects of feed ingredients still appear to be consequent upon direct incorporation of diet long-chain fatty acids into pig fat, taking with them some elements of their degree of unsaturation and flavour.

Entire males, being so much less fat than castrates, may suffer from disadvantages of extreme thinness. Their fat tends to be wetter, softer and more liable to split (the entire male has about 20% of water in the subcutaneous fat, as against 15% for castrates). These small problems may be remedied by limiting unsaturated fats in the diet and by increasing feed levels in order to avoid over-leanness. However, even at equal fatness, the fat of boars tends to be a little less firm than that of gilts. Butchers have complained that lean pigs are difficult to handle and cut, being floppier; this is mostly a characteristic of leanness, but entire males do seem to be more prone than females.

Current prejudice against entire males by some meat wholesalers seems to relate to problems with those few animals that are very lean, or to the equally few others that have been grown slowly. Boars should only be purchased by meat abattoirs and processors under tight contractual arrangements from known producers who can guarantee fast, uninterrupted growth and offer high-level feeding to the animals in a healthy production environment. Pigs failing to meet these minimum criteria should be diverted from the bacon and fresh pork market and used elsewhere in the meat chain.

Carcass quality and grading standards

Pig carcasses may be divided in many different ways, consistent with the huge variety of bone-in joints and bone-out products that come from the pig. Primary breakdown may follow a scheme such as that shown in Figure 2.17. Although it is characteristic of the pig carcass that the value of the cuts does

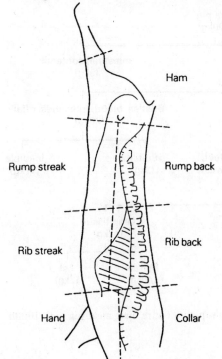

Ham

Rump streak

Rump back

Rib streak

Rib back

Hand

Collar

Figure 2.17 One of the many ways of jointing a pig carcass.

not differ according to position in the body nearly to the same extent as for beef and lamb, because of a lesser tenderness gradient, the most valuable parts are in the loin (rump back) and ham areas.

The chemical composition of the empty body of a 100kg entire male meat pig with a P2 backfat measurement of 10mm will be around 67% water, 17% protein, 13% lipid and 3% ash. The physical composition in terms of muscle, fat, bone, offal and so on is shown in Figure 2.18.

Usually payment for pig carcasses is on the basis of their weight, adjusted for some assessment of carcass quality, as defined in a grading scheme.

Individual grading schemes may contain many or few criteria. The more criteria, the greater the likelihood of a carcass found acceptable on the basis of one criterion being deemed unacceptable on the basis of another. Some example criteria are given in Table 2.4. Standards of over-fatness can be matched by standards of under-fatness. Minimum fat levels (often around 8mm P2) are required for the eating quality of lean meat as well as to help maintain the quality of the fat itself. In some schemes a single factor, such as level of fatness, may be the prime controller of value and payment received by the producer. Often, criteria of considerable importance to carcass quality, such as the quality of the lean and fat, may play no part in grading standards and the payment schedules. In some places a flat-rate payment is made for the carcass mass regardless of quality or grade attainment, while in other places payment is made strictly per kilogram of lean meat produced.

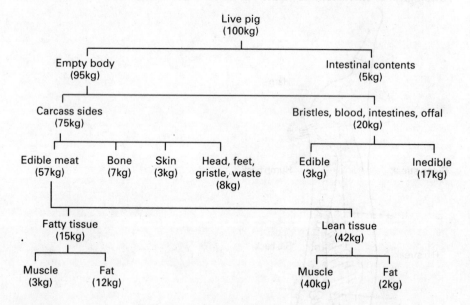

Figure 2.18 Physical composition of a high-quality entire male meat pig with 10mm P2 fat depth.

Table 2.4 Example criteria for grading of pig carcasses

Carcass dead weight (kg)	'Pork' 60
	'Bacon' 60–80
	'Manufacturing' > 80
Fat depth (mm)	P2
	P1 + P3
	Mid-line at: mid-back
	loin
	shoulder
	Ham fat
Muscle depth (mm)	P2 site
	Mid-back site
	Ham
Length (mm)	Carcass
Shape (visual or	Hams
objective score)	Eye muscle area
	Eye muscle shape
Distribution of joints	Fores/hams/loins/belly
Sex	Entire male/female/castrate
Muscle quality	PSE
	DFD
	Colour
	Flavour
	Juiciness
	Texture
	Tenderness
	Taint
Fat quality	Wetness
	Firmness
	Colour
	Tendency to split
	Flavour
	Taint
	Texture

For some types of meat production it is the breed of pig and the method of rearing and feeding which is the vital criterion of quality. For others, such as for the production of Italian Parma hams, the first necessity is for a ham size of 12kg; in effect, a pig of 160kg or more live weight.

Where live weight is the only effective determinant of producer reward, the industry tends towards low efficiency and the product tends towards low quality, there being no alternative incentive. Some aspects of the pig market in the Americas operated in this manner until quite recently. It may be said that it was the demands of the European farmers for greater efficiency, and the demands of the largely pig-meat-eating European consumers for less fat pigs, which brought about the creation of the genetically improved modern strains of European hybrid pig which are now used for meat production world-wide. Such demand was not appreciated at that time by farmers in the Americas who paid less for their pig feed, and whose consumers were mainly eaters of beef and therefore not discerning toward pork quality issues.

As has traditionally been the case with cattle and sheep, in some countries the shape of the live pig, or the pig carcass, may be paramount as a criterion of carcass quality. The fullness of the muscle either side of the backbone and the roundness of the ham are often much sought after as indicators of lean muscle mass. In some breeds of pig, however, roundness of the ham and fullness of shape merely indicate fatness. But when these characteristics are found in the European meat-line strains and breeds, there may indeed be a very positive relationship between the shape of the pig and its lean content. These important differences in the correct interpretation of shape for different genotypes can be separated out only by the objective measurement of lean muscle mass (as well as the measurement of fat depth). Lean percentage can be measured reasonably objectively and directly from the depth of fat and muscle in the middle region of the back, for example around the P2 position (Figure 2.8). Comprehensive automatic systems can now be created which measure shape, while simultaneously probing at many different points down the back and on the ham for determination of fat and lean depth. The same equipment is likely to be able to take automatic measurements of muscle colour and pH. The possible presence of taints such as may be caused by skatole require either the human nose, following application of a hot wire to the fat, or sophisticated on-line chemical analysis techniques.

Where grade is assessed on fat depth alone, good carcass shape may not be adequately rewarded. At equal fat depth the meat-line breed types are likely to have more lean than the conventional Large White and Landrace breed types. In consequence, although some types of high-conformation pigs may contain a higher percentage of lean meat in the carcass, these animals being fatter at the P2 site are liable to be downgraded. The importance of the muscle measurement, in addition to the measurement of fat for the effective and objective prediction of percentage carcass lean in mixed pig populations, is exemplified in Table 2.5. At equal P2 fat depth, homozygote halothane-positive pig types may be estimated to have some 4% more lean meat than the Large White types (as has been shown in Figure 2.13).

Most grading standards and payment schedules relate to fatness (often as backfat at the P2 site, Figure 2.8), within a given weight band. A scheme may, for example, limit head-on carcass dead weight to between 60 and 75kg and pay a premium price for pigs of less than 12mm P2, imposing a

Table 2.5 Prediction of lean meat content of pigs with different muscle depths

Carcass weight	Fat depth (mm P2)	Muscle depth (mm at P2 site)	Predicted percentage of lean meat content	
			By fat depth alone	By fat depth and muscle depth
64	13	60	55	58
66	12	64	57	59
67	12	52	57	57

See also MLC (1990) *Pig yearbook*. MLC, PO Box 44, MK6 1AX, UK.

maximum price penalty for pigs of more than 16mm. Equivalent schemes might pertain for pork carcasses of below 60kg carcass weight, and cutter and heavy pig carcasses of above 75kg carcass weight (Table 2.6). Stepped grade schemes using fat depth as the criterion of assessment fail to utilize known relationships between fatness and lean yield, and the EC grade scheme helps to improve this position by presenting each class in terms of the percentage of carcass lean meat (Table 2.7). Lean meat percentage may be predicted from equations of varying degrees of complexity and efficacy. A simple prediction of percentage lean in the carcass side would be:

$$\text{Percentage lean in carcass side} = 68 - 1.0P2$$

(Equation 2.1)

or:

$$\text{Percentage lean in carcass side} = 65.5 - 1.15\ P2 + 0.076\text{carcass weight}$$

(Equation 2.2)

Most UK pigs of 65–70kg carcass weight have a P2 measurement of 12mm or less. However, in many other parts of the world, particularly those where improved European hybrids are not being used, pigs may still be found with carcass fat depth at the P2 site in excess of 20mm.

Stepped grading schemes, such as described in Tables 2.6 and 2.7, fail to take immediate advantage of the calculation of the absolute amount of lean meat provided from knowledge of the estimated percentage lean in the carcass side and the carcass weight. It would be more logical for payment schedules to be continuous rather than stepped, and based on the absolute amount of lean meat yielded. It follows from the natural growth pattern of the pig that the lighter animals within any allowed weight band have a lower

Table 2.6 Examples of stepped grade schemes for 55 and 80kg carcass weight pigs with emphasis on leanness

P2 fat depth (mm)	Price/kg dead weight (% of average price)[1]
50–60kg dead carcass weight	
< 8	120
8–12	110
12–16	100
> 16	80
75–85kg dead carcass weight	
< 10	90[2]
10–14	120
14–18	100
> 18	80

[1] The average price for 60kg dead-weight pig carcasses would be likely to be some 10–20% higher than for 80kg dead-weight carcasses depending on market demand locally for the two types.

[2] Notice that the imposition of a minimum as well as a maximum fatness can reduce problems associated with overleanness.

Table 2.7 The European Community grade scheme based on lean meat percentage

Percentage of lean meat in carcass[1]	EC grade class	Percentage of UK pigs in each EC class (1990)
>60	S	17
55–59	E	58
50–54	U	21
45–49	R	3
40–44	O	1
<40	P	

[1] Lean meat percentages may be estimated from agreed equipment and equations, for example:

Estimation by Intrascope probe measurement of P2 fat depth:
Lean meat % = $65.5 - 1.15P2 + 0.076W_c$
where W_c is the carcass weight

Estimation by Hennessy probe measurements:
Lean meat % = $62.3 - 0.63P2 - 0.49$Ribfat $+ 0.14$Ribeye muscle depth $+ 0.022W_c$

Equations valid over range W_c = 30–120kg

Measurements including eye muscle depth more accurately credit pigs with superior shape.

fat thickness. Thus, as the potential value of the carcass increases with increasing weight, the likelihood of it being down-graded because of over-fatness is also increased. The (unjust) consequence is a diminished price paid for each and every kilogram of lean meat provided, whilst the total mass of lean meat would, of course, be greater. Further illogicality follows from the stepping of the scheme itself. Whilst an 80kg carcass of 10mm P2 may indeed be worth 20% more than one of 15mm P2 (Table 2.6), it is not logical that the same 20% difference in value also pertains for the 1mm difference beween 13.5 and 14.5mm P2. These inconsistencies would be ameliorated by a grading scheme that paid for each kilogram of lean meat yielded. Within a given breed type, as already described, this could be estimated from a knowledge of the carcass weight and the P2 backfat measurement. Given the calculation of percentage lean in the carcass side, a payment schedule such as depicted in Figure 2.19 might be envisaged. The slope of the line is critical to the cost-benefit of producing leaner pigs, and such a slope may well need to represent a 2% increase in price paid per kilogram carcass dead weight for each percentage unit increase in lean meat content.

Effective determination of percentage lean, however, requires not just knowledge of fatness, but also knowledge of lean muscle mass.

While predictions in the form, 'Percentage lean = k − nP2', where k is around 70 and n around unity, may be adequate for Large White breed types, or for use within given pig populations, the equation fails to allow for the meatier pig strains which carry more lean at any given P2. If only P2 fat measurements are available, a more universal equation form may be, 'Percentage lean = $bP2^{-k}$', where k is around 0.21 and b indicates the degree of blockiness of the strain of pig concerned, and ranges between 90 for unimproved Large White and Landrace breeds and 100 for pure-bred Pietrain and Belgian Landrace types.

While shape can be a helpful guide to improving predictions of leanness made primarily on the basis of fatness, it is liable to errors from subjectivity and, of course, gross errors where a rounded ham is not associated so much with muscle mass as with fat mass. Lean mass can really only be effectively determined by a combination of measurements of fatness and of leanness

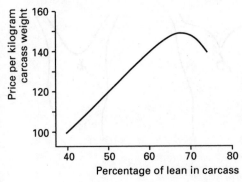

Figure 2.19 Payment schedule based on percentage of lean meat and price per kilogram of carcass weight. Carcass lean may have been estimated from fat depth or fat + muscle depth. The schedule encourages leanness up to but not beyond an optimum level.

itself. This is helped if muscle depth is also measured, usually by probe (see Figure 2.8), and the use of prediction equations such as:

Percentage lean meat in carcass side = $59 - 0.90$P2 fat depth + 0.20Eye muscle depth.

(Equation 2.3)

A pig with 10mm P2 fat and 55mm of muscle depth is estimated to contain, by use of this equation, 61% of lean meat. A poor pig with 45mm of muscle depth and 20mm of P2 fat would be estimated to contain 50% of lean meat in the carcass side. Carcasses with greater than 65% of lean meat may be judged as over-lean and prone to reduced muscle quality.

Such niceties as the exact prediction of carcass lean are not considered necessary in many countries where shape, judged by visual assessment, is considered to cover all the required criteria for carcass and meat quality. Thus Table 2.8, typical for parts of Spain, emphasizes the importance of ham shape within relatively broad categories of fatness.

Table 2.8 Example of a stepped grade scheme for 100kg live weight pigs, giving emphasis to ham shape

Mid-back fat depth at the mid-line (mm)	Ham shape score (1–3)	Carcass grade given (1–5)
10–20	1	1
10–20	2	2
10–20	3	3
20–30	1	1
20–30	2	2
20–30	3	4
30–40	1	2
30–40	2	4
30–40	3	5

Ham shape scores:

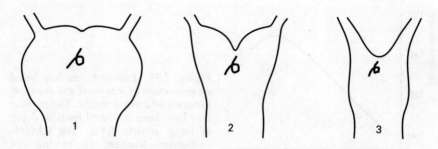

34

Carcass yield: killing-out percentage (ko%)

The carcass weight of a pig is usually between 70% and 80% of its live weight. The loss is mostly blood and internal organs, as shown in Figure 2.18. The carcass weight of the pig conventionally includes the head, feet, tail and skin.

The gut content in pigs is usually around 5% of the live weight, being somewhat greater with high-fibre diets due to both elevated quantities of digesta in the caecum and colon, and the additional water that the presence of fibre attracts to the gut lumen.

Boars have a lower carcass yield than gilts or castrated males, due mostly to the removal of the testes and, in the case of older males, the need to remove a greater amount of cartilaginous tissue in the fores.

Transportation distance and the time between removal from the farm and the moment of slaughter each have additive effects on carcass yield, primarily through stress and disturbance, causing progressive water loss from the body and a consequentially 'drier' carcass at the point of slaughter. Carcass yields of greater than 75% are difficult to achieve if transport distances are much greater than 100km and the time between departure from the farm and stunning much greater than 6 hours. Provision of water sprays and drinkers, and satisfactory reception and holding pens at the slaughterhouse, may help to maintain carcass yield but positive effects tend to be limited.

The major influences upon carcass yield are live weight, fatness and genotype. At live weights of 50–60kg, killing-out percentage may often be little more than 70, whilst at live weights of 100–120kg values approaching 80% may be possible. There are two reasons for this: (i) the carcass grows relatively faster than the gut and the latter therefore comprises a progressively lesser proportion of the whole animal as it increases in size; (ii) heavier animals tend to need more fat, and fatter animals kill-out better than leaner animals at any given weight. This latter is because in pigs two-thirds of the fat is deposited subcutaneously, and somewhat less than one-third is therefore discarded with the internal organs.

Killing-out percentage may be estimated for Large White and Landrace type pigs, all other factors being equal, from equations such as:

$$ko\% = 66 + 0.09W + 0.12P2$$

(Equation 2.4)

35

where W is the live weight and P2 is the measurement of backfat depth (mm).

Equations to estimate killing-out percentage (ko%) tend not to be universal as is already apparent from the foregoing discussion of the benefits of the halothane gene. Pigs of the Pietrain and Belgian Landrace type will kill-out some 2% higher than pigs of the Large White and Landrace type. The probable causes of this improvement are a lower proportion of the offal per unit live weight and a tendency to carry more of the total body fat in the subcutaneous depots. Pigs with lower mature size will be, at any given weight, of a higher degree of maturity. The proportion of gut to carcass will be less than for heavier maturing types, and the killing-out percentage will be greater.

Choice of carcass weight and carcass fatness

The positive relationship between fatness and weight would indicate that the correct slaughter weight for sexes and strains which tend to be fatter should be toward the lighter end of the market, whilst that for sexes and breed types which are naturally thinner should be towards the heavier end. It is germane to production tactics that, unless controlled, fatness will increase faster than body weight, fatness accelerating disproportionately rapidly. This is well illustrated by the prediction of subcutaneous fat (kg) (Y) from carcass weight (X) in one population of pigs:

$$Y = 0.0002X^{2.54}$$

(Equation 2.5)

Total protein mass (Pt) accumulates at a slightly slower rate than live weight and at a distinctly slower rate than P2 fat depth. Thus for groups of cereal-slaughtered unimproved well-fed pigs grown at Edinburgh from 20–220kg live weight (LW) in the early seventies:

$$P2 = 0.149\,LW^{1.093}$$ (exponent >1; P2 gains relatively faster than live weight)

(Equation 2.6)

$$Pt = 0.192\,LW^{0.941}$$ (exponent <1; Pt gains relatively slower than live weight)

(Equation 2.7)

$Pt = 1.33 \, P2^{0.788}$ (exponent much <1; Pt gains relatively much slower than P2)

(Equation 2.8)

While the rule is clear that heavier pigs are invariably fatter than lighter pigs, and as carcass weight increases so does fat depth, the relationship is not so strong for modern genotypes of lower fatness and greater mature size.

1980 (unimproved) P2 (mm) = 0.38 head-on-carcass weight − 10

(Equation 2.9)

1990 (improved) P2 (mm) = 0.15 head-on-carcass weight + 2.0

(Equation 2.10)

This is a logical consequence of the modern improved pig being of lower physiological maturity and a lower proportion of final weight at the given slaughter weight.

Given ultrasonic techniques for the measurement of carcass fatness in the live animal, pigs could, if required, be sent off to slaughter at any given target P2 fatness. But targeting to sale within a given backfat range can be antagonistic to targeting to sale within a given weight range; and the narrower these ranges become the more difficult it is to meet both simultaneously.

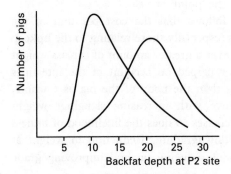

Figure 2.20 The distribution of back-fat depth in the pig population becomes progressively more skewed as average fat depth is reduced.

Fatness is considered as normally distributed through the pig population. However, as average fatness reduces, so the possibility of normal distribution lessens and that for skewness increases (Figure 2.20). Given the biological distributions for fatness and for weight at any given age, it is often the case that there is greater benefit to the producer to sell at preselected calendar dates rather than to try to meet arbitrary and narrow standards for weight and fatness. Such beneficial marketing arrangements would be particularly easily handled if the producer and the retailing organizations were effectively integrated.

Presently, most pigs are slaughtered within the live-weight range 90–120kg, and the head-on-carcass dead weight is usually around 75% of the live weight. Particular markets demand slaughter weights lower and higher than the conventional range (such as for light, fresh butcher pork at one extreme, and for very large, naturally slow-cured gourmet hams at the other extreme). With increasing mature size of modern selected pigs, and with their genetic improvement toward faster rates of lean growth and higher percentages of lean in the carcass, there is a natural (and necessary) tendency for slaughter weights to increase. The current European norm is for live weight at slaughter to be around 110kg.

The need to minimize transportation distances from point of production to point of slaughter will limit opportunities for choice of meat-packer outlets, but locally there usually exists a range of grade and price schedules from which an optimizing choice must be made. But within-schedule choice of production tactics may be even more vital than between-schedule. Pigs may be placed at will somewhere within a range of carcass dead-weight limits, and equally the pigs may be placed at will within the range of fat limits described (together with their appropriate penalties and bonuses). For example, it may be more profitable to elect for a reduction in price per kilogram in return for an increase in the weight of pig at the point of sale.

As slaughter weight increases, it follows that the cost per unit of pig product reduces due to the fixed costs (especially those relating to the upkeep of the breeding herd) being defrayed over a greater amount of carcass weight sales. This particular trend is further helped on account of the intestines being proportionately slower growing than the body of the pig as a whole, thus causing carcass yield to improve with increasing slaughter weight (Figure 2.21). However, as weight increases so does the likelihood of fatness (Figure 2.22) and of consequential downgrading. Reduction in weight at dispatch is often an effective means of reducing fat and improving grade returns.

Heavier weights at slaughter are also associated with reduced efficiency of feed conversion (Figure 2.23). This is partly due to the fatness effect, (1kg of fat gain requiring about three times as much food as 1kg of lean gain), but in lean pig strains is mostly due to the maintenance effect, maintenance representing a non-productive on-cost. At 70kg live weight a growing pig uses about 0.75kg of feed just for maintenance alone, whilst at around 120kg feed usage for maintenance has risen to 1.25kg. Cost per kg of carcass meat

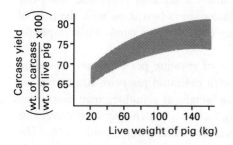

Figure 2.21 Improvement in killing-out percentage with increase in live weight.

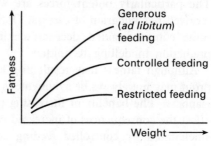

Figure 2.22 As pigs grow, fatness increases. The strength of the relationship relates to the level of feeding.

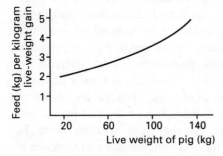

Figure 2.23 Instantaneous feed usage per kilogram of live gain for growing pigs.

Table 2.9 Influence of weight at slaughter on costs of producing 1kg of pig carcass (all costs scaled to those at 100kg live weight)

Live weight of pig (kg)	Carcass weight (kg)	Fixed costs per kg carcass	Feed costs per kg carcass	Total costs per kg carcass
50	35	188	58	125
75	55	133	75	104
100	75	100	100	100
125	95	88	167	101
150	117	83	150	113

in the conditions pertaining in Table 2.9 appears to be minimized at around 75–80kg carcass weight. In general, the tendency is for optimum weight to be increasing year on year. Reductions in variable costs (feed) relative to fixed costs will force optimum weight upwards. If small-sized pigs are required at slaughter, profitability can only be attained if the higher production costs are adequately balanced by greater returns per kilogram.

The number of factors interacting upon optimum slaughter weight are legion, variable and highly dependent upon fluctuating market conditions.

The particularly potent forces are weaner costs, feed costs and the price received per kilogram of carcass. Optimization is best done with the help of some computer-aided decision-making process associated with response-prediction modelling techniques.

Although fatness may result in a loss of revenue per kilogram, optimum profit may not always be synonymous with premium pig price and maximum leanness. The benefits of increasing pig weight at slaughter may more than offset the consequences of increased pig fatness – which may in any event be ameliorated by controlled feeding such as described in Table 2.10. The extent of feed intake control needed is much dependent upon the grade scheme and the penalty which is imposed for fatness. In many circumstances these penalties are inadequate to dissuade producers from opting for a low-input cost system (low feed quality), a high slaughter weight and a fat pig. Mixed groups of entire male and female pigs fed *ad libitum* will result in the females being fatter than the males. Fat levels can be reduced by controlled feeding, but the growth rate and efficiency of the males will be curtailed. Once sufficient benefit is shown to ensue from reducing fat levels in the female, then controlled feeding appears to be called for; otherwise action to improve grading may actually be contra-indicated. Just as for choice of weight at slaughter, the complexities of optimization of level of fatness will require computer-aided decision-making if the right solution is to be determined. Sets of standardized tables for optimum weights and fatnesses at slaughter will obfuscate rather than enlighten.

Table 2.10 Feeding levels required to prevent undue fattening and the production of unacceptably fat carcasses

	Entire males	Females	Castrated males
Improved strains	*Ad libitum*[1]	*Ad libitum*[1]	Slight restriction
Commercial strains	*Ad libitum*[1]	Slight restriction	Medium restriction
Utility strains	Slight restriction	Medium restriction	Heavy restriction

[1] Many 'improved' strains of pigs have been bred for leanness by means of selection for reduced appetite.

Problems of variation

In any production system there is variation; the less well-managed the production process, the greater the variation. In Figure 2.24, both Farm A

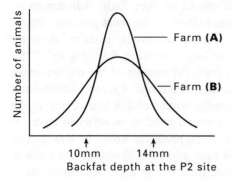

10mm   14mm
Backfat depth at the P2 site

Figure 2.24 Natural variation in fatness experienced on two farms.

and Farm B have the same mean P2 backfat depths, but low-variation Farm A will have relatively few over-thin and over-fat pigs, and proportionately more at around the average. If premium prices are paid for pigs of between 10 and 14mm, then despite both farms having the same average P2 backfat depth, Farm A will be much more profitable, with fewer outgrades than Farm B. Common causes of increased variation are disease, poor feeding management and poor housing.

Additional difficulties arise when more than one standard has to be met simultaneously. Often there are four main categories of carcass assessment:

● carcass weight;
● carcass fatness;
● carcass shape;
● meat quality.

Shape and meat quality are primarily determined by pig type, and pig and meat handling facilities at the abattoir; but fatness and carcass weight are both under the short-term control of the producer. Given a weight range of, say, 70–80kg dead weight, and a maximum fatness of, say, 14mm P2, there are more likely to be outgrades with carcasses at the heavy end of the weight range than at the light, because as pigs get heavier they also get fatter. While pigs are often individually weighed before dispatch, they are rarely checked for fat depth. There is increasing pressure on fat, and therefore some logic in measuring fat depth on individual pigs before dispatch, and sending the fatter pigs away at the lower end of the allowed weight range in order to ensure that a high proportion of the pigs meet the fat depth target. Due to the counteracting forces described above, it is not easy for producers to handle simultaneously a decreasing window size with regard to fat depths and a decreasing window size with regard to weight. The production cost of manipulating the diet and management of individual pigs in order to produce all pigs of similar fatness and weight would be very high. Adjustment of carcass character after slaughter by action taken by the meat processor might often be cheaper than trying to produce carcasses ex-farm that are individually styled to the needs of the pig-processing line (Figure 2.25). Greater overall efficiency might result from the strategy of the pig processor and pig producer accepting the presentation of a range of carcass weights and a range of carcass fatnesses. This strategy would cost a little extra at the processor end of the business, and this would need to be offset in terms of producer price. However, the benefits to the producer would be legion. The need to weigh and measure individual pigs would be obviated, and a whole fattening house could be cleared at one time and at a predetermined date. Such proposals as these require close integration between pig producer and pig processor.

Variation in pig fatness at any given weight has also exacerbated the problem of very lean pigs (those with less than 8mm P2). These problems have become particularly apparent since grade scheme requirements have progressively reduced the level of maximum fat allowed (Figure 2.26). The position is further inflamed by the additional complication that, as average fatness decreases, the distribution for fatness within the population of pigs ceases to be normal and becomes skewed (Figure 2.20). This skewness makes the system highly sensitive at around the point of minimum fatness.

Manipulation of carcass quality by control of feeding

Whilst for many pigs there are few or no grading standards, and for others the standards include measures such as conformation, in many environments pig grading can be considered as primarily dependent upon backfat depth. This is strategically most influenced by breed and genetic merit but tactically most influenced by nutrition. Over the past 20 years of genetic selection in the Large White breed, some 10mm of backfat has been removed from meat pigs. The primary determinant of grade is therefore the quality of the pig. Genetically fat pigs will tend to be always fat within the feasible range of nutritional and environmental variation.

Classical growth analysis, such as proposed by Hammond, presents a view of waves of growth passing through the body as weight and age progress, concluding with the fattening-out process and tissue deposition in the later maturing body parts. It now appears that fatness, far from being related simply to age and weight, is more a function of the level of nutrient supply in

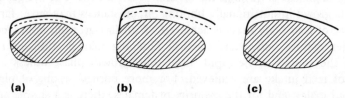

(a)   (b)   (c)

Figure 2.25 Pigs can now be produced which are very lean. The bacon rasher (a) has 3mm of skin and 4mm of fat, giving a P2 of 7mm. An alternative strategy is slaughter at a heavier weight (or use a blocky type, or both) which will give a bigger eye muscle (b) but also predisposes to more fat. The undesirable fat can be trimmed back to 10mm depth by removal of the skin and the first layer of fat (c). This latter pig is likely to have had a P2 fat measurement in excess of 16mm P2. Both types of product (a) and (c) are readily available. Many customers prefer the trimmed rasher with the larger muscle, although the pig giving rasher (a) was by far the most lean.

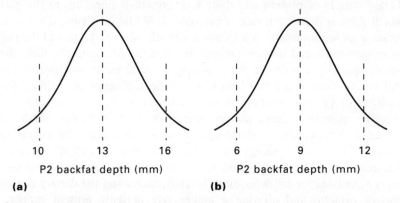

Figure 2.26 With a mean of 13mm (distribution (a)) few, if any, pigs have problems of meat quality. Such a mean would give an acceptable proportion of top grade pigs if the maximum fat measurement was 16mm P2. With a mean of 9mm (distribution (b)) many pigs have problems of meat quality if an acceptable proportion are to achieve top grade in relation to a maximum fat measurement of 12mm.

relation to the level of nutrient need (maintenance + maximization of potential daily lean tissue growth rate). Modern strains of meat pigs may never contain more than the 150g lipid per kg live weight which was already evident in extreme youth at 3 weeks of age. There is therefore little ground for counting age and weight as the major, or only, contributors to body fatness in pigs grown for meat.

Although the long-range strategy for fatness reduction and the meeting of grading standards must be genetic, the tactics, for animals of any given genetic composition, will depend upon the knowledge that fatness is greatly influenced by the quality and the quantity of feed. The major tactical mechanism open to producers to manipulate grade and to achieve target grading standards is through the control of the nutrition of the growing pig. Animals grown in nutritionally limiting circumstances may never fatten until maturity of lean body mass is achieved. With fast-growing improved breeds of larger mature body size, classical fattening may be difficult to obtain during the course of the rapid (immature) growth phase unless very high levels of feed intake are achieved. For more normal strains of pig, for all castrated males, and for the majority of females, there is a simple and direct relationship between the amount of feed eaten (or given) and the level of fatness in the slaughter pig.

Protein

Lean tissue comprises protein and water. The relationship between daily dietary supply of protein and daily lean growth is linear up to the point at which protein deposition rate is maximized. While increasing diet protein is creating an improvement in lean tissue growth rate, the fatness of the pig will be progressively and proportionately diminished. Alternatively, diets that do not adequately provide for the requirement of absolute amounts of protein will fail to allow maximum lean tissue growth. Energy not utilized for the business of protein synthesis will be diverted to fatty tissue growth. But excessive protein reduces dietary energy level. The energy yielded from protein by deamination is about half of the assumed digestible energy of the protein. The effective energy value of a diet containing excess of protein will therefore fall as the deamination rate rises, with the resultant diminution of energy available for fat deposition. Overall, increasing the dietary concentration of protein, and offering a higher rate of daily protein supply, will increase the percentage of lean and decrease the percentage of fat in the carcass; this response is, however, curvilinear in terms of its cost-benefit at higher levels of protein supply. The role of protein concentration upon carcass quality is shown in Figures 2.27 and 2.28.

Level of feed (energy)

As a 10kg piglet may usually contain about 15% lipid, and as modern improved strains of pigs may reach slaughter weight with the same 15% of body lipid, it may be surmised that the balance of fat to lean in the growth will be rather constant over the growing period. An increase in the ratio of dietary energy to dietary protein over the growing period is only required to the small extent needed to cover for the increase in maintenance requirement as the animal grows. Where variation in body composition occurs, the greatest single determinant of that variation in growing pigs is the quantity of

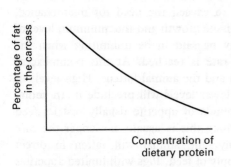

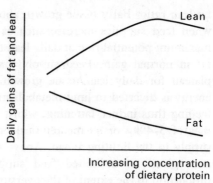

Figure 2.27 Increase in diet protein will reduce fat first by increasing lean growth and second by reducing the amount of available energy

Figure 2.28 Growth responses of lean and fat to increase in diet protein concentration. The response for lean is linear until the plateau for lean-tissue growth potential is reached. The response for fat continues downward even after lean-tissue growth is maximized, so improvements in *percentage* lean would be seen even after the response in lean-tissue growth *rate* has been maximized.

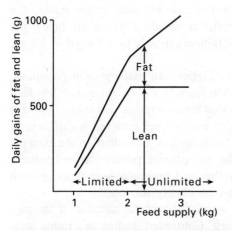

Figure 2.29 Fat and lean growth responses to increase in supply of feed. The actual levels of response achieved will differ from the example shown here according to the sex and genotype of the pig. It is assumed that the feed used is balanced for all required nutrients.

feed consumed (Figure 2.29). In normal growth the linear response of lean tissue to increase in feed supply is coupled with a minimum level of fat. In the limited feed supply phase there is a relatively constant relationship between daily gains of fat and lean over a range of growth rates. The minimum fat:lean ratio is thus an important determinant of body composition during this phase. This minimum fat:lean ratio is a characteristic of sex and breed type. Males and improved breeds have a lower minimum fat:lean ratio. Fatty tissue growth above this minimum can only be achieved when feed supplies increase such as to exceed the need for maintenance, maximum potential rate of daily lean tissue growth and the minimum level of fat in normal gain. Feed supply may be said to be unlimiting when the plateau for daily lean tissue growth rate is reached. At this point excess energy is diverted to lipid metabolism and the animal fattens. High levels of feeding thus induce fattening, whilst lower levels will preclude it. In young pigs of 5–40kg or so the physical bounds of appetite usually restrict feed supply to the limiting range. Above 40kg this is usually no longer the case and, given an unlimited feed supply, the animal will fatten in direct proportion to the extent of the oversupply of feed. Pigs with limited appetites may never be able to eat enough to fatten before slaughter weight is attained; whilst pigs with lower potential for lean tissue growth rate, or with higher appetites, will be likely to fatten for a considerable proportion of their growth, and these animals will need their feed intake to be closely controlled if fattening is to be avoided (see Table 2.10).

Even while feed supply is inadequate to maximize lean tissue growth, there is also some deposition of a minimum level of fat in normal growth. If this minimum level of fatness in normal growth gives backfat depths at slaughter weight which are in excess of the premium grade standard, then only draconian reductions in feed supply and growth rate would bring about a significant improvement. Equally, in the case of some entire males, the minimum ratio may give backfat depths at slaughter which are below the minimum. In this latter case adequate fatness can only be achieved by luxury levels of feed intake.

Pigs of high genetic merit may have higher lean tissue growth potentials, lower minimum fat ratios or both. Such animals will be thinner at low feed intakes and more difficult to fatten as feed level increases. Excessive fatness is feasible for all pigs, but only if the feed intake level achieved is sufficient to maximize lean tissue growth rate and be defined as 'unlimited' in the example in Figure 2.29. It is evident, in order to optimize production and achieve grading targets, that the feed supply that will maximize lean tissue growth without generating excessive fat must be closely identified.

Overall, feeding *ad libitum* will be associated with fat carcasses if the pigs are of high appetite or low genetic merit. Controlled feeding to a ration scale will reduce fatness, but may also reduce daily gain. Given that the daily lean tissue growth potentials are greatest for entire males and worst for castrates, the recent results from the Meat and Livestock Commission Stotfold Research Station given in Table 2.11 illustrate most satisfactorily the relationship between feeding level and fatness, and the way in which sex and feed level interact with regard to carcass quality. Example feeding scales for controlled feeding are shown in Figure 2.30.

Table 2.11 Influence upon carcass quality and growth performance of two levels of feeding and three sexes of pigs. From the Meat and Livestock Commission Stotfold Research Station

	Ad libitum feeding			Controlled feeding		
	Male	Female	Castrate	Male	Female	Castrate
Daily feed intake (kg)	2.1	2.1	2.3	1.7	1.7	1.7
Daily live-weight gain (kg)	0.86	0.79	0.82	0.72	0.68	0.64
Killing-out percentage	75	77	76	75	76	76
Backfat depth (mm P2)	11.6	12.0	14.7	10.3	10.2	12.3
Carcass lean (%)	57	56	53	59	59	55
Lean growth (g/day)	390	360	340	330	320	280

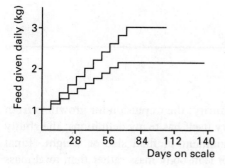

Figure 2.30 Controlled feeding scales, incremented weekly towards a given maximum. The higher scale is appropriate for pigs with higher lean-tissue growth rates, or in circumstances where fatness is not heavily discriminated against.

CHAPTER 3

GROWTH AND BODY COMPOSITION CHANGES IN PIGS

Introduction

The purpose of growth is to reach maturity; the impulsion for growth is from both age and nutrient supply. Maturity itself has many definitions; the ability to breed is reached well before attainment of final size or weight. Final weight may be best judged in terms of lean body mass rather than total mass due to variable, and highly nutritionally dependent, fatty tissue levels being possible in the mature pig. Expected mature lean mass and the time of its being reached is the basis of growth analysis, and it is evident that with selection for increased daily lean tissue gain there has been an increase in mature lean tissue mass. Lean tissue comprises around 22% of protein, and the mature protein mass of modern pigs selected for fast lean tissue growth rate appears to be in the region of 35–55kg. Mature live weights of pigs may vary from 150 to 400kg. Improved hybrid females may reach 300–350kg without being unduly fat, but unimproved sows may become no heavier than 250kg at which weight they are obese.

Growth is usually understood to relate to gain in weight brought about by cell multiplication (as in prenatal cleavage) and cell enlargement (as in postnatal muscle growth). As a part of cell enlargement there may also be simple incorporation of material directly into cells (as in the inclusion into fatty tissue of lipid).

Development relates to changes in the shape, form and function of animals as growth progresses. The blocky Pietrain, Belgian and German Landrace-type pigs develop in a way that results in meatier hams and larger loin muscles. Fatty strains of pigs will develop in such a way as to be a different shape from lean strains.

The study of changes in body proportions and body shape during progress to maturity has not received as much attention as that of overall tissue growth. As a pig develops the differential deposition of fatty tissues in the various parts of the body is not very evident. Most pigs lay down some two-thirds of their fat as external subcutaneous fat, and about one-third internally, the latter mostly as intermuscular fat and fat around the kidneys and intestines. Subcutaneous fat tends to be laid down in a way such that, although different parts of the body have different amounts of fat, these do not develop in marked sequence. A young fat pig will carry a similar balance of fat in its various depots as an older fat pig. Perhaps another reason for development (rather than growth) receiving scant attention is because pigs differ from beef and lamb in having a smaller gradient in eating quality and acceptability between the various parts of the body.

In contrast, the absolute rate of growth is a prime determinant of the efficiency of conversion into meat because of the savings in feed used for maintenance (Figure 3.1). Maintenance costs occur daily, utilizing feed but yielding no product. A pig growing slowly will incur the same feed maintenance costs as one growing fast, but will have less product to offset the fixed cost of that feed. The ratio of lean to fat in the growth is second only to the growth rate itself as a controller of the efficiency of feed use, because the feed energy cost of fatty tissue growth is about four times that of lean tissue growth on account of the differing water content of these two tissues (Figure 3.2).

The relative rates of growth of bone, muscle and fatty tissues are important considerations in the provision of meat for human food. The ratio of lean to

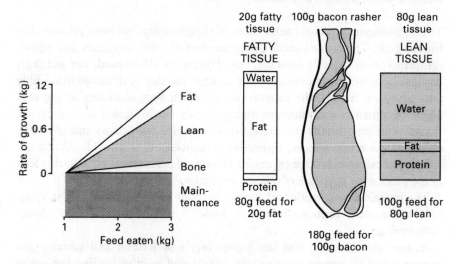

Figure 3.1 As growth rate increases, so does efficiency because the fixed cost of maintenance feed is spread over a higher level of productive output.

Figure 3.2 Composition of lean and fatty tissue.

49

bone contributes to carcass value whilst the amount of fat on the meat is a vital indicator of product quality, with subsequent ramifications into meat processing and retailing practices. Relationships between bone, muscle and fat growth will, of course, also affect the development of the animal. At any given weight, animals grown slowly and being of greater age will tend to have a higher ratio of bone to fat. Animal body shape is strongly influenced by the extent and the position of the fat cover. The rounded shape of a sucking pig is likely to be due to the high level of covering subcutaneous fat. However, this is not invariably the case as, for example, with pigs of the Pietrain and Belgian Landrace type, in which a well-rounded rear quarter does not indicate fatness but, rather, high lean meat yield.

Growth occurs through the medium of the accretion of bone, fatty and lean tissues in the body. It is a result of a positive difference between the continuous anabolic and catabolic processes associated with tissue turnover. Because most fatty tissues (but not brown fat) are turned over slowly, there is a close association between the absolute amount of fatty tissue anabolized and the absolute amount accreted. On the other hand, lean tissue turns over rapidly, such that accretion may be only 5–20% of the total protein anabolized; the less mature the pig the higher that proportion.

General composition of the body

The approximate physical composition of slaughter pigs has been presented in Figure 2.18. The crude carcass, after removal of offal, intestines and blood, comprises about 75% of a meat pig, but of this some 11% is head, feet and skin, leaving about 64% of carcass side. The carcass side may be dissected into edible fatty and lean tissues, the proportions of which vary according to pig sex, genotype, nutrition and slaughter weight. A lean, head-off, skin-off and feet-off carcass side may usually comprise 10% bone, 23% fatty tissue and 66% lean tissue (of the crude carcass, respective percentages would be 11, 30 and 58; leanness of carcasses is often expressed in terms of percentage of dissectable lean in the carcass for purposes of quality assessment).

The chemical composition of the whole body of the growing pig is about 64% water, 16% protein, 16% lipid and 3% ash, with a trace of (liver) carbohydrate.

At any given weight and sex higher levels of feeding will increase the percentage of fat, while at any given weight and level of feeding the entire male is considerably leaner than the castrated male with the female intermediate (60% carcass lean for the entire male, around 56 and 58% for the other two sexes). As already appreciated from Chapter 2, increase in body weight is usually associated with a progressive increase in fatness unless feed intake is limited. Table 3.1 shows the chemical composition of the body of pigs as they grow and demonstrates the increase of fatness and the decrease in water:protein ratio, as well as showing the dramatic effect that feed intake control can have upon fatness. Protein content is found to be relatively much more stable a proportion of the total body than fat content; the former usually varies between extremes of 14 and 18% (if in doubt, assume 16), while the latter may vary from 5 to 40%.

The composition of pigs as they grow is best expressed in the form of an allometric relationship $Y = aX^b$ where Y is the component and X the empty body weight. Values in Table 3.2 indicate the general direction of developmental changes in body components through growth. Pigs fed less and/or of improved genotype would have relatively less fat and more lean at any given empty body weight.

Changes in the general composition of the body with the progression of live

Table 3.1 Chemical composition of pigs (%)

| | Birth | 28 days | 100 kg | | 150 kg |
			full-fed	severe limit-fed	
Water	77	66	60	68	63
Protein	18	16	15	17	16
Lipid	2	15	22	12	18
Ash	3	3	3	3	3

Table 3.2 Growth of dissected carcass and of chemical components of full-fed pigs (20–160kg live weight) described using the allometric relationship $Y = aX^b$, where Y is the component and X the empty body weight (kg)[1]

	b	a	Values for Y when X = 100
Dissected edible lean	0.97	0.41	36
Dissected edible fat	1.40	0.030	19
Dissected bone	0.83	0.16	7.3
Whole body protein	0.96	0.19	16
Whole body water	0.86	0.93	49
Whole body lipid	1.5	0.020	20
Whole body ash	0.92	0.049	3.4

[1] Live weight $\simeq$ 1.05 empty body weight.

weight and time are shown graphically in Figure 3.3 (chemical components) and Figure 3.4 (physical components).

The relationships between bone and muscle and ash and protein are relatively constant (ash = 0.03 live weight or 0.20 protein) due to the use of bone as the support structure for muscle. In contrast, although the lean content of the body varies somewhat inversely with fatty tissue, the relationship is not especially good and this is due in part to the variable water content of lean. Usually dissected lean tissue (muscle) contains 70–75% water, 5–15% fat and 20–25% protein. In a very young pig the water content of lean can be as high as 80% but in a mature animal the water content of lean may well be less than 70%. Although carcass dissected lean may be estimated by the rule of thumb: carcass dissected lean = 2.4Pt, where Pt is the total body protein mass, which is acceptable for meat pigs, the relationship between body water and body protein is exponential. Calculation of total body water from an expression such as $4.1Pt^{0.89}$ allows for the reduction in the amount of water associated with each unit of protein as the animal grows.

Fatty tissue contains 10–25% water, 2% protein and 70–80% lipid. Because the growth of fat is so variable a proportion of the total growth, fat is poorly predicted from the knowledge of either muscle mass or live weight. Most of the variation in fat growth is, however, due to the level of nutrient supply; the more feed, the fatter the pig. But where feed is limited the growth of fat

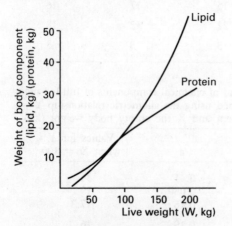

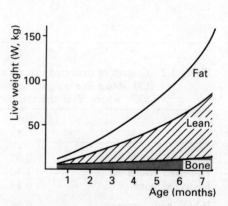

Figure 3.3 Weights of lipid and protein body components of *ad libitum*-fed pigs in relation to increase in live weight. Notice the relative constancy of the lipid : protein ratio ≃ 1 : 1) between 25 and 100kg W (Edinburgh data).

Figure 3.4 Composition of growing pigs.

in relation to lean is much more predictable and may be represented as a constant proportion of protein over much of the growth period. Although it is generally agreed that in the course of normal growth pigs become fatter as they grow larger, the point at which fattening begins is highly dependent on sex, genotype and feed level. It may no longer be taken as axiomatic that the percentage of fat in the body of the pig will change positively during growth. Improved pigs may contain no more than 12% of lipid over the totality of a 10–110kg growth phase (see Figure 3.5).

It is therefore generally helpful to consider the principles of growth primarily in terms of the growth of protein and lean, and to allow fat to follow protein in the form of some given relationship when nutrition is limiting, but to be independent of protein, and wholly dependent on feed supply, when the latter is ample.

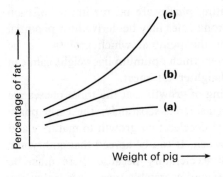

Figure 3.5 Relationship between percentage of fat in the body of the pig and pig live weight. Curve (a) is appropriate to entire male pigs, pigs of improved genotype or pigs whose feed intake is limited. Curve (c) is appropriate to castrated male pigs, pigs of unimproved genotype or pigs whose feed intake is unlimited. Curve (b) is intermediate.

Growth curves

Samuel Brody, in *Bioenergetics and growth*, published by Reinhold (New York) in 1945, writes:

> The age curve of growth may be divided into two principal segments, the first of increasing slope, which may be designated as the self-accelerating phase of growth, and the second of decreasing slope, which may be designated as the self-inhibiting phase of growth. The general shape of the age curve may thus be said to be determined by two opposing forces: a growth accelerating force and a growth retarding

force. The former manifests itself in the tendency of the reproducing units to reproduce at a constant percentage rate indefinitely, when permitted to do so. In the absence of inhibiting forces, the number of new individuals produced per unit time is always proportional to the number of reproducing units. That is, the percentage growth rate tends to remain constant.

These remarks are made in the context more of population growth and growth of cells, rather than of whole-animal organisms; but Brody also makes the point that he believes that the two can be fitted in the same conceptual frame (see Figure 3.6 – *The 'sigmoidal' assumption*). A vital part of the interpretation of Brody's principles is the particular points in the curve at which the various events occur. If, for example, it is wrongly assumed (i) that the curve begins its acceleration at birth (rather than conception), (ii) that puberty is always associated with the point of inflection (which it is not) and (iii) that the accelerating and decelerating phases are mirror images of each other (which they are not), then the wrong rules may be derived for potential rate of growth in young animals, for the point at which growth can be maximized, for the period of growth over which optimum liveweight gain can be made and for the best time of slaughter for meat.

It is fundamental to the understanding of growth that the three phases are dissociated one from the other and considered separately: the early phase of acceleratory growth, the late phase of deceleratory growth to maturity and a middle phase of linear growth in between. It can be agreed that embryonic and early foetal growth must be self-acceleratory because there must be some relationship between current mass and accretable mass (it is ridiculous to suppose that a 1kg animal can grow at a rate of 1kg/day), and 'the number of new individuals produced per unit time being proportional to the number of reproducing units', and 'percentage growth rate tending to remain constant', are entirely consistent statements for the cell multiplication activities of early conceptus growth. Equally, it can be agreed that growth to

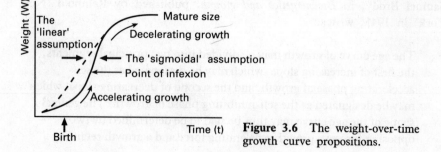

Figure 3.6 The weight-over-time growth curve propositions.

maturity must indeed be deceleratory if mature size is to be rationally attained (if the system does not slow down, then how may it ultimately stop?). But the intervening period need not necessarily be any mathematical continuation of these two phases nor, being so differently driven, need the acceleratory and deceleratory phases be of similar constitution. The intervening period is of prime importance for the production of human food from pigs as it spans the major proportion of the time between weaning and slaughter (the former usually occurring at around 28 days of age, and the latter usually at or below the attainment of 50% of mature size). In the context of Brody's proposition, there can be no *a priori* assumptions about constancy of the percentage growth rate or the number of new cells being proportional to the mass of the existing cells; postnatal growth is more a function of increase in cell size and of cell filling than of cell number, and there may be more self-evident truth in assuming constancy of *absolute* growth rate postnatally rather than constancy of percentage growth rate. In short, the growth of populations of animals, or of bacteria, is not an appropriate analogy for the growth of the body of an individual pig.

Although sigmoid growth curves may be readily drawn, for example for chicks and for calves weaned on the day of birth, it is not immediately evident why the self-acceleratory phase of early postnatal growth should not be attributed in large part to failure to supply adequate nutrients. Indeed, where nutrition and environment are unlimited, there is evidence that absolute rate of growth can be extremely high in the young animal. Piglets given milk *ad libitum* in early life may gain on average more than 300g daily in the first postnatal days and up to 500g daily by day 6. This not only shows the potential is above the 200–400g possible from the milk yield of the sow alone, but also that the potential in a young pig is a remarkably high percentage of its body weight (nearly 15%). Twenty-one-day weights above 7kg *average* for litters of pigs sucking productive and well-fed sows are not exceptional, and individual piglet weights of 12kg at 28 days (rather than the normal expected 7–8kg) again indicate the difference between potential and commercial achievement. Given unlimited nutrition and an excellent environment, it may be that the maximum absolute rate of growth can be achieved early in life and be relatively constant through most of the growth period. Such a proposition denies the idea that achievable growth rate is somehow related in a constant way to the mass of body already attained – as might be assumed from a strict interpretation of Brody's original statements. Also denied is the corollary to the sigmoidal growth curve which is that daily gain, as a function of age or weight, is a quadratic response, peaking at around 0.5 of mature weight.

It is realistic to support a view that maximum gains may be reached early

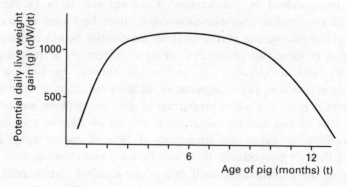

Figure 3.7 Daily gain against time: growth curve for pigs of high genetic merit.

in life and be maintained over a considerable sector of the growth phase. This presents a broadly linear assumption of potential growth once the pre- and immediate postnatal phases are past and before the approach to maturity is begun (Figure 3.6 – *the 'linear' assumption*). Expressed in terms of the influence of time (t) upon daily liveweight gain (dW/dt) a flat top response may therefore be described as in Figure 3.7. However, Brody's propositions remain firm as descriptors of exponential growth *in utero* and of the maturing processes of the later stages of life (Figure 3.8).

Lean and protein growth

The foregoing description of potential growth rate follows somewhat that of a Gompertz curve (Figure 3.9), whose point of inflection at time t* occurs at 0.37 of mature size and which is characterized by a substantial period of effectively linear growth, particularly from 0.1 of maturity (about 30kg live weight for a pig) to 0.6 of maturity (about 180kg live weight for a pig). The upper limit, or maximum potential, for protein growth is unknowable but may be implied from its being approached under nutritionally and environmentally unlimiting conditions. As protein growth is both highly desirous and highly efficient, its maximization is of intense interest. The description of protein growth potential as a function of pig weight, age or stage of maturity is prerequisite to the estimation of nutritional requirement and the derivation of best production strategies. The use of a Gompertz curve to describe potential growth in unlimiting conditions has proved useful.

The curve to give weight W at time t is expressed as:

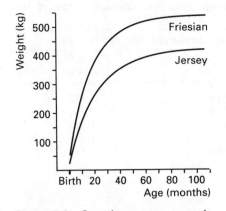

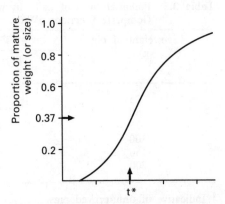

Figure 3.8 Growth curves to maturity for two breeds of dairy cow, one larger maturing than the other (adapted from Brody, S (1945) *Bioenergetics and Growth*, Reinhold, New York).

Figure 3.9 Weight (expressed as a proportion of mature weight) against time, expressed in the form of a Gompertz function. The point of inflection occurs at time t* and at 0.368 of final (mature) size.

$$W = A.e^{-e^{-B(t-t^*)}}$$

(Equation 3.1)

where A is the liveweight at maturity (kg), B is the growth coefficient and t* is the point of inflection (days). Maximum growth rate is achieved at (A.B)/e, but the curve is helpfully rather flat topped. For improved white genotypes appropriate values for A appear to range between 300 and 400kg, whilst values for B may be around or in excess of 0.01. t* is usually about 180 days of age. Unimproved genotypes of different breeds may have lower values for both A and B. (A.B)/e readily calculates to greater than 1kg daily liveweight gain for improved genotypes. Daily liveweight gain (dW/dt) at any given value for W is calculable as:

$$dW/dt = B.W.log_e(A/W)$$

(Equation 3.2)

Potential daily liveweight gains using different values for A and B are given in Table 3.3.

The most important component of the daily liveweight gain is protein, as this controls both production efficiency and product quality. Daily protein retention (Pr) measured by slaughter experiment at Edinburgh for unimproved entire male pigs was found to be 122g between 20 and 105kg, 144g between 20 and 150kg, and 101g between 20 and 200kg. The relative constancy of the rate of protein retention over much of the growth phase was confirmed and the Gompertz curve derived from the experimental data is presented in Figure 3.10.

Table 3.3 Potential rates of daily liveweight gain (g) calculated by use of the Gompertz function: $dW/dt = B.W.\log_e(A/W)$

Liveweight of pig (W, kg)	Daily liveweight gain (dW/dt, g)	
	$B = 0.009^1$ $A = 250^1$	$B = 0.011^2$ $A = 300^2$
25	518	683
50	724	985
75	812	1,144
100	825	1,208
150	690	1,144
200	402	892

[1] Indicative of unimproved pigs.
[2] Indicative of improved pigs.

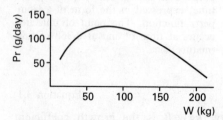

Figure 3.10 Daily rate of protein retention (Pr) in relation to the pig body weight (W) (data from experiments at Edinburgh).

Values for A and W expressed directly in terms of protein mass (rather than liveweight) allow calculation of the daily rate of protein retention at any given body protein mass. Different pig genotypes and sexes can be described in terms of their values for B and for A. By the simple determination and presentation of two numerical values the potential for protein retention at any given weight can be estimated. With this knowledge optimum nutritional and management strategies for any given pig type can be elucidated. These two values for B and for A are pivotal to the effective production of meat from pigs. Figures 3.11 and 3.12 show curves calculated for the equation $dW_p/dt = B.W_p.\log_e(A_p/W_p)$, where W_p is the protein mass at any given time (t), and where A_p is the protein mass at maturity. In shortened nomenclature, dW_p/dt is referred to as $P\hat{r}$ (for the ultimate or potential daily rate of protein retention at any given point in growth), A_p is referred to as $P\hat{t}$ (for ultimate total body protein mass) and W_p is referred to as Pt (for current total body protein mass). Figure 3.11 describes values for B and A (as $P\hat{t}$) of 0.0105 and 37.5kg, indicative of castrated male pigs of unimproved genotype, while Figure 3.12 describes values for B and A (as $P\hat{t}$) of 0.0125 and 47.5kg, indicative of entire male pigs of improved genotype. Other sexes and types would fall with intermediate values. The shape of both curves argues well the possibilities for rapid early growth, and argues

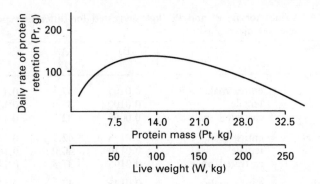

Figure 3.11 Daily rate of protein retention for growing pigs calculated using the equation $P\hat{r} = B.Pt.ln(P\hat{t}/Pt)$, where $B = 0.0105$ and $P\hat{t} = 37.5kg$.

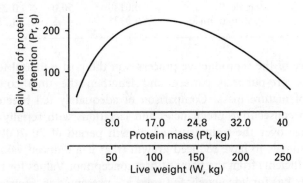

Figure 3.12 Daily rate of protein retention for growing pigs calculated using the equation $P\hat{r} = B.Pt.ln(P\hat{t}/Pt)$, where $B = 0.0125$ and $P\hat{t} = 47.5kg$.

adequately the tendency toward a relatively stable value for potential protein growth over much of the commercially active growing phase to slaughter, namely between 40 and 140kg live weight. Table 3.4 suggests values for B and for Pt̂, and for maximum attainable rates of protein deposition (Pr̂), as appropriate to different types and sex of pigs. Values offered range from 0.0095 to 0.0135 for B, from 32.5kg to 52.5kg for Pt̂, and from 115g to 260g for Pr̂.

Growth in all three sexes of the meat pig, with the exception of entire males used for breeding, is interrupted by slaughter, usually at about 30–50% of mature size (80–140kg). Growth in the female breeding sow is interrupted by the incursion of conception and pregnancy. Pregnancy dramatically affects the rate of protein retention, creating a marked downturn in the rate of maternal growth. Nevertheless, derivation of nutrient requirements for a growing sow requires knowledge of the pattern of accretion of body protein (and lipid) and the Gompertz function (Equation 3.1) is as adequate a descriptor of later growth as early growth, but the factors used must allow for

Table 3.4 Values for B, Pt̂ and Pf̂ (kg) suggested for different types and sex of selected meat pigs

		B_p	A_p (Pt̂)	(B.A)/e (Pf̂)
Utility:	entire male	0.0105	37.5	0.145 (0.145)
	female	0.0100	35.0	0.129 (0.130)
	castrated male	0.0095	32.5	0.114 (0.115)
Commercial:	entire male	0.0115	42.5	0.180 (0.180)
	female	0.0110	40.0	0.162 (0.160)
	castrated male	0.0105	37.5	0.145 (0.145)
Improved:	entire male	0.0125	47.5	0.218 (0.220)
	female	0.0120	45.0	0.199 (0.200)
	castrated male	0.0115	42.5	0.180 (0.180)
Nucleus:	entire male	0.0135	52.5	0.261 (0.260)
	female	0.0130	50.0	0.239 (0.240)
	castrated male	0.0125	47.5	0.218 (0.220)

the presence of the reproductive processes in the post-pubertal female which disrupt the pre-pubertal pattern and lengthen the time course for the approach of mature mass. Comparison of adequately fed breeding animals involved in consecutive pregnancies and lactations with serially slaughtered growing pigs over the uninterrupted growth period of 20–200kg W shows breeding animals to have reduced growth (that is a reduced value for B, the growth coefficient) from the point of first conception. Values for body protein content (Pt, kg) for adequately fed sows are presented in Figure 3.13. With an initial value for Pt at first conception of 16.6kg, respective gains from

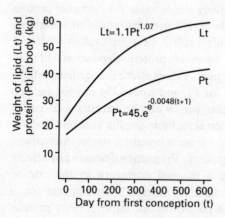

Weight of lipid (Lt) and protein (Pt) in body (kg)

$Lt = 1.1Pt^{1.07}$ Lt

Pt

$Pt = 45.e^{-e^{-0.0048(t+1)}}$

Day from first conception (t)

Figure 3.13 Increase in maternal body protein and maternal body lipid from the time of first conception. The point of inflection (t^*) was coincident with conception ($t^* = -1$), indicating conception to be the point of diminishment in daily rate of growth.

conception to conception over the first four parities were 11, 8, 5 and 2.5kg. The B value, at 0.0048, is at best only half of that likely to have pertained prior to conception. It is evident that the rate for Pr measured for these sows can only be regarded as observed performances under the duress of the reproductive cycle and not as any indication of what may have been a preferred pig growth target. It is also equally clear that the initiation of the reproductive processes is the cause of the inflection of the growth curve, and not its consequence.

It may often be convenient to estimate Pt in live sows from knowledge of W and backfat depth at P2:

$$Pt(kg) = -2.3 + 0.19W - 0.22P2$$

(Equation 3.3)

Changing the shape of the potential growth curve

The shape of the actual (achieved) growth curve is readily changed by nutrition, environment and indeed pregnancy. For any given individual, however, the shape of the *potential* curve is, by definition, an intrinsic and unchangeable character. Genetic selection may, however, be imposed upon a population to artificially move the intrinsic characteristics of that pig population. If individuals in an existing population grow according to case (a) in Figure 3.14, and that population is then selected for improvements in lean tissue growth rate, then the slope of the regression of weight (W) upon time (t) will be steepened in the direction of the arrow and towards case (b). If time at maturity is delayed only a little (which appears so), then the undeniable consequence is an increase in A, the mature size. Because enhancement in growth rate is a basic goal in pig production, as in other branches of animal production, then it should not be too surprising to note that genetic selection and breed substitution has consistently resulted in the creation of larger animals, not only in the case of pigs, but also sheep, cattle and fowl. But perhaps the greatest change in mature size over recent years has indeed been with the

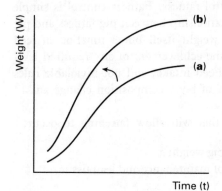

Figure 3.14 Genetic selection for increased liveweight gain steepens the slope for weight-over-time and also increases final mature body weight.

domestic pig, consequent upon the extremely high selection pressure that has been placed upon lean tissue growth rate in genetic improvement schemes.

Populations of animals conforming to case (b) in Figure 3.14, as well as growing faster and maturing bigger, have higher maintenance requirements in their adult breeding herds and fatten less readily. As may be interpreted from the Figure, at any given weight pigs conforming to case (b) will be less mature than those of case (a). Conventional slaughter weights for case (a) animals, maintained for use with case (b) animals, will ensure a meat of lower physiological maturity derived from animals at a lesser proportion of their potential final size. Increased size at maturity brings with it an increase in optimum slaughter weight.

Fat and lipid growth

Little pigs gain fat rapidly in early life, with special help from mothers' milk with 8% content of fat. The most rapid phase of fat growth may well have occurred in the pig before 4 weeks of age, whereas maximum lean impulsion may not be reached until rather later. Whilst there is less than 2% of fat in a pig at birth, by 21 days of age there is usually more than 15%. Modern pigs may never be fatter than when 21 or 28 days of age at the point of being weaned. It is possible that maximum fatness has been attained within the first month of a 4–7 months' growth time to slaughter. Carcass pigs can achieve slaughter weight at 100kg readily with less than 20% of fat in the carcass. Pigs now rival broiler fowl as being the leanest meat available for human consumption. Over-fat pigs are no longer excusable, for although pigs do tend to get fatter as they get bigger (as appetite outstrips lean growth potential) modern requirements for lean meat at relatively high carcass weights have invoked the need to control fatness. Fatness control is simple with modern pig strains because the relationship between pig fatness and pig weight is not primarily a function of weight itself but a function of feed intake. Feed intake control will cause the achievement of any required level of fat at any required carcass weight. Feed intake, and not inviolable rules relating to time and weight, is the crux of body composition change and of carcass fatness in meat producing pigs.

The following conditions are those that will allow fattening to occur:

1. when the diet is imbalanced (at any pig weight);
2. when feed intake exceeds the needs of maintenance and lean tissue growth (at any pig weight);

3. when, for sound physiological reasons, the body places fat accretion above lean accretion in its order of priority (as is the case postnatally between birth and weaning, or during pregnancy when preparing for lactation, or in expectation of times of feed shortage);
4. when mature lean mass is achieved and ingested feed has no other function than fat growth to satisfy.

At this point it is helpful to differentiate total body lipid (Lt) into three components: essential fat, target fat and surplus or depot fat. A minimum level of fat is essential to normal metabolic function; this is probably about 5% lipid in the whole body (Lt = 0.05W). Target fat is the minimum level of fatness at which, having achieved that target, the animal feels sufficiently physiologically comfortable to partition available nutrients and to maximize metabolic effort toward the primary aim of reaching the potential for lean-tissue growth rate. At levels of fatness below the target, the achievement of target lipid levels will detract from the achievement of potential rates of protein retention, as the physiological priority will be to the former and not the latter function. Target fat gains may be conventionally expressed in terms of a given relationship with lean or protein gains. Because protein gains will be restrained until target levels of fat are reached, then at all times prior to the lean-tissue growth rate reaching its maximum potential, the ratio of fat to lean will reflect directly the proportion of fat in the gain which is the target level (see Figure 3.15).

While target fat is a physiological necessity for normal growth, surplus or depot fat is quite differently a means of either (a) dealing with surplus energy present in an imbalanced feed, or (b) creating a depot store of energy in the

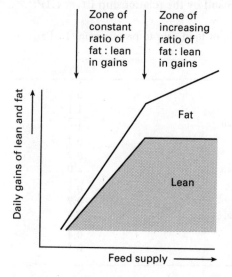

Figure 3.15 The ratio of lean to fat in the gain will tend to be constant until lean-tissue growth rate reaches maximum potential.

body in preparation for some prospective feed shortage. The full extent of depot fat storage in the pig is largely unknown, although Plate 6.6 may give some indication! Pigs with more than 50% of the body present as lipid are entirely credible but highly uneconomical. It would be unusual to find even obese contemporary pigs carrying more than 30–40% body fat. By definition, depot fat should probably be excluded from a quantitative description of growth on the grounds of the four conditions which allow fattening to occur, and which have been given above, perhaps with the singular exception of the fattening of a very young pig, which in any event may be justified by the use of a higher target fat value.

In comparison to the daily gains of protein (Pr), it would appear that for the sucking pig minimum target gains of lipid (Lr) are around or above equivalence:

$$(Lr:Pr)_{min} \geq 1$$

(Equation 3.4)

Target fat levels during normal growth between weaning and puberty, expressed in terms of $(Lr:Pr)_{min}$, are given in Table 3.5. It will be evident from comparison with Table 3.4 that selection has resulted in reduction of $(Lr:Pr)_{min}$ as well as an increase in Pr. This, however, is consequent upon simultaneous selection *against* Lr as well as *for* Pr. There is no particular physiological reason why high values for Pr should be associated invariably with low values for Lr.

After the attainment of puberty there is a new dimension – that of pregnancy. Pregnancy creates an enhanced level of target daily lipid gain which is probably greater than quantitative equivalence to the daily rate of protein gains, as shown in Figure 3.13 and by the relationship $Lt = 1.1Pt^{1.07}$.

Table 3.5 Values for the minimum ratio of lipid gain to protein gain $(Lr:Pr)_{min}$ in selected meat pigs

		$(Lr:Pr)_{min}$
Utility:	entire male	0.9
	female	1.1
	castrated male	1.2
Commercial:	entire male	0.7
	female	0.9
	castrated male	1.0
Improved:	entire male	0.5
	female	0.7
	castrated male	0.8
Nucleus:	entire male	0.4
	female	0.5

Most of the extra fat laid down in pregnancy is used to fuel lactation. Breeding sows with greater than 30% of total lipid in the body ($Lt > 0.3W$) are likely to have reproductive problems due to over-fatness. Breeding sows with less than 17% body lipid are liable progressively to lose reproductive efficacy in terms of increased time taken to return to oestrus after weaning, reduced ovulation rate, reduced success for embryo implantation and survival and reduced piglet birth weight.

While for full-fed unimproved genotypes a body fat mass twice that of the protein mass, or more ($Lt > 2Pt$), used to be unremarkable, it would appear that the body content of lipid in the contemporary adult sow should preferably be maintained at 1.5 times the protein content ($Lt = 1.5Pt$), or approximately 20–25% of the total body weight as lipid ($Lt = 0.2W$). The lipid content should never drop below equivalence with the body content of protein ($Lt_{min} = Pt$).

The pattern of lipid gains in Figure 3.13 suggests that an initial value of Lt of 22kg at first conception may be followed by conception-to-conception gains for parities 1–4 of 16, 12, 7 and 4kg Lt respectively. Gains of Lt are thus approximately equal to 1.5Pt. With each parity, pregnancy gains of Lt were, of course, greater than this to allow for lactation losses.

It may often be convenient to estimate Lt in live sows from knowledge of W and backfat depth at P2:

$$Lt(kg) = -20 + 0.21W + 1.5P2$$

(Equation 3.5)

The special case of early growth

Having been born with only 1–2% of lipid, less than the 'physiological minimum', and considerably less than any conceivable target minimum (which would be 7–8% of lipid), the neonate pig is being entirely reasonable when it puts lipid gains above protein gains in its order of priority for nutrient partitioning. By so doing the baby pig can achieve a body lipid content of 15–20% by 21–28 days of age. To achieve these dramatic changes in body lipid, Lr in early life is self evidently much in excess of Pr. The trauma of weaning, however, rapidly amends total body lipid content to below 10% of body weight before more normal positive growth rates are resumed.

The remarkable ability of the young pig to grow rapidly in early postnatal life is not restricted to lipid growth. Total body growth rate potential is perhaps even more outstanding. The impressive nature of postnatal live-weight gains possible in a young pig is only exceeded by the apparent inability of pig producers world-wide to exploit fully the potential that is on offer. But the growth potential of young pigs is remarkable only in comparison with conventional agricultural expectation, not in relation to many other animal species such as the newly hatched pigeon, the young rabbit, the seal (seals double their weight in the first 3 days of life) and the whale, all of which grow, in proportion to their current size, much more rapidly than the fastest growing young pig. In the first week of life the baby pig consumes about four times its maintenance requirement, the baby seal six times maintenance, while the baby whale consumes eight times maintenance. Even the human baby conforms to the pattern; between the fourth and seventh week of age R.J.W., when 4kg live weight, gained 290g weekly, a rate of growth which (if maintained) would have resulted in a 24 stone (154kg) 10-year-old.

In terms of their potential, healthy pigs of 5kg live weight can readily be shown to grow 600g daily in circumstances of unlimited nutrition. Growth increments of 10% or more of current body weight are achievable between birth and about 8kg live weight, B values within Equation 3.2 having been measured up to 0.02 for young pigs at Edinburgh. Figure 3.16 (as measured) presents only conservative possibilities for the growth of sucking pigs, the particular animals tracked here having no access to supplementary feed of any kind. Loss of growth impulsion toward the later stages of the period of measurement shows the consequences upon potential growth of a food supply source (mother's milk) unable to keep up with pig requirement and growth potential.

The tremendous impulsion found in the early growth of mammals is likely to be associated with survival mechanisms, and may constitute an exception to the Gompertz assumption which seems to hold so well elsewhere throughout growth to maturity. When analysing *achieved* growth rates of young pigs, however, gains made do appear to fit the Gompertz assumptions of slower growth in early life with subsequent acceleration; but this is not reflective of any natural law, it is more likely to be accounted for by the following three phenomena:

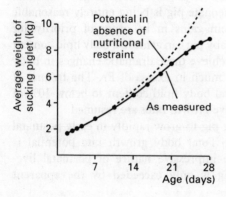

Figure 3.16 Growth of piglets in a litter without access to creep feed and without symptoms of ill-health.

1. Expression of growth curves over a substantial period of the whole life fail to account for perturbations in one small sector; that is, the curve is dominated by the majority of the growth pattern and cannot properly express the exceptional early growth phase.

2. Young animals are not offered unlimited feed supply, and growth is thereby restrained below potential through the medium of nutrient limitation.

3. The young pig is usually weaned in the middle of the early growth period (at around 21–28 days of age), and the consequences of weaning are often both a compromised health status and a severely reduced nutrient intake. As the trauma of weaning is recovered from, an acceleratory phase of growth emerges (Figure 3.17).

Young pigs would grow considerably faster than they do on commercial production pig farms if conditions were more close to ideal, and the relatively slower rates of growth actually experienced are consequent upon the imposition by man of a variety of restraints, rather than any natural growth pattern. Figure 3.16 can be taken as indicative of reasonable and conservative expectations for pre-weaning growth performance of young pigs, and Table 3.6 shows some regularly achieved post-weaning performances of pigs at Edinburgh maintained under unexceptional conditions.

Weight loss by negative fat growth

Negative growth is not the reverse of positive growth, and is not open to a similar mathematical approach. Fat loss is invariably the consequence of an imbalance between energy demand and energy supply, the difference being made up from internal nutrient resources. The anabolism of fatty tissues may cease entirely before protein anabolism even slows down, while the

Table 3.6 Post-weaning growth performance of pigs at University of Edinburgh

Live weight at start (kg)	Live weight at finish (kg)	Days	Daily feed intake (g)	Daily live weight gain (g)
6	12	13	500	450
6	24	31	800	581
8	16	14	650	590
12	24	16	900	760

breakdown and utilization of body lipid and the build-up of body protein can proceed simultaneously (the former fuelling the latter). Negative fat growth happily utilizes the surplus fat in body depot stores, the purpose for the original deposition of body lipid depots being to protect against the very imbalance between nutrient demand and nutrient supply such as often occurs after weaning and during lactation, when pig appetite for ingested energy fails to measure up to metabolic requirements. Fat catabolism, however, may often bite deeper into target fat if demand for energy is strong enough. In this way the absolute lipid content of the pig may be found to fall below target levels. While fat reduction by utilization of the depot stores down to target body fat content levels may be considered as a normal aspect of pig bodily function, incursion into target fat will undoubtedly bring with it negative consequences and optimum body function cannot be expected thereafter, until such time as target lipid levels have been restored.

Negative growth therefore has an important contribution to make to normal growth and reproductive processes, but in so doing may dramatically influence body composition, and is only a normal phenomenon provided that it relates to depot fatty tissues. Thus, according to level of feed intake in lactation, lactating sows may, during the suckling period, lose lipid up to levels even as high as 20kg. Fat loss in lactating sows is closely related to weight loss:

Fat loss (kg) during 28 day lactation = 7.5 + 0.3 live weight loss (kg)

(Equation 3.6)

which indicates that even at sow body weight stasis there is some 7.5kg of fatty tissue catabolized during lactation, or 250g per day. More usually sows will lose about 10kg of live weight in lactation, creating normal fatty tissue loss levels of around 10kg in the course of a lactation, or around 500g of fat daily. The pregnant mammal anticipates impending lactation and lays down depot fat during pregnancy in order to lose it subsequently through the milk. The baby seal will grow 1.5kg per day – putting farm livestock of any age or weight to shame – on milk containing 50% fat produced by the dam which at this time may well be eating nothing. Daily losses of fat by lactating dairy cows can be more than 1kg per day. In terms of metabolic body weight, these amounts are all similar, and are found in the pig at about 10g per kg metabolic body weight. The same rates of loss also occur in lactating ewes. Because fat losses can be counterbalanced by water gains, the energy value of weight loss can rise to be considerably in excess of the gross energy value of pure fat (40MJ); energy values for live-weight loss for lactating animals as great as 90MJ/kg live weight have been recorded. Body fat losses during lactation must be made back during the subsequent pregnancy or progressive fatty tissue depletion will occur as the breeding sow ages.

Fat losses by newly weaned baby pigs will relate to the extent of the pre-weaning fat stores built up and the physiological need to maintain fat in the face of post-weaning nutritional stress. Weaned baby pigs preferentially lose subcutaneous fat rather than internal fat (the ratio of the rates of fat losses being about 9:1). Immediately after weaning pigs will usually show weight stasis, live growth impulsion picking up about 1 week later (Figure 3.17). The extent of this post-weaning check may vary from 2 days to 2 weeks, depending upon both pre-weaning and post-weaning management, housing and nutrition. It is apparent that weaning is a traumatic time for young sucking animals. Pigs of 28 or 21 days of age at weaning have body compositions approximating to 15% protein and 15% lipid, but only 7 days after weaning, whilst the protein content may have remained relatively stable at around 15%, the lipid content may well have fallen to well below 10%. Sometimes the percentage fat in the total body of the young weaned animal can be halved within the week. The apparent ability of little pigs to circumvent the laws of proportionality (losing lipid but not gaining protein) is due to the water in the empty body which increases to counterbalance fat losses.

Figure 3.18 shows data from four separate experiments. In experiments 1 and 2 pigs were weaned at 28 and 21 days of age respectively, and the compositions shown at these ages therefore relate to the sucking pig. In experiment 3 the pigs were weaned at 14 days of age, and the composition given therefore pertains to 7 days post-weaning. It is apparent that weaning has not only brought about a great reduction in the percentage of fat, but fat losses have been compensated for on a proportional basis by the addition of water. At 55, 53 and 42 days of age, pigs in experiments 1, 2 and 3 had failed to make good their post-weaning fatty tissue losses (gaining lipid at only half the rate of protein [$Lr \simeq 0.5Pr$]), but pigs on experiment 4 had done so. These latter pigs were given an unconventional diet with a high energy:protein ratio. Over the period of the experiments the composition of the post-weaning live growth in experiments 1, 2 and 3 was about 50% protein and 7% fat, but in experiment 4 there was more fat than protein in the gain (about 15% protein and 19% fat).

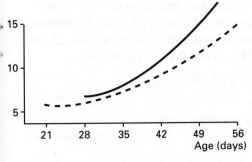

Figure 3.17 Growth of commercial pigs weaned at 21 (broken line) or 28 (solid line) days of age. The post-weaning lag phase of little or no growth may last for as short a time as 2 days or as long as 2 weeks.

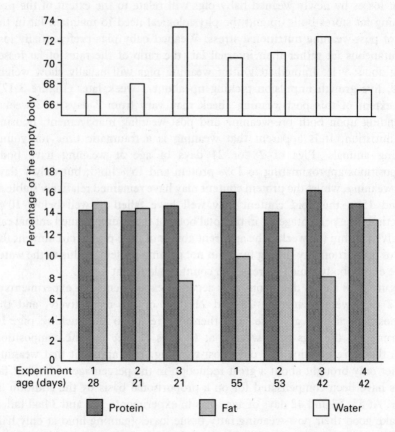

Figure 3.18 Percentage composition of pigs. Pigs on experiments 1 and 2 at 28 and 21 days of age were taken at the point of weaning. Pigs on experiment 3 were weaned at 14 days of age (from Whittemore, C T, Aumaitre, A. and Williams, I H (1978) *Journal of Agricultural Science* **91**:681).

In a further experiment negative lipid growth was studied in greater detail. Control sucking pigs were left on the dam, whilst those on the experimental treatment were weaned. Piglets were taken at 2-day intervals, during which time the weaned pigs only managed to maintain weight stasis. The suckled pigs, however, gained around 300g per day (Figure 3.19). For the weaned piglets, water, lipid and protein relationships drawn up between gains of the chemical components and gains of the empty body weights were:

$$WG(g/day) = 0.56EBWG + 53$$

(Equation 3.7)

$$LG(g/day) = 0.29EBWG - 56$$

(Equation 3.8)

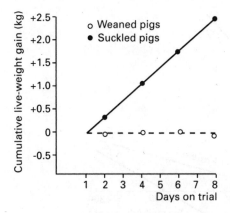

Figure 3.19 Cumulative weight gains of suckled and weaned pigs from 21 to 28 days of age.

$$PG(g/day) = 0.15EBWG - 4$$

(Equation 3.9)

These relationships are illustrated in Figure 3.20 which presents the compositional proportionality with respect to total gain. The slope of the protein line is 15% and it resolutely passes close to zero. At body-weight stasis, some 50g of lipid are lost from the body daily and this from pigs that are themselves only 5kg in weight (that is, 15g/kg metabolic body weight, or about 50% more than from the lactating cow). The counterbalancing chemical component was, of course, water, for which there were positive gains of 50g at weight stasis.

At low daily live-weight gains – below 200g – lipid losses are seen to support protein gains, and there is quite a range of gains over which lipid loss and protein retention occur simultaneously. Physical dissections of the tissues showed that, at zero live-weight gain and rapid subcutaneous tissue loss, simultaneous tissue gains were taking place, mainly in the heart, lungs, liver, intestines and other essential body components. Patterns of negative growth associated with the utilization (and subsequent recovery) of depot fat stores are shown in Figure 3.21, which demonstrates lipid losses both post-weaning and during lactation. Failure to allow recovery of depot stores by the provision of energy supply in excess of the immediate needs of protein growth or pregnancy anabolism would inevitably lead to progressive decline of fat levels, and ultimately the reduction of total body lipid below target level, with inevitable unfortunate consequences for pig health and productivity.

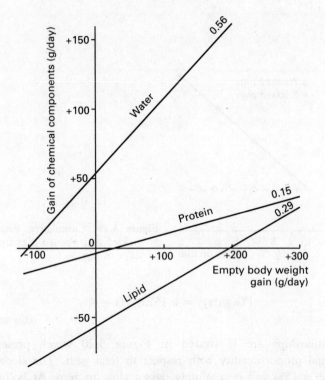

Figure 3.20 Relationships between total gain and the components of the gain (calculated from Whittemore, C T, Taylor, H M, Henderson, R, Wood, J D and Brock, D C (1981) *Animal Production* **32**:203).

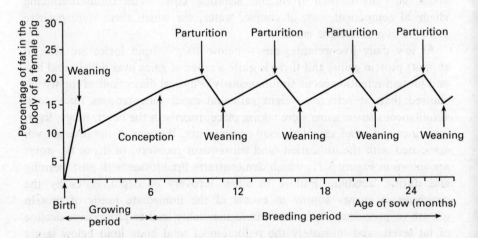

Figure 3.21 Changes in lipid content of the body of pigs consequent upon fatty tissue losses after weaning and during lactation.

Weight loss by negative protein growth and compensatory or catch-up growth

Rates of protein deposition less than the potential (Pr<Pr̂) are natural consequences of normal growth limitation such as failure to optimize nutrient supply and the imposition of higher priority demands for target fat; but a return to maximization of protein retention (Pr=Pr̂) and rates of lipid retention above (Lr:Pr)$_{min}$ may be welcomed by the growing pig.

Whilst a conservative shortfall in protein retention rate below maximum is unlikely to elicit any special response outside the principles of normal growth adumbrated earlier, it is reasonable to expect that if the shortfall is serious the pig may sense an increase in difference between achieved protein mass and potential protein mass for any given temporal age, and may wish to reduce such a difference by making some effort to 'catch up'.

Negative protein growth (rather than reduced positive growth) is an unusual and metabolically disruptive phenomenon indicative of severe nutritional imbalance or deficit. It would appear that catabolism of muscle protein, with resultant degeneration of essential tissue, may occur during lactation in sows suffering feed-protein deprivation and if possible for whom milk synthesis for the good of the offspring has priority over the good of self. It is also reasonable to suppose that the pig will volunteer its body protein tissue to support its life (maintenance needs) before choosing to expire from starvation with a full complement of musculature. Such may indeed be necessary in conditions of highly seasonal nutrient supply patterns which can occur in the wild. For purposes of the science of the domestic pig, however, it can be taken that negative protein growth is indicative of severe management failure, and will have equally severe repercussions for productivity. The discussion of compensating growth is therefore limited to its possibilities subsequent upon reduced (and not negative) rates of protein retention and, as such, the redressing of lost potential gain is deserving of attention:

First, it may be proposed that the pig has no interest in compensatory fat gain beyond reachievement of target fatness, for which purpose lipid retention will have preferential call upon nutrients over protein retention until (Lt:Pt)$_{min}$ is attained.

Secondly, it may be proposed that any pig will maximize protein growth in times of nutritional adequacy. If adequate nutrition follows a period of nutritional deprivation, then the rate of protein retention will be seen to increase in the normal course of events. If control animals were themselves performing below potential, but assumed to be performing at potential, then the realimented group will have all the appearances of 'catching up'; but again they will not be functioning outside the accepted principles.

Thirdly, it may be proposed that any pig making true compensatory gains need not invoke the idea of supermetabolic efficiency. It is merely enough to have an elevated feed intake over normal expectation for pigs of that given weight. If protein retention was restrained previously by limitations to feed intake, then enhancement of intake will naturally lift protein retention rate.

Fourthly and last, it follows from the above that true compensatory growth should be defined strictly as occurring: (a) when previous failure to achieve rates of $Pr = P\hat{r}$ are resultant from a nutritional shortage which may be overcome by a compensating elevation of appetite above previously found maximum levels; and (b) when the previous potential rate of protein retention ($P\hat{r}$) is exceeded as an integral part of the true catch-up process.

There is no reason to believe that compensation of the type covered under conditions (a) and (b) above could be in any way advantageous to the production process. The period of growth during which Pr is less than $P\hat{r}$ can only result in a loss of efficiency which, if ever made good, will incur increased feed usage overall in the attainment of any given final weight. Growth restraint in one period of life to exploit compensatory gains in a subsequent period could only be an economic consideration if feed was particularly expensive in the period chosen for restraint, and particularly cheap in the period chosen for catching up.

The concept of compensation is nevertheless most intriguing from the point of view of growth analysis because if it exists, then its existence shows that the pig has a particularly clear view of its actual position in comparison to its optimum position on the preferred growth track; and, furthermore, it is ready to adjust the situation if harsh reality and target expectation fall too far adrift from each other.

From experimental data available to date it is evident that pigs do recognize deviations from both a preferred weight for age and a preferred lipid:protein ratio. Thus, growing pigs created by nutritional imbalances to be excessively fat or lean will, given the chance, readjust body composition to the preferred Lr:Pr ratio. Further, previously restricted pigs may, upon re-alimentation, show short-term levels of protein retention greatly in excess of levels previously understood to be the maximum possible. Most interesting of all, young growing pigs placed into luxury energy status by the encouragement of generous levels of depot lipid stores may achieve, upon being presented with a diet of exceptionally high protein, the most remarkable rates of daily live-weight gain. The experimental pigs of Dr Kyriazakis of Edinburgh, prepared in this way, achieved daily gains of 0.925kg at a live weight of 13kg.

The fascination of science for compensatory growth should not delude those concerned with commercial production into the mistaken belief that the pig, through mechanisms of compensatory growth, can make amends for deficiencies in production or nutrition. This is not the case, and there has been no evidence forwarded by either science or experience to suggest that optimum growth efficiency would be achieved by any other route than allowing the rate of protein retention (Pr) to approximate to the maximum potential available (Pr̂) at all times.

Growth response to feed supply

The frame for growth analysis often requires an assumption of adequate nutrient supply, although circumstances presumed adequate are frequently to be found wanting upon closer examination. It is a self-evident truth that growth can only occur when there is provision of sufficient nutrients. Feed intake is therefore a prime determinant of the rate of weight gain, of body composition and of carcass quality in meat-producing animals (Table 3.7).

In young pigs the response of protein retention to feed intake is linear up to maximum appetite (Figure 3.22); while in slightly older animals (Figure 3.23) protein retention reaches a plateau at higher levels of feed intake. In these latter studies (which were from Australia) high intakes were achieved. 36MJ of digestible energy per day is more usually associated with 70kg pigs than with those of 35kg. It is evident from this work, and other studies

Table 3.7 Energy value of the live-weight gains of bacon pigs and beef cattle. The energy content of fatty tissue is about 36MJ and that of lean tissue about 5MJ per kg. The fatty content of meat is therefore well indicated by the energy content

	Feeding level			Sex		
	High	Medium	Low	Male	Female	Castrate
Pigs	11	9	7	8	12	16
Cattle	22	18	12	14	20	17

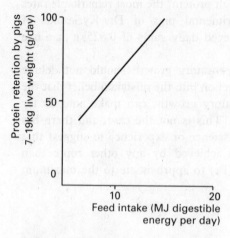

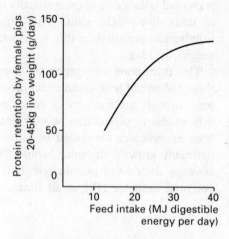

Figure 3.22 Showing the linear response of protein retention to feed intake, with no curvature, even in these very young pigs (note the high intake achieved) (calculated from Campbell, R G and Dunkin, A C (1983) *Animal Production* **36**:185).

Figure 3.23 Showing the linear response of protein retention to feed intake followed by a plateau at about 35MJ digestible energy and 125g per day of protein retention (note the high feed intake achieved) (calculated from Campbell, R G, Taverner, M R and Curic, D M (1983) *Animal Production* **36**:193).

undertaken elsewhere, that protein growth responds linearly to feed intake up to a maximum point, at which it plateaus.

Given high feed intakes, the plateau may be attained relatively early in life. Gompertz assumptions are confirmed by experimental data showing that maximum limits to daily protein retention are rather constant over a wide range of live weights covering 10–60% of mature weight (or about 20–120kg in the case of pigs) (see Figure 3.24).

The point is amply further demonstrated with data relating to male turkeys (Figure 3.25). Growth rate accelerates rapidly to a plateau early in life, at about 7 weeks or 15% of mature weight (which is around 20kg). This rate does not further increase, but rather it holds at a relatively constant 120g or so daily through to 70% of mature weight.

The linear/plateau response for protein retention to feed intake shown in Figure 3.26 relates to pigs of around 50kg live weight. A plateau occurs at the maximum growth potential for the animal, which in case (a) is 500g of lean growth daily. In case (b) the maximum potential is 750g lean growth daily. (a) and (b) may be different genotypes, or perhaps different sexes. In pigs the entire male has a much higher potential for lean tissue growth than either the female or, particularly, the castrate (Table 3.8).

During the linear phase for lean growth, fatty tissue growth will be

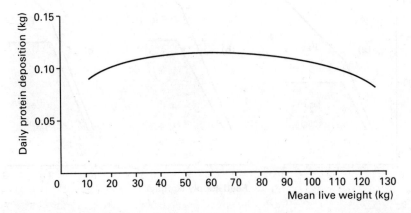

Figure 3.24 Daily protein deposition rate in relation to *mean* liveweight. The experiment began at 20kg and progressed by serial slaughter at increasing live-weight points through to about 180kg. Mean live weight is the mean between 20kg and the serial slaughter point (from Tullis, J B (1982) *Protein growth in pigs*. PhD Thesis, University of Edinburgh).

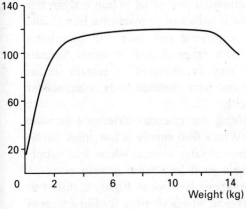

Figure 3.25 Growth rate of turkeys as a function of live weight. The mature weight of these animals is in the region of 20kg (unpublished results from Emmans, G C, Edinburgh School of Agriculture).

Table 3.8 Lean-tissue growth potentials of pigs (g/day total lean in whole live body)

	Potential lean tissue growth rate (g/day)
Entire male	700
Female	600
Castrated male	500

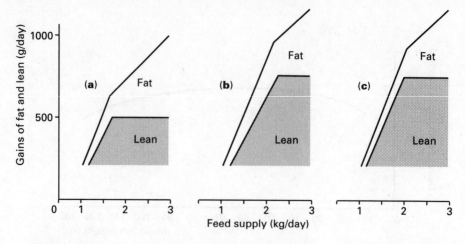

Figure 3.26 Effect of increase in feed level upon the growth of lean and fatty tissues in the whole body of pigs of 50kgW.

restrained to a minimum level (Lr:Pr)$_{min}$ on the assumption that under conditions of normal growth the animal prefers to target for lean whilst maintaining some minimal level of fat in normal live weight gains (Figure 3.26). In these cases ((a) and (b)) the minimum ratio of fat to lean is about one of fat to four of lean. Until feed intake is sufficient to maximize lean tissue, the animal will not fatten whilst it is growing and there will be relative constancy in the composition over a wide range of body weights. Animals given diets of low nutrient density may be expected to remain in the nutritionally limited phase of growth and have constant body compositions over much of their growth to maturity.

Figure 3.26 is also useful in clarifying the essential difference between leanness and lean-tissue growth rate. Where feed supply is low, pigs can be lean while having a poor lean-tissue growth rate; whereas where feed supply is high, pigs can be fat while having a good lean-tissue growth rate.

One of the consequences of the hypothesis laid out in Figure 3.26 is that animal breeders must provide a test regime which supplies feed in excess of the requirement to achieve maximum lean-tissue growth in the best candidates, or differences between the improved and unimproved genotypes will fail to be distinguished. In the case of the example illustrated for pigs of around 50kg live weight, minimum feed supply for distinguishing between genotypes is around 2.3kg. In case (a), 2kg of feed per day will bring about fattening, whereas in case (b) maximum lean-tissue growth rate has yet to be achieved. As soon as the feed supply fully satisfies maximum potential for lean-tissue growth, rapid fattening occurs, the rate of daily live weight growth increase as a function of feed supply reduces, and feed conversion efficiency worsens. To the left-hand side of this break point (the point where the linear response becomes a plateau) growth may be described as nutritionally limited; to the right of this break point it may be described as nutritionally unlimited. Therefore 2kg of feed represents unlimited nutrition for the pig in case (a) but limited nutrition for case (b).

It should come as no surprise to animal breeders that selection against fat so often brings about a reduction in appetite; *ad libitum* intake is pushed to the left along the feed supply scale in Figure 3.26 (that is, feed supply is reduced), and thus the lower appetite of the pigs moves feed supply into the nutritionally limited (low fat) rather than nutritionally unlimited (high fat) area of growth response. Pigs tested on an *ad libitum* feeding regime, and selected only for live-weight gain, would be expected to make substantial positive improvements in daily feed intake and daily lean-tissue growth rate, but the pigs may also become fatter.

Strain and sex differences may also be revealed in the ratio of fat to lean in nutritionally limited growth. In pigs the ratio can vary from less than 0.5 of fat to 4 of lean for entire males of improved strains, to more than 2 of fat to 4 of lean for castrates of unimproved strains. Selection against fat in meat is more likely to diminish this minimum fat ratio – as the example in case (c) (Figure 3.26) – than to increase lean-tissue growth rate. Fat reduction can therefore be achieved either by selecting for animals of type (c), or ensuring that nutrient intake remains limited.

The propositions put forward here are fundamental to pig production strategies. They state that there is a maximum lean-tissue growth rate, and that this is largely independent of animal weight and age. They further assume linear responses up to the plateau. They govern the circumstances in which animals cannot become fat, and give rules for feeding to maximize efficiency of feed use and growth rate whilst minimizing fatness. Animals with higher lean-tissue growth potentials can consume greater amounts of feed with consequently improved feed efficiency and no increase in fatness. Entire males of improved genotype do not fatten because appetite is low in relation to potential growth rate. Selecting for greater appetite would initially enhance growth and not fatness, but as lean potential is approached, further selection for appetite will bring about fat pigs.

Under normal circumstances young animals, because of limited size, will have low appetites and will therefore find themselves in nutritionally limited growth (Figure 3.27). The same would apply for older animals that have poorer appetites or that are given low-quality feeds, such that within appetite their achieved nutrient intake is low. Older animals with higher inherent appetites have a better chance of finding themselves in nutritionally unlimited growth. This also goes for animals given higher nutrient-density diets. This is the major reason why animals fatten as they get heavier.

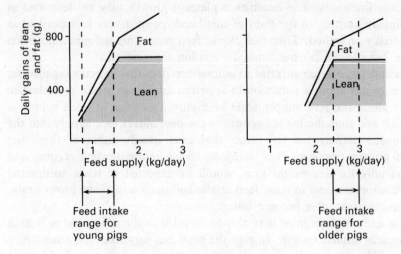

Figure 3.27 As animals get older and bigger their ability to eat more feed increases. This means that older pigs are more likely to become fat than younger pigs. However, if diet nutrient density also diminishes with age, the likelihood of fattening decreases.

Whilst growth potential may therefore be satisfactorily described in terms of biological constants and mathematical functions, the achieved rates and composition of gains in pigs are most directly influenced by nothing more complex than the level of nutrient intake achieved at any given moment.

Hormonal manipulation of growth

The endocrine system is the intermediary between the intrinsic (genetic) instructions relating to the quantity and quality of growth that is required of any individual, and the enactment of those instructions through the biochemical processes accreting protein, lipids, ash and associated water.

Growth can therefore be manipulated at three levels:

1. through genetics – by artificial selection;
2. through the endocrine system – by reduction or enhancement of body hormones;
3. through the environment – by increase or decrease in the supply of nutrients for metabolism.

Normal growth is regulated through a hormonal complex including, amongst others, growth hormone (or porcine somatotrophin [PST]), insulin, somatomedins (which are insulin-like growth factors), thyroid hormones, glucocorticoids, epinephrine, androgens and oestrogens. Growth manipulation can be achieved by increasing the concentration of anabolic hormones in the system, by increasing the sensitivity of target organs to existing hormone concentrations or by diminishing the effects of feedback control systems (this latter diminution may be achieved either by compromising the feedback system or by desensitizing the primary system to the effects of the feedback by immunization).

Growth hormone, somatomedins, insulin and thyroid hormone interrelate within a hormonal complex which is itself subject to a layer of controller and feedback mechanisms. Growth hormone releasing factors from the hypothalamus impinge upon the pituitary, influencing output rate. Somatostatin is released as a feedback response to growth hormone and somatomedins; somatostatin controls (restricts) growth hormone release, together with some degree of lesser restraint upon insulin, glucagon and thyroid hormone (ameliorating the strength or effect of somatostatin, for example by immunization against it, will enhance growth by reducing somatostatic inhibitory effects). The somatomedin hormones themselves increase the rate of energy metabolism, stimulate fatty tissue growth to some extent but, most importantly, act directly to increase the rate of protein accumulation into body tissues. The somatomedins stimulate many of the primary hormones involved in promoting growth, and represent the major hormonal influence upon daily gains of lean and fatty tissues. Faster-growing animals will show elevated blood levels of insulin, growth hormone and somatomedins.

The main effects of some of the major hormones associated with growth are presented in Table 3.9. Fat growth is usually enhanced by positive support or conversion of glucose to fatty acids, while it is restrained by encouragement of fatty acid oxidation. Lean growth requires active glucose metabolism for creation of the high levels of energy needed, together with a

Table 3.9 Main effects of some major hormones associated with growth

	Fat growth	Lean growth
Growth hormone	−	+
Somatomedins	+	+
Thyroid hormones	−	+
Insulin	+	+
Catecholamines	−	+
Oestrogens and androgens	+	+
Glucagon	−	+

high level of stimulation of amino acid anabolism and protein accretion. The catecholamines act through beta-adrenergic receptors and may also stimulate insulin, growth hormones and thyroid hormones.

Androgens and oestrogens have been available as exogenously administered anabolic agents acting both directly on lean tissues to enhance their growth, and also indirectly by encouraging release of growth hormone, insulin and thyroid hormones. In-feed androgen/oestrogen growth promoting agents were available for use formerly as growth-promoting agents.

Present interest in hormonal growth enhancers and manipulators is concentrated primarily on the readily available manufactured growth hormone and also upon the highly effective beta-adrenergic agonists.

It is evident that in normal and natural circumstances, without any exogenous interference, fast-growing animals show elevated levels of growth hormone. DNA recombinant technology has enabled the industrial manufacture of 'man-made' pig growth hormone, or recombinant porcine somatotrophin (r-PST) whose composition and action is *almost* indistinguishable from natural pig growth hormone, as secreted from the pig's own pituitary gland. (It appears that r-PST itself may have one amino acid different and one more amino acid in the chain and, of course, the different methodology of delivery and the inability to also manipulate the interrelating hormones is bound to influence mode of action.) When delivered into the hormonal complex via the blood system, r-PST brings about dramatic increases in growth rate (about +10%) and carcass meat content (about 5% of lean in the carcass side), and reductions in fatness (about −25%). There will also be some negative effects upon appetite which is usually (unfortunately) reduced. It would appear that, at normal levels of circulating hormone, there are no serious adverse consequences of the use of r-PST to pig health, but, if physiological dose rates are used, neither would any negative consequences be expected in view of the exogenous hormone being similar in form or action to the endogenous. The possibility of negative health effects must, however, increase as significant performance change upon dosing is achieved. In some trials, evidence of respiratory and locomotor disorders have been seen. Following r-PST treatment, there may also be some loss of muscle tenderness over and above any direct consequence upon meat quality of a large reduction in fat.

Beta-adrenergic agonists are in the family of catecholamines including endogenous norepinephrine and epinephrine. The catecholamines (in the naturally occurring forms norepinephrine and epinephrine) stimulate adrenergic receptors with consequences not only for heart rate and bronchodilation but also lipolysis. Beta-adrenergic agonists which are analogues of epinephrine have been manufactured as cimaterol, clenbuterol and ractopamine. These substances decrease fat mass (about −10%), increase leanness (about +10%) and may also improve growth rate and feed efficiency (although the latter is unfortunately likely to be through a reduction in the level of feed consumption).

In addition to being different in mode of action, the beta-adrenergic agonists are also importantly different from r-PST in that they are orally active (whereas PST must be delivered into the bloodstream). This facilitates administration, but increases human hazard.

Exciting as the responses of pigs to exogenous hormones might be, it still remains to be ascertained:

1. whether the consumer will accept pork products as being of the highest quality in the broadest sense when that consumer has knowledge that the pigs concerned have been treated with exogenous hormone preparations (however safe these may be shown);

2. whether the same benefits of increased growth rate, efficiency and leanness cannot be achieved more economically and more simply through conventional genetic selection for lean-tissue growth rate.

These comments are made in the knowledge that whilst in some countries the pigs are over-fat and carcasses can be made acceptable by r-PST administration, many European genotypes are now as lean as is compatible with minimum standards of meat juiciness, flavour and tenderness, and the need to reduce fat content by exogenous hormone intervention no longer exists.

CHAPTER 4

REPRODUCTION

Introduction

Reproductive activity begins in earnest with puberty at which the oestrous cycle begins and ova are made available for fertilization. Pregnancy and lactation follow, with a return of the oestrous cycle soon after milk flow ceases. The efficiency of the breeding-pig herd is heavily dependent upon the level of biological productivity – the number of piglets produced annually from each female. The main events in the breeding life of the sow are shown in Figure 4.1. Some of these are of relatively constant duration while others are variable (Figure 4.2). Key factors under production control are evidently the age at first conception, the lactation length and the interval between weaning and the next conception. With regard to the latter, two elements of

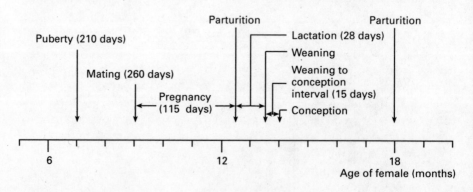

Figure 4.1 Events in the life of a breeding female pig.

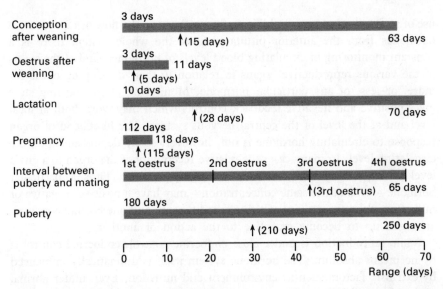

Figure 4.2 Aspects of reproductive performance. Time-scale ranges found in practice – ↑ marks the average positions.

the system are relevant: first, the weaning to oestrus interval, which governs the *possibility* for a fertile mating and secondly, the frequency with which an oestrous event *actually* culminates in a sustained and worthwhile pregnancy.

In the mammal, reproduction involves a remarkable complexity of interacting events, controlled with great precision by secretions of the endocrine system. Individual organs of the body, such as the ovaries, produce hormones which act directly on other organs, such as the uterus, while overall control emanates from the central nervous system. The most convenient route to deliver instructions to body organs is via the ubiquitous blood system which can carry to target tissues the hormones synthesized in various glands around the body, such as the anterior pituitary gland, the posterior pituitary gland, the adrenal glands, the pancreas, the ovaries, the testes, the uterus and, indeed, embryos or foetuses themselves. The central nervous system joins with the blood circulatory system via the major hormone-producing sites of the anterior and posterior pituitary glands, the latter extending from the hypothalamus – a part of the brain located on the brain's underside. The posterior pituitary is joined with the hypothalamus by nerve fibres which ramify through both bodies and provide a direct link with the brain and central nervous system. There are no nerve connections with the anterior pituitary, which must be communicated with through an internal hormone-carrying blood supply whereby the hypothalamus can, by use of secreted releasing or inhibiting factors, control the flow of reproductive hormones from the anterior pituitary. Over the whole system there is a constant monitoring of circulating blood levels of hormones, and of the status of the various reproductive organs in relation to their developing needs for more, or less, of any particular hormone. Monitoring of the existing state gives positive and negative feedback controls which may work both at local level and at the level of the central nervous system. The likelihood of organ response to circulating hormone is not, however, solely dependent upon the circulating level. In many cases the relative receptivity of the organ to a given level of circulating hormone may be highly volatile. Thus a particular hormone – even at the same concentration – may have a positive, negative or null effect upon the activity of a single body organ; and one hormone may set an organ up to become receptive to the action of another.

While reproduction is under close endocrine control, endocrine control is by no means absolute. The hormone system itself is dramatically influenced by extrinsic factors such as environment and nutrition. Even under normal operating conditions there is constant interaction with the whole spectrum of physiological, genetic, biochemical, behavioural and physical factors that pervade the life of the breeding sow. Reproductive success requires not merely that the highly complex endocrine system operates at *its* optimum, but also that this is combined with the simultaneous creation of an optimum internal and external environment in the broadest sense.

Hormones of the oestrous cycle

The brain links with the anterior pituitary gland through releasing hormones (RH) and inhibiting hormones (IH) which control – by positive and negative actions – the flow of the major reproductive hormones into the bloodstream, and thereby to their sites of action upon the reproductive system. The major hormones from the anterior pituitary gland are somatotrophin (STH), thyrotrophic hormone (TH), adrenocorticotrophic hormone (ACTH), prolactin, follicle stimulating hormone (FSH) and luteinizing hormone (LH), the latter two often being referred to together as the gonadotrophins. LH and FSH invariably act synergistically, and by a change in the ratio, one to the other, different responses may be elicited in target organs. Prolactin levels are controlled by the anterior pituitary reacting to both releasing hormone and inhibiting hormone coming down from the hypothalamus. LH and FSH are stimulated by gonadotrophic releasing hormone (GnRH) coming in pulses (every hour, or thereabouts) from the hypothalamus to the anterior pituitary. The pulsatile nature of GnRH release results in LH also appearing in the circulation in pulses which also occur approximately hourly in the pre-ovulatory situation. LH and FSH release are negatively controlled by a feedback mechanism in which the amount of GnRH released from the hypothalamus depends upon the brain monitoring existing circulating levels of FSH and LH in the blood; and also monitoring the circulating levels of their secondary hormones, the steroids androgen, progesterone and oestrogen (of which more later), in relation to the control of the progressive steps of the reproductive process.

In passing, there is an inhibiting effect of melatonin upon the release of gonadotrophins from the anterior pituitary; melatonin is released in response to the hours of darkness and melatonin production is suppressed by daily provision of 16 hours of light.

Pituitary FSH, LH and prolactin – with FSH initially in the ascendancy – encourage the development and maturation of ovarian follicles. Developing follicles, in their turn, produce oestrogen. When ready, the ova are released from the ovarian follicles due to stimulation by elevated levels of pituitary LH released into the circulation. This (substantial) pre-ovulatory surge of LH is composed of a progressive build-up of pulses of increasing frequency. Corpora lutea, initiated by the action of LH, form in the ovary at the points from which the ova left the follicles; and from the corpora lutea, the hormone progesterone is secreted. The primary pituitary hormones FSH and LH (the gonadotrophins) are therefore responsible for the secondary release of the steroid hormones oestrogen and then progesterone from the female gonad.

In the absence of a pregnancy, the oestrous cycle of the pig occurs every 21 days or so. In the intervening period the active corpora lutea secrete increasing levels of progesterone which rise to a maximum at around days 12–15 of the cycle (Figure 4.3). Pregnancy failure having been registered by the absence of embryos in the uterus, maintenance of the corpora lutea and secretion of progesterone become futile. The corpora lutea are therefore attacked by the hormone prostaglandin F2-alpha which is produced by the (disappointed) uterus at day 15 or 16 of the cycle, and which brings about luteolysis, that is the regression of the corpora lutea and a consequent reduction in circulating progesterone. The oppression of GnRH by progesterone is lifted at this time with resultant release of LH and FSH, and oestrus can be expected about 4–5 days later. As progesterone concentrations fall, the steroid hormone oestrogen rises, being produced from the follicles of the ovary under gonadotrophic (both FSH and LH) stimulation. The oestrogen peak occurs approximately a day before oestrus.

The role of the hormone prolactin is somewhat confusing at this time.

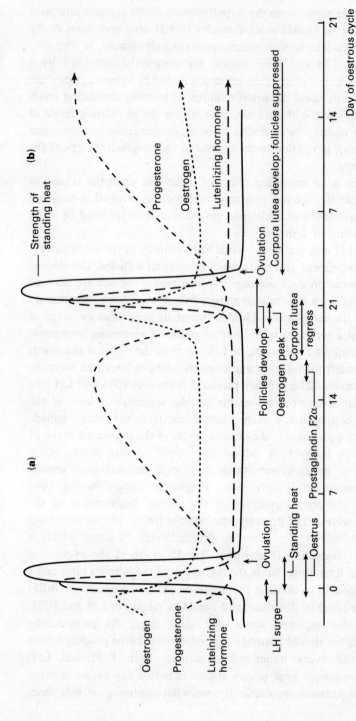

Figure 4.3 Relative changes in the blood concentrations of hormones of the oestrous cycle (a) in the absence of pregnancy, (b) in the presence of pregnancy. The rise in oestrogen follows proportional changes in the development of the ovarian follicles from which the hormone arises, while similarly the changes in progesterone levels follow proportional changes in the development and regression of the corpora lutea.

Prolactin levels are elevated 4–5 days before oestrus, encouraged apparently by the secretion of prostaglandin F2-alpha. Prolactin may be involved with prostaglandins in the breakdown of the corpora lutea (luteolytic), a complete reversal of its role in corpora lutea support during pregnancy and lactation (luteotrophic). Upon luteolysis prolactin levels fall, then – again with prostaglandin in attendance – rise during oestrus. This latter rise is enhanced by the presence of the boar and the act of mating. In this case, an association between oxytocin released during copulation from the posterior pituitary, and prolactin released from the anterior pituitary, foreshadow a similar partnership which occurs during lactation.

The recognizable signs of oestrus are standing heat and acceptance of the male (vulva reddening and swelling is particularly apparent in the gilt, but may occur less or even covertly in the adult sow). As a consequence of the oestrogen peak there is an LH surge from the anterior pituitary gland, and a coincident but smaller rise also in FSH. LH takes the ascendancy and peaks half a day later, closely coincident with the start of oestrus. In fact, it would appear that the first action of the initial oestrogen release is to reduce basal LH release. A store of LH builds up in the anterior pituitary. As oestrogen levels build ever higher, the hypothalamus encourages a GnRH pulse and a now greatly enhanced pre-ovulatory surge of LH leaves the pituitary and enters the bloodstream. This massive dose of LH stimulates local intrafollicular prostaglandin secretion which initiates rupture of the follicles (that is, ovulation), and causes ova release from the follicle in the course of the next 3–6 hours (Figure 4.3). Oestrus in the pig will normally last 1–3 days, with an expected average of 2 days. In the first day, LH and FSH levels will fall, the fall in LH being far the greater which now returns to base. While LH seems to have a low base with a dramatic pre-ovulatory surge, FSH has a more steady level of secretion through the cycle, with a smaller fall at this time and a smaller rise assisting LH at ovulation. Oestrogen will have returned to its base level by around the same time and the other steroid hormone, progesterone, will begin to rise, being secreted from the corpora lutea which have now been formed in the collapsed follicles from which the ova were released. If the pig is not mated, or if a reasonable number of fertilized ova (4–5) are not registered as present in the uterus, then the rise of progesterone will be halted at days 14–16, and the cycle will be reinitiated so that oestrus should occur at intervals of around 21 days. The initiation of follicular growth on day 16, and following, of the oestrous cycle is not only consequent upon a rise in circulatory levels of gonadotrophins, but also upon the ovary itself becoming more sensitive to the existing levels of circulatory hormones.

Elevated gonadotrophin levels are found in the blood some 40 hours before ovulation. Oestrus begins soon after blood oestrogen levels are elevated. Thus the 48-hour (usually 1–3-day) oestrous period begins around 6–8 hours after the specially raised levels of FSH and LH release from the anterior pituitary (Figure 4.3). In some cases (if the heat is short) oestrus may be terminated before the ova are present in the tract, whilst in other cases (if the heat is long) mating may occur some time after ovulation. Pigs will usually enter standing heat around 6 hours after the onset of oestrus; some 12 hours after gonadotrophin release. The gonadal hormones oestrogen and progesterone have, amongst their other roles, action upon the brain which reacts by creating appropriate sexual behaviour patterns to encourage and allow the singularly unusual behavioural event of copulation. The achieving of a successful pregnancy requires not only an ample number of fertilizable eggs in the right part of the tract, but the simultaneous presence there of male gametes, the spermatozoa. This delicate timing requires remarkably precise changes in the behaviour of the female pig, and often the reversal of many normal agonistic behaviours, in order that she will stand to accept the boar during exactly the right period of time, neither too soon nor too late.

Once a pregnancy is established, then oestrogen, FSH and LH remain relatively quiescent under the dominant influence of a high level of progesterone which suppresses hypothalamic GnRH release and thereby inhibits release of the gonadotrophins FSH and LH from the anterior pituitary gland. Recent research has also implicated the opioid peptides in this scheme. Opioids (morphine-like compounds) are released from the central nervous system and they appear to be antagonistic to GnRH. Progesterone will maintain the pregnancy through to parturition so long as the number of foetuses in the uterus is adequate to be considered (by the pig) as a minimum worthwhile and viable litter (four or five). One of the first requirements of the pregnancy is that the new embryos, floating free in the uterine fluids, should be well nurtured by uterine secretions and should be, in due course, firmly attached to the uterine wall whence their cells will proliferate into foetal tissue. Progesterone acting on the uterine wall is essential to these processes.

Oestrus will be accompanied by ovulation only when an appropriate LH surge has also caused the collapse of the follicles. The combined events of oestrus and ovulation require the combined effects of elevated and sudden release of both gonadotrophins (FSH and LH) from the anterior pituitary, the former to stimulate oestrogen production from the follicle and standing heat, and the latter – which has the greater elevation of surge – to collapse the ovarian follicles and free the ova. Given an inoptimum FSH:LH ratio, it is possible therefore to have oestrus without ovulation, or (perhaps worse) with only a limited number of eggs released. This sort of oestrus is more common when the cycle is first initiated in the pubertal gilt, or when there is some suppression of ovarian function such as by prolactin and the stimulus of a sucking litter during lactation which may bring about a partial anoestrus.

Attempts to control the oestrous cycle so that groups of pigs may be mated at the same time (particularly useful to achieve batch farrowing or batch use of artificial insemination) have frequently failed. One notable exception was the product methallibure which totally inhibits oestrus and ovulation by differentially influencing the secretion of pituitary gonadotrophins, especially FSH. When used with cyclical females, methallibure allows normal fertility 4–9 days after withdrawal. However, the material proved unsafe to both pigs and humans, and is not available in most countries.

Puberty

Puberty will normally occur at ages in excess of 190 days and weights in excess of 100kg. The initiation of breeding in pigs, which will occur usually one or two oestrous cycles after the first pubertal oestrus, requires a combination of most or some of: an appropriate age, greater than 220 days; an appropriate weight, greater than 120kg; and a fatness level of something in excess of 14% of the body as lipid, but preferably a level in excess of 16% of the body as lipid; that is, a lipid:protein ratio in the body of 1:1 or greater. The lower fat level is probably sufficient for puberty itself, but the higher level is optimum in view of subsequent pregnancy and lactation loadings.

Puberty may lie quiescent in females kept away from males, kept at higher temperatures, given a short day length, or allowed peace and seclusion. Puberty is aroused and forwarded, however, by the presence of males (especially those which are older and more experienced), by mixing together competitive female groups, by transportation, by placing into new surroundings and by other such general excitements. The maximum effect of boar presence and other puberty-positive stimuli may be achieved if introduced at around or after 180 days of age. First mating may best take place some 42 days after the pubertal heat. The number of ova released from the ovary and available for fertilization increases over the first two or three or so oestrous cycles as the reproductive system and its associated hormones mature. This results in a commensurate increase in potential litter size (Figure 4.4). Conception rate should also rise from around 80% at the pubertal oestrus to around an expected 95% at third oestrus. There is good reason therefore to mate not at the first pubertal oestrus but rather at the second or, even better, at the third.

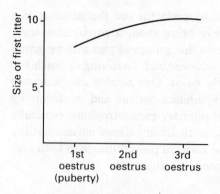

Figure 4.4 Influence of time of first mating after puberty on litter size.

It is the increasing levels of oestrogen from the developing follicles of the ovary, stimulated by FSH and LH from the anterior pituitary gland, which help to stimulate growth of the reproductive tract and which finally give the hormonal surge which produces signs of the first pubertal oestrus in the form of standing heat. Meanwhile, as the first pubic heat approaches, hourly pulses of luteinizing hormone (LH) are evident, culminating in the first experience of an LH surge immediately prior to first oestrus. Pregnant mare serum (PMS – high in FSH), followed by human chorionic gonadotrophin (HCG – high in LH), may stimulate oestrus and ovulation, provided the female is in a suitably receptive state. These systems are of limited use to stimulate puberty, or to induce early puberty, because the rest of the system may not be receptive. Exogenous FSH and LH hormones, in the form of serum gonadotrophin and chorionic gonadotrophin, should only be used to trigger puberty as a last resort, that is after all other methods have been tried, and after an age of 265 days has been reached.

Ovulation

Ovulation occurs 40 (usual range, 20–60!) hours after the beginning of oestrus. Once initiated by the LH surge, however, ovulation itself is a relatively compact process. The upper limit on ultimate litter size is set by the total number of viable ova released from the follicles of the ovary and despatched to the oviducts for fertilization.

In cyclical, non-pregnant pigs, each ovary will produce some 6–14 ova at every ovulation. The breeding female shows improvement in ovulation rate with age up to a peak at the fifth parity (Figure 4.5); while gilts usually produce up to 14 or 15 ova, sows will usually produce 20–25. The number of follicles developing at each cycle, and the consequent number of ova released, are under the control of FSH and LH and it is not surprising that exogenous hormonal administration in the form of pregnant mare serum (PMS) which is rich in FSH and human chorionic gonadotrophin (HCG) which is rich in LH, will increase ovulation rate. However, with breeding sows there is not so much an ovulation rate problem, as a problem of embryo loss subsequent to fertilization. Increasing the ovulation rate from some 20 to some 25 ova shed is somewhat pointless when litter size is adjusted anyway to around 12–14 in the normal course of events.

Ovulation rate is greatly influenced by breed type and is open to improvement by genetic selection. Cross-bred and hybrid pigs also tend to have higher ovulation rates than pure-bred.

There may be some negative effect of under-nutrition on ovulation rate, although converse positive effects of especially good nutrition are difficult to discern. In the gilt, effective ovulation is dependent upon an adequate level of body fatness, and a harmonized physiological system not stressed by under-satisfied demands for continuing body tissue growth. Reduced feeding levels will both diminish body lipid level and deny optimum muscle growth. In such gilts increased feeding levels will improve upon the situation and help to lift what would have otherwise been a depressed ovulation rate. Such a position, however, would best not be entered into in the first place and the appropriate nutritional strategy is one of maintaining adequate growth and fatness over the whole of the period up to puberty and to continue that strategy between the post-pubertal heats leading up to first mating.

There is a helpful (flushing) effect where the ovulation rate is increased when gilts previously underfed are generously fed for a short period before mating. This phenomenon is only likely, however, when the level of previous feeding has been below normal standards and ovulation rate is expected to be

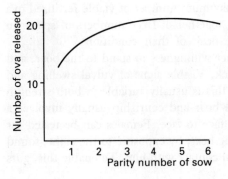

Figure 4.5 Ovulation rates are higher for multiparous sows than for primiparous gilts, and will increase up to parity 4 or 5. Numbers of piglets born alive will also increase from around 9.5 in parity 1 to 11.0 in parity 3 and 12.0 in parity 5, but will decrease thereafter to 11.0 in parity 7 and 10.0 in parity 9 due to the greater numbers of foetal losses and piglets born dead, which occur in older sows.

depressed. Again, therefore, in properly managed pigs neither the longer-term nor the shorter-term nutritional effects upon ovulation rate are relevant, responses only occurring in animals which have been inadequately fed in the first place. The need to raise the number of ova released by the sow is highly questionable in the normal course of events due to an already demonstrable oversupply. But ova are not oversupplied in the gilt (see Figure 4.5) and attention to nutrition before the first heat and between heats is clearly of importance if the number of ova released is to be maximized.

After weaning, ovulation will normally occur within 7 days. Given the existence of oestrus it is only in exceptional cases of disease, nutritional failure, mycotoxin contamination of the feed or inadequacy or excess of body lipid that ovulation rate will be curtailed so that litter size is adversely affected. Between weaning and mating, multiparous breeding sows will probably be fed generously in any event to counter body-condition loss during lactation. Litter size in breeding sows is more usually a function of embryo loss than numbers of ova shed, while low productivity per sow per year is more usually a function of failure to come into oestrus post-weaning in the first place. In the absence of reproductive disease, return to heat following mating due to release of insufficient ova to support a patent pregnancy would be considered unusual.

Fertilization

The introduction of the male to a female in standing heat will elicit copulation. Copulation is highly likely to result in fertilization providing it occurs before, but not too long before, the time of ovulation. Given that this is readily achievable by the simple expedient of mating at 12-hourly intervals through the oestrous period, it is the proportion of eggs fertilized which is most critical to pig fecundity. The maximum number of viable fertilized ova will only be achieved by the coming together of large numbers of sperm in peak condition with ova also at the peak of their condition.

The female will show oestrus by her willingness to stand to the boar, and also to human pressure upon her back. Visible signs of vulval swelling and reddening may also be apparent but this is usually variable in both strength and incidence. She will seek out the boar and courtship usually involves a preliminary exchange of pleasantries face to face. Females can be tested for heat, and encouraged to show a strong heat, by exposure to the sight, sound and smell of the boar, and by physical contact with him. To enable this, gilts and multiparous breeding sows approaching oestrus should be kept close to the boars, either penned together or separated only by open-spaced bars.

The particular moment of copulation involves a complex of behavioural and hormonal interactions. The hormone oxytocin is released from the posterior pituitary gland in a dramatic pulse, or series of closely spaced pulses at the behest of the central nervous system. The action of oxytocin is to contract smooth muscle and this will occur within less than a minute of its release into the bloodstream. The half life of oxytocin is short and the effects may be lost again within a single, or a few, minutes. Because the posterior pituitary is linked directly to the hypothalamus of the brain, no releasing hormone is needed. Rather a direct neuro-endocrine reflex is enacted whereby the brain itself triggers release. The target organ for oxytocin is the uterus where it causes contractile movements which help to wash the semen from the body of the uterus into which it was ejaculated and up the female tract so that sperm encounter the ova descending the oviducts at this time. Oxytocin release in the male may be an accessory to the act of his ejaculation. Oxytocic effects are antagonized by the hormone adrenalin which is released at times of stress (Figure 4.6). (In passing, oxytocin will be shown as being involved in the contraction of uterine muscles as an essential part of foetal expulsion at birth, and again later in the contraction of mammary alveoli muscle in the course of milk ejection.)

Puberty in the boar occurs at live weights in excess of 100kg and ages in excess of 200 days. Fertility is low to begin with, rising to working levels at around 15 months (450 days). Pubertal spermatogenesis, and for that matter all subsequent sperm formation through mature breeding life, is initiated and supported by FSH, LH and prolactin. LH stimulates testicular steroid (androgen) production, which itself is needed as part of the hormone complex supporting spermatogenesis. Sperm production is a continuing process and sperm may be stored before ejaculation for periods of weeks. The process of spermatogenesis and the full maturing of viable sperm may also be prolonged

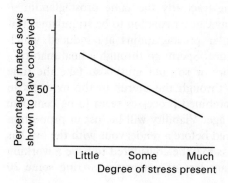

Figure 4.6 Putative influence of adrenaline upon oxytocin activity as perceived in the relationship between stress during mating and conception rate. Inasmuch as such a response may be observed, the cause is not clear and the effect variable. Physical disturbance and disruption at the time of mating does, however, appear to reduce reproductive performance in sows.

over days or weeks, much according to the frequency of copulation. But because the ejaculate is produced at one moment in time, whereas its sperm content is produced continuously, it follows that semen will contain sperm which are old and which have lost their virility, and also sperm which are immature, as well as fully competent sperm. Some sperm will also exhibit defects and deformities of diverse sorts. Normally, about 75% of sperm will be seen microscopically as motile, and about 85% as morphologically normal.

In the pig the sperm-rich fraction is joined during the act of mating by copious quantities of seminal (that is, accessory) fluids. Sperm and seminal fluid together create the ejaculate. The boar will produce between 100 and 500ml of semen per ejaculate with an overall concentration of around 0.3×10^9 sperm per millilitre or a total sperm count of about $30-150 \times 10^9$ sperm. This level of production can be achieved three or four times per week. On each occasion the boar will copulate for 5–15 minutes, with sperm flowing in waves during the full course of the duration of the mating, which should therefore remain undisturbed. The combined physical length of the male and the female (one behind the other), the need for courtship and the time taken for copulation itself, all mean that generous space must be provided in the mating area. If a boar works three times weekly and sows need to be mated two or more times at each oestrus, then the number of boars needed will approximately equal the average number of sows coming into heat in any week. In a herd of 100 sows, farrowing 2.3 times yearly, five sows will, on average, come on heat every week. Some further allowance should be made for unevenly spread weanings and young boars not yet fully in use.

The passage of as few as 2×10^9 sperm through the cervix and into the uterus is usually sufficient to ensure adequate fertilization. There is, therefore, a high degree of overkill in the delivery of $>30 \times 10^9$ sperm, but many of these, although delivered through the cervix, do not pass to the top of the uterus.

Upon delivery of the sperm into the female, they move through the uterus and up into the oviducts. Under the influence of seminal oestrogens and uterine prostaglandins, as also the influence of oxytocin, the uterine lining contracts to help move sperm up the tract. By the same prostaglandin + oxytocin activity, the act of mating may cause ovulation to be stimulated (if it has not already taken place); follicular prostaglandins are induced which foster enzyme action and follicle rupture. Sperm go through a final maturing in the female (capacitation) of 4 hours or so, and must also take the time (about 2 hours) and effort to travel through the uterus to the oviducts in order to meet with the ova. These combined processes seem to be complete in 8 hours. After this the sperm will age. Viability will be lost in proportion to the extent of ageing that has occurred before a rendezvous with the ova has been effected. Overall, the sperm mass may be considered to have a working life of about 36 hours, with significant loss of potency occurring some 20

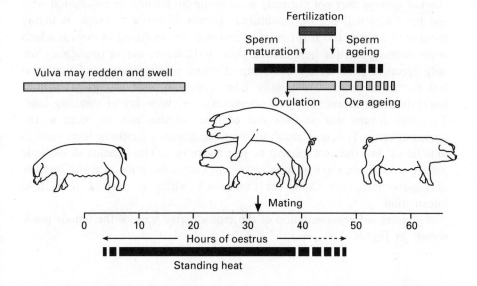

Figure 4.7 Synchronization of oestrus, mating and ovulation to optimize fertility.

hours after original deposition into the female tract. There is, therefore, an optimal window of maximum viability for sperm between maturation and ageing of about 12 hours, and this will occur about 8 hours after mating. Although only one sperm nucleus enters each of the 10–25 fertilizable ova, which may be fertilized, it is helpful if semen volumes from more than one mating can be contributed into the female tract over the period of standing heat.

Meanwhile, the ova also require to mature in the female tract subsequent to their expression from the follicle. Release of some 15–25 ova from the two ovaries is achieved over a period of about 4 hours. After release the maturing process is achieved in around a further 2 hours or so. Ova are then at optimum viability for the next 8 hours. Subsequent to this time maturity deteriorates into ageing, which occurs rather rapidly, and loss of viability is apparent from around 12 hours after ovulation. Ova released later are viable, but the endocrine and uterine environment relates most closely to the earlier released ova, and so the latecomers may find themselves out of phase with their support systems. About 30% of ova released might be under threat in this way. Fertility may be expected to fall rapidly as a consequence of egg population ageing any time after 10 hours post-ovulation, and from sperm population ageing any time after 18 hours post-mating. Optimum time for mating is toward the middle of the period of standing heat (which is of course of unknown duration), and around 6–12 hours before ovulation. The necessary synchronization of events is depicted in Figure 4.7. From the Figure it will be apparent that matings taking place before 30 hours after the onset of oestrus may not normally be of optimum fertility or conception rate, and litter size will not be maximized. However, because oestrus is highly variable in length, and it is ovulation and not the beginning of oestrus which is the target point for best mating time (6–12 hours before ovulation), the only practical protocol for achieving effective synchronization of ovulation and mating in order to maximize litter size is to mate at regular 12-hour intervals (early morning or late afternoon), on every day of standing heat. This will ensure that at least one or more of the matings occur at the optimum time. This may result in sows being mated anything from once to four times, but the average will be just above two. The benefits of multiple mating over single mating, with respect to both conception rate and number of piglets born, are shown in Figure 4.8 which is derived from field information.

A generalized representation of the reproductive tract of the female pig is shown in Figure 4.9.

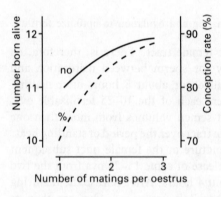

Figure 4.8 There are significant improvements for conception rate and litter size when sows are mated twice rather than once only. Three-times mating may achieve some further marginal improvements. Increased mating improves the chances of achieving optimum mating time when this is difficult to know or predict. There is also some (lesser) benefit of increasing sperm volume in the tract.

Artificial insemination

As the boar may ejaculate $>50 \times 10^9$ sperm at any one time, but as only 2×10^9 sperm placed through the cervix are necessary for effective fertilization, then one ejaculate may be used for 25 or more matings through the practice of artificial insemination. In a working boar, 500ml or so of ejaculate will contain about 100ml of sperm-rich semen containing up to 0.5×10^9 sperm per millilitre. Something more than 70% of these should be motile and something less than 15% should show physical defects.

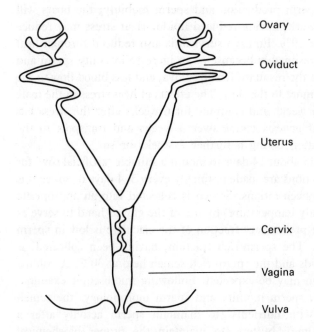

Ovary

Oviduct

Uterus

Cervix

Vagina

Vulva

Figure 4.9 Reproductive tract and gonads of the female pig.

Artificial insemination (AI) is used mostly in genetic improvement programmes, but also at commercial level to extend the boar base and help to handle peaks of boar requirement. In the latter case, often about 25% of the matings may be by AI. The distribution of use of AI world-wide is highly variable, as may be seen from Table 4.1. As a general rule, AI gives conception rates about 10% worse than natural mating and results also in a litter size some 10% lower; but when properly carried out by trained and experienced staff may be equally as effective as natural service for both conception rate and litter site. Hotter climates are perhaps the least appropriate to AI operations as a high environmental temperature (above

Table 4.1 Extent of use of Artificial Insemination

	Percentage of total breeding females bred by AI
The Netherlands	60
Germany	30
Denmark	25
China	25
UK	12
USA	7
Canada	2

25°C) decreases both sperm production and sperm mobility; the boars will also be less willing to work, having reduced libido. Heat stress may reduce pregnancy rate by some 20%. Embryo survival is also reduced through heat stress by about 20%, presumably because of an increase in faulty sperm and genetic incompetence of the resultant fertilized ova, and less blood flow to the reproductive tract, but more to the skin. The effects of heat stress in the male may be delayed by 2–6 weeks and continue for 6 weeks after the stress has passed, due to spermatogenesis taking over 4 weeks and transport to the testes and final ejaculation taking a further 2 weeks or so.

Boars can be trained in about 14 days to mount a suitable 'artificial sow' for semen collection. Collections are made optimally every 3–4 days or so, as this gives the best sperm concentrations. Semen is collected into an appropriate container (at around body temperature) by use of the gloved hand to serve as a surrogate cervix. The preliminary fraction of the ejaculate is low in sperm and may be discarded. The sperm-rich fraction, having been collected, is filtered of accessory fluids and the sperm-rich semen held at 30°C. A volume of 100ml per ejaculation may be expected. Following microscopic examination for sperm density, sperm motility and sperm morphology, the semen may now be diluted with nutrients (to maintain sperm activity after a reasonable period of time), buffers (to maintain the proper biochemical medium for sperm viability), chilling media (to protect the sperm from cold shock) and antibiotics (for disease prevention). The diluent is not only to allow much wider use of a given ejaculate, it is also an essential part of ensuring the proper activities of inseminated sperm so that effective fertilization may be achieved at reduced dose rate. Equally, it must not be assumed that excessive sperm density or sperm volume will improve fertilization by artificial insemination; for the nutrients and buffers to operate to their full effect the concentration of sperm must be close to optimum. Each ejaculate may result in the making up of around 25 individual doses of 50ml, having been mixed with diluent to contain around 2×10^9 sperm per dose. The diluted semen can remain potent for 3–5 days at room temperature (20°C). This is considerably longer than undiluted semen which, in the absence of nutrients and buffers (but with antibiotics), will only keep for 1–2 days.

For best results, two doses of semen should be given at a 16-hour interval. To maximize efficiency, it is essential that the artificial insemination catheter is locked securely into the cervix of the sow, and that the 50ml dose is allowed to flow slowly at its own rate through the cervix and directly into the uterus with the minimum of backflow and loss of semen.

One of the major problems with AI is recognition of the best time to mate, as sows show varying lengths of oestrus, degrees of vulval reddening and willingness to stand to human pressure on the back. The best time for AI is the same as that for natural mating (see Figure 4.7). Ironically, the most helpful aid to successful AI and correct insemination timing is the use of a teaser boar.

Boar semen can be deep-frozen, but the method is more laborious than for cattle, and the results much less effective. Boar-to-boar differences in the success of freezing are striking. Frozen boar semen remains only of interest for purposes of long-term storage of genes, to maintain stocks of genetic diversity for future breeding programmes.

Pregnancy

Most of the ova released (Figure 4.10) will be fertilized (say 22 of 23), but imperfections in some 10% of the early embryos will identify them as unfit and cause their rapid demise. Rising progesterone levels dilate the oviducts, and after 2 days in the oviducts the embryos move into the uterus where – again under the influence of progesterone – the environment is improving for them. Cell division brings logarithmic growth of the embryo and at 8 days the blastocyst begins to elongate dramatically. These elongated embryos are pushed around both uterine horns by contractions of the uterine wall. Some

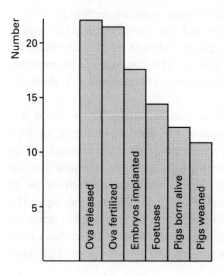

Figure 4.10 Losses of potential piglets during the course of pregnancy. Only half of the ova released complete the pregnancy as live-born piglets. Of total losses, 10% fail to be fertilized, 45% fail to implant, 30% fail to reach term and 15% are born dead; a further 10% die between birth and weaning.

embryos may well migrate from one uterine horn to the other. The objective of this exercise is to achieve even spacing throughout the female reproductive tract. As a result, however, a further 10% are lost (Figure 4.10) so that only 75% of fertilized ova actually implant (say 17 of 22). Attachment of the embryos to the uterine wall begins on day 13 and is complete 10 days later. The embryos are evidently sufficiently competent by day 12 to register their presence and prevent the uterus from dispatching the luteolytic hormone prostaglandin into the bloodstream, which would regress the corpora lutea and remove pregnancy support.

Attachment to the uterine wall is in fact aided by secretions of oestrogen from the embryo itself, the embryo at this point beginning to control its own endocrine environment. The embryo releases oestrogen from around day 12 and this prevents luteolysis, which would normally occur on day 15 or 16. It seems that embryo oestrogens redirect prostaglandin secretions away from the vascular system (via which they would reach the ovaries) and into the uterine tissues where the consequences are positive. Embryonic oestrogen production is paramount for the maternal recognition of pregnancy, and it is these same secretions which encourage insulin-like growth factors with the particular objective of supporting early pregnancy growth of the uterus.

It is understood that progesterone, emanating from the corpora lutea, is the prime hormone maintaining pregnancy. Prolactin also has an important role in cherishing the corpora lutea, and helping to maintain circulating levels of progesterone, especially in late pregnancy when progesterone levels may begin to decline. Opioids from the central nervous system are involved in the operation of the anterior pituitary gland, and these act negatively on the GnRH pulse (thus suppressing peaks of the gonadotrophins FSH and LH), as well as having some regulatory activity upon prolactin secretion.

This is a fraught time for the embryo and any disturbance, additional to that already existing in the internal milieu, will add to the likelihood of embryo loss because of implantation failure. The comfort of the maternal body – psychological, physical, nutritional and environmental – is therefore paramount at this time.

For the pig, pregnancy is not an all or nothing affair. There are at this point degrees of pregnancy. If at any time the mass of embryos, or later foetuses, becomes too small to yield a viable litter, then the strength of the signal will be inadequate to prevent uterine prostaglandin secretion into the bloodstream, a luteolytic attack on the corpora lutea and self-termination of that pregnancy. An adequate mass of embryos by day 14, and subsequently, will however prevent luteolysis by active secretion of luteotrophic hormone from the embryos themselves, which will positively support the corpora lutea and encourage the output of pregnancy-sustaining progesterone. This local activity also requires the active cooperation of an adequate balance of prolactin, FSH and LH from the anterior pituitary to strengthen the corpora lutea. This balance appears to comprise a 'discrete presence' of the gonadotrophins, for progesterone must suppress FSH and LH to a considerable degree if the oestrous cycle is to be firmly put into abeyance. Luteotrophic encouragement *from the centre* can be lost if the system is disturbed, or if the central nervous system demands it. In this event, regardless of the adequacy of the embryonic or foetal load in the uterus, or its luteotrophic output, the pregnancy will be lost.

The vital nature of progesterone in pregnancy support, and the serious problem of embryo loss in early pregnancy in pigs, has led to consideration as to whether litter size might be increased if the early pregnancy levels of circulating hormones were to be enhanced, or at least any progesterone diminishment avoided. The possibility has been raised that high levels of feed intake in the first 3 weeks of pregnancy might reduce the effectiveness of circulating progesterone; first by encouraging a higher level of body lipid (which absorbs progesterone) and secondly by enhancing blood flow rate through the liver and thereby accelerating progesterone clearing. These notions lent credibility to the idea of restricting the feed allowance of the sow in early pregnancy. Whilst over-fat sows are certainly likely both to be less fertile and to have smaller litters, the idea that feed reduction in early pregnancy will prevent embryo loss has not been satisfactorily proven. More reasonably, sows tending to be low on body lipid following lactation are likely to be less disturbed by adequate feed levels than by restricted ones. Whilst it appears clear that excessive energy intakes of above 40 MJ DE daily (3kg of cereal-based diet) are likely to be detrimental in early pregnancy, normal feeding should properly be related to the condition of the sow which is probably more important for progesterone support than factors such as blood flow and metabolic clearance rates.

Once implanted, the development of the embryo into a foetus may yet see further losses and neither is the foetus safe through to term, nor even at parturition when often one apparently perfectly good pig may be found dead at birth.

Of the 17 embryos that may be supposed to have implanted, three will shortly be lost to leave 14 foetuses (Figure 4.10). Most of these (13 or 14) will be likely to survive to term; barring, that is, catastrophic under-nutrition or the intervention of disease or toxic substances, all of which would result in reduced numbers born, with dead, partly decomposed and reabsorbed foetuses at birth; or even abortion and the complete loss of the pregnancy. The chronic losses of perhaps four pigs between early embryo implantation and live birth may be due to failure in the particular part of the uterus at which the young was attached or later perhaps due to overcrowding. It is reasonable to suppose that where spacing as a result of early pregnancy migrations in the uterine fluids has not been properly effected, then crowded implantations could result. The weaker of the brethren would then labour at progressive disadvantage. Overcrowding of the uterine horns is however considered to affect more the birth weight of individual piglets rather than litter size – at least for litters up to 14 or so in number. The negative influence of litter size on individual birth weight is unsurprising in view of finite resources in the uterus with regard both to space and to nutrients, and is shown in Figure 4.11.

The dramatic difference between ova fertilized and live pigs born is certainly a major problem area yet to be satisfactorily handled by scientist and pig producer alike. That concentration on numbers born is more important than either of the more popular whipping boys of pre-weaning mortality or lactation length is shown in Table 4.2.

The time spent in the uterus is around half of the total life-span of a pig destined for meat production. During this time the majority of foetal growth is made in the last 30 days of pregnancy (Figure 4.12). Meanwhile the dam herself is preparing for lactation by growth of mammary tissue (Figure 4.13). It is apparent from measurements of the maternal backfat depth that sows on constant feeding level during the course of the pregnancy feel the increasing nutritional demands in late pregnancy; and maternal body lipid catabolism may be observed subsequent to day 90 of the pregnancy. Calculation of nutrient requirement (Chapter 11) confirms the significant difference in the

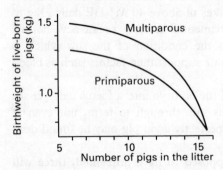

Figure 4.11 The birth weight of piglets is greater for multiparous than primiparous sows. As litter size increases, the average birth weight decreases.

Table 4.2 Factors influencing number of pigs weaned per breeding sow per year

Factor	Action	Result
Pre-weaning mortality	Reduce by half (from 15 to 7%)	+ 2.25
Lactation length	Shorten by 10 days (from 30 to 20)	+ 1
Numbers born	Increase by quarter (from 10 to 12.5)	+ 6

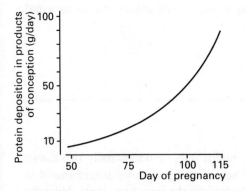

Figure 4.12 As pregnancy progresses towards its conclusion, the growth of the foetuses and other products of conception is exponential.

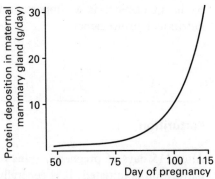

Figure 4.13 As pregnancy progresses towards its conclusion, the growth of maternal mammary tissue is exponential.

way the nutritionist should view early and late pregnancy. An increased level of feeding may be helpful in the later stages of the pregnancy in order to prevent losses of maternal body stores and also to encourage growth of the foetus at this time and to maximize weight at birth. Unfortunately this proposition may be interpreted as a means of reducing the necessary feed allowance in mid-pregnancy, which would be unwise. Also, the relationship between food given to the sow in pregnancy and the final weight of the newborn is weak, and neither must it be assumed that there is benefit in making an already big foetus even bigger. Only in the event of a low birth-weight problem being specifically identified need enhanced late pregnancy feeding be resorted to. It is conventional to accept that piglets with birth weights of below 1.1kg will have lower viability and poorer postnatal growth rates, whilst those below 0.9kg will have only a poor chance in postnatal life. Where W_b is the weight at birth, expected percentage survival rate of an *individual* piglet may be approximated as:

$$\text{Percentage survival} \simeq 75W_b^{0.8}$$

(Equation 4.1)

The relationship between growth rate from birth to 90kg and original birth weight may similarly be approximated as:

$$\text{Growth rate (g, birth to 90kg)} \simeq 550W_b^{0.3}$$

(Equation 4.2)

The appropriate nutritional strategy is to feed adequately for maternal growth and body lipid storage during the whole course of the pregnancy as well as for the growth of the foetal products of conception and the mammary glands. As in all cases it is as important to avoid nutritional excess as to avoid nutritional insufficiency.

Parturition

After 115 days of pregnancy (plus or minus one, or occasionally two days), parturition is initiated. It is generally agreed that pregnancy is terminated at the express wish of the foetal load, which gives the initiating signal. How the extremely low coefficient of variation about the mean of pregnancy length is achieved remains mysterious. It is apparent, however, that given the exponential growth occurring in late pregnancy (Figure 4.12), the environment will be becoming rapidly less acceptable after day 110 of pregnancy and the message of readiness to leave the womb is likely to be accepted with alacrity by the maternal system which by this time will be stretched to its limit. It seems that the foetal hypothalamus responds to stress by producing adrenocorticotrophic hormone (ACTH). In response to ACTH there is a flush of foetal corticosteroids. These stimulate uterine secretion of luteolytic prostaglandin, and progesterone secretion from the corpora lutea declines in response (Figure 4.14). Prostaglandin is central to the effective precipitation of parturition, and the prostaglandin/progesterone antagonism feeds upon itself logarithmically such that progesterone levels dive as the termination of the pregnancy is approached, and the maternal corticosteroids take up the call from the foetal load to rise into a dramatic spike peaking at parturition.

Meanwhile oestrogen levels from the placenta have been climbing at a more conservative rate and from a somewhat earlier starting point, reaching a peak before parturition and declining rapidly to base levels, together with progesterone and the corticosteroids, soon after. The role of the oestrogens is primarily to prepare the uterus for contractile activity (dramatic) and presumably also to encourage nest building and other such maternal responses needed to achieve within the space of a single day the cataclysmic change in the dam from being merely pregnant to being acutely maternal.

In preparation for giving birth, the hormone relaxin allows expansion of the pelvic ligaments, the cervix dilates, and with the crucial and central help of the muscle-contracting hormone oxytocin from the posterior pituitary gland and in conjunction with a further release of uterine prostaglandins, the (oestrogen) sensitized uterine walls and abdominal muscles contract to expel the foetal load. While all these acute activities are underway in the maternal

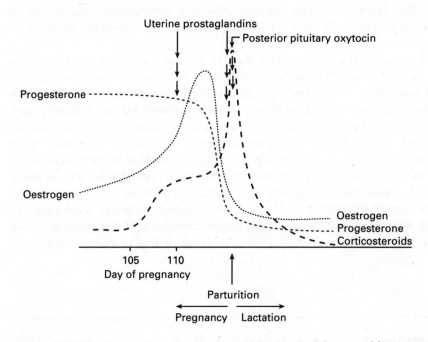

Figure 4.14 Showing various hormonal activities involved in parturition.

body, the mother herself having prepared a nest settles down. Although she may rise during the course of parturition she will normally lie with abdominal and uterine muscles straining. Some sows may appear to fall into a state of conscious quiescence which encourages passivity and avoids stress or panic. The creation of this passive state is a probable consequence of combined hormonal and central nervous system involvement, possibly involving opiate-like activity.

Parturition may take 1–5 hours in order to deliver 9–14 newborn piglets; the time between individual births may vary from a few to 60 minutes. Protracted parturition may result in infection causing MMA: metritis in the uterus, and mastitis and agalactia in the mammary gland. Administration of oxytocin at this time can be helpful in shortening an arduous farrowing, and antibiotics can help reduce the risk of post-parturient infection. The young are not delivered in any particular posture, nor usually with any mechanical problem. About 10% of foetuses brought to term will be born dead, but these will be born without difficulty. The placenta will be expelled partly along with the delivery of the foetal load but mostly towards or at the end of parturition. Neonate piglets will release themselves from the placenta by breaking the cord and within minutes will make their way to the mammary gland to seek colostral milk. With the single exception of being crushed by mother, failure to obtain colostral milk and subsequently succumbing to disease organisms in early life is the primary cause of postnatal mortality in pigs.

Provided that it is certain that day 112 of a pregnancy is reached, injection of prostaglandin, or an analogue, will bring about parturition almost exactly 24 hours later. Prediction of the time of farrowing in this way may simplify farrowing-house management and ensure day-time births. Although such preparations have been readily available for some time, their application has by no means become universal and their cost benefit remains highly controversial. The reason for this is that the relationship between supervision at farrowing and high postnatal piglet survival is not always clear on many units. Some managers achieve low piglet mortality with a policy of non-attendance at farrowing. Delivery of the exogenous hormone earlier than day 112 will, of course, bring about undue shortening of the pregnancy and premature birth, with consequent likelihood of loss of neonate viability.

Lactation

Mammary structure

The structure of the mammary gland is shown in Figure 4.15 (general structure) and Figure 4.16 (alveolal structure).

The post-pubertal mammary growth surge leaves 12 or 14 (or occasionally more) semi-developed mammae. The mammary gland at this stage has a nipple with a ring (the teat areola) of touch-sensitive tissue. Each nipple has for its orifice two canals exiting on a flattened plane just below the tip of the nipple. The internal structure of the virgin gland has a complete, though poorly developed, blood and nerve supply system; while the gland cistern, sinuses, large ducts, smaller ducts and finally the fine ducts are all patent. Nevertheless, the majority of the mamma is fatty tissue together with connective tissue of undifferentiated body cells and some structural collagen. The undifferentiated cells are destined to develop into the active milk secretory cells with their support tissues when, in the fullness of time, pregnancy intercedes.

The purpose of milk is to feed the newborn young; it is hardly surprising, therefore, that it is pregnancy which is responsible for initiating the milk-manufacturing process. With the onset of pregnancy, the internal elements of

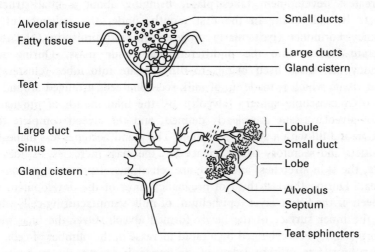

Figure 4.15 Internal structure of the mamma (diagrammatic representation, alveoli not to scale).

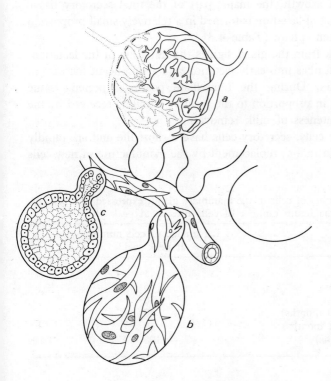

Figure 4.16 Structure of the alveolus showing (**a**) blood capillary supply, (**b**) myoepithelial cells and (**c**) lumen epithelium of secretory cells lying on a basal membrane.

109

the mammary gland become frenetically active. It is during pregnancy that the greatest development takes place, bringing about a final structure complete by full term. In the first month, under the influence of the pregnancy hormones (particularly progesterone and prolactin), the ducts proliferate further into the undifferentiated tissue mass. During mid-pregnancy this mass itself begins to differentiate into lobes (clusters) of alveolar tissue which is made up of milk-secreting cells arranged around the inside of microscopic spheres (alveoli). By the final month of pregnancy, lobes of alveolar tissue are clearly defined, and the alveoli complete their formation and fill with a syrupy secretion. The gland becomes less malleable, more ducts and secretory tissue replace fat, the mass increases, as does the volume, the skin stretches, and there are sensations of pressure, tenderness and heat. During the last third of pregnancy most of the development will have been of the active layer (epithelium) of milk-manufacturing cells which lie on the inner surface of the newly formed alveoli. Over the final week, secretory activity is accelerated by a rapid increase in the number of cells and their synthesis rate. Within 3 days of the impending birth, the mammae are full of milk, and only back pressure (because there is no milk being withdrawn) is preventing the secretory cells from continuous synthesis of milk. As a result of the very rapid development of the active mammary tissue in these last stages of growth, the major part of the total secretory tissue present at the beginning of lactation is formed in a relatively small proportion of the total development phase (Table 4.3).

The removal of milk from the gland signals the initiation of the lactation. Mammary growth continues into early lactation and a significant increase in secretory tissue occurs. During the immediate postnatal period, tissue development proceeds in proportion to the positive stimulus received by the frequency and completeness of milk removal.

Like all active body cells, secretory cells have a short life and are rapidly turned over. In early lactation, replacement by the manufacture of new cells

Table 4.3 Possible number of cells in the mammary gland, expressed as a percentage of the maximum number of cells present at peak yield

	Percentage of cell number at peak yield
Birth	<1
Puberty (200 days)	5
Mating (240 days)	10
Half-term pregnancy (10 months)	20
Full-term pregnancy (12 months)	60
Peak lactation (13 months)	100

greatly exceeds the rate of loss, and there is a net gain. After peak yield has been reached, the balance of gain and loss is tipped in favour of a gradual decline in total cell number. In the pig this occurs from the third to fourth weeks after farrowing, and from that time on the lactation is inescapably programmed into decline toward the end of natural lactation at 10–12 weeks, when the active secretory cells of the alveoli regress and degenerate. Withdrawal of the sucking stimulus, at any time in the lactation, will cause rapid build-up of secretory milk products in the alveoli, severe back pressure and the termination of milk synthesis by the cells, which is quickly followed by degeneration of the active secretory layer of alveolar epithelium. The residue of the alveolal and duct structure remains intact. After the respite of pregnancy, the maternal body reawakens to the inevitable approach of the next parturition, and some 2–3 weeks pre-partum a new layer of secretory epithelial cells is formed within the alveolar framework. This comprises the beginning of the next acceleratory growth phase for the gland as it prepares for the next lactation.

The 12–14 mammary glands of the sow are formed into the udder, the tissues of which may hold their own weight again of milk. Limits to the rate of milk production are a combined function of the mass of tissue actively synthesizing milk, the frequency and completeness of milk withdrawal and the capacity of the cisterns, sinuses, ducts and internal alveolar spaces (lumena) to store the milk once secreted. Secretory products from the epithelial cells first fill the lumen of the alveolus and then, under the pressure of gravity and continued synthesis, the newly formed milk passes down the ducting system. Immediately prior to suckling, about 10–15% of the total milk is held in the gland cisterns, sinuses and large ducts while the other 85% is held in small ducts and alveolar lumina (Figure 4.15).

The nervous system is not directly involved in milk secretion or milk removal, but is essential to the suckling process which requires the active, conscious and willing participation of the central nervous system of the lactating female. The mammae themselves are supplied with nerves which lead *to* the central nervous system *from* terminal endings in the glandular tissue and the skin and around the base of the nipple. These termini are particularly sensitive to touch, stretching and pressure, and convey to the central nervous system both that the mamma requires to be emptied, and that the means of milk withdrawal (the litter) is present at the udder.

It is a remarkable attribute of mammary glands, and a characteristic central to understanding the phenomenon of suckling, that the alveoli which hold the newly secreted milk, and the small ducts which transfer milk to the gland cistern, are not themselves in active contact with the nervous system. The activity of these elements of the mammae are controlled by hormones via the blood supply. The mechanism for hormone release is triggered via the nervous system at the level of both the brain and the nerves present at the skin surface of the mamma and in the region of the nipple. At the microscopic level, the alveolar cells are seen to possess myoepithelial cells lying lengthwise (Figure 4.16). These myoepithelial cells, in common with muscle cells, have the property of contraction. Myoepithelial cells are unusual, however, in that their contraction is initiated not by a nerve, as for muscle, but by hormone action.

Milk synthesis and production

The biosynthesis of antibody-rich colostrum (Table 4.4) gets under way in the last quarter of pregnancy and builds exponentially with mammary tissue growth (Figure 4.13). The mammary gland development will be under the general control of somatotrophin (STH), while the rise in oestrogen at the end of pregnancy is also likely to identify for the mammary gland the impending parturition and the likely presence of young whose survival will depend upon nutrients in the form of milk. The colostral milk is readily and immediately available for withdrawal by newborn piglets. Indeed, milk may be expressed manually from the nipples up to 24 hours before parturition occurs. Pressure in the mammary gland is positive, and the early milk flows easily from the mammae. Oxytocin circulating at this time as a part of parturient events also acts upon the mammary gland helping the outward movement of milk. The first few hours of lactation involve the withdrawal of lactation products actually formed before parturition. It is parturition itself which triggers the need for a full-blown lactation involving the production at hourly intervals of some 250–500g of milk and a total yield of some 6–12kg (or more) daily.

Feedback from the hormonal changes at parturition (Figure 4.14) causes the secretion of the lactogenic hormonal complex comprising prolactin, STH, TH, ACTH, insulin, oestrogen and progesterone. All these are important for sustained lactation in pigs but STH does not have the same dramatic milk

Table 4.4 Composition of pig milk (g/kg)

	Colostrum	Normal milk
Water	700	800
Fat	70	90
Lactose	25	50
Protein	200	55
Ash	5	5

production enhancement effects as is the case with dairy cows. Prolactin secretion from the anterior pituitary is normally inhibited by the presence of an inhibiting hormone secreted by the hypothalamus. The synthesis or secretion of the inhibitor is suspended, with the result that the prolactin flows uninhibited to initiate and sustain lactation. Prolactin first appears at elevated levels 1–3 days before parturition, rising dramatically in conjunction with prostaglandin, and being a contributory party to parturition itself. Prolactin present at the time of birth may also be seen as the initiator of the lactation at this time. The metabolic demands of milk production are immense, as will be shown in Chapter 11, so hormones involved in the biosynthesis of milk constituents (ACTH, STH, TH, insulin, corticosteroids and so on) are central to adequate milk yield.

Prolactin maintains the ovaries relatively dormant throughout lactation, cherishing the corpora lutea, suppressing the growth of ovarian follicles and inhibiting oestrogen secretion, taking over these roles from progesterone in pregnancy. Although gonadotrophin levels in the anterior pituitary gland itself can be quite high during lactation, little or none is released, and there is full and effective suppression of both FSH and, equally importantly, of LH. Oxytocin is released approximately hourly as a result of suckling, and circulating oxytocin is likely to have an additional feedback role in enhancing prolactin and suppressing the gonadotrophins. Prolactin levels are particularly high in early lactation, but fall gradually over the course of the lactation, especially subsequent to 6 weeks. The lactation will fail due to inadequate prolactin support at any time subsequent to 8 weeks post-partum. Abrupt weaning at any time will bring about a sudden and massive prolactin drop. Prolactin and oxytocin are assumed to be the main lactogenic hormones in the pig, with TH, adrenocorticoids (which also inhibit GnRH) and STH giving support. Prolactin increases rapidly during suckling, and remains enhanced for about 40 minutes post-suckling before levels fall away over the next 20 minutes prior to re-enhancement. Whilst piglets suck approximately hourly during the first 4 weeks of lactation, subsequently the suckling frequency reduces as the young become progressively less dependent upon milk as their main nutrient source.

The sucking stimulus maintains lactational anoestrus and quiescence of the ovaries by oxytocin and by prolactin inhibition of GnRH release, and thereby the withholding of FSH and LH secretions. FSH appears additionally suppressed at this time by inhibin. During lactation the positive feed to the central nervous system given by the sucking stimulus also encourages the release of opioids from the CNS, which are positive to prolactin but negative to GnRH, and thereby negative to FSH and LH. Days 1–14 after parturition are required for effective uterine involution and membrane repair and refurbishment. Subsequently, however, follicles may develop, depending upon the strength of the lactation. The administration of GnRH, or of LH and FSH gonadotrophins themselves, can give rise to oestrus during lactation, but it is evident that the effectiveness of such treatment is directly and strongly related to the number of days which have passed since parturition.

During lactation there will be low-level, gradual and chronic increases in FSH and LH. At weaning, however, through the actions of hypothalamic GnRH pulses, there is a flood of pulsatile LH and a rapid rise in the FSH and LH basal levels. The pituitary gonadotrophins stimulate follicular growth, initiate ovulation and bring about oestrus through the previously described course of events (Figure 4.3) some 5 days after weaning.

The level of milk yield is in part a function of the lactational ability of the dam (her body size, her body reserves and her nutrition), and in part a function of the sucking stimulus of the young (litter size, piglet weight and piglet vigour). Lactation peaks at around 3 weeks post-partum (Figure 4.17). The yield of milk varies greatly between individual females, and there is much dispute as to the true level of milk production possible from modern hybrid sows. However, simple calculation using expected conversion efficiencies of milk into sucking pig reveals a total lactation yield in 28 days which is unlikely to be less than 320kg, or an average of 11.5kg daily. The influence of the size of litter upon milk yield of the sow and the milk intake per sucking piglet is shown in Table 4.5. Milk yield also increases with parity number as may be seen from Table 4.6, but so also does litter size – so yield per piglet tends not to increase with parity number. The intake of milk solids by sucking pigs naturally follows the lactation curve, and will peak at 3 weeks of age with a daily consumption of some 0.32kg of solids per piglet per day. Individual piglets starting with a birth weight of 1.3kg and a fat content of around 2% will 3 weeks later have achieved a body weight of 6kg or more with some 15% of lipid.

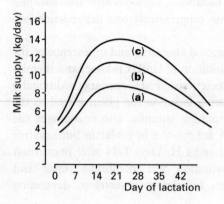

Figure 4.17 Milk production of (a) unimproved sows, (b) modern hybrids with smaller litters, (c) modern hybrids with larger litters.

Table 4.5 Milk yield in relation to number of sucking young

Number of sucking young	Milk yield of sow (kg/day)	Milk intake of piglet (kg/suckler/day)
6	8.5	1.4
8	10.4	1.3
10	12.0	1.2
12	13.2	1.1

Table 4.6 Increase in expected milk yield with parity number

Lactation number	Average daily milk yield (kg)
1	8
2	10
4	11
6	12
8	10

Lactation finally ceases in natural circumstances around 10 weeks post-partum as a result of progressive loss of maternal willingness to suckle and a reduction of any need for the young to suck, as by this time they can be fully nutritionally independent of their dam. Optimum economy of production is, however, achieved by weaning much earlier than this: rarely as early as 14 days post-partum, often at 21 days post-partum, optimally at 28 days post-partum, often at 35 days post-partum and in many countries even yet at the traditional times of 6–8 weeks. The major consequence of imposed weaning is that, in comparison to natural weaning, it is early and it is abrupt. Thus the sucking stimulus is lost instantaneously at a time when potential milk synthesis rate is high. Loss of the sucking stimulus and the absence of oxytocin give negative feedback to prolactin, while the failure of milk to be removed from the gland brings about resorption of biosynthesized milk constituents back into the bloodstream. This reverse flow also provides negative feedback to the hormones supporting lactational metabolism. Within 2 days the lactation is irretrievably lost and circulating levels of supporting hormone have fallen back to base. Two days may appear to be a short time for milk flow to cease and for the pig to register the need to return to the activities of the oestrous cycle, but even within the first half day after weaning the sow will have noted a dozen missed feeds. There is no need to remove either feed or water from weaned sows in order to enhance cessation of milk production and the return of the oestrous cycle. The loss of sucking stimulus and the pile-up of secreted milk in the mammary glands is quite sufficient to send all the necessary signals to the hormonal and nervous systems. Neither will the mammary gland suffer any damage as a result of weaning, nor is the removal of feed or water likely to do anything other than increase stress (rather than diminish it).

The pig is considered lactationally anoestral; that is, the presence of lactation suppresses completely the oestrous cycle. This is however only partly true. First, some sows may show transient oestrus early in lactation, probably due to residual levels of circulating oestrogen; but there is no ovulation. From 3 weeks post-partum there is a real possibility of providing sufficient positive stimuli to encourage both oestrus and ovulation whilst the sow is still lactating. Such stimuli would be a reduced level of sucking stimulus, reduced daily milk yield, a high level of physical activity on the part of the sow, the placing of competitive sows in groups and the presence of a boar. There is a developing interest in delaying weaning due to the benefits that accrue to the young pigs when they are left with their dam. But for delayed weaning to be economic, pregnancy by day 40 post-partum is a requisite target. This may be realistic in some production systems. In the vast majority of circumstances, however, lactational oestrus is of academic interest only, and it is the effective presence of lactational anoestrus which brings with it the need to wean piglets before the oestrous cycle can return and the sow be made pregnant again.

Weaning before 21 days post-partum may cause a reduction in ovulation rate, and an increase in embryo loss due to the uterus not yet being returned to a state of appropriate fitness to receive a new pregnancy after the exertions of the previous one. Full uterine involution is considerably assisted by the sucking stimulus and the uterus of a sucked sow will return to normal quicker than that of a sow weaned very shortly after parturition. There may also be an increased proportion of early weaned sows which do not come into heat within 5 days of being weaned, and there may be a further proportion of sows which will return to heat, having been mated but failed to conceive. Taken together these failures would result in an increase in the (unproductive) weaning to conception interval.

In the case of weaning after 21 days post-partum, however, the weaned sow will bounce back rapidly into an effective oestrous cycle with the rapid initiation of follicular development under FSH, and with the subsequent surge of LH release driving the sow into oestrus and ovulation.

Although the sow is not seasonally anoestrous, and may be considered as a regular breeder throughout the year, there may be seasonal effects influencing the readiness of weaned sows to re-breed. It would appear that the days between weaning and oestrus may lengthen if the temperature is high, if the temperature is low, if the atmosphere is excessively humid or if day length is shortening.

Natural suckling and milk ejection

The neuro-endocrine milk ejection reflex is a short-term physiological system of great elegance and particularly essential to the pig. Mammary cisternal storage is small (around 10–15% of total milk volume in the gland) and most of the milk is held between sucklings in the alveoli and small ducts. While milk may be removed passively from the cistern by sucking action of piglets, the mass of milk cannot be sucked from the mammary gland, and must be actively expressed from the alveoli and ducts by positive decisions on the part of the dam. Piglets will initiate the suckling and will arrange themselves along the udder, each to a mammary gland nipple. When all are present and the sow is content, the sucking stimulus received through the mammae from the piglets informs the brain via the nervous system that milk release is now appropriate. The direct link between the hypothalamus and the posterior pituitary is used for the almost instantaneous release of a substantial 'belt' of oxytocin into the blood. In 20 seconds the oxytocin reaches the target tissue which comprises the myoepithelial cells surrounding the alveoli (Figure 4.16). Oxytocin causes these to contract, the alveoli to implode and the milk is ejected. The milk flows immediately from the gland to the waiting throats of the sucking litter – passage being complete in 15–25 seconds. Oxytocin has a short half-life of less than 1 minute and is rapidly metabolized by the targeted myoepithelial cells. The milk ejection reflex is depicted in Figure 4.18.

Mammalian young are suckled at relatively even intervals through the day, the time elapsing between feeds depending upon species and upon the age of the young. Piglets suck approximately hourly during the first 4 weeks of lactation, but progressively less frequently thereafter. During the interval between sucklings, milk is secreted from the epithelium into the lumina of the alveoli, and also flows to the ducts, sinuses and cistern. As milk accumulates, pressure in the mammary glands rises. The level of pressure, and thereby the frequency at which milk should be removed, will depend upon the degree of cisternal filling. The sow has a proportionately much smaller gland cistern than the cow, and it is significant that piglets suck much more frequently than calves.

During normal suckling sessions in the pig, oxytocin is usually released as a single dose, although a double dose is not impossible. Oxytocin release by a series of drips appears to occur in the immediate period after birth, when the newborn young are not sufficiently strong to suckle at regular intervals but

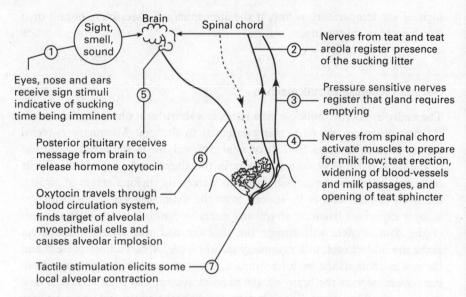

Figure 4.18 The elements involved in the milk ejection reflex.

rather require the supply of milk to be continuously available from the mammae.

As milk withdrawal sessions are readily entered into by mammals, it is sensible to assume that the suckling female derives some positive and acceptable stimulation from the experience. The positive nature of the experience may be associated both with the reversal of the unpleasant sensations connected with the build-up of pressure and tenderness in the gland as it fills with milk and also with the pleasant sensations of the various hormone effects.

Initiation

Nerves run from endings situated around the teats (which are touch-sensitive), at the gland surface (sensing tension, temperature and pressure) and in the gland tissue (registering intramammary pressure). Impulses from these nerve endings join the spinal column from where reflex responses may, within 5 seconds, cause teat erection and squeezing of ducts and cisternal walls (see also Figure 4.18). In this way, smooth-muscle action moves milk from the ducts to the sinuses, controls the flow of milk into the cistern and at the same time prepares for the free flow of milk from alveoli when the myoepithelium contracts. Other pulses pass from the tissue of the mammae along the nerve fibres to the hypothalamus in the brain, where the need for milking is registered and the presence of the means to do it is noted. This results in behavioural responses allowing sucking by the young. Information then passes from the hypothalamus to the posterior pituitary which results in the release of oxytocin (Figure 4.18).

Nerves running from ears, nose and eyes may relay information to the brain about the imminent presence of what the animal has come to learn are the means of achieving milk removal. By such conditioned stimuli, the nervous link between the mamma and the brain is short-circuited; the resultant release of oxytocin is, however, the same. Nerves also relay information that acts to impede oxytocin release. For example, milk release may be held up by pain in the mammae, or by the presence of factors prejudicial to milk removal – such as unaccustomed noises or predators. But, most important of all, milk release in the pig is disallowed by the squabbling and squealing of piglets competing for a nipple. Only when all the piglets are simultaneously settled and happy, each with a nipple in its mouth, and when fighting has ceased, will milk let-down occur. Relay to the brain of negative information will result in nerve pulses passing from the brain to mammary smooth muscle, causing constriction of the blood-vessels, closure of the small ducts and constriction of the milk passages. A reduction in the size of the blood-vessels prevents blood-borne oxytocin reaching the myoepithelium in effective quantities, while closure of small ducts and milk passages results in a hold-up of milk in the glands. Secretion of the hormone adrenalin would also cause constriction of the blood-vessels. The most significant effects occur at the level of the brain itself which, when conditions are unsuitable, simply prevents oxytocin release from the posterior pituitary and so terminates at source any chance of milk ejection.

Passive withdrawal

This phase of natural suckling is so called because the *active* participation of the mother is not required. The young approach the dam and begin to massage the mammae. Both the presence of the young and pressure in the gland are needed simultaneously to trigger behavioural responses conducive to suckling. The mother–young coupling may be initiated by either party, depending upon the relative urgencies of gland discomfort (mother) or hunger (young). The mother becomes progressively less tolerant of the attentions of her offspring as they get older. When sucking begins, the teat sphincter at the tip of the nipple opens and milk may drip out. Although the teat is taken into the mouth of the sucking young by means of negative suction pressure, this is primarily to maintain the position of the teat in the mouth and not to withdraw milk. The tongue is usually partially wrapped around the teat and extends from the base to the tip, even including some of the mamma itself. The tongue makes a rippling, stroking, movement so as to drag milk down from the base (top) of the teat to the teat sphincter at the tip (bottom). In the case of the pig, and also the human, where the gland cistern is small (holding only about 10% of the total milk), little milk is obtained during this phase. Continued butting and sucking by the young may also stimulate by mechanical action contractions of a few of the alveoli and small ducts. Whilst this passive withdrawal of cisternal milk is taking place, nerve pulses pass from the teat in response to the stimulation of sucking, together with impulses picked up from the ears, nose and eyes of the sow. These pulses enter the hypothalamus which then passes the information on to the posterior pituitary, and oxytocin is released into the bloodstream.

Milk ejection

The milk ejection reflex takes place 20–40 seconds after the initial release of oxytocin, and the oxytocic effects last only for 15–30 seconds. Upon contraction of the myoepithelial cells, the alveoli collapse and milk is forcefully ejected into the small ducts. The ducts shorten and widen and the milk rushes through into the sinuses and gland cistern. Only alveolar and ductal milk is expelled by the action of oxytocin on the myoepithelium; there is no contraction of large ducts or cistern. In the sow and the human female, a small cisternal capacity removes the need for a strong sphincter; most of the milk in the gland is held in the alveoli and ducts, and is only likely to leave the gland following collapse of the alveoli when the young are present at the teat.

During ejection, milk flows freely into the area of lower pressure in the mouth of the young. The function of the mouth during this phase is not to suck milk from the gland but rather to facilitate the active flow of milk from the gland, which occurs under pressure created by milk forced down into the cistern from imploded alveoli. After ejection, the alveoli expand back into shape and milk flow ceases. If all the milk is not yet withdrawn from the gland, reverse flow occurs from the sinuses into the ducts, and from the ducts into the alveoli. Milk flow may only be reinitiated by another release of oxytocin.

The general nature of oxytocic action on all mammae simultaneously has its implications for animals with multiple young, as all the young must be present at the udder at the same time. With pigs, the first phase of suckling may be a protracted affair, the mother being unlikely to allow milk release before the whole litter is settled down, each piglet to its own teat, ready to receive. The ability of the mammal to release oxytocin from the posterior pituitary is strictly limited – perhaps because the trigger mechanism becomes jaded, or because the store of hormone is depleted. In any event, a refractory period follows each session of suckling, during which time milk ejection by oxytocic activity is precluded. If milk is not removed during the time when oxytocin is active, it will not again be available until the system is reloaded and primed.

For the pig, milk ejection is vital to effective milk removal as only 10–20% of total milk is held in the gland cisterns. This characteristic can militate against the use of mammals with small milk storage capacities as commercial dairy animals, particularly if there is also any unwillingness to participate freely in the release of milk for purposes other than feeding their own offspring.

Termination

At the end of milk removal, the alveoli relax (expand) and the sphincters close. The sucklers will attempt to draw the last drops of milk from the gland by returning to massage, butting and vigorous sucking to try to spill out any milk which may have been held up in inaccessible pockets and in the more remote sinuses of the mammary tissue. Between 5% and 20% of total milk synthesized will unavoidably remain in the gland, milk removal never being completely effective. This residual milk is not any loss to production, as it joins the new milk synthesized for the next milking session. Ineffective gland-emptying resulting in more than the minimum of residual milk, however, does have serious repercussions if the extra milk causes pressure build-up before the next milking sufficient to reduce the synthesis rate. This latter is most likely if the refractory period for milk ejection prevents bringing forward the next suckling session. Natural circumstances allow demand feeding by mutual consent of mother and offspring, which may overcome problems of incomplete gland emptying at the previous session.

Lactation length

Natural weaning takes place over a period of 3–4 weeks, beginning around the eighth week. The switch from the dominance of the lactogenic complex to the dominance of the gonadotrophins and the hormones of the oestrous cycle is probably rather gradual, but once follicular growth is underway there will be an acceleration towards oestrus. In practical production systems natural weaning rarely, if ever, occurs because whatever lactation length is imposed upon the sow it will certainly be shorter than a grossly uneconomic 12 weeks.

The choice of lactation length is contingent upon the interests of both weaned young and weaned mother, and is resultant from the simple physiology of anoestrus in the sow. Until the sow is weaned, she may not return to oestrus and become pregnant again. Thus, it can be calculated: if pregnancy is 115 days, if a sow shows oestrus and conceives in 5 days after weaning and the lactation length is 56 days, then with 10 pigs weaned per litter she will produce 21 piglets per year. But if all other circumstances remain the same, a lactation length of 21 days will give 27 piglets per year.

However, all other circumstances do not remain the same. At lactation lengths of less than 21 days, the uterus of the sow has not completed involution (Figure 4.19) and such is the strength of the lactogenic hormonal complex at this time that weaning the litter imposes a trauma that takes as long to recover from in terms of the return to oestrus as would have been the case if the litter had been left sucking. From 21 days onward, however, weaning opportunities appear realistic.

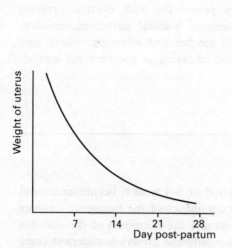

Figure 4.19 Progress of uterine involution in the sow post-partum.

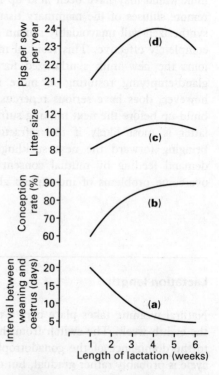

Figure 4.20 Effect of lactation length (age at weaning) on (a) interval between weaning and oestrus, (b) conception rate, (c) litter size, and (d) pigs reared per sow per year.

Figure 4.20 shows the influence of lactation length on various aspects of reproductive performance. First (a) oestrus does not snap back post-weaning quite so rapidly for 2- or 3-week weaning as when the litter is weaned at 4 weeks or subsequently. This is probably a combined effect of the state of the uterus and, most importantly, of the strength of lactogenic complex because yield and sucking stimulus rise to their peaks at 3 weeks post-partum. Secondly (b) conception rate is poor if the pig is weaned before 3 weeks. Some of the sows showing oestrus do not conceive, and they return to oestrus 21 days later thus extending the weaning to conception interval. Thirdly (c) it appears that fewer ova may be fertilized and fewer embryos survive to term as lactation length is reduced. These failings are serious for lactation lengths below 18 days but, as Figure 4.20 suggests, the extent of production loss is not large between 21 and 28 days of lactation length. The final number of pigs produced per sow per year (d) does not change greatly between 21 and 32 days of lactation length.

The decision as to when to wean should not be based on the interests of the sow alone, however, but also on those of the sucking young. The digestive tract of the neonate, and its ancillary enzyme systems, is attuned to receiving milk as the sole nutrient source. Normally piglets would begin to eat solid food from 10 days of age at the investigatory level and with real, if modest, nutritional purpose from 15 to 21 days or so. The pattern of consumption by a litter, however, will not measure that of individual piglets within the litter; some of whom will be eating enough solid food by 14 days of age to provide for up to 200g of body growth daily, whereas others will be eating none at all. The capacity of the digestive tract and the ability of the enzyme system to digest non-milk sources of carbohydrate and protein appears to be critical in development terms at around 3 weeks post-partum (Figure 4.21).

The immune system of the piglet is heavily dependent upon mother's milk; again until 3–4 weeks of age when active immunity builds up in the piglet and competency to deal with invasive disease organisms develops. There is some justification in setting a piglet weight, rather than age, as definitive of readiness to be weaned. Weights in excess of 6.5kg have been proposed. It is possible that a litter could be weaned sequentially – the bigger pigs at around 21–24 days and the smaller ones around 26–30 days by which time they would have reached target weight. Over the latter part of this time the sow will devote a higher proportion of her yield to the smaller and weaker piglets, and so experience a more gradual diminishment of sucking stimulus. This apparently most acceptable of management practices is, however, not often adopted due to its inconvenience and the need to split up litters upon weaning and placement into nursery pens.

Thus it is that at 21 days post-partum developments are underway which will give the piglet an ability to survive on solid food and to be competent to

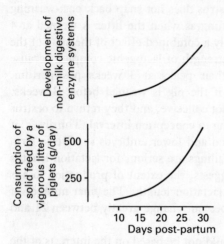

Figure 4.21 Development of non-milk digestive enzyme system and consumption of solid feed by litters of sucking pigs. Individual pigs may well consume at least twice or half of the litter average.

thrive independent of its dam. Weaning at 21 days of age will accelerate these developments, but will also test them at the very time when they are at their most fragile. Sometimes the test is not survived and weaned piglets enter a phase of inappetence, slow or negative growth and susceptibility to disease, especially in the intestinal and respiratory tracts. The ability of the pig to cope with the trauma of weaning improves with age at weaning and so in consequence does the post-weaning performance of the piglets. It is usual to be able to show that piglets weaned at 28 days of age will grow to slaughter weight in fewer days overall than piglets weaned at 21 days of age, there being already some 5kg advantage, on average, evident by 12 weeks.

Because of the relative incompetence of early-weaned pigs, the degree of management care needed and the cost of feeds and equipment are directly related to the earliness of weaning. On the grounds of sow biology, weaning between 21 and 28 days appears optimum. On the grounds of piglet biology, the optimum is somewhere between 28 and 42 days. Overall, it is difficult to justify lactation lengths of less than 24 days or greater than 35 days (Figure 4.22).

The interval between weaning and farrowing

If lactation length is 28 days or longer most (90% or more) of the breeding sows will show oestrus within 7 days of weaning. There will be a normal distribution around day 5, with few sows showing oestrus as soon as 3 days or

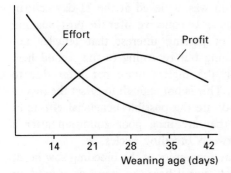

Figure 4.22 As weaning age increases, the effort involved in post-weaning piglet care decreases. Overall profitability of the breeding herd appears to be maximized when weaning is around 28 days post-partum.

as late as 7 days after weaning. Remaining sows, not barren, will come into heat some time in the next 20 days or so. The weaning to oestrus interval should, on average, therefore be about 6 days ($[0.9 \times 5] + [0.1 \times 15]$). Modern production practice rarely achieves this ideal, weaning to oestrus intervals being more usually around 8 days. Amongst others, this is due to some sows being weaned in poor body condition which has the effect of lengthening the weaning to oestrus interval, due to the presence of disease or sometimes to a failure for the oestrus to be observed and acted upon until the next cycle.

At oestrus there will be conception failures. These will result from predictable shortcomings such as inadequate mating frequency, a pregnancy initiated with insufficient fertilized ova to sustain it, poor sow body condition, disease, infertility in the male and the like, which are to be accepted as a normal part of any biological production process. Conception rates are often high, however, and perhaps 85% of sows mated will conceive with the remaining 15% conceiving at the next heat or the one after that (say 12% and 3% respectively). On this basis the 8-day weaning to oestrus interval will now be extended to a 12-day weaning to conception interval ($[0.85 \times 8] + [0.12 \times 29] + [0.03 \times 50]$).

Prior to pregnancy term, there will be pregnancy failures, again as a part of the normal course of expected biological events. Even in the best-run herds some sows will be assumed pregnant, when in fact they are not. Such sows may reach expected term before the absence of pregnancy is diagnosed. Returns to oestrus in cases of long-term non-pregnancy may usually add another 5 days on average to what may now be termed the total 'empty days'. Testing for pregnancy by simple ultrasonic techniques can confirm the presence or absence of pregnancy at 20–40 days post-mating (depending upon the sophistication of the machine and the skill of the operator). Many herds operate a culling policy on the basis that any sows not confirmed pregnant by 50 days after weaning should leave the herd. This will pick up anoestral (barren) weaned sows but would allow for a conception failure at the post-weaning heat, provided that conception was achieved at the 21-day return. Other managers allow the sow two chances to conceive after the post-weaning oestrus before she is culled. It is of passing interest that females not conceiving at the first heat after weaning but returning to the second heat some 21 days later, will have about 0.5 piglets more per litter due to enhanced ovulation rate (Figure 4.23). This is not enough to offset the loss of 3 weeks of cycle time, but it does indicate the possible beneficial effects of giving some sows, especially those that are in a poor condition after a strenuous lactation, some respite between breeding cycles.

Days empty are a vital element in the economics of the breeding sow herd. Days empty may be expected in conventional herds to stand at around an average of 17 days in each cycle. If this number is exceeded then a frank reproductive problem – or management problem – may reliably be suspected. With 17 days empty per reproductive cycle, together with 115 days of pregnancy (a 132-day interval between weaning and farrowing), and 28 days of lactation, the reproductive cycle is completed in 160 days. This allows 2.3 litters per sow per year.

Individual sows failing to show oestrus after weaning, and also chronically anoestral or suboestral sows, may respond positively to treatment with exogenous serum and chorionic gonadotrophins (FSH and LH).

About 40% of the sow herd is replaced annually. But infertility and failure to breed are not the major cause of culling in herds which are kept under modern intensive conditions and in which disease is controlled and nutrition is good; that distinction goes to locomotor problems connected with joints, legs and feet.

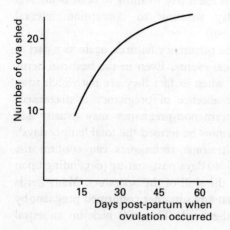

Figure 4.23 As the interval between parturition and follicle rupture increases, so also does the number of ova released.

Productivity summary

On most units between 18 and 26 pigs are produced annually from each breeding female. As the female eats feed and takes up space and labour regardless of her productivity, it costs almost as much to produce 18 young pigs as 26. Some major factors influencing productivity of breeding females are shown in Figure 4.24.

Mortality amongst newborn pigs at or soon after birth represents a major loss. Deaths at birth comprise about 2–20% of pigs born, with a further 2–20% dying after birth itself but within the first week of life. Clearly there

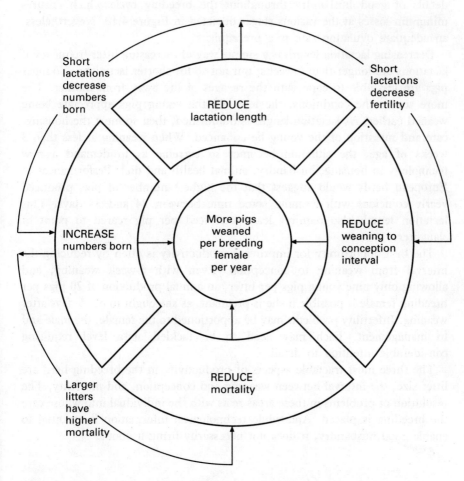

Figure 4.24 Increasing productivity in the breeding herd.

are significant benefits to be derived from keeping mortalities to the lower end of the range.

There are three approaches towards reducing the predilection that little pigs have for suicide: first, the provision of a warm, acceptable and hygienic environment to reduce pathogenic organisms; next, adequate care and nutrition of the mother to encourage high individual pig birth weight; and last, a large number of young pigs born would ensure that, despite unavoidable loss, sufficient young nevertheless remain alive.

A large number of pigs born alive in each litter is crucial, but not easy to attain. It would appear that feeding during pregnancy has less effect on numbers born than on birth weight. There are breed and strain differences in prolificacy, but between-farm differences are substantial. It is attention to the details of good husbandry throughout the breeding cycle which ensures minimum losses at the various stages depicted in Figure 4.10. Nevertheless, an adequate ovulation rate is a prerequisite.

Decreasing lactation length is a simple way of increasing litter frequency if lactations are longer than 6 weeks; but not so for shorter lactations. Younger pigs are less able to cope with the ravages of life away from mother. The more severe the conditions, the less will the young pigs appreciate being weaned earlier. As lactation length is diminished, then so must the housing, care and nutrition of the young be enhanced. When weaning at less than 3 weeks of age, the situation becomes so extreme as to demand a new technology in housing, husbandry, animal health and diet. Performances of European herds would suggest that the highest number of pigs produced yearly coincides with weaning some time between 24 and 35 days. This lactation length also requires least total feed per pig reared to point of slaughter.

The best opportunity for improving productivity is often by reducing the interval from weaning to conception. Even with 6-week weaning, and allowing only nine young pigs per litter, an annual production of 20 pigs per breeding female is possible if she is pregnant, as she ought to be, 5 days after weaning. Infertility problems may be apportioned to the female, the male and to management. Each may need to be tackled at a level requiring considerable attention to detail.

The three most tractable aspects of productivity in the breeding herd are litter size, the interval between weaning and conception, and mortality. The resolution of problems in these areas rests with the individual into whose care the breeding is placed. And while technological information is essential to enable good husbandry, it does not necessarily bring it about.

CHAPTER 5

ASPECTS OF PIG BEHAVIOUR AND WELFARE

Introduction

In those parts of the world where human welfare can be measured in terms of the quantity of food that people receive, the welfare of animals is usually found fairly low on the priority list of agricultural activities. First and foremost comes the need to improve the quantity of animal products available to the human population, and next to reduce the price of those animal products so that there can be an increasing substitution of meat into what would otherwise be an involuntary vegetarian diet.

However, in countries where the well-being of human society tends to be examined in terms more sophisticated than the basic provision of an adequate diet (the nutritional needs being already adequately or excessively met), society will tend to place the welfare of animals kept for the production of meat high on the list of agricultural priorities. Well-kept humans may perhaps be the more ready to consider the lot of their animals.

Animal behaviour science requires, first, an accurate description of behaviour, secondly, an understanding of the control of behaviour and thirdly, a quantitative assessment of how behaviour can be accommodated. All three elements impact upon good husbandry and optimum production, as well as upon animal well-being.

Animal welfare attempts to deal with both the *actuality* of animal health and the *concept* of animal well-being. Broadly, 'well-being' comprises two elements: the human view of the animal's state and the animal's view of its own state. The first aspect falls into the purview of human sociology, the second into the discipline of various animal sciences, perhaps the most important being that of the study of animal behaviour. Animals may suffer from purposeful or accidental ill-treatment and there is no difficulty in identifying such abuse as being contrary to both acceptable ethics (and often national laws) and production efficiency. But well-being implies freedom from deprivation; and here there are degrees of freedom and levels of deprivation, particularly with regard to the importance of the freedom to express behavioural needs.

The modern science of farm animal behaviour, much of which we owe to Professor David Wood-Gush, not only allows a sound foundation for the assessment of welfare, it also provides the basis of good husbandry and stockmanship, and a major means of improving both the level and efficiency of pig production.

Self-choice feeding

One of the basic rights of an animal is to the provision of adequate amounts of feed of proper nutrient content. It is often taken for granted that a pig should be fed a mixed diet, balanced for nutrients by control of ingredient content, and that the decisions regarding appropriate feedstuff mixtures and nutrient contents are best made by nutritionists, feed compounders and production managers. This, however, is rather a recent idea and may not be in the best interests of the animals. Pigs must have balanced their own diets quite satisfactorily for many thousands of years of evolution by choosing from their environment those feedstuffs that were best suited and most needed. There is no reason to presume that the modern pig has not retained the abilities of its ancestors to choose a satisfactory diet from a range of ingredients.

In order to satisfy its nutrient needs by self-choice feeding, the pig must be broadly aware of the following precepts:

1. The pig must have some mechanism for identifying – in quantitative terms – its needs for maintenance, for growth of lean and fat, for lactation and for the relative balances between those needs. The pig must have some means to recognize that different needs can be satisfied only by different nutrients; energy for maintenance, protein for lean growth, calcium and phosphorus for bone, trace minerals and vitamins for metabolic processes and so on. These different nutrients must be differentially distinguishable in the feedstuffs on offer.

2. Choices must be made within the boundaries of appetite and within the ability of the pig to excrete unwanted nutrients unavoidably ingested.

3. Choices should not be limited by narrowness of range of ingredient supply or obscured by the confounding of needs. Problems of choice will occur when a wanted nutrient is only to be found in a feed ingredient which already contains an excess of an unwanted nutrient, or in a feed ingredient which is unpalatable. Pigs will discriminate with considerable precision against feedstuffs they find unpleasant to eat. They will also avoid fungal toxins, some plant toxins (such as those in cruciferous and leguminous plant seeds), and extraneous non-feed solid particles such as stones, glass and metal. There can be aversion to specific tastes, such as bitterness or unpleasantness which may be associated in plants with toxic or anti-nutritional factors.

4. The pig should be granted an opportunity to learn; to experience the metabolic consequences of different ingestion choices, to remember which feedstuffs were associated with the various possible consequences and to recognize either the feedstuff or the feedstuff component (or nutrient) in terms of its metabolic consequences.

Despite prehistorical evidence to the contrary, the idea that pigs could select and balance their own diet is often met with scepticism; not least because (a) it was assumed, in contrary to the market demand for lean pork, that the pig's idea of bodily perfection was obesity, and (b) many experiments testing the possibility for effective self-choice feeding have been negative. With recent improvements to genotype the first assumption is no longer universal. The second may have been due to the experimental design being directed to testing for principle (1) above, but failing to accommodate principles (2), (3) and (4). Indeed, while principles (2) and (3) may have been denied unwittingly through poor experimental design, principle (4) – the opportunity to learn – was in some experiments purposely obfuscated by devices such as frequent alteration of the geographic location in the pen of the different diet types.

Self-choice feeding toward the possibility of effective energy and protein balance is of paramount interest not only because this is the most important economic aspect of nutrient supply, but also because discrimination in this particular aspect of diet formulation may be presumed to be of special concern to the pig. Successful self-choice amongst pigs with regard to optimum energy and protein balance would have the following benefits:

1. Allowing for individual variation in nutrient requirement. In this way the requirement of all pigs is exactly satisfied, and the diseconomies are

avoided of over-providing for some members of the population, under-providing for others and optimally providing for only a few. This is the inevitable consequence of compounding a single diet to the average requirement of a variable pig population. This benefit prevails both with regard to the use of a diet by a group of pigs at any one time, and with regard to the change in diet specification which occurs for every individual pig over time as it grows. Not only do pigs of similar age or weight differ in their rates of protein and lipid growth, but the ratio of diet protein needs to diet energy needs changes as the pig grows. This is because the need for energy becomes increasingly important in comparison to protein as maintenance costs and lipid growth become progressively greater relative to protein accretion.

2. Determination of the optimum balance of diet energy and protein by the nutritional wisdom of the pig rather than the nutritional wisdom of the human diet compounder. If the pig is able to follow the four principles outlined above (intrinsic knowledge of nutrient need and recognition of nutrients in feedstuffs, adequate appetite, absence of confounding factors and the opportunity to learn), then it may be that the pig is more skilled in determining its individual optimum than the human.

3. Achievement of maximum growth rate unrestrained by nutrient limitation imposed by inadequacy or imbalance of the nutrient supply.

There is also the possibility of *dis*benefits:

1. The pig cannot be forced to eat what it may not wish to eat because of factors relating to unpalatability or (human-imposed) imbalance. Often a production manager may wish the pigs to grow at a different rate, or with a different lipid:protein ratio in the growth than the pig might consider optimum. Also, it may be necessary for the pig to consume ingredients less than optimally palatable. These things can only be achieved if the pig is given a simple compounded diet with no choice amongst the ingredients.

2. A period of learning or training is likely to be required and this takes time and management effort.

3. At least two sources of feed must be available to the pig in order for choice to be exercised. This doubles the feeder equipment requirement.

None the less, experiments at Edinburgh conducted over many years have shown that *if* pigs are given at the same time both a high-protein and a low-protein feed and allowed to choose freely between them, and *if* the feeds are of ingredients all of which are similarly acceptable to the pig, and *if* a short period of accustomization to both feeds by alternative provision is allowed, *then*:

- the pig has 'nutritional wisdom' and can recognize the relative protein concentrations of the feed;
- the pig can balance its diet with respect to protein and energy;
- optimum growth rate can be consistently achieved.

Table 5.1 shows responses of young pigs given single diets of various protein content from 12 to 30kg live weight. Daily live-weight gain is seen to increase with crude protein content to a peak at 217g CP/kg diet, but then at 265g CP/kg diet the growth rate diminishes, showing (i) the classical response to diet protein and (ii) the optimum diet crude protein content. In the second part of the experiment pigs of similar weight were offered choices between pairs of these diets, and the results are shown in Table 5.2. The pigs offered a choice between two diets of differing crude protein content appeared able to discriminate between them and select sufficient of each to create a balanced

Table 5.1 Influence of diet content of crude protein upon daily live-weight gain of pigs given a single diet (from Kyriazakis *et al.* (1990) *Animal Production* **51**: 189)

	Crude protein content of the diet (g/kg)			
	125	171	217	265
Growth rate (g/day)	492	627	743	693

Table 5.2 Daily live-weight gain of pigs given a choice of two diets of different crude protein content (from Kyriazakis *et al.* (1990) *Animal Production* **51**: 189)

Crude protein contents of diet pairs (g/kg)			Crude protein content of final diet chosen (g/kg)	Live-weight gain (g, 15–30 kg)
125	and	171	160	682
125	and	217	208	752
125	and	265	204	768
171	and	217	202	769
171	and	265	205	763
217	and	265	218	764

diet; which in this event appears to be between 202 and 208g CP/kg. The precision with which this final diet was independently chosen despite widely different source feed pairs is quite remarkable. Pigs offered the two lowest and the two highest crude protein concentration diet pairs were, of course, prevented from achieving the preferred balance, the optimum solution being infeasible. Table 5.2 also demonstrates that the choice volunteered by the pigs was indeed optimum, achieving rates of daily gain rather superior to that determined with the best single diet in the first part of the experiment. This experiment also showed that pigs chose to reduce the protein level in the diet that they selected as the pigs grew. This is consistent with expected changes in the balance of nutrient demand. The crude protein content of the final diet voluntarily chosen from pairs of feeds with adequately widely differing crude protein content diminished smoothly from 230g CP/kg at the beginning, to 210g in the middle, to 190g CP/kg at the end. Dr Kyriazakis and his group in Edinburgh have again demonstrated the latter phenomenon with older pigs offered a choice of a pair of feeds at 119 and 222g CP/kg when grown from 45 to 105kg live weight. Gains in excess of 1kg daily were achieved, while the diet selected by self-choice began at 195g CP/kg and then diminished smoothly over time to 170g CP/kg when pigs were 60kg, and 155g CP/kg toward the experiment's end.

Behaviour associated with reproduction

Mating

Successful intromission requires the female to be in standing heat and receptive to the attentions of the male. In this condition the female will accept willingly indignities that would otherwise attract either an aggressive or an aversive response.

Usually the female entering oestrus will seek out (by smell, sound and sight) and approach the male for the purpose of head-to-head contact. Vocalization, jaw-champing and frothing at the mouth will be evident from both parties, but especially from the male. The female may nudge the male. In response, the boar may vigorously bunt and nudge the female and investigate the genitalia with his snout, while she will stand relatively immobile. After the niceties of courtship are completed (which, if the sow is in full oestrus, may take 1–5 minutes), the boar will mount (Figure 5.1).

Figure 5.1 Mating. Note the ample space given in the pen and the presence of straw on a dry, concrete floor.

More than one attempt is likely to be necessary, often accompanied by (fruitless) protrusion of the penis and pelvic thrusting. Achievement of the correct position enables insertion. Copulation takes the form of waves of thrusting followed by periods of relative quiescence. Passage of semen is usually associated with especially deep thrusts. The deeper action is associated with the ultimate locking of the spiral glans into the folds of the cervix and ejaculation. This element of the affair may last between 2 and 10 minutes, and its duration is highly variable. The female may move around somewhat during the course of the events. In natural circumstances, boars may mate three or four times daily over the period of the sow being in standing heat, each mating being followed by a refractory period.

The following may be surmised if optimum conception rates and fertility are to be achieved:

1. Courtship is a necessary part of fertilization.

2. Females entering oestrus should be in sight, smell and sound of the male, and be able to move freely into physical nose-to-nose contact (normally through the open bars of a pen), see Figure 5.2(a) and (b).

135

Figure 5.2 (a) Group pens of weaned sows entering oestrus and awaiting mating. Note that the pens for sows alternate with pens for boars, and only open bars separate the two. (b) A boar investigates from his pen the possibility of a gilt in the adjoining pen coming into oestrus and being ready to mate.

3. Mating quarters should be of adequate size to allow (i) bunting and investigation, (ii) the possibility of escape for the female from the attentions of the boar if she is not in full standing heat, (iii) effective mounting and mating. These three activities require a minimum area of $10m^2$, and preferably more (Figure 5.1).

4. The floor of the mating pen should be clean and dry to give a good grip and it may also be helpful if bedding is provided (Figure 5.1).

5. Standing heat itself is best identified in the presence of a boar.

6. Attention should be paid that correct intromission has taken place.

7. Mating should not be hurried, nor interrupted at any time.

8. Upon dismounting after successful coitus, the boar should be moved quietly away from the sow before the sow is moved to her separate quarters.

9. Given adequate space, there is no reason why a boar should not run freely with a group of sows, provided he is not over-used. Evidence of the female having been mated is conspicuous by red abrasions and bruising behind the shoulders of the female resultant from the activity of the boar's front legs.

Parturition

Choice of nest site need not be immediate. Feral pigs (domestic animals allowed to run wild) will often build at the margins of woodland, constructing the nest from tree branches, saplings, new woody shoots and bark but especially favoured are grasses, sedges and rushes which are carried in large mouthfuls and with great enthusiasm in the period prior to farrowing. The nest will comprise a wall of material built by rooting movements of the snout and is likely to be asymmetrical. Pigs, of course, build nests for resting in the normal course of events by the same means. Given an area of coppice and ground vegetation, together with some form of shelter, pigs will actually clear the plant material which may be eaten, chewed and discarded, or incorporated into the nest site (or shelter) after some degree of mastication or manipulation by the mouth. Pigs kept in groups in open straw-yards will create nests and lie together in subgroups. Nest building may take some time, with much fetching and carrying of straw and subsequent manipulation and chewing. Sites, once chosen, may be quite long-lasting. The need to lie communally extends from a very young age, immediately after birth, through all classes of pigs as they grow, and includes adult sows and boars. Only when about to farrow will the female seek a degree of isolation from other members of the group.

In the pre-parturient female, nest-building begins about 24 hours before birth and is again associated with vigorous activity of jaw and snout. There is frequent turning around and a deal of moving to and fro, often with the specific purpose of fetching and carrying nest-building materials. Pre-parturient sows have been observed under feral conditions to travel between 100 and 200m in an hour and a half or so, going backwards and forwards with nesting material. Some of this frenetic activity may continue during parturition, usually within the confines of the nest area. Modern husbandry systems may frustrate nest-building activity by confining the pre-parturient sow into a farrowing crate (Figure 5.3). However, this also prevents turning and moving during parturition which may otherwise lead to squashing of the

Figure 5.3 A farrowing crate which restrains the sow before and during parturition, and throughout lactation.

newborn underfoot. Thwarting the nest-building drive does not appear to adversely affect the process of parturition itself. Some amelioration of the extent of nest-building thwarting may be achieved by the provision of straw at the head of the confined sow, which at least facilitates mouth manipulation of nest-building material, and also allows nest-building movements with the forelegs whereby material is raked back into a mound. There remains considerable argument as to whether sows should be allowed freedom to come and go, and to turn around, in the place of parturition. While some authorities point to research evidence under controlled conditions of low piglet mortality in 'yard farrowing accommodation', others point to the history of mortality of 20% or more associated with open-pen pig house designs, and the progressive reduction in perinatal mortality as first farrowing rails, and next the farrowing crate itself, were introduced – the beneficiaries of the farrowing crate being not farrowing sows but neonatal piglets.

During parturition the sow will exhibit periods of passivity when awareness of extraneous activity in the environment is apparently low. This semi-comatose state is interspersed with periods of activity, when she is restless and may rise and attempt to turn. No attention is given to those of the litter which have been born, and the dam does not assist in cleaning and drying. When lying, the sow will usually be on her side, exposing the udder for frequent access by newly born piglets who will suck to obtain the benefits of colostrum as soon as possible after birth. Meanwhile abdominal and uterine muscle contractions expel foetuses, with little trauma, usually at the rate of one foetus for each deep muscle effort, sometimes with vigorous tail movements, and over a period of 2–3 hours, with variable lengths of time between each delivery (usually around 15–20 minutes). Some placental material is voided during the course of farrowing, but most at or near the end. The umbilical cord may be broken at birth, or by the piglet struggling to make its way forward round the rear legs of the sow to the mammary glands. The young are up, active and mobile, without assistance, within a minute of birth.

In the natural course of events the dam will attempt consumption of placenta and piglets born dead. Cannibalism of live-born piglets is rare.

Teat-seeking behaviour by the neonate is aided little by the mother. Fore and rear leg movements, which have been interpreted as giving guidance to the young looking for the mammae, are usually counter-productive and cause more disturbance than assistance.

During the first few hours post-partum milk is continuously available at the udder, and a teat order begins to be established at this time. Within 5–10 hours of birth suckling frequency occurs approximately half-hourly, and is established at hourly intervals within the first day. The hourly frequency continues up to around 35 days of age, after which suckling bouts occur progressively less often.

Nursing and sucking

The establishment of the teat-order in piglets is renowned. It consists of the development of each individual piglet's preference for a particular nipple, and that piglet's willingness to fight for possession of its nipple, with much squealing and biting. If each suckling were to be the occasion for a litter of 10–12 piglets to squabble anew over which mamma was to be occupied by each piglet, the process of feeding would become unconscionably inefficient. The teat-order therefore represents the means by which the members of the litter may come to a general agreement over the first 2 days or so of life as to which nipple they will suck. Teat-order is usually established within 3 days of farrowing. Quarrels tend to ensue over those nipples which are placed in the middle of the udder, where mistakes are more readily made. There is also confusion when the sow changes the side that she usually lies on, for this requires the piglets to adapt to a change in their vertical position if they are to maintain their preferred teat. It is thought that the anterior mammae are most favoured, and won by the bigger piglets, but the evidence is weak. Piglets fostered on to sows having farrowed 1–2 days previously will have little problem finding/winning a place at the udder, but after this time problems will arise as those mammae not regularly sucked will have begun to dry up, and the interloper will cause serious disruption to the established order.

Nursing may be initiated either by the sow, or more commonly by the piglets. Prior to this, either or both parties are likely to be resting or sleeping. After a lapse of time from the previous suckling of about 1 hour, the sow may be roused from slumber by the attentions of one or two members of the litter who have awakened early. If the sow is willing to be sucked, she will roll on her side and expose both rows of teats. For effective and equitable feeding of the whole litter, it is necessary for co-ordination of sow and piglet behaviour to ensure all members are present, correct and ready to receive before the sow allows milk to flow from her mammary glands to the sucking young. The pattern of milk flow from the mammae of a sow is shown in Figure 5.4. Phase 1 of suckling comprises the piglets jostling for position on the udder for purposes of identifying and retaining their own preferred nipple. This phase may last for 20–60 seconds, or more if there is quarrelling. Nosing and butting the mammae with vigorous up and down movements of the head herald phase 2, which lasts for about 30–40 seconds. Phase 2 will not begin until the disputes amongst the piglets at the udder have been resolved, and all quarrelling and squealing has ceased. This phase stimulates oxytocin

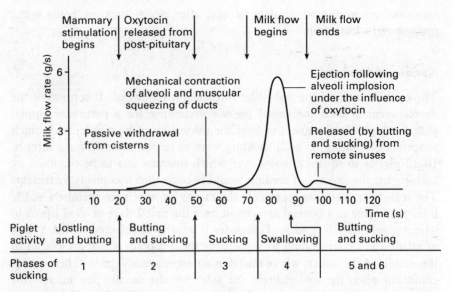

Figure 5.4 Pattern of milk flow from mamma of a sow. Total yield/sucking ≃ 600g.

release, and the piglets may get a little milk by passive withdrawal from the cistern. The piglets may then be observed to go quiet for about 20 seconds during phase 3, sucking on the teat with slow mouth movements of about one 'draw' of large amplitude upon the teat every second. Milk flow (phase 4) lasts for a mere 10–20 seconds, and may be recognized by the piglets' rapid mouth movements (about three per second) of smaller amplitude while they stretch their necks, flatten their ears and swallow 40–80ml of milk. After this, there may be a short period of quiet sucking, like phase 3, but often interrupted by noisy sucking and 'smacking' of lips (phase 5). The final phase (phase 6) returns to vigorous butting of the mammae. This phase is usually terminated by the sow rolling on to her udder and covering her nipples. The six phases of sucking are depicted in Figure 5.5 (i)–(vi).

The sow herself has meanwhile been grunting in a characteristic rhythmic vocalization pattern of about one grunt every second. At around the beginning of phase 3 of the suckling, this rate of grunting escalates dramatically to about two grunts per second, and then subsides down to one grunt per second by phase 4; and finally the sow will stop grunting some time during phase 5 (Figure 5.6). Sometimes two peaks in grunt rate may be heard, the second occurring around the beginning of phase 4. It has been observed that not every nursing given by a sow will necessarily result in the release of milk. Unsuccessful sucklings are characterized by the absence of phase 4 from the routine, and a failure on the part of the sow to increase discernibly her rate of grunting. It would appear from the timing and

Figure 5.5 The six phases of suckling: (i) jostling for position.

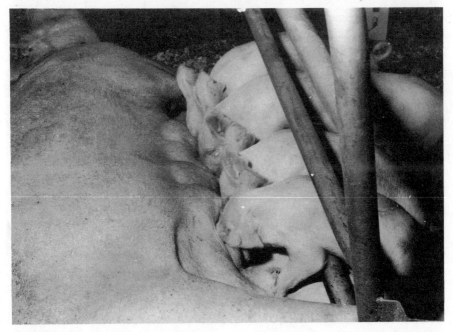

Figure 5.5(ii) Massage of mammae.

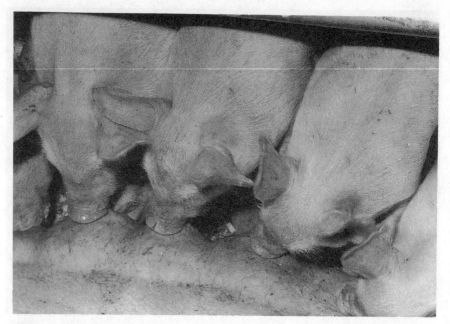

Figure 5.5(iii)   Quiet sucking.

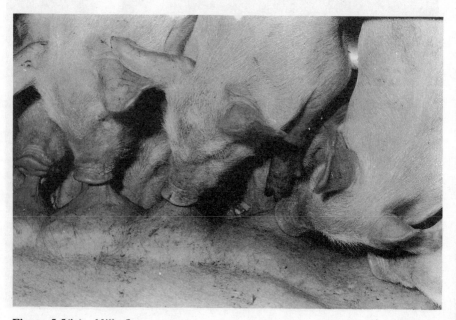

Figure 5.5(iv)   Milk flow.

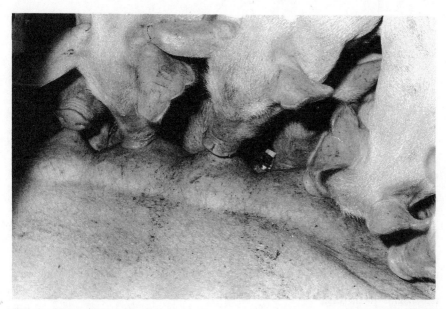

Figure 5.5(v) Quiet sucking.

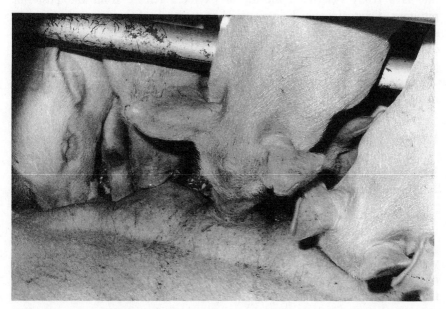

Figure 5.5(vi) Final massage.

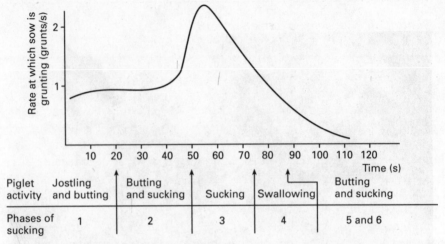

Figure 5.6 Rhythmic grunting of sows during the course of a bout of suckling. The solid line shows a single peak in grunt rate. The timing of the increase in rate of grunting in relation to milk flow (Figure 5.4) would suggest that the peak may indicate the release of a burst of oxytocin from the posterior pituitary into the bloodstream. Piglets may be seen to swallow milk about 25 seconds after the grunt rate begins to increase.

characteristics of the vocalization pattern of the sow that the increase in grunt rate is a ready external indicator of oxytocin release from the posterior pituitary. Oxytocin may be presumed to have an elapse time of about 20 seconds between release from the pituitary and action upon the gland. The duration of the effect of oxytocin upon alveolal implosion appears to be patent for a further 15–20 seconds.

Nursing and sucking will be seen as having been disrupted or abnormal in nature if:

1. piglets do not stop jostling for teats within a reasonable time from the initiation of the suckling;
2. the rate of sow grunting does not increase;
3. phase 4 of sucking (rapid mouth movement) is not evident.

Because of the limited amount of milk held in the cistern of the mammary gland of the sow, adequate levels of milk will only be supplied to the sucking young when the proper conditions of teat-order, increased grunt rate and phase 4 sucking behaviour prevail. With the singular and brief exception of the first hour or two post-partum, the presence of the young sucking at the mammae is not an adequate positive indication of receipt of nourishment.

Competitive and aggressive behaviour

Pigs normally live in family groups. Competition amongst them is restricted in the most part to sibling disputes in the first weeks of life, and otherwise to aggression with other families when there is an encounter, particularly if there is also competition for food or space. There is also the normal fierce, aggressive behaviour amongst competitive males. By and large, however, the family group is not a threatening, but rather a warm, local environment for pigs to live in. This results from early definition of the hierarchical social (dominance) order amongst siblings, amongst less-well-related pigs, amongst breeding females and also males. This hierarchy may need to be reaffirmed, or indeed altered, by occasional acts of aggression, but in the normal course of events vocal and postural threats, with the occasional physical challenge, are usually adequate to maintain an ordered social structure amongst pigs living in groups. Once the dominance order has been established, however, and provided groups are not destabilized by the addition of strange pigs so that a new order must be established, the vast majority of aggressive acts between pigs take place at the time of feeding, or at other times when there is real competition for a scarce resource – such as newly provided straw.

This relatively harmonious state of affairs does not pertain in many domestic pig-keeping situations. Aggression arises in modern pig husbandry systems as a result of:

- Lack of adequate space to respond submissively to a threat, and to achieve full avoidance action and movement away. A situation that would otherwise have been averted is thus caused to develop into the need to defend and fight.

- Competition for food which creates a situation of constant reassessment of the hierarchical order in the face of hunger and unwillingness of low-ranking animals to comply readily with their more lowly position when to do so would be, effectively, to fail to eat.

- Frequent disturbance of social hierarchies occasioned by the mixing of pigs of different social groups. This occurs at weaning, often again on one or two occasions while growing, on despatch for slaughter and in the slaughterhouse lairage. Adult sows kept in groups are mixed into disturbed social structures at least once every parity while awaiting mating or while pregnant.

Threats usually take the form of vocalizations, or movements of the head, and may be followed by head-butting and shoulder-pushing, although lesser body positions and the mere physical presence of a dominant pig may elicit the necessary avoidance reaction on the part of submissive pigs of lower rank. Most fighting amongst pigs is with the mouth open and teeth bared. Upward thrusting movements and bites inflict elongated bruising, grazes and gashes, mostly to the flanks, shoulders and face. A pig may charge at another, usually targeting the head (if head-on) or the flank and belly. Bunting and rooting is common, especially at the underside and between the hind legs.

The consequences of fighting amongst pigs are severe. Scarring to the faces of sucking pigs due to fighting at the udder may facilitate entry of bacteria and the development of exudative dermatitis. Aggression amongst newly weaned pigs mixed together into broken family groups can reduce the growth of all the pigs in the pen for a period of up to 2 weeks while the new social order is established. If pigs of equal strength and dominance meet repeatedly, individual animals can be severely damaged. Pigs pushed down the social order by fighting may subsequently fare less well, be denied full access to feeding possibilities and fail to thrive. Not only are these pigs inefficient in their growth, they are also soft targets for disease organisms and therefore a threat to the whole group.

Because much aggression starts with competitive feeding, and its consequences are often most severe when expressed in terms of diminished feed intake, it is essential that adequate trough space is available for all pigs to eat as frequently as they may wish from *ad libitum* hopper systems. If meal fed, the trough length must be adequate for all pigs to eat at the same time and avoid aggressive interactions while doing so. This will require a trough space allowance per pig which is greater than the width of the pig across its shoulders. Optimally twice shoulder width is an appropriate guide to trough space allowance per pig if a competitive group is to be fed from the same trough simultaneously and if pigs of lower social rank are to receive their full and fair share of food provision. Divisions also help considerably, even just between the pigs' heads.

Adult sows may fight vigorously when placed into novel social groups as may often be the case after weaning for group-housed sows. As such fighting may occur around the time of oestrus, and during the first 3 weeks of the next pregnancy, losses of embryos can result which reduce litter size. Sow groups will often form with one dominant animal which eats greedily and may become large and fat. There are also likely to be one or two subservient animals which actively avoid competitive feeding situations and, in addition to being wounded and bruised, may become emaciated. This unacceptable behaviour in group-housed sows can be ameliorated by individual feeding. Placing pelleted food scattered amongst the bedding, or on the ground in piles, separated by a substantial distance, is helpful in allowing all sows equality of access, and is often practised in large straw-yards or in outdoor pig units. For smaller pens, however, it is necessary to feed each member of the group individually. This is best done with an individual crate into which each pig can be shut and protected from others while feeding takes place (Figure 5.7). There are also available electronic feeding systems (Figure 5.8) where the feeding station is continuously accessible to pigs wearing electronic identification tags. One stall is adequate for 15–20 sows, but is often used for twice that number. *Ad libitum* feeding of sows is unusual in pregnancy as consumption is likely to be considerably in excess of both need and an economic allowance. In any event, it should not be assumed that the presence of an *ad libitum* feed hopper will ensure freedom of access by all pigs in a group. Dominant animals may well guard the feeding station and prevent access to pigs of lower social rank. This can also happen with electronic feeding stations.

Tail- and ear-biting amongst growing pigs should not properly be considered as aggressive or competitive behaviour; this phenomenon is abnormal, and the behaviour aberrant. A pig may bite another's tail or ear, which will first be bruised and then bleed. The troubled pig will often be passive to these proceedings. The bleeding may then attract further attacks over the next hours or days. At this time significant amounts of tissue will be lost and there will be copious bleeding. The bitten pig will lose condition rapidly and become distressed. A more general outbreak can occur. Frequent observation of growing pigs is necessary in order to control the problem. The pig attacked and the pig attacking should be removed from the pen at the first sign of disturbance, and placed alone in individual accommodations. It can now be safely asserted that although some strains of pig might be more prone to tail- and ear-biting than others, the cause of this problem is environmental and managemental. Predisposing factors are:

- inadequate space allowance;
- inadequate temperature control;
- inadequate air movement;
- inadequate nutrient content of the diet;
- inadequate amounts of feed provided;
- unpalatable feed;
- inadequate trough space and access to feed;
- inadequate water supply.

Dr Peter English of Aberdeen University makes the point that there is an interaction between the importance of social rank in aggressive and competitive situations and the general standards of management on the

Figure 5.7 Individual feeding facilities for group-housed sows.

Figure 5.8 A continuously available electronic feeding station for use by group-housed sows in straw yards.

production unit. Thus pigs of low social rank, submitting to aggressive pen mates in competitive situations, will suffer greatly in circumstances where, for example, there is limited living space or limited trough space. This concept is a vital part of good husbandry and adequate welfare.

Locomotion, ingestion and elimination

The notions that pigs (i) spend most of the time sleeping, (ii) eat greedily and (iii) have dirty habits are misconceptions (at least in part) arising from modern husbandry systems.

In the semi-natural environment of the Edinburgh Pig Park, domestic pigs were kept in groups of about six adults, together with young offspring, in areas of 1.2ha. Feed was provided to a level adequate for a protected environment, and therefore likely to have been somewhat inadequate to provide fully for dietary needs in the semi-natural environment. During the day, the pigs spent half their time grazing and rooting, and a further third walking about, and generally manipulating, working over and interacting physically with their environment (Table 5.3); only 6% of the time was spent lying. The males spent 4% of their time in agonistic behaviour; young adults spent rather more (6%), but the sows hardly any (1%).

In the Edinburgh Family Pen System of Pig Production four sows and their litters had access to defecation areas, rooting areas (peat), activity areas (straw) and nesting areas (straw). The growing pigs – which were housed together with the adults – were fed *ad libitum*, while the sows were individually fed to ensure that the full feed allowance was available to each.

Table 5.3 Comparative daytime activities of pigs in extensive, semi-intensive and intensive production systems (percentage of total daylight hours)

	Pig Park	Family pen	Sow tethers/ stalls
Eating-related and rooting	50	15	2
Moving and interacting physically with the environment	30	25	–
Lying recumbent	6	60	76
Standing	–	–	22

The overall space available for some four or five adults and 40 young (through to finish) was generous at 110m². Despite opportunities for ranging more widely, the pigs spent more than 75% of their time in the nest area, being found in the activity area for only 10% of their time; of the remaining time more was spent in the feeding area than in the rooting area. Pigs spent more than 50% of their daytime sleeping, a further 25% in general activities, 5% eating and 10% of their time rooting (mostly in the nest area). Table 5.3 shows clearly the dramatic reduction in foraging activities (from 50 to 15%), and a dramatic increase in time spent lying (from 6 to 60%) compared to the Edinburgh Pig Park.

In stalls or tethers, pregnant sows will spend about one-fifth to one-quarter of the daylight hours standing and over three-quarters lying. Eating only takes 10–15 minutes per day (see Table 5.3). Some of the time spent standing will be used for repetitive activities such as bar biting and chain chewing. There appears to be an inverse relationship among pigs between the incidence of such repetitive activities and the time spent lying. It is not surprising that standing usually occurs before, during and after meals and drinking, and whilst other activities (mating and human-related) are going on in the house.

Locomotion

As has been seen, feral pigs spend a significant amount of time active and moving about the range area (more than three-quarters of the daylight hours; Table 5.3). Much of this movement is in association with foraging. Pigs also rapidly explore new territory to a wide extent. They will wander, walk purposefully, trot significant distances (hundreds of metres) and scamper (tens of metres). Very rapid movement may be accompanied by sharp, excited grunts and barks. It would appear that young pigs especially enjoy rapid movement, and will leap with agility when startled. Chasing behaviour is an important part of play – and aggression. Healthy pigs play more often and more vigorously than sick ones. This general description is not a familiar pattern in intensive production systems where there is little need for locomotor activity; indeed, little opportunity. Undue degrees of restraint on movement may exacerbate locomotor problems in joints and may also predispose to frustration. However, when given freedom to move widely, but where the need to do so is absent, pigs will usually volunteer to spend a high proportion of the day inactive. This was clearly observed in the Edinburgh Family Pen System of Pig Production (Table 5.3) where, despite ample opportunity for activity and exercise, the pigs spent some 60% of their time lying. Well-fed young pigs when not eating or eliminating will usually be found sleeping. The lactating sow with a ready provision of ample food will lie with equanimity in the nest with her piglets. It is self-evident that sows in tethers, stalls or crates are denied locomotor activity other than moving to and from the lying position, and moving backwards and forwards one or two steps. Such pigs spend rather more time lying, even than in the family pen. However, evidence for the need to do something other than sleep is provided by the significant amount of time that pregnant sows in tethers and stalls will spend standing (Table 5.3).

In intensive production systems, movement between houses and at the time of dispatch can be a stressful event for the animals. Pigs are best moved in limited file down a race with solid sides and an open top (Figure 5.9). Contrasts between intense light and dark should be minimized. Pigs readily accept a change of slope, but not more than 15°.

Ingestion

Feral pigs will spend more than 50% of their day foraging; this they tend to do in groups. Foraging involves rooting up and eating pasture grasses, small trees and shrubs and seeking other feed such as plant seeds, fruits, berries and leaves and shoots of larger trees. Insects and small animals, such as beetles, worms and grubs, are commonly consumed whilst larger animals, and the young of larger animals which are not swift enough to avoid capture, may also be killed by pigs. Pigs will eat available carcasses of animals which have died, or been killed by other predators. The pig is an omnivorous and opportunistic feeder. Having such a wide variety of feeding opportunities, it is not surprising to learn that they have both efficient fore-gut and hind-gut digestive systems. The 'monogastric' digestive system operates in the stomach and small intestine, which is well-adapted to dealing with lipids, sugars, starches and proteins. The highly efficient caecal and large intestinal digestive system operates through the help of a substantial bacterial population, which will deal with nutrients in plant tissues high in non-starch polysaccharides. Neither is it surprising that the pig can readily discriminate the different nutritive values of a wide variety of feedstuffs which enable self-choice feeding and the automatic balancing of the diet. The pig derives satisfaction from seeking feed and is well-equipped to do so with a high sense of taste and smell and a sensitive snout. The pig's snout offers a capability for rooting and moving substantial quantities of inert material in search of feed. Many of the feed sources sought in nature require substantial manipulation and mastication before swallowing.

The pig therefore possesses considerable drives both to forage for feed over long periods of time (and distances) and to manipulate thoroughly what it

Figure 5.9 Pigs being moved quietly down an open-topped, solid-sided race.

finds in its jaws. The natural situation can be contrasted with modern systems of production where the full day's nutrient requirement is provided in a trough and can be consumed in a feeding time of around 20 minutes. Being comminuted already, the need for feed mastication is minimal. Competition at the trough creates a positive relationship between the rate of eating and individual success in terms of growth. It is also noticeable that the less pigs are allowed to eat in the day, the more rapidly they will eat (the more greedy they will appear). The digestive processes also differ greatly between feeds which may be ingested in feral situations (these are high in structural non-starch polysaccharides) and the nutrient-dense starch/oil/ protein mixes which form the major part of an intensive compound diet. The latter of these diets is digested for the most part in the stomach and small intestine and fills a much lesser volume within the gut. Concentrated feeds may frequently fail to satisfy hunger while fulfilling nutrient requirements. Hunger will lead to apparent greed as the time of the next feed approaches. It is evident that the strong and inherent drives to select feedstuffs amongst alternatives, to forage for feed, and to manipulate feed and non-feed matter with the jaw, are all left entirely unsatisfied in modern production systems. However, the need for eating-related activities appears to be influenced by individual environmental circumstances, and it must not be assumed that the time spent in feral conditions foraging for feed is necessarily a behavioural need in circumstances where ample feed is readily available. Where there is both ample straw for manipulation and a rooting area in addition, pigs spend only 15% of their time in eating-related and rooting activities. Nevertheless, this is considerably more than the 2% time spent eating by sows in stalls and tethers, and pigs in richer environments with bedding and rooting areas available will spend a quarter of their time moving and interacting physically with the environment, much of which will involve work with snout and jaws.

Elimination

Pigs are fastidious in their elimination patterns. Defecation is usually in special chosen places in the pen, often in corners or alongside walls or other barriers. In the Pig Park defecation places were selected and used with care, there being little or no random elimination over the area. Only when this (clean) behaviour is prevented or confused will pigs soil lying areas and defecate without consideration for location. Purposeful soiling may occur at high temperatures, either indoors or outdoors, when pigs will wallow in faeces and urine or (preferably) in mud. This is entirely logical behaviour as the pig has a very limited ability to sweat and a predisposition to heat stress and sunstroke.

Choice of lying area and dunging area may differ between the designer of the pig's house and the occupant. This difference of opinion will lead to pen soiling. At lower stocking densities, 'lying areas' may become used for elimination. In the Family Pen system the peat rooting area and the straw-bedded activity areas were frequently used for defecation and urination, despite the provision of areas designed specifically for elimination. This may have been because the pigs did not recognize the defecation area designed for them, or they found the need for the rooting and activity areas superfluous and therefore utilized them for elimination. The frequent inability of the pig and the pen designer to fully agree about the 'correct' defecation area, but the need to keep pens clean of accumulation of excreta, has led to pen designs with fully slatted floors. Faeces will fall through the slats to the slurry chamber below wherever in the pen the defecation has occurred. Unfortunately, although solving one problem, the full-slatted pen is a barren environment which does not allow the use of bedding materials. It is questionable whether such pens are satisfactory for adequate pig welfare.

Injuries and mutilations

Damage to the body can be inflicted through pigs fighting each other at any stage of life. Damage may be particularly serious in newly grouped weaned pigs, and growing pigs and adult sows upon regrouping; all in consequence of the need to re-establish a dominance order and the unwillingness of one or other of two equally matched aggressors to submit. Occasionally the introduction of a single animal into an otherwise stable group will elicit multiple attack to the severe detriment of the interloper. Pigs' teeth can inflict cuts and slashes, but equally as damaging can be rooting and thrusting movements of the head which will bruise deeply, damage limbs and cause internal injury.

Second only to abrasions from fighting, the most common damage to growing pigs results from tail- and ear-biting.

The third most common mutilation is damage to legs, which occurs especially in growing pigs on slatted floors and in pens without bedding. Here, although removal of the skin and physical injury to limbs and joints may cause locomotor problems, more significant perhaps is the high frequency of swollen joints, the presence of lumps and bumps around the legs and especially swellings on knees and hocks. The same floor types can also predispose to damage to the sole of the foot through abrasions, to infections between the toes, to cracked and damaged toes, and to uneven or aberrant toe growth. Where the skin is broken, and where there is any limb injury, there is the added likelihood of the incursion of infective agents which will cause further local inflammation and occasionally systemic toxicity and disease. Secondary infections may enter through any damaged limb to cause a general septicaemia.

Damage to leg joints and other locomotor problems may also commonly be found in sows as a result of fighting when straw-yards are used for loose-housing. However, more frequent damage to bones, joints, muscles and tendons arises in tether and stall systems where pregnant and lactating sows may have difficulty in rising and lying down and may lose their footing on slippery surfaces. Sows in stalls or tethers are also prone to swellings, abrasions and other injuries to the feet, ankles, hocks, knees and also to the shoulders and thighs resultant from concrete floor surfaces and slats. Such surfaces are often also dirty and a ready source of secondary infection.

Tethered sows are susceptible to damage from the tether, especially neck tethers, but also those used around the girth. Metal-bar and chain neck tethers, even when protected with plastic coating, can on occasion rub, inflame and cut into the flesh, creating a deep wound. Only by substantial modification of production systems, and by constant observation and monitoring of young pigs and sows, can the consequences of these injuries be lessened.

At least four categories of mutilation are inflicted by man in the normal course of husbandry practice and common custom. These are:

1. The sharp baby teeth of piglets are clipped, usually within 1 or 2 days of birth at the same time as iron is administered. Piglets' teeth are clipped in order to avoid (a) damage to each other while fighting for position at the udder, (b) damage to the nipples of the sow and (c) reduction of sow milk yield. There is some dispute about the need for this procedure, and in many cases neither (a) nor (b) may occur if piglets' teeth are left unclipped. However, the consequences of either when the piglets are older, when the damage is already done and the task of cutting the teeth so much more difficult, are so considerable that most managers opt for routine teeth clipping. The clippers used should be sharp and the teeth clipped, not crushed, to the level of the gum.

2. In production systems where tail biting is a problem, the tail may be removed, again within 2 days of birth. Often the teeth-clippers are used. Perhaps more satisfactory may be the use of a high-temperature blade, such as that developed for beak trimming in poultry. The tail may be removed in its entirety, half may be removed or just the last quarter at the tip. The removal of the major part of the tail is on the assumption that aberrant tail-biting behaviour cannot occur if there is no tail. The removal of a smaller part creates a more sensitive tail tip; thus encouraging bitten pigs to move rapidly away before the damage is done, rather than to remain passive to the attentions of a tail-biting aggressor. The appropriate solution to tail biting is, of course, the removal of its original cause, not the removal of the pig's tail – either in whole or in part. However, where tail biting is a real problem on a unit, it is in the best short-term interests of the animals to dock all tails, not to await trauma in the knowledge of its likelihood.

3. Ear-notching. Pig identification is a part of good husbandry. Plastic coloured and numbered identification tags can be inserted into the ear by the creation of a single hole with a sharp instrument in a part of the ear where there is some presence of cartilage but few or no blood-vessels. The blood-vessel of the ear will, if severed, bleed copiously, and this is to be avoided. Such ear tags have an adequate record of security, but some will be lost. Many traditional pig identification systems rely on removing notches from the periphery of the ear and this can be done at the same time as iron is given, the teeth are clipped and the tails are docked. Up to four notches can be accommodated on each ear, allowing ambitious systems of numbering or coding according to week of birth, litter of origin and ultimate destination, or whatever. The system is fail-safe unless the ear becomes further mutilated by the attentions of other pigs, or by being caught up and torn. In general, the more comprehensive the numbering system the more mutilated the ear, and the more tattered its appearance. The ear may also be used as the site of a permanent tattoo mark; this may often be of four or five letters and/or numbers, and requires the use of a punch made up of many needles and blacked with paste. The area of damage is extensive in a small ear but the discomfort appears to be short-lived, and the mutilation is not evident. Tattoos on the inside of the ear cannot, of course, be read from a distance in the same way as a large plastic tag or notched ear.

4. Castration. The removal of the male's testicles serves a number of purposes:

    (i)   those males not considered of good type and fit to sire the next generation are prevented from mating and producing offspring;
    (ii) pigs that would otherwise be lean will be more fat;
    (iii) the flesh of the castrate will not be subject to tainting, as can occur with entire boars as they mature.

    These evident benefits, amongst others, have created in some countries a need to castrate male animals if they are to be used for meat.

    Castration is best completed before 21 days of age, and preferably before 10 days of age. Castration is achieved by the use of a sharp knife to make two incisions in the scrotum through which the testicles can be drawn and the attaching cords broken or severed. The area should be cleaned with antiseptic before and after treatment. Piglets being castrated complain vociferously, but this is probably due mostly to being held upside down while the operation is carried out. There is clearly discomfort after the event as a result of the breaking of the cord joining the testicles to the body, the general trauma in the scrotal and groin areas and, of course, the incisions. There should only be slight risk of subsequent infections.

Although a universal practice only a few years ago, many production systems have now dispensed with this mutilation. There are adequate alternative ways of preventing unwanted breeding (for example, separate penning). The higher growth rates, leanness and efficiencies of the entire male over castrate are of great benefit to the economy of the production system, and provided the entire male is slaughtered at a young age (below 160 days) the likelihood of taint is low. Besides, tests for tainted meat may now be incorporated into the quality control system on the slaughterhouse line to ensure 'rogue' carcasses are identified and withdrawn from use for meat.

Behavioural response to stress

The major stressors are disease, injury and subjection to aggression. These may be addressed by an effective health programme, sound housing, good management and avoidance of placement of the pig in situations where it will be subject to aggressive interactions which result in the deprivation of feed, water or lying facilities.

Stress may also occur as a result of other, specific environmental problems, for example on moving pigs from place to place, whenever they are goaded, during transportation, under conditions of heat (or cold), in slaughterhouse lairage and at the point of slaughter. Such stress may be manifest as porcine stress syndrome (PSS), and pale, soft, exudative meat (PSE). Susceptibility appears to be affected by genetics, since some strains are more prone than others. Some of this effect is due to the occurrence of the halothane gene, since pigs which are halothane-positive (nn) are susceptible to stress and the occurrence of this trait varies between strains. A level of stress-related sudden death between weaning and final slaughter of around 1% can be expected on production units. At levels of above 2% there should be investigation as to the cause.

In behaviour and welfare terms, stress is often perceived as the consequence of the restraint which occurs on intensive production units. Particular restraints are seen (i) with tethered and stall-housed pregnant sows which cannot turn around and have limited (less than 1m) forward and backward movement (Figure 5.10), (ii) with lactating sows similarly

Figure 5.10 Pregnant sows restrained in stalls.

restrained in farrowing crates and (iii) with common stocking densities among growing pigs (Figure 5.11) where there may be both many pigs in a group (more than 10) and limited space per pig (less than 1m² per 100kg of pig). The outward expressions of these stressors may be seen as stereotypic behaviours, extremes of placidity and apathy and aberrant behaviours such as tail biting. All types of stress, disease and injury represent an inability to cope with the environment, cope with other pigs or cope with the frustration of desired behaviours. Stress will bring about a depression of the immune system and an increased likelihood of disease. There will be reduced appetite and growth, and decreased breeding efficiency. The degree of stress required

Figure 5.11 Weaned pigs at high stocking density and with large group size.

to bring about any given and quantified reduction in performance, however, is complex and not yet clear.

Stereotypic (repetitive) behaviour may typically take the form of one or other, or a combination, of repetitive tongue sucking, bar biting (Figure 5.12), bar sucking, chain chewing, head waving and excessive water drinking. The presence of stereotypic behaviour has sometimes been suggested to indicate that the pig is coping with the stress, but is now more generally accepted as showing that normal control mechanisms for behaviour

159

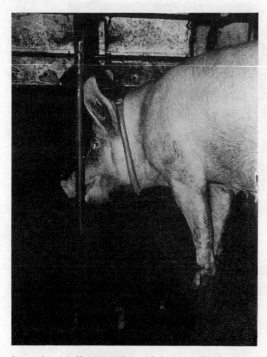

Figure 5.12 Tethered sow bar-biting.

have been disrupted, and is usually accepted as indicative of a low level of welfare.

Recent investigations by Dr Mike Appleby, Dr Alastair Lawrence and their colleagues at Edinburgh have challenged the presumption that stereotypic behaviour is a response only, or even mainly, to restraint. It is true that it is seen most commonly in pregnant sows in stalls/tethers on intensive units; but there is a strong relationship between feed level and stereotypic behaviour (Figure 5.13), and this may be found in both restrained and loose-housed sows. At high-level feeding, stereotypic chain chewing has been found to be virtually absent from sows in tethers, and in loose-housing provided with chains for pigs to chew on. In both systems generously fed sows spent 80% of their time resting. At low-level feeding, however, resting behaviour was reduced to 70% of daytime hours for tethered sows and only 40% for loose-housed sows. The incidence of stereotypic chain chewing was found to occupy 10–20% of the daytime hours of both tethered and loose-housed sows. It appears here that the frustration resulting in stereotypic behaviours is that of a desire to forage, and to use the manipulative potential of snout and jaws. Also, there may be a frank hunger problem describable not in terms of nutrient shortage but of inadequacy of feed bulk.

Care is needed, therefore, in the interpretation of the causes of stereotypic

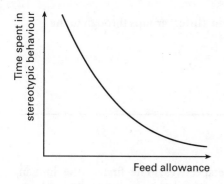

Figure 5.13 Relationship between incidence of stereotypic behaviour and the amount of feed given to pregnant sows housed in tethers.

behaviour. It is further the case that increased freedom levels may themselves bring stress, particularly when adult sows are placed into groups which lead to competition for feed and to the aggression which comes from the establishment and maintenance of the dominance hierarchy. Individual sows in group-housed conditions may become heavily stressed, and this may be ameliorated by separation from the group and individual penning and feeding (which implies an increased degree of restraint).

Outdoor sow-breeding, pig-keeping systems are rated highly in terms of reduction of stressors associated with extremes of confinement, but these same systems bring with them different stressors related to group living: variable feed intakes; wetness, coldness and hotness; poor ground conditions; an increased possibility of inadequate housing. There is reason to accept the view that it was the inadequacies of outdoor and loose-housed systems which encouraged first the introduction of the individual sow feeder, next the indoor farrowing crate, next the sow stall and, finally, the various attempts which have been made to provide housing with a fully controlled environment.

There is much present interest in the possibilities for systems of production which increase potential movement and space whilst avoiding some of the new stressors which may be associated with it. Such systems will usually need to include:

- avoidance of injury and disease;
- generous provision of food and water;
- small group size;
- solid flooring for at least a part of the pen;
- use of bedding (for example, straw);
- provision of material for oral manipulation;
- low-density stocking;
- provision of ample trough space for growing pigs;
- individual feeding for breeding sows;
- stability of animal groups by obtaining constancy of individuals in the group;

● maintenance of growing pigs in family (litter) groups through to the point of slaughter.

Welfare

Welfare is best examined from two perspectives. The first is the human perception of pig welfare; the second is the pig's perception. The first is likely to be variable, shifting as the precepts of society change. It will be geographical in its interpretation, depending upon local custom and culture. It will be motivated through a political will and may be enforced by legislation. The second perspective may accord in some or many respects with the first, but has a different base. The pig's perception of welfare will tend to relate to its state of health and well-being, freedom from injury, adequacy of supply of feed and water, freedom from unacceptable acts of aggression, freedom from stress and the ability to pursue those behaviours necessary to create an agreeable life within the context of the environment in which the pig finds itself. These elements have been discussed in the preceding sections of this chapter.

Some environments are irredeemably disagreeable and the behaviours performed in them therefore tend towards the aberrant. Early weaning, before 21 days of age, cannot be readily coped with by piglets because of the abnormal and traumatic nature of this situation, and the degree of deprivation caused by it. The negative consequences are legion in terms of production inefficiencies, but also included must be behavioural depression, aberrant fighting, belly-nosing and bullying. Again, it is true that the behavioural needs of a tethered sow are different from those of a free-ranging one, and in the former case many behaviours shown under free-range are not needed. But if that tethered sow is inadequately fed, on a low grade of flooring, and also cold, then there is no behavioural repertoire that could make the position agreeable to her. It is not logical to say that such a sow, having been rendered negatively apathetic to her plight, has no behavioural needs because she has given up trying to show any.

It is perhaps important that scientific investigation should bring together the human and the pig perceptions of welfare and ensure that legislation or guidance through the former does indeed enable improvement in the latter. It would be misguided to assume that reduced welfare is present when (i) a situation arises which is unacceptable to humankind but not necessarily also to the pig, or (ii) there is absence from one environment of the behaviour seen in another, which latter environment is perceived by the human to be 'better'. Thus pig welfare in some conditions of restraint may be improved over that in some conditions of freedom; for example, when the former allows adequate provision of feed but the latter allows only competition and aggression. In outdoor and feral conditions, pigs spend much time in rooting activities but this is not the case when there is adequate feed supply and straw in a more-protected environment. This was noticed when pigs maintained in the high-welfare and spacious Edinburgh Family Pen System of Pig Production failed to use the area provided for rooting in peat and soil, but rather abused it as a place for defecation and urination.

In so far as welfare reflects the human idea of optimum animal care, generalizations about the circumstances which should prevail on pig units are impossible. Legislative guidance may be given in terms of:

1. Codes of conduct describing housing systems, environmental control, feeding, management, health and disease control.

2. Approval of particular pieces of equipment, pen and house construction.

3. Licensing of premises for holding pigs and of persons to keep pigs.

All would require an inspectorate (who may or may not have powers in law), while the last implies also education and demonstration of a given level of knowledge and competence. Licensing may override the need for welfare codes and for approvals of equipment if it is considered that the degree of satisfaction of the welfare needs of the animal may best be judged by a trained inspector. The ability of the pig-keeper to maintain a high level of welfare on a pig unit is probably more dependent upon his own knowledge, skill and degree of caring than upon a strict interpretation of codes, or the use of (or avoidance of the use of) particular approved systems or pieces of equipment. There is the added dimension that the desire to legislate against given equipment and systems of production may exceed the strength of the information that the enactment of the law would require. In the wish to remove the use of one piece of equipment, inadequate consideration may have been given to the welfare implications of the (often relatively untried) alternatives.

Overriding in the interests of pig welfare is the quality of stockmanship – the knowledge, ability and willingness of the person in charge of the stock to care fully for their needs. From the point of view of the pig, the quality of the person looking after it must take first priority over and above the quality of production systems, housing and equipment.

Some authorities would wish to base their requirements for adequate welfare on the extent to which the pig can express basic rights:

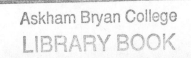

- to the provision of water, adequate amounts of feed and feed of proper nutrient content;
- to freedom from disease and injury;
- to a comfortable environment;
- to freedom from aggression and fear;
- to express normal behaviours and satisfy normal behavioural needs.

Such a list is laudable, and apparently unarguable, but its movement into laws or guidelines for practical pig production is fraught with difficulties. Nutrient supply for what purpose? What if a pig desires to be grossly fat? How free from disease? Health itself requires a degree of subjection to organisms causing disease and therefore contains the possibility of disease. Sometimes one freedom denies another: freedom to express the normal behaviour of the creation of a dominance order requires a degree of aggression and fear. The definition of a normal behavioural need is greatly dependent upon circumstances; a normal behaviour on free-range may not be required in a protected environment.

In many countries three levels of formal activity relating to welfare can pertain: first, laws relating to the prosecution of persons who cruelly ill-treat or abuse animals; secondly, laws relating to the banning of certain production systems which are considered inappropriate on grounds of the distress caused being excessive in terms of both human and animal perceptions (such as, perhaps, stalls and tethers for pregnant sows, and very early weaning of piglets at less than 3 weeks of age); thirdly, guides or codes which are not themselves law but failure to observe them may be used as evidence to support convictions under the first two activities. A fourth possibility, the licensing of stock premises and stock-keepers for pig production enterprises, has not yet been invoked. In the UK there are Codes of Recommendations for the Welfare of Livestock: Pigs (1988), which are published by the Ministry of Agriculture, Fisheries and Food, and which include the following within their preface:

> The basic requirements for the welfare of livestock are a husbandry system appropriate to the health and, so far as practicable, the behavioural needs of the animals and a high standard of stockmanship.
>
> Stockmanship is a key factor because, no matter how otherwise acceptable a system may be in principle, without competent, diligent stockmanship, the welfare of the animals cannot be adequately catered for. The recommendations which follow are designed to help stockmen, particularly those who are young or inexperienced, to attain the required standards. The part that training has to play in the development of the stockman's awareness of welfare requirements cannot be overstressed. Detailed advice on the

application of the Code in individual circumstances is readily available through the official advisory services and in advisory publications of which a selection is listed at the end of the Code.

Nearly all livestock husbandry systems impose restrictions on the stock and some of these can cause an unacceptable degree of discomfort or distress by preventing the animals from fulfilling their basic needs. Provisions meeting these needs, and others which must be considered, include:

- *comfort and shelter;*
- *readily accessible fresh water and a diet to maintain the animals in full health and vigour;*
- *freedom of movement;*
- *the company of other animals, particularly of like kind;*
- *the opportunity to exercise most normal patterns of behaviour;*
- *light during the hours of daylight, and lighting readily available to enable the animals to be inspected at any time;*
- *flooring which neither harms the animals, nor causes undue strain;*
- *the prevention, or rapid diagnosis and treatment, of vice, injury, parasitic infestation and disease;*
- *the avoidance of unnecessary mutilation; and*
- *emergency arrangements to cover outbreaks of fire, the breakdown of essential mechanical services and the disruption of supplies.*

Pig husbandry systems in current use do not equally meet the physiological and behavioural needs of the animals. Nevertheless, within the framework of the statutory powers under which the Code has been prepared, an attempt has been made, on the basis of the latest scientific knowledge and the soundest current practices, to identify those features where the pig's welfare could be at risk unless precautions are taken. The Code sets out what these precautions should be, bearing in mind the importance to the pigs of their total environment and the fact that there is often more than one way in which their welfare can be safeguarded.

The Codes of Recommendations themselves then spell out, in 56 paragraphs, particular aspects of housing, equipment, feeding and management, including different provisions for indoor and outdoor pigs. The Recommendations aim to enhance welfare by guiding toward, amongst others: (i) provision of proper housing and effective equipment; (ii) avoidance of injurious structures and surfaces; (iii) provision of a clean, dry bed and provision of bedding materials and proper flooring; (iv) protection against fire and power failures; (v) proper environmental temperature, humidity, air movements and lighting for all classes of pigs; (vi) adequate provision of water and feed at proper levels and of proper nutrient composition, and proper feeding arrangements whereby competition amongst pigs for feed and aggression during eating can be avoided; (vii) a high level of animal care and individual attention, especially at critical times, including frequent animal inspection; (viii) avoidance of the creation of unstable pig groups; (ix) proper conduct of mutilations, such as ear-marking (where necessary), castration (to be avoided where possible), tail-docking (to be avoided where possible), tooth-clipping (to be avoided where possible); (x) the protection of newborn piglets from crushing by the sow and the special provision required for the comfort and cleanliness of farrowing sows; (xi) the age of weaning (not less than 3 weeks in normal circumstances); (xii) proper allowances for floor space and the provision of feeding, sleeping, exercise and dunging areas; (xiii) prevention of aggression amongst pigs, especially sows in groups, and adequate provision of individual sow feeding facilities; (xiv) avoidance of stalls and tethers for pregnant sows, but the use rather of open-pen yard systems with straw which allow a greater degree of exercise and expression of behavioural needs; (xv) proper boar accommodation and mating areas, together with needs for bedding and dry and suitable floor surfaces.

These Codes are helpful in that responsible persons could not find them in any way negative to pig welfare, but strict adherence to the Codes does not itself ensure welfare. There remain vexed questions: for example, that of the relative extent of welfare of sows in or out of sow stalls, which depends much on the circumstances prevailing for the sow kept in an alternative. Nevertheless few would suggest that current systems for keeping pigs, either intensively, semi-intensively or outdoors, are in any way ideal. Any steps toward the improvement of pig welfare should be welcomed, not only by the pig but also by the pig manager, who will be aware that a high-welfare system is not only likely to be efficient and productive but also to create a more pleasant environment in which people may work rewardingly with pigs.

CHAPTER 6

DEVELOPMENT AND IMPROVEMENT OF PIGS BY GENETIC SELECTION

The early years

Where the Englishman Robert Bakewell of Dishley Grange succeeded in the creation of the improved Leicester Sheep, he failed with pigs. But the principles applied by Bakewell are germane to all the farm animals, including pigs, and he was conspicuous in having made farm animal breeding commercially successful through the sale of the stock he improved.

Bakewell farmed in the latter half of the eighteenth century, during which time the number of people in the British Isles doubled to about 10 million by around 1800. The Industrial Revolution created a population of town dwellers and factory workers separated from their agricultural origins and requiring to be provisioned from an agricultural industry which was itself being revolutionized to cope with the new social environment. Slaughtered pigs were able to provide lard, meat, leather and bristle. Some 200 years on, it appears that both the Industrial and Agricultural Revolutions have run their course. Having created for the industrial nations a society where much of the physical toil has been taken out of factory work, the plentiful nature of food has become an embarrassment and not a comfort and, rather than simply assuaging the hunger of the people, agriculture in northern Europe and the US must now satisfy strident demands for *quality* in food, and also accommodate such additional diverse sociological aspirations as animal rights and environmental care.

At the start of the eighteenth century, Townsend's and Tull's development of root brassica and potato crops for winter keep, together with the more plentiful availability of dairy by-products, millers' offal and waste from butter, cheese, bread and beer manufacture, had allowed a higher proportion of stock to be over-wintered rather than slaughtered in the late autumn. Prior to the 1700s most meat was either from castrated males surplus to breeding herd needs, from pigs slaughtered and salted down at the year's end because of lack of winter keep or from sows which had served their time breeding. Year on year continuity greatly increased the potential size of breeding herds and encouraged the development of breeds of livestock with the sole purpose of providing meat. Successive years of breeding life enabled individual dams and sires to be maintained long enough for their offspring to be assessed and a breeding value for the parents to be quantified. Selection of parents shown to have produced offspring with beneficial characteristics (or traits) enabled the pursuit of positive and directional selection, through the use of proven parents in the breeding herd nucleus over subsequent years. The nineteenth century saw the setting up of Agricultural Societies (The Highland in 1822 and the Royal Agricultural Society of England in 1838) and the opening of pedigree stud books. Such activities helped the planned promulgation of proven sires. It was also at around this time that agriculture was forwarded as a worthy topic for research and academic study, Edinburgh being the first University to appoint a Professor of Agriculture (Andrew Coventry) in 1790.

David Low, Esq., FRSE, second Professor of Agriculture in the University of Edinburgh, contributed substantially to the science of pig breeding by preparing for publication (by Longman) in 1842 *The breeds of the domestic animals of the British Islands*. Low points to the many and varied genera (sic) of hog, all feeding 'on plants, but especially on roots, which their strong and flexible trunk enables them to grub up from the earth. They devour animal substances, but they do not seek to capture other animals by pursuit . . . they delight in humid and shadowy places'. Low describes three genera of the family Suidae; Hogs, Wart-hogs and Peccaries. The Hogs include (a) the Wild Boar of Europe – considered to be the progenitor of the domesticated pig (Plate 6.1), (b) the Babyrousa pig of the East Indies (Plate 6.2) and (c) the smaller Asian or Siamese pig (Plate 6.3). The admixture of the Chinese pig, assumed to be derived from the Asian and Siamese roots, with the European domestic breeds throughout the nineteenth century, has been identified as the primary means by which the indigenous European breeds received the perceived benefits of reduced age and size at maturity and increased fatness. Indigenous breeds (such as the Old English; Plate 6.4) throughout much of northern Europe were considered as being too large and too lean. Only very few of the European domestic breeds present at that time escaped importation of Chinese genes. When David Low wrote, it was clear that Europe contained very many domesticated breeds of various sizes, colours, shapes, ear position and degrees of prolificacy and fatness. Low's illustrator depicts two, the Neapolitan (Plate 6.5) and the Berkshire (Plate 6.6). Although both of these are coloured it is evident that saddleback and wholly white breeds were also commonplace, especially the Yorkshire (to be subsequently called the Large White), the Lincolnshire, the Norfolk and the Suffolk. These same counties still espouse by far the greater part of British pig production. White pigs with lop ears (Landraces) were also evident at that time in France, The Netherlands, Belgium and Germany. In Russia it is asserted that pigs were wilder and more akin to the Wild Boar type than to the European domestic types. The American breeds are supposed to be derived from mixtures of all types, that continent being subject to immigrants (and their livestock) from most other continents of the world, including Africa, Asia and Europe.

Despite the fatness of the wild boar and its influence at that time on some of the European breeds – especially those in the east – the European indigenous breeds tended on the whole to be large, late maturing and rangy. Those breeds receiving substantial quantities of genes from pigs of Asian origin became, by contrast, small, fatty and early maturing. Intermediate types were evident, depending upon the extent of influence of the Chinese pig. However, apparently unique amongst the farmed pigs of the nineteenth century appear to have been those of the Duchy of Parma in Italy, which combined characters otherwise seen as incompatible – large size together with an aptitude to fatten. These characteristics still remain the essential prime requirements for the creation of top quality Parma ham.

Robert Bakewell's idea that animals could be bred especially for the improved production of meat followed from the acceptance that a given animal genotype was not, as had been previously assumed, fixed, but could be changed by the imposition of a selection programme through the hand of the animal breeder. Characteristics such as size, fatness, rate of growth, prolificacy and efficiency could be assessed and progressively improved generation by generation, to create a new type of farm animal, one whose specific and primary purpose was the production of meat. The notion of artificial selection in farm animals pre-dated that of natural selection forwarded by Darwin (*The origin of species* was published in 1858), who learned much from the agricultural practices of the times. The 'wrong' assumption of a static animal type has the corollary that individual animal size, fatness and rate of growth are only affected by health, nutrition and the environment, and that the choice of breeding sire and dam does not imply the possibility of progressive changes in the characteristics of the next generation. The basic 'correct' principles of selection for improvement of type, as perceived and practised by Bakewell, were, in contrast:

- Breed type can be improved by the progressive selection and fixing of characteristics.

- Animal populations show variation in their characteristics, and this

169

phenotypic variation can be observed and quantified. Although much of the variation will be due to health and nutrition (environment), some will be passed on to (inherited by) the next generation (genotype). The genetic quality of the parent can thus be judged by assessment of the progeny, and potential sires can be ranked on the performance of their offspring.

- If selection for improved characters is made in a single place under a single set of conditions, then the chance of the improvement appearing in the offspring is higher. This is because what can be seen and measured (the phenotype) is the sum of genotypic and the environmental variations and, wherever the environmental variation can be reduced, then the relationship between the phenotype and the genotype strengthens.

- Beneficial characters can be fixed into a line of animals if particular individuals (or families of individuals) are used frequently within the development of that line (line breeding). Especially, an individual parent of high quality was frequently used in the 1800s in both the maternal and paternal side of the family in order to fix some special quality into subsequent generations. In the classic case of the famous Shorthorn bull, Comet, his sire, Favourite, was mated to his own (Favourite's) mother, to sire Comet's dam. These principles of inbreeding had been learned from thoroughbred racehorse studs after the introduction of Arab blood primarily from only three sires. It was also noted, however, that inbreeding could lead to a lessening of reproductive ability.

- The choosing from the population of very few male animals of extreme beneficial type, that is to say a high intensity of selection, followed by their concentrated use as sires in a nucleus breeding herd, will bring about the most rapid rate of change. For females, such a high intensity of selection is not possible due both to the sex ratio needed in a breeding population (many more females than males) and to the large percentage of females required for retention in the herd as replacements. Because of imposed selection of less desirable types on the female side, progress is unavoidably slower than in the male, where only the best types can be selected. There is, however, benefit from at least minimal selection amongst the females so that all animals showing adverse characteristics contrary to the direction of selection can be removed: in effect, the rejection of females not true to type.

- Having created an improved animal type in the breeding herd, and fixed that type so that it breeds true generation upon generation, the male offspring from the herd can be considered as appropriate for sale as top-crossing sires to commercial meat producers for use on their female stocks. In this way the improved character (or characters) now secured in the nucleus may be spread through the rest of the population by the sale of males. Bakewell demonstrated particular entrepreneurial skills in this direction. To facilitate the marketing exercise the new improved type requires to be named, and to be readily identified both by eye and by subsequent performance. The potential purchaser must be able to see that he is buying the new improved product; and having bought it, the improvement must be demonstrable in added value terms which, at the time of Bakewell, was seen as reaching market at a younger age and a higher degree of fat cover.

The term 'improvement' from 1750 through to 1900 was associated with the creation of special breeds of pigs with an increased penchant for fatness. In pigs, fat was seen as a beneficial character. Fat was a healthy provider of energy to a population of manual workers, and also provided an effective cooking lard for both meat itself and also for other foods in the absence of readily available vegetable oils. If there was a tradition of British livestock breeding to increase fatness, then this was less the case for Continental breeds where lean meat remained a primary interest of the consumer and where vegetable oils were more readily available for cooking. In general, while Continental breeds of meat-producing animals remained relatively lean and of large size, many British breeds became smaller and fatter.

Prior to the creation of the new 'improved' breeds, the incorporation of fat into meat had been achieved by the fattening of animals which had reached their mature size, and were ready to partition nutrients away from lean growth and toward storage fat. It is evident from records relating to the unimproved breeds that adult size was considerably greater than the subsequently 'improved' types which superseded them. The original Yorkshire breed of pig was so large that some could reach the height of a pony and half a ton in weight (at least double the size and weight of the improved Large White). The most notable character common to all 'improved' meat-producing livestock was that of earlier maturity, which was defined as adequate fatness and finish at a younger age and lighter weight. This was achieved by selection of types which incorporated high levels of fat into the body tissues *during growth* and reached lean-mass maturity at a smaller size – both relatively highly inherited characteristics. In the pig population selection for early maturity and fatness was certainly closely associated with reduced final size. It appears that these 'improvements' were also negatively correlated with reproductive characteristics. The reproductive problems were likely to have been caused by the incorporation of new genes from Asian pigs. These pigs (including some Chinese types) were small, chunky and often coloured. They were chosen to cross with the indigenous European breeds because of their propensity to reach both sexual and size maturity at a younger age and to be fat. (It is well to remember that in the nineteenth century, as now, there were very many breeds and types of Chinese pig, and it may not be presumed that they were the same 'Chinese pigs' that are presently of interest for their putative possession of genes for especially high prolificacy.) The problem of the creation of a sire with excellent characteristics for meat, but with less excellent characteristics for reproduction, is readily solved by a crossing programme following the differentiation of sire and dam lines. This simple expedient appears to have been rejected through the nineteenth and early twentieth centuries for the pig breeds. During this time there was much pressure to create pedigree breeds which were complete in themselves, the male and the female lines being of the same type and breeding true.

The creation of particular pedigree breeds, and the hope that they would excel in both meat and reproduction characteristics, was a common attribute (failing) of pig farming practice through the nineteenth and twentieth centuries, but it was not any part of Robert Bakewell's breeding plans. Bakewell was quick to exploit the separate development of sire lines with specially excellent carcass and growth characteristics, and of dam lines with specially excellent reproductive characteristics. Most rapid genetic progress was made overall when these different sorts of qualities were single-mindedly pursued in quite separate populations which were only brought together at the point of creation of the slaughter generation.

Robert Bakewell counteracted reproductive shortcomings in his improved meat line of Leicester sheep by the use of the improved breed only as a top-crossing sire upon highly prolific dams of a different breed (or cross-bred) for the production of the slaughter generation. The sheep industry of England utilized, two centuries ago, commercial cross-bred breeding flocks showing the added benefits of hybrid vigour in the reproductive traits. Thus, while the sire line was formed through the use of in-breeding and type-fixing techniques, the dam line was formed by a two-step programme: pedigree selection in the nucleus, followed by a stratification programme involving out-crossing of prolific breeds. This lesson was to be well learned – or rather re-learned – by the modern pig breeding companies only relatively recently. In the early years, no such sire-line or cross-bred dam line breeding schemes had appeared to the pundits of the day to be necessary for the pig, breeds being developed from the same root for both meat and reproductive efficiency. Thus the same pedigree breed was assumed to be proper to provide sire, dam and slaughter generation. This assumption was, of course, false.

The desire to make 'coarse' indigenous pigs more obese seems to have run amok in the late nineteenth century with show-ring exhibitors priding themselves in grossly over-fat breed types. These, it appears, were simultaneously being rejected by the purchasing populace who looked for their pork and bacon from *un*improved breed types such as the Yorkshire (Large White) and the Tamworth, these latter having apparently escaped the more serious ravages of the 'improvers'. The Tamworth had previously been expressly criticized for its over-leanness! (A clear lesson in the importance of conserving genes which, whilst appearing useless to contemporary breeders, become central to future improvement programmes.) In the event, the market for pork and bacon products exceeded the remaining availability of reasonably lean pig types to provide for it. The fancy breeders had failed to listen to the messages from the market-place. There arose, in consequence, a market for leaner bacon and ham imported into Britain from the Continent. Importations of pig meat to Britain rose from 300 to 300,000 tons in the latter part of the nineteenth century. Continental pigs, however, were not only of leaner genotype, it is likely that they were also grown with superior production skills.

The breeds which majored in the early twentieth century in Britain were the Large White, the Tamworth and the Saddleback. The Danish pig industry developed at the same time a base population from European Large White and local Danish Landrace pigs. The Danes created a white breed, the Danish Landrace, with lop ears (as opposed to the prick ears of the Large White) whose success was resultant not from the genetic base itself so much as from the logical improvement of that base in the direction of customer requirements for lean bacon. The improvement programme for the Danish pig was driven through the medium of the central progeny test. Classical Robert Bakewell breeding theory and entrepreneurship! By the 1920s half the pig meat consumed in Britain (the presumed cradle of the science of livestock improvement) was imported, mostly from Denmark.

Thus the great opportunity afforded to the British to apply the teachings of Bakewell was, for the pig, largely missed in that country. It was however relentlessly pursued on the Continent with beneficial consequences especially for the Danish, German and Dutch pig industries.

The creation and improvement of the twentieth-century breed types

The century through to 1950 saw the consolidation of the breed types that had been originated in the 1800s. The agricultural industry producing meat from beef and lamb developed stratification schemes, with the use of specialist meaty sire lines on cross-bred dam lines strong in reproductive traits. Meanwhile, in stark contrast, the production of milk from dairy cows, and pork from pigs, pursued the use of the same breeds for both parents of commercial stock. During the latter half of this period the extreme Middle White type of pig, which produced a small fatty carcass seen as ideal for the pork trade of the nineteenth century, was supplanted by the Large White and the Scandinavian Landrace types, more suited to cured bacon production. By the middle of the twentieth century production efficiency required faster growth rates, while consumer preference demanded not just less fat, but almost its absence.

Perhaps because there was evidently so much to be done in the case of the fatty pig, it is the pig industry which has championed the breeding of contemporary lean meat strains of farm livestock. Two hundred years on, it is the pig industry which leads in the application of the science of genetics to the production of the commodity demanded by the market, and it is the sheep industry (the pioneer of two centuries ago) which has this time been tardy to understand either the elements of the application of genetics or the basic rules of market demand and customer satisfaction.

In the creation of lean growth and the avoidance of fat, the propensity to fatten at a lighter weight and a younger age becomes a negative and defunct character. Selection for lean growth brings with it an increase in final adult lean body mass and final body size dimensions, such as used to be shown by some of the unimproved pig types and indeed as preserved in some of the Continental breeds of cattle.

Although selection for growth, meat and reproductive characteristics created improved breed types also in Russia and Asia, little if any progress which may have been made early in the twentieth century now remains; perhaps with the singular exception of the Chinese Meishan type. This is small, slow growing and extremely fat, like other Asian types, but sexual maturity is early and litter size is up to 30% greater than for the European white breeds.

Presently, the world pig population is of two distinct kinds. First, there remain indigenous Asian types in village communities and on traditional farms throughout China, South-East Asia and parts of Africa. But for commercial pig meat production as a business-based farm enterprise, the pigs of Europe and North America predominate. Pig breeding companies, based in the European Community and the US are presently exporting genes to every part of the world where pig meat is eaten – to the exclusion of the indigenous types.

The genetic base for the contemporary pig breeding businesses around the world has been derived from relatively few breeds. Most of these have

Plate 1.1 Part of a 600-sow intensive pig unit under construction in Japan

WILD BOAR & SOW.

Plate 6.1 European Wild Boar. From Low D, 1842, *The breeds of the domestic animals of the British Islands.* Longman

Plate 6.2 Babyrousa pig (East Indies). *Courtesy*: Dr A A Macdonald

Plate 6.3 Siamese pig

Plate 6.4 The Old English breed. From Low D, 1842, *The breeds of the domestic animals of the British Islands*. Longman (Note the obvious ancestry of the Saddleback and Hampshire breeds)

NEAPOLITAN BREED.

PROFESSOR LOW'S ILLUSTRATIONS OF THE BREEDS OF THE DOMESTIC ANIMALS.

Plate 6.5 The Neapolitan breed. From Low D, 1842, *The breeds of the domestic animals of the British Islands*. Longman

Plate 6.6 The Berkshire breed. From Low D, 1842, *The breeds of the domestic animals of the British Islands*. Longman

Plate 6.7 Improved pig of the Large White or Yorkshire breed type. Contemporary. *Courtesy*: Cotswold Pig Development Company

Plate 6.8 Improved pig of the Scandinavian Landrace cross-breed type. Contemporary. *Courtesy*: Cotswold Pig Development Company

Plate 6.9 Improved pig of the Belgian Landrace breed type. Contemporary. *Courtesy*: Pig Improvement Company

Plate 6.10 Improved pig of the Duroc breed type. Contemporary. *Courtesy*: Cotswold Pig Development Company

Plate 6.11 The Chinese Meishan breed type. Contemporary. *Courtesy*: Pig Improvement Company

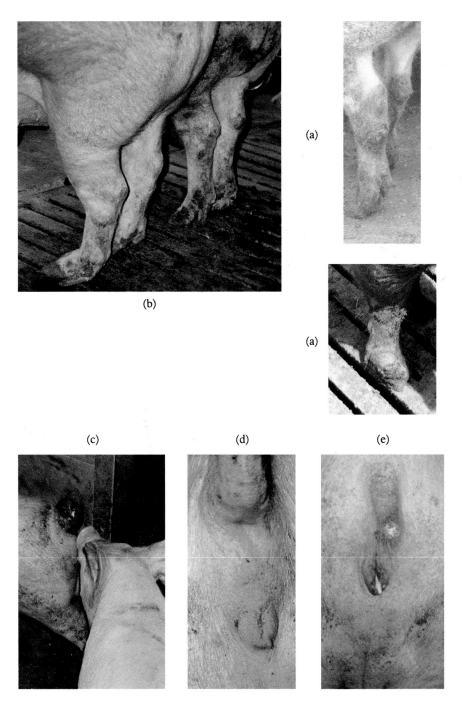

Plate 7.1 (a) Lameness and limb infections, (b) dropped pasterns, (c) rectal prolapse about to lead to cannibalism, (d) atresia ani, (e) vaginal discharge

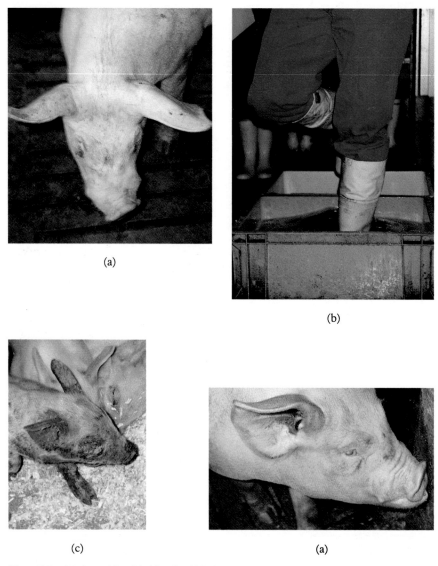

Plate 7.2 (a) Atrophic rhinitis, (b) disinfectant foot bath placed at door of farrowing pen, (c) exudative dermatitis

emerged in Europe and North America as populations breeding reasonably true to type in the early 1900s. The pig populations of Australia, New Zealand, South America, South-East Asia and Japan have developed significantly toward the European and North American patterns over the last two decades. These countries have been able to accept the advantage of the developed white European base of the sixties and seventies, and with it have recently achieved some remarkable rates of further genetic improvement.

The **Large White** or **Yorkshire** breed (Plate 6.7) is, world-wide, the most widespread of modern pigs. Easily recognized as wholly white and with prick ears, the breed is renowned for being superior to all other types in its growth rate and beaten only by the Meishan for its litter size. Modern strains are of large size, the females maturing above 300kg live weight. Growth rate may readily exceed 750g daily from birth to 100kg producing a carcass with 55–60% lean meat. Puberty is at around 180 days, but litter size will be around 11–13 piglets with an average birth weight of about 1.25kg. Now accepted as a lean and meaty pig, it was at one time rather fatter. The Large White has however been subjected to intensive selection against fat over the last 30 years, and is now one of the leanest of all pig types. The breed is seminal to the foundation of most white pig stocks and as a pure breed is fundamental to most crossing programmes involved in the production of the present-day hybrid female. In many countries of the world, and including northern Europe, it is also widely used as a top-crossing sire.

The **Landrace** is a general description of a white lop-eared type, in effect 'the race found most abundant nationally'. It is helpful if this large and variable group of pigs are considered as falling broadly into two quite different varieties. The original Scandinavian Landrace types (Plate 6.8) are long, quite lean, and acceptably prolific, but not especially muscular. Selected for bacon production and farmed as a single pure breed for many years in Denmark, the Danish Landrace is perhaps the most famous of all breeds as an exemplar of the success of progeny testing and selective breeding. Now, however, its main use world-wide is in crossing with the Large White. Nevertheless, given its dominance of the quality pig meat market for more than half of the twentieth century, it is not surprising that the Danish industry showed some reluctance in accepting that the pure breed could indeed be improved upon. Such acceptance has however now resulted in the Danes of the 1990s pursuing cross-breeding and top-crossing programmes with characteristic vigour. In addition to Denmark, this Landrace type is found in, and holds the name of, Norway, Sweden, France and many other nations. The size and performance of the Scandanavian Landrace approaches that of the Large White, some strains bettering the latter.

The other Landrace variety is typified by the Belgian (Plate 6.9), Dutch and German Landrace breeds. Also white and lop-eared, these are, however, slightly less prolific, but are specially noted for being muscular. The depth of muscle in ham and loin, the lightness of bone, the high percentage of lean carcass meat and the full and rounded shape of the hams and eye muscles, characterize these breeds as popular for use as a top-crossing sire; indeed they are used widely as such throughout Europe. They are also sometimes used as commercial females in their native countries, producing in the progeny an extremely muscular carcass, but a slaughter generation which tends to be stress susceptible, to grow rather slowly and to have arisen from a modestly sized litter. In some European pig-keeping systems the Scandinavian Landrace type is used in the female line, whilst the blockier Belgian, Dutch or German Landrace types are used as the top-crossing sire.

The **Duroc** pig (Plate 6.10) is coloured, ranging from gold through red to dark brown, and probably stems from the Berkshire. This robust and heavy-boned breed is as numerous in the US as the Large White or Yorkshire. Whilst neither especially prolific nor lean, it is deeper-muscled (but fatter) and faster growing than the Scandinavian Landrace. Its popularity in the US is said to stem from its ability to thrive in a simple farming system. It is germane that the European white hybrid females found it difficult at first to make their mark in conventional North American farming systems until they were crossed with a Duroc. The use of the Duroc has increased in Europe over recent years. It has found favour as a top-crossing sire in Denmark (amongst other countries) being said to impart (by having 4% rather than 2% *intra*muscular fat in the eye muscle, and by having a deeper red colour in the meat) improved meat quality to the Danish Landrace. Used in the female line the Duroc is purported to offer robustness to the hybrid mother and disease resistance, particularly in the context of outdoor and extensive pig-keeping systems. These last qualities of robustness and disease resistance may be especially important in the context of latter-day welfare requirements for production systems, which may nevertheless be offset by recent evidence that increasing the proportion of Duroc genes into a Large White/Landrace pig type is likely to reduce growth rate, increase fatness and reduce litter size.

Hampshire pigs are black with a white belt. Commonplace in the US, Hampshires were derived originally from the Essex and Wessex saddlebacks of England (Plate 6.4). Now with a reputation for meatiness rather than reproductive performance, the Hampshire has been used in the creation of some hybrid sire lines. While it is hard to see what particular characteristic it may possess not found in the other types already mentioned above, today's Hampshire pig does have the advantage of no longer being stress susceptible. The **British Saddleback**, of the same root as the Hampshire, is perceived to have rather different qualities, motherability and adaptability to outdoor and extensive pig-keeping systems. The progeny tend, however, to be rather fatter and slower growers than white pigs. It remains to be seen whether the Duroc cross will ultimately oust the Saddleback cross as the main source of females for outdoor pig-keeping systems.

Pietrain pigs from Belgium are piebald (white and black spotted), rather small and stocky, and of extreme blocky shape with large hams, large loin muscles, a high percentage of lean meat in the carcass, light bones and high carcass yield. Nervous in disposition, they are readily stressed and prone to sudden death. Their reproductive efficiency is low. In comparison to the Large White the Pietrain produces two less piglets per litter, but can have $10cm^2$ greater eye muscle area (54 v. 44) at 100kg live weight. Whilst relatively few in numbers of pure-breeding herds, these pigs have had a dramatic effect on the muscling qualities of many European pigs, and their genes have been used to improve the shape of many of the leaner white types, especially the Landraces. Pietrain blood is considered by many to be highly useful in top-crossing sires, but it is counter-productive in the female line.

The European and North American axis to modern pig-breeding initiatives has created a narrow breeding base which emphasizes growth rate, efficiency and leanness, while maintaining prolificacy often by the expedient of hybrid vigour rather than by positive selection. Some Chinese types, exemplified by the **Meishan** (Plate 6.11) may offer increased litter size, but have the very severe shortcomings of extreme fatness and slowness of growth. A Meishan pig has an adult weight of no more than 150kg, and grows at a rate of 400g daily to produce a carcass with only 45% lean meat, but it also reaches puberty at less than 100 days of age, has 16 nipples and an average litter size of around 14 piglets with a birth weight of about 800g. The 'extra' three or four piglets born are a result of both a greater ovulation rate (20 v. 17 for the Large White), and better embryo survival (70 v. 65 for the Large White). At least as important is the genetic diversity to be found in the many varieties of Chinese and Asian pigs which comprise the indigenous populations of many countries and still provide a significant part of the world's pig-meat supply. Such genes might not be useful now, but may well have unforeseen qualities which could be needed in the future, particularly if advances in genetic engineering were to allow the extraction of beneficial genes while avoiding transfer of negative traits.

Many old European and North American breed types have fallen by the wayside along the road of the creation of the modern pig. These may, with care, be preserved as rare breeds. While many of the British breeds – too numerous to catalogue – are already extinct, others survive as extremely small breeding groups, including the **Berkshire, Lop, Gloucester Old Spot, Large Black, Middle White** and **Tamworth**. American breeds being left aside of today's mainstream breeding programmes include the **Poland China** and

Spotted, the **Lacombe** and the **Chester White**, although each of these may yet be found in substantial numbers. The classic nineteenth-century European breed types of France, Italy and Iberia are now, sadly, lost.

Pig improvement strategies

One of the most remarkable books in Animal Science in this century has been Douglas Falconer's *Introduction to quantitative genetics*, first appearing in 1960, and the third edition of which was published by Longman in 1989. This is the classic source text for all animal breeders and has had a considerable influence upon pig-improvement strategies. The three ways of altering the genetic make-up of pigs, and thereby affecting their improvement in a given direction, are well known and simple. They are:

1. the use of hybrid vigour;
2. the incorporation of new genes from pigs of different breeds;
3. positive selection for a given character (or characters) within a breeding population.

While the strategies themselves may be well accepted, the methodologies employed to pursue them are many and various, and the subject of differences in both scientific and practical opinion.

First principles and basic tenets

Animals in a given population vary in the degree to which any particular character is expressed. Usually the variation is normally distributed so that most animals appear to be about average, but a few are really exceptional (Figure 6.1). Conventional breed improvement relies upon the notion that use of exceptional individuals as parents of the next generation will bring about a positive shift in the population mean (Figure 6.2).

The variance seen amongst individuals in a population (σ^2P) is caused through variance in genetic make-up (σ^2G), variance in the environment (σ^2E) and σ^2GE, the interaction between genotype and environment.

$$\sigma^2P = \sigma^2G + \sigma^2E + \sigma^2GE$$

(Equation 6.1)

The degree to which σ^2P (what is seen and measurable) is influenced by σ^2G (that part of σ^2P which is genetic in origin) will be increased if σ^2E (the influence of the environment) is minimized.

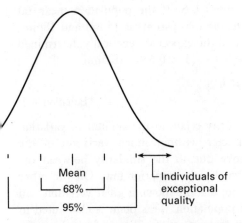

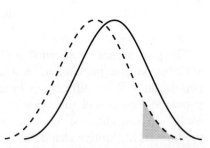

Figure 6.1 Normal distribution of a character in a population. Nearly 70% of the population are not greatly better or worse than the average. Exceptional animals are found in the top 2.5%.

Figure 6.2 Selection in the parent population (broken line) of superior individuals (shaded area) will result in the mean of the offspring population being to the right of that of the parent population: that is to say, improved.

The genetic variance is itself the sum of the additive (σ^2A) and non-additive gene action (dominance and epistasis, σ^2D, σ^2I).

$$\sigma^2G = \sigma^2A + \sigma^2D + \sigma^2I$$

(Equation 6.2)

Additive genetic variance (σ^2A)

The proportion of σ^2P represented by σ^2A (that part of total variance which is due to the additive genetic variance) is the heritability (h^2). Heritability (σ^2A/σ^2P) values range between 0 and 1. The additive genetic make-up expresses the breeding value of a parent, and is passed on to the offspring in a simple and additive way.

Heritability (h^2) is estimated by quantification of a character in the parents and in their progeny, and calculating the relationship which exists between the two in the expression of that character: that is, the regression of offspring on parent. By definition, measurements of a character can only be of the phenotype which – depending on h^2 – may or may not bear a close relationship to the genetic constitution or genotype.

Selection for inherited characters within a single population of animals allows generation-by-generation improvement. The amount of change of the population (the response, R) depends upon the heritability and the extent of the superiority of selected parents over the population mean (S). The heritability for fat depth in pigs is about 0.5. If the population mean fat depth is 20mm, and breeding value of the two parents is 16mm and 18mm, then the parental superiority is 3mm and the expected genetically determined fat depth of the offspring would be $20 - (3 \times 0.5) = 18.5$mm:

$$R = h^2 S$$

(Equation 6.3)

The possibilities for parental superiority relate to the amount of variation available in the population. A character without much variation in the population will be difficult to improve due to the similarity between the population mean and the value shown by the superior few. On the other hand, selecting the best few from a population showing great variation will produce rapid positive change in the population. This point is elaborated in Figure 6.3. In population (a), the best few pigs are far distant from the mean in terms of their leanness, but, in population (c), the best few pigs are only slightly superior to the population mean.

To select only the fewest possible of the most excellent of individuals, and to spread their genes through the maximum number of less excellent animals, is fundamental to any animal-breeding programme. The fewer chosen, the higher will be the intensity of selection (i), the greater the extent of parental superiority and the greater will be the response:

$$S = i\sigma P$$

(Equation 6.4)

The amount of genetic improvement per generation (R) can now be expressed as the product of the variance of the character in the population, its heritability and the intensity of selection employed:

$$R = ih^2 \sigma P$$

(Equation 6.5)

The *rate* of genetic improvement *per year* (ΔG) depends upon the generation interval (GI); the quicker the turnover of each generation, and the shorter the interval, the more rapid will be the annual rate of genetic change:

$$\Delta G = ih^2 \sigma P / GI$$

(Equation 6.6)

The maximum rate of genetic improvement will always be more rapid in large breeding populations than small ones because of the influence of i. Increase in the population size will (a) increase the variation available to select from, (b) allow a greater intensity of selection without increasing the degree of inbreeding and (c) allow improved genes to be spread through a greater number of prospective parents. Large populations are self-evidently

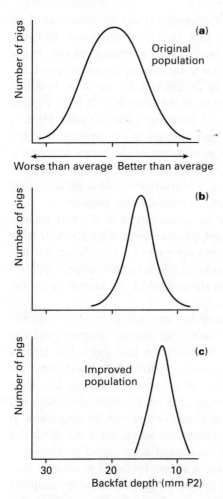

Figure 6.3 Distribution of backfat depth in a population selected for low fat.

found as breeding herds with many females, but modern biotechnology may extend the role of a single female of high quality by multiple ovulation and embryo transfer. Gene distribution from excellent males is also extendible by use of artificial insemination.

Correct identification of 'better' parents for the next generation from amongst a number of candidates is obviously crucial to realizing a high value for the intensity of selection (i). This is not so easy as may appear. Some characters are not available for direct measurement, such as litter size or milk yield in a male, or weight of lean tissue mass in a pig which must remain alive if it is subsequently to breed. In these cases measurements must be made in relatives, or predicted from a correlated character which can be measured in the live animal. Even measurements made on the candidate himself may not be all that they seem. A fat pig with a big appetite is not the same sort of fat pig as one with a small appetite. The character measured and the environment in which the character is assessed are importantly related, so the method of testing chosen and the character to be selected are interdependent.

The **progeny test** is an assessment of an individual's worth or breeding value by measurement of the performance of the offspring. It is an explicit requirement for assessment of worth and breeding value for milk yield or prolificacy in a male. It was used with great success in the early years of the rise of the Danish pig industry to select against fatness. Although accurate, it is expensive. For traits of low heritability, such as reproductive performance, it is now accepted that information on any individual requires to be derived from as many close relatives as possible, especially the progeny.

The **performance test** assesses worth by measurement of the performance of the individuals themselves and is most effective for use with more highly heritable traits such as growth rate and carcass quality. In the latter case this is predicted from correlated measurements such as fat depth in the candidate and from direct measurements made in siblings which are sacrificed for the purpose.

The performance test and the progeny test are not, of course, mutually exclusive. One candidate's performance test is his parents' progeny test! The breeding value of an individual can therefore be best predicted from a combination of information from his own performance test, from the performance of his sibs, from progeny test information and by data from ancestors and a wider forum of relatives. All such information naturally accumulates as a breeding programme progresses generation by generation. If all these data are put together, and appropriate adjustments are made for differences in time and place, then an enhanced accuracy of prediction is possible. This is especially useful for traits of low heritability, such as milk yield or litter size selected for in the female lines. Considerable computer power is essential to deal with the huge mass of quantitative data that may usefully pertain to a single candidate for selection. This technique is now accepted practice in large breeding companies and is known as Best Linear Unbiased Prediction, or BLUP.

It is invariably the case that selection must be based on more than one character. Each additional character slows down the rate of progress in any one, but maintains a fitter overall population. Thus, selection may need to be for growth rate *and* carcass quality, or feed conversion efficiency *and* litter size. To select simultaneously for more than one character, an **index** is required. The index will include, and also 'weight', the various elements according to their relative importance. Where the negative correlation is not strong, the index can accommodate, but problems can arise if two desired characters are highly negatively correlated, such that positive change in one causes deep negative change in the other. In this case the usual solution is to choose separate selection lines for each character.

Non-additive genetic variance

Non-additive genetic variance ($\sigma^2 D$ and $\sigma^2 I$) represents contributions from parents which do not add together simply. Characters showing non-additive genetic variance cannot be selected for progressively within a breeding population. However, such variation is responsible for heterosis, and is extremely valuable in breeding plans. Heterosis is usually shown most strongly when two pure-bred breeding populations are crossed. The causes of non-additive genetic variation are generally understood to be dominance ($\sigma^2 D$) and epistasis ($\sigma^2 I$). Dominance relates to interactions at a given locus (A over a), while epistasis relates to interactions across loci (A or a alleles with B or b, or C or c, etc.). Dominance is seen to occur when the gene complement of one parent suppresses that of the other, as would be the case when a homozygous halothane-positive pig is mated to a homozygous halothane-negative pig and the resultant heterozygous offspring is halothane negative. Were there to be no dominance and the genetic variation was additive, the heterozygote would be expected to show an intermediate value of halothane response. However, this is not the case (Table 6.1). This particular example also shows that often dominance is not complete, and the heterozygote may continue to show some small tendency towards characters such as PSE associated with the halothane gene. The heterosis that results from dominance effects is best seen when individuals, homozygous for different characters, some positive and some negative, are crossed and as a result show in the heterozygous offspring all characters positive. Where traits A and B both show dominance, the homozygote AAbb (showing excellence for A, but poor quality in b), crossed with the homozygote aaBB (showing excellence for B, but poor quality in a), will give progeny AaBb, which because A and B are dominant show excellence for both A and B.

Table 6.1 Incidence of pale soft exudative muscle (PSE) in two populations of pigs and the resultant cross. The halothane-negative gene is dominant over the halothane-positive gene

		Incidence of PSE in pigs stressed before slaughter
Halothane-negative homozygote	(NN)	5
Halothane-positive homozygote	(nn)	60
Heterozygous offspring	(Nn)	10

Epistasis ($\sigma^2 I$) occurs where the genes from the two parents complement each other and build together a whole that is greater than the sum of its parts. This would be the case where A, B, C, etc. were all contributing elements of a complex productive trait, and had the opportunity of positively interacting *the one with the other*. Such a situation might readily be envisaged for litter size where ovulation rate, fertilization rate and embryo survival all contribute to the number of piglets born. Typically non-additive gene action of this sort results in heterozygote performance that is superior to either of the parent homozygotes.

Using hybrid vigour

Non-additive genetic change, demonstrated as heterotic effects and brought about by dominance and epistasis, is by nature once-for-all. Progressive improvements are not possible, and the phenomenon must be recreated at each generation when two particular pig strains are crossed. It neither occurs for all crosses, nor in all characters. The consequence of a particular crossing regime for a heterotic effect in a given trait cannot be predicted before it has been tried out. However, once heterosis is found it will reappear for the particular character each time those two particular pig strains are crossed. Heterotic phenomena in pigs are primarily observed in the reproductive traits, which are of low heritability and difficult to improve by conventional selection amongst additive genetic variation within a breeding population. The inverse relationship between heterotic effects and the extent of additive genetic variance follows in part from Equation 6.2, the genetic variation ($\sigma^2 G$) being the sum of the additive genetic variance ($\sigma^2 A$), and the heterotic ($\sigma^2 D$) and epistatic ($\sigma^2 I$) effects.

Table 6.2 considers the case of three breeds and two crosses between them. For growth rate, the crosses may be seen to perform exactly at mid-parent values. For example, $(650 + 750)/2 = 700$. This is the null expectation. But for litter size there is heterosis. The mid-parent value for Case 4 (Large White $\times$ Landrace B) would be ten piglets born, but performance levels of 10.75 are noted. One may question the benefit of such a cross as the performance of the Large White breed (Case 3) is already better with respect to both growth rate and litter size. However, a third character such as lean-meat percentage or ham-shape score may be superior for the Landrace B, and the mid-parent value for this quality, together with the mid-parent value for growth rate and the above mid-parent value for litter size may make a worthy cross-bred generation. Case 5 (Large White $\times$ Landrace A) shows the best hybrid vigour response, producing a litter size of 12, one whole piglet above the best parent (the Large White). The Case 5 cross certainly maximizes litter

Table 6.2 Performance of three pure-breds and two cross-breds showing the presence and absence of heterosis (hybrid vigour) in the cross-bred; (a) progeny superior to average of the parents, (b) progeny superior even to the better parent

	Growth rate (g/day)	Litter size (numbers born)	Ham size (1–5 score)
Case 1: Pure-bred (Landrace type A)	650	10	3
Case 2: Pure-bred (Landrace type B)	700	9	4
Case 3: Pure-bred (Large White)	750	11	3.5
Case 4: Cross-bred (Large White × Landrace B)	725	10.75[a]	3.75
Case 5: Cross-bred (Large White × Landrace A)	700	12[b]	3.25

Although this example shows no heterosis and mid-parent values for growth rate, this is not always the case. There is often some heterosis (type (a)) shown for growth rate in crosses between the Pietrain and Large White, and between Duroc and Large White.

size, although not growth rate or ham shape, which will have to be handled by a different stratagem.

True heterosis had, of course, been early demonstrated by maize breeders, upon out-crossing two in-bred, low yield, homozygous lines. The resultant F1 heterozygote produced a yield of maize which was greater than either of the two original lines even before the in-breeding regime commenced. To create hybrid vigour in pig populations it was at first thought that performance depression by in-breeding might also be necessary. This, however, is not the case, and there is no such prerequisite. Nevertheless, to show hybrid vigour it is necessary that the two breeding lines used should be quite distinct from each other.

Heterosis in the reproductive traits was noted and exploited by L M Winters of Minnesota, and later by Smith and King in the early 1960s when working at the Animal Breeding Research Organization at Edinburgh. The Large White and Scandinavian Landrace breeds remain the two breeds most often used in cross-breeding schemes to enhance reproductive traits in pigs. Heterosis is, however, equally as evident in crosses between the Duroc and the Landrace and Large White breeds. It has been consistently demonstrated that there is heterosis at two levels. First the cross-bred litter out of the pure-bred dam is of greater size and vigour, resulting in higher numbers weaned and a greater weaning weight; cross-bred progeny may have at least a 5% lift in litter size and a 10% lift in litter weight at weaning. Secondly, the use of the first generation F1 hybrid females as mothers for the next generation gives a further boost (5%, at least) to performance due to the greater reproductive vigour of the dam (Table 6.3). Litters from F1 mothers may confidently be expected to show 1.0–1.5 more piglets weaned as a result of greater numbers born and subsequent piglet viability. Due to improved milk production and piglet vigour there is also likely to be an enhancement of up to 1kg in individual piglet 28-day weaning weights.

It will be recalled, however, that these gains are not progressive. They have been of this same order for 30 years and only occur upon the recreation of the F1 by the out-crossing of pure-bred lines. The pure-bred lines must therefore be maintained as independent pure-breeding populations for the specific purpose of creating generation after generation of new heterozygous F1 progeny. In the meantime progressive selection can also be made in the pure-bred populations as exemplified in Table 6.4.

One of the characteristics of the F1 generation, in addition to heterosis in the reproductive traits, is the uniformity of the heterozygote, which is particularly standardized in physical appearance and performance. This has considerable commercial advantage in itself. If, however, F1 heterozygotes *from the same parental sources* are themselves crossed to create an F2 generation there is a disassociation of the genes which had previously

Table 6.3 Effect of cross-breeding upon reproductive performance

	Percentage improvement over pure-bred for pigs weaned/dam/year
First cross	5–10
Back-cross or three-way cross	10–15

Table 6.4 Showing an example of improvement made in the pure breeding populations by selection, together with exploitation of heterosis in the cross-bred offspring

	Litter size (numbers born)
Pure-bred line A	8.5
Pure-bred line B	9.5
Cross-bred (A × B)	10.0
After n generations of selection including litter size	
Pure-bred line A	8.75
Pure-bred line B	9.75
Cross-bred (A × B)	10.25

complemented each other so beneficially. The result of crossing heterozygotes may be unfortunate in two respects. First, there is less hybrid vigour resultant and secondly, the progeny may be more variable. Variability may show in form, function and some aspects of performance.

There may therefore be some disbenefit in choosing either:

1. to use the same F1 construct for both the sire and the dam of the slaughter generation; or

2. to use the slaughter F2 generation as a source of replacement breeding females.

It is important that the male used on an F1 parent female should be either pure-bred (one of the pure lines which created the F1 is satisfactory), or a cross-breed of different origin which includes at least 50% of novel genes. Equally vital is that producers forgo the temptation to draw replacement female stock from out of their finishing pens of slaughter pigs but, rather, the female breeding stock must be recreated at each and every generation from crossing the 'pure-bred' lines.

Hybrid vigour in the reproductive traits of pigs is also evidenced in the male side. In this event the phenomenon shows itself through hardiness, earlier sexual maturity, longevity, increased libido and improved sperm concentration and viability. As a result, cross-bred boars can be easier to use, are more ready to mate, will improve conception rates and will produce litters of greater size in consequence of enhanced semen characteristics and libido. Between 0.25 and 0.75 piglets more per litter have been claimed from the use of hybrid boars. But producers value particularly their increased libido as compared to the 'pure-bred' types.

Breeding regimes planning to utilize hybrid vigour in the female line tend to take one or other of the forms in Table 6.5. Option 1 is a typical two-way cross which may be handled with either line as male or female (i.e. a variant on the scheme would be for L to take the place of LW and vice versa). Sometimes the same breed may be used in the male and female line, but a different strain is chosen for those different purposes. Option 2 is a three-way cross where P is always the top-crossing sire, but LW and L may be reciprocated. Option 3 utilizes a cross-bred male line from sources different from the female line thus maximizing hybrid vigour in both sire and dam, as well as offspring. Option 4 is a variant continuing on from Option 1 where females from the slaughter generation are drawn out for replacements in the breeding herd. This scheme is called criss-crossing and some heterosis is preserved by alternate use of the two source breeds for the sire of the next generation. Such a plan fails to optimize heterosis and requires both growth and reproductive traits to be dealt with by use of an index and simultaneously

Table 6.5 Crossing plans which encourage hybrid vigour in the female line

1.
$$LW\sigma \times L\female \text{ (or the reciprocal)}$$
$$\downarrow$$
$$LW\sigma \times (LW \times L)\female \text{ (F1 hybrid dam)}$$
$$\downarrow$$
slaughter generation

2.
$$LW\sigma \times L\female \text{ (or the reciprocal)}$$
$$\downarrow$$
$$P\sigma \times (LW \times L)\female$$
$$\downarrow$$
slaughter generation

3.
$$D\sigma \times LW\female \qquad LW\sigma \times L\female$$
$$\downarrow \qquad\qquad \downarrow$$
$$(D \times LW)\sigma \times (LW \times L)\female$$
$$\downarrow$$
slaughter generation

4.
$$L\sigma \times (LW \times (LW \times L))\female$$
$$\downarrow$$
$$LW\sigma \times (L \times (LW \times (LW \times L)))\female$$
$$\downarrow$$
$$L\sigma \times (LW \times (L \times (LW \times (LW \times L))))\female$$
$$\downarrow$$
etc.

selected for in both populations. Once popular, it is now largely discredited. In the third option, breed 'D' may be any suitable sire-line sire, as indeed may breed 'LW'. However, crosses between the Large White or Landrace, and the Duroc, the Hampshire, the Pietrain and Landraces of the Belgian and German types are all feasible, as are other possibilities.

The incorporation of new genes by importation

Importation of genes for specifically required characters into an otherwise satisfactory population is second only to complete population replacement as the simplest form of genetic improvement.

Where the alternative gene source is better overall than the existing population, then the latter's complete replacement is indicated. This could be done either at a single stroke, or by 'grading up'. The one-step option for breed replacement is usually employed in pig production, rather than the more tedious grading-up process. Either a unit is requiring stocking *de novo* or the opportunity is taken to depopulate an existing unit, disinfect and rest and restock with a replacement pig type of improved quality and higher health status.

Examples of instantaneous replacement therapies would be the substitution of the modern white hybrid female for local indigenous pig stocks, a decision to change breeding company or a policy change in choice of top-crossing sire.

Grading up is a tiresome process by which the replacement breed is used upon the base population to sire five generations, until the population base has 97% of the genes of the sire breed and may therefore be considered to be that breed. The appropriateness of grading up would be limited to where the availability of the new breed is restricted to few individuals but it is required to create a new nucleus herd. A typical possibility would be the use of semen from a muscular breed to create, from indigenous stocks, a particularly meaty sire-line population.

Often, the alternative gene source is *not* better overall than the existing population, but holds within it one or two specific characters of value. Wholesale breed replacement is therefore uncalled for. The incorporation of imported genes is required to augment rather than replace the existing population and to create an amended breed. The injection of genes is achieved by one or more initial crosses followed by subsequent selection to retain the wanted genes, but to exclude the unwanted ones. In this way the Meishan has been used to import genes associated with prolificacy into the lean and fast-growing Large White and Landrace types. The first cross shows heterosis above mid-parent in prolificacy giving about +3 piglets per litter, but it is also painfully fat and slow growing. By subsequent back-crossing with the White breed, and appropriate selection, prolificacy may be held in the genome while slow growth and fatness are cast out. The achievement of +1 piglets per litter with no loss of growth rate or meatiness would be considered as a highly beneficial breeding programme. In the event, the Meishan crossing programme has been somewhat disappointing.

The Belgian Landrace is a meaty breed, but also halothane positive with lowered breeding efficiency, easily stressed and whose carcass meat is liable to loss of quality through PSE. For this reason it is usually used only as a top-crossing sire, the normal halothane-negative (NN) gene being dominant and the heterozygote (Nn) being stress resistant when produced from a normal (NN) mother. It is an important point of principle in the use of the heterozygote (Nn) type that while the normal type (N) is dominant for predisposition to stress (and mortality) and to reduced meat quality, and these problems are therefore absent from the Nn heterozygote, the inheritance of lean meat percentage is additive, the Nn heterozygote showing values intermediate between the two parents (about +2.5% of lean meat). But the carcass would be even more valuable if the mother were also to carry the gene for meatiness whilst avoiding all the problems of the associated halothane gene; in effect, a new breed of meaty Large White created by the importation of genes for meatiness, but the exclusion of genes for stress. The disassociation of blocky muscling and stress susceptibility is now proving possible though arduous, and meaty pig types which are halothane negative may shortly be available and suitable for use in the sire line. However, *if a heterozygote* is used in both male and female lines, PSE reappears in a proportion of the (nn) offspring (Table 6.6). It remains an important goal for animal breeders to achieve separation of meat characteristics from stress characteristics in breeds such as the Belgian Landrace and Pietrain. But it may be that meat-line qualities could be achieved equally as well by selection within a population of halothane-negative (NN) pigs.

If the animal breeder perceives that a mixture of genes from both breeds, or equally as often from three or four breeds, could be of benefit, then a completely new breed could be manufactured. First, the source populations are crossed giving a mixed heterozygous population which is then mated back to itself. The next generation has within it the genes from all the source breeds and a greater amount of genetic variation than was ever available to select from within any of the founding populations. The gene mix can now be exploited by the selection of animals showing wanted characters and rejection of those which do not. Matings amongst selected animals bring about concentration of wanted genes until the population is pure-breeding. It is a common difficulty with such schemes that variability in the offspring may continue for many generations after the original admixture has taken place. Nevertheless, this technique was responsible for the Minnesota and Lacombe breeds. Most importantly, gene importation and breed mixing has created the majority of the pure-breeding nucleus dam and sire-line populations of many of today's breeding organizations, which are new or amended breeds created both by gene importation and by selection. Specialized sire lines are invariably made from admixtures of some, or all, of Hampshire, Pietrain, Belgian Landrace, Duroc and Large White breeds. The success of the method depends upon the quality of the source populations used and the efficiency of the subsequent selection programme.

Although the construction of a *pure* breeding population of perfect pigs showing all beneficial characters at levels of excellence seems a laudable and

Table 6.6 The suppression of the halothane gene (nn) by dominance effects, and its subsequent reappearance upon the heterozygote being used in both the male and female lines

NN × nn → Nn:	Intermediate for meatiness, stress-resistant heterozygote
Nn × Nn → NN:	Homozygote, reduced meatiness, stress free
Nn:	Heterozygote
Nn:	Heterozygote
nn:	Homozygote, meaty, stress susceptible

tidy objective, it is by no means the most efficacious route to combining characters of different types. Breed-crossing programmes can be justified without the need to invoke the benefits of hybrid vigour. Achieving mid-parent values can be ample reward in itself. The genes of the Duroc and Large White are well combined by the simple expedient of using the cross-bred (D × LW). Such a cross has found favour both in the sire line when mid-parent values for meat-quality characteristics are appreciated, and in the dam line when better than mid-parent values for prolificacy and for hardiness are useful in lower-input pig-keeping systems. The crossbred is *not* mated back to itself in any attempt to create a new breed. Rather pure-breeding nucleus Duroc and Large White populations are sustained to create the cross-bred product anew at each generation.

Modern breed improvers make frequent use of gene importation to introduce new characters and to produce a greater pool of variation to select from. All three techniques, breed amendment, new breed creation and continuous cross-breed production, are commonplace tools of the trade for national and international breeding companies. Gene addition and use of heterosis are complementary: lines which have been the subject of gene importation can be used in hybrid breeding programmes in both the male and female lines.

The third and equally complementary leg of breed improvement is the achievement of positive change by genetic selection within pure-breeding populations. The continued effectiveness of hybrid breeding and gene importation programmes is dependent upon progressive improvement in the source populations by selection. It is explicit that no breeding programme is sustainable by cross-breeding and gene importation techniques alone.

Breed mixing has sometimes, most unfortunately, been used in substitution for a progressive breeding programme. This dubious practice involves the maintenance of four or five small herds, one for each breed, within which no selection takes place, but whose sole purpose is to maintain various different breed types. (For example, one prolific breed, one coloured breed, one fast-growing breed, one lean breed, one meaty breed and so forth.) Market needs are then satisfied on the basis of mixing and matching the nucleus breeds. This strategy promises much but delivers little because there is no genetic improvement taking place in the source breeds.

Positive selection within a breeding population

Improvement by selection within a population depends upon the presence of additive genetic variance ($\sigma^2 A$) and the shifting of the population mean in the favoured direction by the use of sires, and to a lesser extent dams, of excellence. Proof of a candidate's superiority depends upon measurement; but only the phenotype can be measured. The phenotypic variation for a trait is the sum of the additive genetic effects and the environmental variation, providing the influences of dominance and epistasis (heterosis; $\sigma^2 D$ and $\sigma^2 I$) are taken to be small. Where $\sigma^2 G = \sigma^2 A + \sigma^2 D + \sigma^2 I$:

$$\sigma^2 P = \sigma^2 G + \sigma^2 E$$

(compare also Equation 6.1)

As will be recalled, from Equation 6.5, response to selection (R) depends on the heritability (h^2), the intensity of selection (i) and the extent of variation existing within the population as measured by the standard deviation (σ). While population size impacts upon i, a short generation interval maximizes the rate of response (see Equation 6.6).

Heritability (h^2)

Heritability relates to that part of $\sigma^2 P$ which is accountable by $\sigma^2 G$,

$$h^2 = \sigma^2 A / \sigma^2 P$$

(Equation 6.7)

and is demonstrated by the influence of the parent upon the genetic composition of the offspring. Heritability (h^2) can be determined as the regression of the performance of the offspring on the performance of the parents (mid-parent), but can also be built up from knowledge of resemblances between other relatives – especially sibs. Figure 6.4 shows the estimation of h^2 for growth rate from a population of pigs. Each dot is the

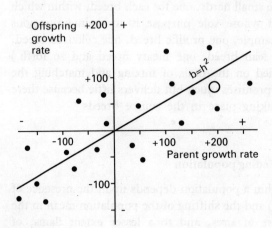

Figure 6.4 Use of regression of offspring on parent to estimate heritability.

value for the mean performance of one pair of parents against the performance of their offspring. Thus, for the open circle, the growth rate of the sire was 900g daily and that of the dam 700g daily. The mean growth rate of all the parents was 600g so the mid-parent value was $(900 + 700)/2 = 800 - 600 = +200$g. The growth rate of four offspring which were measured was 650, 600, 750, 700. The mean growth rate was $(650 + 600 + 750 + 700)/4 = 675 - 600 = +75$g. Over all parents and offspring, the regression has a slope of 0.5.

Heritability values for some important pig production traits are given in Table 6.7.

Characters such as carcass quality which are influenced little by environment are highly heritable (as σ^2E becomes smaller, then σ^2G becomes a higher proportion of σ^2P). Conversely, characters such as reproduction which are much influenced by environmental effects are lowly heritable. Heritability (h^2) is an *estimate* and may therefore vary. If σ^2E is great, h^2 values for reproduction and growth can be less than 0.1 and 0.3 respectively but, where the environment is controlled more closely, h^2 for these characters can readily double. It follows that control of σ^2E will maximize the efficiency of identification of true genotype and thereby increase h^2. The greater h^2 is, the faster will be the rate of genetic improvement made by selection of best parent.

The selection differential (S)

The selection differential is a function of the intensity of selection (i) and the variation (σ). Response to selection is therefore the product of the heritability and the selection differential, as shown in Equations 6.3, 6.4 and 6.5.

The more exceptional the selected parents, in comparison to the population mean, the more exceptional will be their offspring and the faster will be the rate of positive change in the population as each generation passes. Given the regression of offspring on mid-parent shown in Figure 6.5 for growth rate, it is self-evident that the further to the right-hand side of the X axis are the parents, the higher up the Y axis will be the progeny. It should also be evident that the steeper the slope of b, the better the progeny response and the closer the progeny performance to selected parent levels. Breeding plans should aim to maximize movement in the direction shown by the arrows in Figure 6.5.

Where some 40% of the population has to be selected as parents for the next generation, then the selection differential will not be great. A low differential and a high proportion selected will be typical for selection of breeding females. A high selection differential, with usually less than 10%, and often less than 1%, selected, would be typical for sire selection, especially

Table 6.7 Heritability estimates for pigs

Reproductive characters (lowly heritable)

Ovulation rate	0.10–0.25
Embryo survival	0.10–0.25
Numbers born	0.10–0.20
Survivability of the young	0.05–0.10
Readiness to rebreed	0.05–0.10
Milk yield	0.15–0.25
Milk quality	0.30–0.50
Longevity	0.10–0.20

Growth and carcass quality characters (moderately heritable)

Daily live-weight gain	0.30–0.60
Lean-tissue growth rate	0.40–0.60
Appetite	0.30–0.60
Backfat depth	0.40–0.70
Carcass length	0.40–0.60
Eye muscle area	0.40–0.60
Ham shape	0.40–0.60
Meat quality	0.30–0.50
Flavour	0.10–0.30

if artificial insemination is utilized to facilitate maximum coverage of females by the minimum number of quite exceptional males. The influence of the proportion selected upon degree of excellence of selected parents is illustrated in Figure 6.6.

The selection differential is dependent upon both i and σ. Selecting the top 10% of animals for a character with low variation in the population gives a lower value for S than selecting the top 10% for a character with high variation, as is shown in Figure 6.7, for characters of (a) low and (b) high variation. The influence of heritability and selection differential upon the extent of improvement in three characters is depicted in Figure 6.8.

It has often been wrongly assumed that as a selection programme progresses then the extent of variation available to select from must diminish, and the response must slow ultimately to a plateau; and it *is* certainly reasonable to expect some limit to exploitable variation for characters which threaten health. For traits being selected downwards, such as backfat depth, it is unavoidable that variation must lessen as the population mean approaches 0 (Figure 6.3) but for traits selected upwards, such as reproductive characters and lean tissue growth rate, no limit is evident. Common sense suggests that litter size must be limited by uterine capacity at some point, but even this expectation will not be met if there is a correlated positive response in maternal size. There is no reason to suppose that a plateau to response is likely to be reached in the foreseeable future for characters associated with lean growth and reproduction.

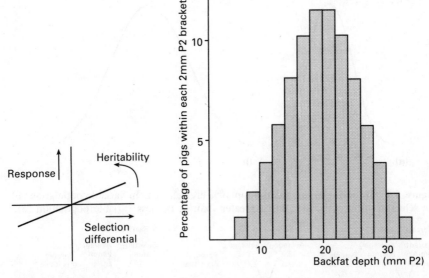

Figure 6.5 Maximization of selection differential and heritability will maximize response. The better the parents, the better the response; the higher the heritability, the better the response for any given selection differential.

Figure 6.6 Example of the variation in P2 measurements taken from a population of pigs out of which males are to be selected. If the number of candidates is 100 and five require to be selected, the highest P2 measurement in the selected pigs will be about 11mm, and the average about 10mm. If the number of candidates is 100 and 25 require to be selected, the highest P2 measurement in the selected pigs will be about 16mm, and the average about 14mm.

The extent of genetic variation in meat quality traits is the subject of present discussion; it could be that whilst some aspects of meat quality such as flavour and tenderness might be reasonably heritable, there could be little additive genetic variation ($\sigma^2 A$) available to select from, because $\sigma^2 P$ is small.

Generation interval (GI)

For the animal breeder it is the annual rate of response that counts, not simply response per generation:

$$\Delta G = h^2 i\sigma/GI$$

(Equation 6.8)

The fewer years per generation the faster the rate of response. Pigs become sexually mature between 6 and 9 months of age, and pregnancy takes a further 4 months. A pig generation can therefore be turned around in about 1 year. In practice, it may not always be economical to use a sow for only one

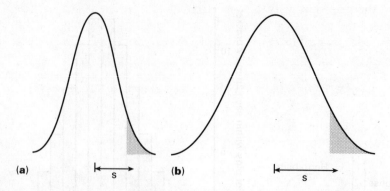

Figure 6.7 In both populations the top 10% are selected. In variable population (**b**) the selection differential S is much greater than in the less variable population (**a**).

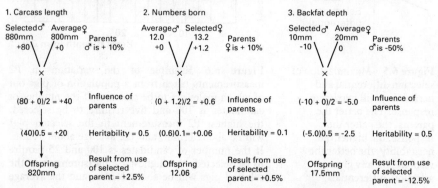

Figure 6.8 Influence of heritability and selection differential upon rate of improvement following use of a selected parent. In the first example the selected male has a carcass length 10% better than the average of the breeding herd. If the heritability for carcass length is 0.5, a 2.5% improvement would be found in the offspring. Some 12 months are likely to elapse between the birth of the selected male and the birth of his offspring. In a less variable population of pigs a male with a carcass length of 880mm might not exist, and the longest animal be only 5% above average (840mm), so the offspring would only be improved by 1.25%. In the second example a breeding female has been selected consequent upon her producing litters with 10% higher than average number of pigs. With a heritability of 0.1, the offspring are likely to show an improvement of only about 0.5%. In this case 24 months or more elapsed between the birth of the selected female and the birth of her offspring. In the third example a male with P2 10mm, used on a population with average P2 20mm, will produce offspring whose P2 measurement will be about 17.5mm (heritability assumed to be 0.5). A male with P2 measurement of 14mm would, by the same calculation as example 3, produce offspring P2 measurements of around 18.5mm. The size of difference between the average of the selected parent and the average of the population will depend upon the number of animals which have to be selected from amongst the candidates. It is, for example, much more probable that the mean P2 of those selected will be 10mm if one is selecting only five individuals from 100 candidates; and conversely, more probable that the mean P2 will be 14mm if one was selecting 25 individuals from 100 candidates (Figure 6.6).

litter; she may remain in the herd for up to 4 years, which considerably lengthens the generation interval. There is benefit in keeping to a rigorous regime of culling sows out of the nucleus herd – perhaps to a multiplier herd – after a given number of litters (often two). But if reproductive traits are a part of the selection programme it is impossible to accurately predict breeding value for reproductive characters in only a few litters. This is where information from relatives becomes so useful.

Much the same argument pertains to the male. The most accurate assessment of the breeding value of a sire can be made by study of his progeny, but there is a necessary delay while the progeny demonstrate their qualities. Measurement of the performance of the sire himself may be a less accurate determinant of breeding value than measurement of the progeny, but it reduces generation interval, performance test data being available even while sexual maturity is yet being attained. The shorter generation interval allowed by performance test far outweighs any loss of precision in breeding-value determination. Where the character is sex-limited, like litter size, the progeny test might be considered obligatory to assess the breeding value of a candidate male; but information from relatives, if they be numerous enough, can do much to allow the allocation of a breeding value for reproductive traits to a potential sire even before he is sexually mature.

There is a temptation to keep an especially good sire or dam in the breeding nucleus. In the time when line-breeding techniques were used to fix a particularly desirable quality, the use of one particular individual, or family, over many generations, was an essential part of pedigree breeding strategy. Individuals of outstanding merit were sought and could remain in a herd for 4 years or even longer and set in place the future destiny of that population (and its owner). But this practice fails to maximize rate of genetic change. The advances in growth rate and leanness over modern times have only come about since breeders have been willing to rapidly replace their herd boars, maintain a high rate of turnover of breeding stock, and reduce generation interval to the minimum. The realized generation interval for pigs in progressive breeding companies is presently about 400–450 days. The effective lifespan for the use of a dam in a pure-breeding nucleus which is the subject of intensive genetic selection should normally be from two to four litters, and the effective working life of boars in a nucleus should be measured in weeks or months rather than in years.

Pure-breeding (nucleus) herd size

Pedigree breeders used to have relatively small breeding herds of around 100 sows and five boars. Effective herd size is increased by provision of a testing facility to which can be sent candidates from many herds, followed by the willingness of individual breeders to use only tested sires of the most exceptional quality. Such was the basis of many national pig improvement schemes, but these have mostly fallen into low usage or disuse since breed improvement responsibilities have passed to the independent breeding companies.

There are many reasons for the demise of the small pedigree breeder, even with access to central testing. Amongst these have been unwillingness to cooperate to create a genuinely operational nucleus of significant size. But equally as important was the fear that central testing might spread disease. Candidates could bring to central test those disease organisms resident on the various breeders' independent units, exchange those organisms in the course of the testing programme, and then redistribute them to the pedigree breeders through the medium of those candidate boars which were selected. The rise of interest in high health status stock demanded by commercial buyers of improved pigs compromised central testing as a means of increasing effective nucleus herd size.

Response to selection has been shown to be positively infuenced by the accuracy of measurement, the intensity of selection (i) and the amount of variation available (σ). All these are facilitated by large population size. In large herds it is also easier to minimize generation interval.

Where heritabilities are high, nucleus herd sizes of 100–500 pure-breeding females may be acceptable, as with the male side of a sire-line population which may need to be open to importation. For moderate heritabilities a nucleus of 200–600 breeding females is required, whilst for low heritability traits, herd sizes in excess of 1,000 are prerequisite for acceptable rates of genetic change. It is unusual (but not infeasible) to consider single-site, pure-breeding herds of such size. The effective herd size would usually be 300–600 breeding sows for proper management and recording of élite animals which are the subject of intensive genetic selection programmes. The nucleus could, if necessary, be spread over a number of separate units; in effect, the nucleus is made up from a group of herds. This is not a significant complication given computer recording and the relative ease with which the genes of a boar can be spread across sites by the use of artificial insemination.

Accuracy of measurement and effectiveness of candidate comparison

The selection of parents of special worth assumes that those selected do indeed possess it; the measurement taken must reflect the character measured. But misrepresentation of the facts can be a significant contributor to environmental variation. Many elements conspire to ensure this :

1. Human error at the point of measurement.

2. Recording and computer entry error.
3. Measurements being of necessity indirect, such as ultrasonic measurements for fat depth and the need to predict lean mass in the live animal from associated fat depths and body weights.
4. Measurements being possible only upon relatives rather than the candidate itself, such as for litter size or meat quality.

In addition to the type of errors in effective measurement identified above, there are other errors which are directly attributable to the way the nucleus herd is managed. They are built into the system itself. These should be assiduously rooted out. Examples would be: candidates measured for lean gain having different body composition and/or age at the start of a performance test; candidates measured for growth and efficiency having eaten different amounts of feed over the duration of test; candidates measured for litter size having different management routines at the time of mating; candidates measured for susceptibility to PSE having relatives slaughtered at different processing plants and so on.

Improving the accuracy of measurement is part of reducing environmental variation ($\sigma^2 E$). Many failures are the result of management convenience, and it is axiomatic that a genetic selection programme is *in*convenient for management considerations. The approach of staff and management to a nucleus breeding herd requires to be radically different from that of a commercial production unit.

Genotype:environment interaction

Genes are, perforce, expressed in the environment. A pig cannot grow in the absence of an environment containing feed. An inherited resistance to a disease cannot be expressed in the absence of the disease organism (Table 6.8). This might be thought to mean that candidate testing must take place in the environment to which the genes of the progeny will be exposed.

One problem arising from this analysis is that the commercial environment

Table 6.8 Interaction between genetic competence for resistance to *E. coli* K88 and the presence of the organism in the environment

	Incidence of diarrhoea (%)	
	Without K88	With K88
Litters from sire A	15	35
Litters from sire B	10	15

(the environment of the progeny) is more variable and unstable than that of a nucleus herd where the intensive testing and selection must take place (the environment of the parent). There is disadvantage in a variable environment because where environmental effects (σ^2E) are large, effective heritability will be low due to σ^2A being a small proportion of σ^2P:

$$\sigma^2P = \sigma^2A + \sigma^2E$$

(Equation 6.1)

$$h^2 = \sigma^2A/\sigma^2P$$

(Equation 6.7)

Genotype:environment interaction impacts heavily upon choice of nutritional environment for a programme of selection for lean-tissue growth rate and fatness. The commercial environment may favour restricted feeding of a diet of average nutrient quality. But under these conditions differences between individuals for lean-tissue growth rate may not be readily apparent. It is only at higher levels of energy and protein that differences in lean growth and fatness become evident, as demonstrated in Figure 6.9.

Inasmuch as it appeared evident earlier that candidate testing must take place in the environment of subsequent gene exposure and expression, it is now equally evident that, to the contrary, candidate testing must take place in the environment that best distinguishes between the genotypes.

For the breeder, priority goes to the selection of candidates in an environment which reduces variation and increases the likelihood of successfully distinguishing between candidates of different merit. This will inevitably mean a test regime different from the circumstances of production.

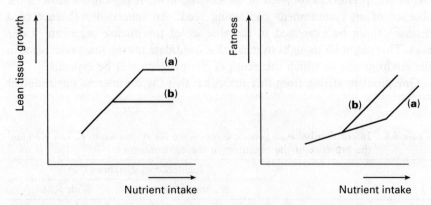

Figure 6.9 Showing genotype : environment interaction for the genetically influenced characters of lean-tissue growth rate and body fatness and the nutritional environment. Note that it is only at the higher nutrient intakes that the two genotypes are distinguishable from each other.

It is necessary in this event that there is no severe genotype:environment interaction of the sort where a candidate chosen as better on the test regime may turn out to be worse on a commercial farm (Figure 6.10).

More usually, the quality of the commercial environment simply fails to allow the full expression of selected characters. Producers with previous experience of pig strains of moderate quality who have upgraded their genetic base by the purchase of improved stocks also require to upgrade their environment (especially the nutritional environment) if the improved genetic complement is to express itself to the full.

The presence of genotype:environment interactions casts doubt upon the wisdom of making comparisons of different genotypes in a common environment. To compare various breed types and breeding company products, and identify which is the best, appears sensible at first sight. However, if each type is tested under a different regime the comparison is confounded; while if each type is compared under the same regime, that identified as 'best' will merely be the one which most closely fits that particular regime. Regardless of the inconvenience or difficulty of the interpretation it is often the case that different pig types can only be properly compared when each is tested in the environment best suited to that type. The corollary is that such comparisons perhaps should not be attempted, and assessments of commercial products are probably best made by the market-place. Pig purchasers should opt for the pig type most likely to perform best in the individual conditions of that purchaser. This will relate to the test environment chosen for selection of parents in the nucleus breed-improvement programme. Breeding companies should therefore expect customers for their stock to be interested not only in selection objectives and selection intensities, but also in methods of testing.

Index selection

Selection strategies rarely, if ever, have only one selection objective. The combination of objectives requires the construction of an index which converts many numerical values for various traits into a single value which allows candidates to be ranked for excellence by use of one single number – or index. Lean tissue growth rate alone is not adequate to ensure a high quality carcass; minimally, fat depth should also be included in the selection programme. Reproductive traits may require active positive selection as well as carcass and growth characteristics.

Notwithstanding the need to pursue more than one objective if a realistic programme is to be maintained, there are negative consequences to overly complex and numerous selection objectives. Most rapid progress is made if one objective only is pursued single-mindedly. But optimum progress in

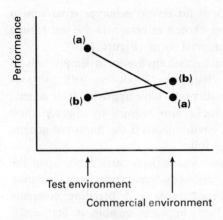

Figure 6.10 Severe genotype : environment interaction rendering performance on test a useless indicator of merit in commerce.

efficiency and economy of the pig production process can equally only occur where more than one trait is the subject of selection pressure. As each additional objective is included into the index then the rate of progress in any one is progressively slowed. Simultaneous pursuit of negatively correlated traits will ensure that ultimately there will be zero progress if too many traits are selected. Only those characters crucial to performance should be the subject of improvement by selection. Breeders ignore this discipline at their peril. Such behaviour led to the demise of the Ayrshire dairy cow and to the poor recent record of the sheep-breeding industry. Equally, heeding the message has helped the pig-breeding companies and also, for example, the North American dairy cattle-breeding industry.

Where two traits are negatively correlated (such as is probably the case for reproduction and carcass quality) it may be helpful to avoid their inclusion in a single index by dividing the population into two lines, a dam line and a sire line. The two lines may then be crossed at the point of creation of the commercial F1 generation. The rationale behind such a scheme is described in Figure 6.11.

There are some threshold characters which may best not be included in an index, but dealt with by simply culling all individuals not reaching the required minimum level (the independent culling level). These would apply regardless of the ultimate objective of the breeding scheme. Meat quality may be adjudged as one; leanness is only beneficial if the muscle is free from PSE. Good legs are prerequisite for boars and sows, regardless of potential litter size. All sows must have at least 12 nipples, all boars must be willing to mate. Only when all the threshold levels have been passed would candidates go forward for calculation of their merit on the basis of an index.

The construction of an index to allow multi-trait selection is complex.

1.   Even when only a single character is the subject of the improvement

	Dam line		Sire line
Initial position	10	Litter size	10
	50	Lean percentage	50
Select for 10 years	Index including litter size and lean percentage		Index including backfat depth and lean percentage
	12	Litter size	10
	55	Lean percentage	65

$$\times$$
$$\downarrow$$

Litter size 12
Lean percentage 60

Figure 6.11 Use of specialized sire and dam lines with differently weighted indexes for each line.

programme, the efficiency of selection of that character can be improved if associated characters are included into the criterion of assessment. The measurement of percentage lean tissue in a live candidate boar is improved if it is determined in combination with ultrasonic backfat depth, muscle depth, ham shape, growth rate and feed conversion efficiency, as well as by use of dissected lean content obtained from slaughtered sibs.

2. When multiple characters are required to be improved in the selected population they may be strongly or weakly positively related. The daily rates of total gain and of lean-tissue gain are closely related; but daily lean-tissue growth rate and fatness are not necessarily so. Under *ad libitum* feeding conditions the relationship between lean-tissue growth and fatness may be positive, whereas under restricted feeding conditions the relationship may be negative. Such positive and negative relationships require to be understood and statistically accounted in the index.

3. Some characters are more important than others and it is sensible to weight the important ones so that improvement is more rapid in those elements than in the less-important ones. Fat depth would be considered of great importance in a country with fat pigs and a human population discriminating against fat. But fat depth would be of less importance in a country which already had thin pigs, and a human population concerned more with flavour and cookability than with the ultimate in leanness.

Weightings are sometimes dealt with by attaching financial values to each character. If it is obvious that a selection objective must have economic benefit, then the index is an ideal means of apportioning more weight to those characters of greater economic benefit. However, even in the context of a single market, economic benefits can be transient and liable to fluctuation. If each fluctuation is accommodated into the index as price changes occur, there is a danger that the index merely ensures a limited rate of progress due to constant direction changes. If, however, fluctuations are ignored, the market situation may drift to a position where the weightings of characters in the index no longer reflect reality. For breeding companies trading internationally, the situation is even more complex as there are large differences in the economic ranking of various characters between countries. Ham shape, for example, is economically vital in Spain, but trivial in the UK. It follows that economic weightings placed into indexes should be viewed with considerable suspicion. It is perhaps best to decide selection objective priorities, settle on their relative importances, consider likely rates of progress, and then weight the index. The weightings will thus reflect clear management decisions and be expressed biologically, rather than being open to the caprice of changing financial circumstances.

The National Swine Improvement Federation of the United States offers, amongst others, the following examples of simple indexes (I).

For sows:

$$I = 100 + 7(N - \bar{N}) + 0.4(W - \bar{W}) - 1.4(D - \bar{D}) - 53(B - \bar{B})$$

where N is the number of piglets born alive, W is the weight of the litter (lb) at 21 days of age, D is days to 230lb live weight, and B is backfat depth (inches). If $\bar{N} = 10$, $\bar{W} = 130$, $\bar{D} = 160$ and $\bar{B} = 1.0$, and for two candidates N = 11 and 12, W = 150 and 140, D = 130 and 135, B = 0.7 and 0.8; then I = 166 and 160 respectively. The index, upon use, shows itself to be weighted in favour of numbers born and maternal growth rate, which is not unreasonable.

For males:

$$I = 100 + 112(DG - \bar{DG}) - 120(B - \bar{B})$$

where DG is daily gain (lb), and B is backfat depth (inches). If $\bar{DG}$ is 1.5 and $\bar{B}$ is 0.6, and for two candidates DG = 1.7 and 1.6, B = 0.5 and 0.4; then I = 34 and 35 respectively. This index appears weighted approximately equally for gain and fatness.

A simple index used in the UK for the selection of gilts from group feeding pens is:

$$I = (100 \times DLWG) - (2 \times P2) + (0.02 \times W)$$

where DLWG is measured from birth in kilograms, P2 is the depth of fat over the eye muscle measured in millimetres, and W is weight in kilograms at the point of measurement. All animals are weighed and ultrasonic backfat determined at the P2 site at an age reasonably close to 150 days. Two candidate gilts reaching 89 and 95kg live weight at 150 days of age with respective P2 measurements of 10 and 12mm would have similar index scores.

The simple boar index:

$$I = (0.08 \times DG) + (50 - P2)$$

is weighted more favourably towards growth rate than the index:

$$I = (0.04 \times DG) + (100 - 2 \ P2)$$

Two candidates A and B, with daily gains of 800 and 900g, respectively, and P2 backfat measurements of 15 and 20mm, respectively, will score on the first index 99 and 102, but on the second index 102 and 96. Thus candidate B is perceived to be better (being faster growing) when the first index is used, but candidate A is better (being less fat) when the second index is used.

These example indexes are, of course, extremely simple, although they are of use. Their main limitation is that they only include information from the candidate. It can readily be understood that when information is coming into an index from many different sources, and the various weightings and relationships have to be coped with, then indexes used by practising breeding companies can be so sophisticated that they are only manageable by computer techniques.

Selection objectives

It is established that the fewer the selection objectives the faster the rate of improvement in those chosen. Priorities must therefore be set at the outset of the breeding programme.

There are three broad areas:

Reproductive traits;
Meat quality;

Efficiency of growth;

and two broad criteria:

Worth;
Potential rate of progress.

Taken separately, the same unit of additional financial return can be obtained by about 20% improvement in daily *live-weight* gain, 4% improvement in feed conversion efficiency, 3% improvement in carcass lean percentage or 7% improvement in litter size. For growth rate, a rate of improvement of around 20g daily live-weight gain annually appears to be readily achievable by breeders and shows no diminution. Concentrated selection for carcass lean percentage appears to be able to accumulate up to 1% of lean annually. Amounts of improvement possible for reproductive traits remain largely unknown. Litter size may be increasable by 0.1–0.2 piglets annually, with the help of BLUP, but it has been speculated that faster rates may be possible.

Achievement of improved carcass lean percentage by reduction in appetite and growth rate with no change in feed conversion efficiency is clearly less beneficial than improvement of carcass lean percentage by increase in lean-tissue growth rate which is also positively associated with feed-conversion efficiency and daily gain.

Table 6.9 shows responses to selection for four characters each selected alone under *ad libitum* feeding conditions. The Table shows that selection for daily gain alone will have positive responses in terms of lean tissue growth rate and appetite but the pigs will get fatter. Selection for reduced backfat alone will bring about reduced daily live-weight gain and reduced appetite and have only a small positive effect on lean-tissue growth rate. Selection for feed efficiency alone has only slight consequences for each of the characters, on account of the conflicting influence of gain, appetite and fatness on efficiency – but appetite is reduced. Selection for lean-tissue growth rate alone has similar effects to selecting for daily live-weight gain alone. Selection for increased appetite alone will produce fast-growing but fat pigs. It will be noted that only by use of a properly constructed index can positive gains be made in all the required characters.

It is entirely sensible that most genetic improvement over recent years has been made in reduction of fat, as this was clearly the most immediately cost beneficial. The position is augmented by the high heritability of this trait which has allowed a rate of change of about 0.5mm P2 (or 2.5%) backfat reduction annually. Potential progress for further backfat reduction in some pig populations is now greatly diminished, however; present levels having been reduced to 8–10mm P2 are now approaching the minimum necessary

Table 6.9 Responses to selection under *ad libitum* feeding

	Character selected					
	Daily live-weight gain alone	Reduced backfat depth alone	Feed efficiency alone	Lean-tissue growth rate alone	Increased appetite alone	Index[1]
Character responding						
Daily live-weight gain	++++	−	++	+++	+++	+++
Lean-tissue growth rate	++++	+	++	++++	+++	++++
Reduced backfat depth	−	++++	++	+	− −	++
Increased appetite	+++	− −	− −	++	++++	++

[1] Index including values for live-weight gain, lean-tissue growth rate and backfat depth.

for adequate meat quality. Further improvements in carcass leanness in the future are likely to be made through positive change in carcass yield and negative change in carcass bone content.

Although the potential for further progress in fat reduction is somewhat limited in many North European pig stocks, there remain large improvements possible for most other pig types world-wide. There is no evidence yet of any diminution in rate of change possible for traits associated with increased lean-tissue gain, and as these are being selected upwards there is no reason to suppose any immediate limit, provided, of course, that restraints are not placed on positively correlated traits; if appetite were restrained, then further improvement in growth rate would be limited.

For optimum production of meat the selection index is likely to take into account:

daily live-weight gain;
fat depth;
percentage lean;
daily lean-tissue growth rate;
eye muscle area;
ham shape;
muscle quality;
fat quality.

The relative importance of meat quality (eye muscle area, ham shape, muscle quality and fat quality) in comparison to lean growth (daily live-weight gain, percentage lean and daily lean-tissue growth rate) varies greatly. In some countries the former group of qualities are unrewarded at the point of slaughter and therefore considered unimportant to the producer. In other cases meat quality is prerequisite in a slaughter pig to the extent of these characters being handled as criteria for independent culling. There is reason to believe that the shape of pig types such as the Large White and Scandinavian Landrace can be improved by selection, and some such types are now coming forward without having had gene importations. Most well-shaped White breeds have, however, been created through gene importations from Pietrain or Belgian Landrace types.

The accumulation of all these various production characters together need not always seriously detract from progress in any one. Rather they may be supportive of each other when positively correlated, as are live-weight gain, lean-tissue growth and lean percentage, and also as are eye muscle area and ham shape. The proposition, as has been shown in Table 6.9, that selection for fat depth alone may bring with it the negatively associated traits of reduced daily gain and appetite, supports the view that these characters should be included in the index in order to stop reversion. Some indexes also include feed-conversion efficiency, but this can be a complicating factor as it is itself a compounded value which is liable to be influenced by a number of disparate characters.

Meat quality is only now receiving a measure of attention as a selection objective. As usual, the first response of breeders to consumer interest in meat quality has been to look for quick solutions through the importation of genes from other breeds. The use of the Duroc in the sire line has been considered (perhaps wrongly) as a way toward rapid improvement of the eating quality of fresh pork products from the Large White and Landrace breeds.

Within-population selection for intramuscular fat may be worthwhile, and although there is close association between intramuscular fat (a possibly positive attribute) and subcutaneous fat (a definitely negative attribute), some strains of the Duroc breed do suggest that this association can be weakened. Many types of Duroc carry almost twice as much intramuscular fat as many modern populations of Large White and Landrace, but only around 25% more subcutaneous fat. The cost in terms of other selection objectives forgone while intensive selection is made simultaneously for intramuscular fat and against subcutaneous fat might, however, be extortionate. Beyond fatness, it is possible that in the future there may be successful selection for other aspects of meat quality, given the present levels of improvement in knowledge about the causes of tenderness and flavour. Besides, there is ample evidence to suggest that tenderness and flavour could be readily and greatly improved by attention to the preslaughter, slaughter, postslaughter, meat retail and household cooking aspects. Perhaps until standards are improved here there is little merit in creating a superior genotype.

Genetic improvement in meat quality is likely to remain for a little while yet at the level of a search for breeds possessing particular attributes of flavour and tenderness, and the incorporation of these into the male lines. But in the long term it *is* possible to envisage a programme for the improvement of meat quality by within-population selection.

Improvement in reproductive performance has a remarkable pay-off, but until recently has been abhorred as a selection objective by managers of breed-improvement programmes. Not only is heritability low, but the trait is also sex limited. Because measurement of the breeding value of a boar for reproductive characteristics can necessarily only be made by measurement of relatives and progeny, many litters are required before a dependable view of a candidate can be obtained. These problems are exacerbated by a long generation interval. In the past within-population selection programmes at nucleus level have been avoided, and the need to incorporate litter size and number of litters per year into a selection index has been given low priority. Instead:

1. New genes have been sought and incorporated from renowned prolific breeds such as the Large White and some Chinese types.

2. Heterosis is employed in breeding programmes by the crossing of White breeds, and use of these F1 hybrid crosses as parents for the slaughter generation.

3. Special herds of super-prolific sows have been created from the results of searches through many thousands of pigs for individuals showing quite exceptional reproductive performance, such as four litters with more than 14 piglets born. The outstanding individuals are bred together by backcrossing to form the core of a hyperprolific dam line which may have up to two piglets more born per litter. To maintain health status, the genes from this line are then moved into the conventional nucleus herds through the medium of embryo transfer, artificial insemination and hysterectomy. The down-line expectation from this tactic is around 0.25–0.75 more piglets per litter at parent level. Theoretically, this procedure could be handled continuously, but it takes about 6 years of backcrossing and testing to secure hyperprolificacy in the genome for a benefit of about one piglet per litter, which is somewhat slower than conventional selection using modern techniques.

All these policies have been, to some degree, useful, but equally all suffer from failing to achieve year-on-year progress. For further improvements in reproductive capacity there is no avoidance of within-population selection (with the possible exception of genetic engineering techniques which are not yet available).

The most appropriate reproductive trait for selection is probably litter size, as this at least is relatively easily measured. Modern breeding programmes can now incorporate litter size, and other reproductive traits if required, into indexes used for selection in the female line. The present acceptance of reproductive characters into the suite of selection objectives by contemporary breeders is due to two recent advances. First, better environmental and management control procedures are allowing a significant reduction to be made in the size of σ^2E and a consequent lift in heritability from around 0.10 up to around 0.20. Secondly, the use of the computer and Best Linear Unbiased Prediction animal model statistics (BLUP), by incorporating additional information from many relatives, times and places, avoids the need for lengthy generation intervals and tedious progeny tests of the prolificacy of a sire's daughters. BLUP is likely to realize significant improvements in litter size and other fertility rates over future years. BLUP technology may allow a rate of improvement of 0.2–0.3 piglets per litter per year if litter size is the only selection objective in a population of 1,000 breeding sows or more. With a mixed model and selection for traits other than litter size, rates of improvement of 0.20 piglets per litter per year have been predicted, but yet to be proven. It is conceivable that selection for ovulation rate alone would result in an early loss of response and a plateau in litter size improvement due to the counteracting forces of embryonic losses and limited uterine capacity. Litter size is a compound trait which requires at least three separate elements to be simultaneously improved. Selection for litter size (rather than its component parts) should ensure progress in all the associated traits required and there is no reason to believe that litter-size response will plateau in the near future. But 14 rather than 12 functioning mammary glands may become an essential component of a breeding sow.

The progeny test

The progeny test post-Bakewell was reinvented by the Danes in the early years of the twentieth century. Central progeny testing, being expensive but accurate where the effects of environment must be minimized, is most preferred for traits which are sex-limited and lowly heritable, such as litter size and milk yield. The progeny test using central testing station facilities for identification of superior sires with regard to growth and carcass quality traits was used with great success in Denmark through the middle decades of the twentieth century and was responsible for creating the Danish Landrace pig type and the excellent lean Danish bacon by Wiltshire cure which was derived from it. The progeny of candidate sires were sent for central testing where in one common environment growth and carcass characters were assessed primarily through an index of feed efficiency and leanness. One of the attributes of the scheme was the large population size made available on account of all nucleus breeders contributing to the scheme and being ready to use the proven boars provided from the scheme.

Although central testing to reduce environmental effects has been used for milk yield in dairy cattle, the exorbitant costs have necessitated a more realistic approach and the evaluation of progeny on their farm of origin. This requires:

1. an effective artificial insemination service;
2. the wide ranging use of the sire being evaluated;
3. sophisticated computer techniques to handle the records;
4. BLUP animal evaluation techniques to cope with the statistical methods

required to generate a sire breeding value from as many relatives as can possibly be accommodated, and also to take into account the differences between farms, seasons, generations and times.

Lack of these facilities in the past led to the demise of the progeny test for the selection of reproductive traits in pigs. Now, however, combinations of performance and progeny testing are back in use in nucleus herds of female lines for purposes of the improvement of litter size (especially) and 3-week litter weight or milk yield (rarely).

The performance test

Performance testing of the modern era probably originated in the US amongst Duroc breeders. Certainly it was the North American pioneers of the application of genetics to pig breeding – Hazel, Lush and Dickerson – who clarified the potential of performance testing. They saw the possibility of examining more individuals and reducing the generation interval. More rapid progress can be made in the growth and carcass traits of moderate/high heritability by performance test than by progeny test. The American breeding plans were developed further by Dickerson, by Fredeen of Canada and by European geneticists, such as Clausen, Jonsson, Johansson and Skjervold in Scandinavia, and Falconer, Robertson, Smith and King in the UK, and by the disciples of these latter, such as Brascamp, Bichard and Webb.

During the 1950s and 1960s, central testing stations were set up throughout Europe and also in North America on the Danish pattern. None of these schemes was as successful as that of the Danes. The Danish scheme was centrally controlled by Danish Slaughterhouses, a single autocratic organization with a clear view of a limited goal (Danish bacon for export) and with the ability to organize the distribution of improved genes through a breeding pyramid which concentrated the best genes in the nucleus herds.

A typical performance (and sib) testing operation on a national scale would involve the setting up of a number of central test stations to which were sent (usually) four sibs: two entire boars (the candidates), a castrated male and a female. The candidates joined as large as possible a population of contemporaries, all to be tested and compared under the same conditions of place, time, environment, nutrition, health and management. Pigs would be entered at around 25kg and a test begun at a given set weight (30kg). After being grown to a second given set weight (100kg) individual daily live-weight gain and ultrasonic backfat depth (and muscle depth) would be determined for individual pigs, and feed consumption and feed conversion efficiency measured for individual candidates or for a sib group. The candidates would be assessed for physical competence, including type, legs, feet, number of nipples and so on. The castrated male and female would be similarly measured for their performance, but would then be slaughtered to yield sibling data on killing-out percentage, lean-meat percentage, fat content, muscle quality, eye muscle depth, shape of ham and other associated carcass and meat-quality characters. Breeding value is estimated from the index score, which would be constructed for each candidate male, built up from all of the information available from the quartet, and expressed as a deviation from the mean of contemporaries on test.

Boars of proven excellence are then available for return to their original owners and use in pure-breeding nucleus herds, or to be sold for use in other pure-breeding nucleus herds. Boars proven to be good, but not outstanding, would be sold to commercial producers for siring cross-bred slaughter pigs. Boars shown to be of less than average quality would be slaughtered.

Alongside the central test for males, the females would be performance-tested on the nucleus unit of their origin. Selection rates would not be high, and this element of the scheme would contribute only a small proportion of the genetic gain. After an initial sorting on conformation, legs, feet and nipples, females would be evaluated for their weight for age and ultrasonic backfat depth and ranked according to a simple index. If required, the index could include litter size of origin, and some assessment of the dam's general prolificacy.

Such a scheme (but without litter traits) operated in the UK through the second half of this century under the Pig Industry Development Authority and later the Meat and Livestock Commission. It was instrumental in creating the rapidly developing nucleus pure-bred herds which later became seminal to the UK breeding company structure. From 1980 onwards it has had to be realistically accepted that the scheme has foundered due to the lack of candidate sires coming forward. The scheme has been the victim of its own success. The breeding companies have themselves adopted similar schemes for use in-house and the performance test remains central to company progress by genetic selection.

The withdrawal of progressive pig-breed improvers from the UK national scheme, and the setting up of their own schemes in the context of independent and competitive trading, was hastened by four further issues.

1. The testing regime was particularly targeted to the improvement of carcass quality and was having no beneficial effect upon daily lean-tissue growth rate. The breeding companies, on the other hand, were already beginning to identify that their grandparent stocks were approaching the minimum levels of backfat depth that could be acceptable within the constraints of carcass quality, and their main interest was now turning

away from lean percentage and towards lean-tissue growth rate, conformation, meat quality and reproduction.

2. The national scheme was naturally biased towards general national needs. But the breeding companies were already embarking on individual objectives. The export market, for example, requires particular qualities not always recognized as priorities nationally, for example ham shape and muscle size.

3. Central testing represented a potential health hazard due to the mixing of animals on test from many farms of origin. As health in their stock is prerequisite to breeding companies selling on to other producers a live pig product, the notion of bringing into nucleus herds sires which had been exposed to a range of diseases during test was an anathema.

4. Breeders recognized that the test regime used was not best placed to make effective selection of characters of primary interest now. A genotype:environment interaction effect had arisen between the national test and the customer base.

There remains a residual service in the UK which may survive to fulfil a role in both the national and the private context. First, aspirant breeders are given an opportunity to have stock tested to a level of efficacy not possible in their own (yet) limited circumstances. This was expected to encourage new entrants into the breeding company scene, foster competition and avoid the development of oligopoly. Secondly, the rump scheme allows for the sires that are candidates to remain at home to be tested on-farm, while the siblings are sent for central testing. This avoids all the health problems, although reduces somewhat the quality of the performance test on the sire himself. But breeders are offered an opportunity to rank their own candidate boars, through measurements on relatives, by use of an independent test.

The scheme can also be used to rank candidates from various different farms and sources. For this to occur the nucleus herd itself must be ranked. Thus, a candidate boar is represented first in relation to other candidates from his own herd but then, secondly, his herd is represented in relation to its ranking against other herds. The ranking of herds is not readily effected with accuracy, but the greater the sample size of animals forwarded from each herd to the central test facility the better. Each herd in the scheme must therefore submit a minimum number of candidate sire sibs to central test annually if they are to remain in the scheme.

Most importantly perhaps, the central test facility can offer to all breeders a cost-effective carcass and meat-quality assessment service. Whilst independent breeders may feel able to pursue an effective in-company performance testing programme independently of a national scheme, few of them would be able to carry out a substantial dissection and meat-quality control programme.

It is true that ultrasonic techniques can measure fat and eye muscle depth, and even assess muscling of the ham. With ultrasonics, an excellent prediction of lean-tissue mass (and hence lean-tissue growth rate) can be provided from measurements on the live animal; the need to slaughter in order to determine body composition is now much reduced. However, these predictions are themselves dependent on regression equations, and these equations must be determined from data sets including information from both live and slaughtered animals. Not only is the central carcass analysis service highly effective at providing data for building these regression relationships, but these relationships are themselves:

    breed type specific;
    company specific;
    liable to change over time as genetic progress is made.

Therefore a forward-looking breeding company needs to maintain a continuous flow of sample animals for full carcass analysis so that (a) the prediction equations that are used in indexes for assessing sire candidates can be kept up to date and accurate, and (b) carcass and meat-quality characters determinable only on the dead animal can be measured and monitored.

If ultrasonic technology has allowed good prediction of body composition of live animals, technology to assess killing-out percentage, cut-out values, balance of high-price cuts, proportion of bone and so on, is less well advanced. Slaughter of sample animals from candidate sires is also unavoidable if any assessment is to be attempted of breeding value for such characters as fat quality, muscle colour, muscle tenderness, water holding capacity, intramuscular fat levels and the like.

National schemes for performance testing at central facilities will have a continuing place in breed improvement especially where:

1. there is a clear need to change the whole population of a nation's pigs in an agreed direction, and where the selection objectives are few;
2. a scheme is needed to encourage new entrants into the breeding business;
3. there is a wish to attempt to rank and compare different nucleus units within a breeding company, or different breeding companies within a given market environment;
4. a carcass dissection and meat quality facility is needed to service pig breeders;

5. there is a national structure which promulgates improved genes in such a way as to ensure unidirectional movement which avoids dissipation of excellent genes to non-nucleus breeding stock: in effect, where there is a determined and controlled breeding pyramid.

The sceptic may conclude that of these five conditions which favour a national testing scheme only number 4 is, in fact, a perceived need!

A pyramidal breeding structure (Figure 6.12) is needed for both national and company schemes. The notion rests on the simple premise that the most excellent genes should be maintained in the nucleus for improving that nucleus. The improved genes are then multiplied up and moved down to commercial level. This does create a degree of genetic lag, as it takes time for improvements to filter down from great-grandparent nucleus level to parent level and the slaughter generation, often 4–5 years. However, once initiated, progress is maintained year on year.

The mainstay of central testing philosophy is the reduction of environmental effects and the placing of all candidates in the same environment to allow equal opportunity to express genetic merit. Outstanding questions remain of genotype:environment interactions and the suitability of one test regime to all candidate types. There also remain serious questions relating to herd of origin effects, where success on tests is rather too closely related for comfort to pre-test environment and treatment. However, this latter can be countered by starting at an earlier age and lighter weight, and by the use of Best Linear Unbiased Prediction animal models. It is disturbing to note that the correlation between conventional test and a modern test with BLUP and earlier start to reduce pre-test effects may be as low as 0.5, rather suggesting that one or other is not specially effective.

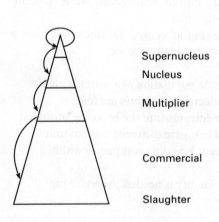

Figure 6.12 Breeding pyramid. Performance and progeny testing, and the selection of individuals of special merit as sires and dams of the next generation, takes place at supernucleus and nucleus level. The stocks improve rapidly due to concentration of testing resources, and to concentration of selected genes from proven sires. Progeny of nucleus parents are moved to multiplier level. Here the genes are multiplied to sufficient number to allow sale to commercial producers. Pigs at the top of the pyramid will usually have the highest health status. Movement of stock is always, and only, down the pyramid.

Supernucleus

Nucleus

Multiplier

Commercial

Slaughter

The loss of the pivotal role of central testing in contemporary pig breed improvement should not be taken to imply any dilution of general national breeding effort, rigour, intensity or accuracy. Rather the reverse, but the activities of the test station are now mostly handled in-company, on-farm. Most breeding companies are pursuing, with even greater effect, the same performance-testing protocols so painstakingly set up by the erstwhile national schemes.

The test regime

Selection objectives (expressed in the form of the index) will interact with the test regime. Great care is required on the part of the breeder to match the selection environment with the objective. The deep influence which the test regime has upon the character actually selected has only recently been fully appreciated. Clear indicators came from the original UK national scheme whose index purported to advance growth rate, feed conversion efficiency and carcass quality, but which advanced only backfat reduction; growth rate remaining unaltered over all the years of the use of the index and there being no efficiency effect occurring independently of that created by backfat reduction.

An index weighted towards growth *would* allow fatness to increase if feed was offered *ad libitum*, but *would not* allow fatness to increase if feed was offered at a restricted level. In populations which are no longer overfat, it is reasonable to select positively for appetite which requires both measurement of individual feed intake and index weighting towards growth. Indexes in which low fat depth and good feed conversion efficiency have high economic weightings will bring about appetite reduction when used in populations of fat pigs. Because positive selection for lean tissue growth rate, together with positive selection for muscle size and ham shape, is likely to predispose the population to a fall in meat quality, then meat-quality characters are worth weighting heavily in an index used for populations of lean pigs with already excellent growth and muscle size characteristics.

Test regimes offer many and varied possibilities, from which may be exampled, amongst others, any combination of the following:

Start: age; weight; weight within a given age range.

End: time from start; weight; consumption of a given amount of food.

Feed intake: severely or moderately restricted; appetite; *ad libitum*.

Diet: high density; low density; average density.

Select for: leanness; lean-tissue growth rate; lean-tissue feed conversion.

Select against: fatness; poor feet and legs; low meat quality.

The earliest possible start weight, within a given range of age, is beneficial only so long as the increased susceptibility of the young piglet to disease and environmental abuse does not adversely influence the test by increasing variation. Start weights of 30–35kg are conventional, but in excellent facilities 20–25kg may now be possible. The test end should be as far distant from the start as possible within the bounds of costs and the need to minimize generation interval and the disposal of culls. The longer the test period, the more accurate can be the measurement of growth characters. The test should not, of course, continue after growth impulsion has diminished, because the sensitivity of the test will be reduced and the separation of candidates into rank order will be less effective (Figure 6.13).

End of test is most conveniently fixed as a given number of days from the start. This helps management and planning, and facilitates the use of a fixed-feed, fixed-time ration scale where all pigs get the same feed allowance which is incremented weekly regardless of pig weight. Most tests are continued for 10 or 12 weeks, targeting for end weights in the region of 90–120kg (the better candidates being heavier).

Termination of test by weight appears convenient from the point of standardization of carcass attributes, but in fact it is unsatisfactory for the interpretation of growth rate and feed intake data due to various candidates finishing at different times and having eaten different amounts of feed. Adjustments to measurements of backfat thickness to allow for the effects of pigs finishing at different weights can, fortunately, be readily and accurately

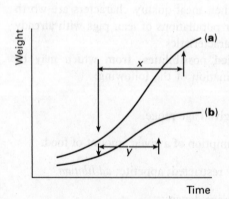

Figure 6.13 Length of test appropriate for two pig types (**a**) and (**b**). x and y indicate respective appropriate test lengths. To test population (**b**) over the same length of time as (**a**) would discriminate unjustly against (**b**); while to test (**a**) over the same length of time as (**b**) would fail to optimize. In both cases the period of testing includes a small part of the acceleratory phase of growth as well as the linear phase. The figure shows how difficult it would be to devise a single test regime suitable for two different pig populations. Tests should be designed to be population-specific and used only to rank candidates from within a single population.

made from knowledge of the relationship between weight and fatness for the population concerned, and this may be obtained by the simple expedient of determining the regression of backfat depth upon live-weight. One such estimate, for example, would predict P2 to rise by 0.15mm for each kilogram increase in live weight for pigs in a particular population finishing test at around 100kg. This coefficient would be expected to be higher for fat populations of pigs and lower for leaner populations.

The amount of feed consumed by candidates on test has interesting repercussions for the interpretation of the test (Figure 6.14). Feed intake during early growth (L) is inadequate to distinguish between the two candidates, (a) and (b). Over the period during which feed intake is adequate to maximize lean-tissue growth rate for candidate (b) (M), the test progressively better differentiates between the two types. Over the period identified as L, pigs with larger appetites will have faster lean-tissue growth rates, thus these pigs will be favoured even though subsequently the maximum lean-tissue growth rate may not be greater for such candidates. If this is the case the test loses accuracy as the two characters tested (early appetite and maximum lean-tissue growth rate) are confounded.

Figure 6.15 shows how if both candidates eat the same amount of food, providing that amount is in excess of n, then candidate 2 will lay down more fat, and fat depth comparisons positively assist in differentiating the two different lean-tissue growth rate genotypes. If, however, candidate 2 eats less than candidate 1, then it will not necessarily be fatter. Optimum feed intake to separate the two candidates is at the beginning of the intake phase identified at n, with both candidates eating the same amount of feed. Pigs with big appetites allowed to eat more feed on test than contemporary candidates may combine fat accumulation with improved lean-tissue growth rate (Figure 6.15). It is because of this sort of response that feed intake should be either fixed and equal for all candidates, or known for each individual.

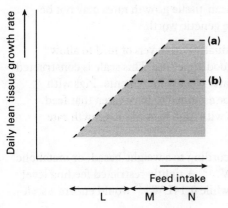

Figure 6.14 Daily lean-tissue growth response of two candidates (a) and (b) to an increase in feed intake. Feed intake in the range identified as L is typical of early growth, feed intake in the range identified as M is typical of the middle period of growth and feed intake in the range identified as N is typical of the later period of growth. For the consequences of this response upon the efficacy of a test regime to select superior genotypes for feed intake level, see text.

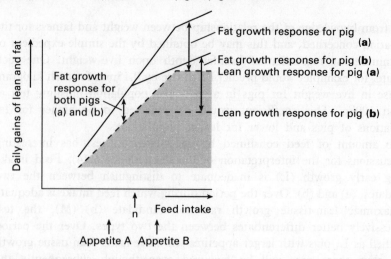

Figure 6.15 Lean and fat growth responses of two candidates (a) and (b) to an increase in feed intake. The lean : fat ratio is relatively constant until the plateau for daily lean response is reached, after which the animal begins to fatten. Note how if pig (b) had appetite A it would be slower growing than pig (a) with appetite B, but it would be demonstrably less fat. A selection index heavily weighted toward fat reduction may favour pig (b) over pig (a).

Ad libitum feeding systems for candidates housed in groups are particularly difficult to interpret.

Contemporary test regimes may often use one or other of the following rationing systems:

1. All candidates are provided individually with the same levels of feed on a time-based feed scale over a fixed time period, for example 1.5kg daily for the first week, incremented weekly by 0.15kg over a 10-week period. If the scale chosen is readily within the appetite of all candidates then feed intake will be guaranteed to be identical for all pigs, but some candidates with particularly high lean-tissue growth rates may not be able to fully demonstrate their true genetic worth.

2. All candidates are provided individually with levels of feed to allow appetite expression for pigs with good appetites. This scale is constructed similarly to 1, above, but with greater weekly increments. Pigs with lower appetites refusing feed will be assumed to have eaten that feed. Such a scheme will favour animals with high lean-tissue growth rate potential and high appetite.

3. Candidates are fed individually according to a weight-based, or metabolic weight-based scale such as $0.12LW^{0.75}$. A tightly restricted feeding level would be ensured at $0.10LW^{0.75}$, while $0.14LW^{0.75}$ would ensure a scale

close to, or above, *ad libitum*. Feeding a group of candidates on a weight-based scale related to the average of the group would, of course, be nonsensical. Weight-based scales have the consequence of according to faster growing, heavier pigs the greater feed allowance for which they have demonstrated a need. This avoids the time-based feed-scale problem in which the smaller (poorer candidate) pigs are offered the same level of feeding (and therefore effectively more feed) than the larger (better candidate) pigs, thus favouring the poor and penalizing the good. The evidence, however, does not confirm the practical reality of this effect. Feeding to a time-based scale pressures the good pig to grow lean from his limited ration, whilst the poor pig is encouraged to grow fat – thus helping to differentiate between them. It is also the case that time-based scales need not be restrictive on the better pigs if that scale is generously allocated as for case 2, in which the needs of the better pigs with the bigger appetites should be the basis for the decision about the size of the weekly increment.

Further, in practice, the weight-based scale may do no more than spend 9 weeks of test reinforcing the first week's performance, which often is the arbitrary result of a range of non-genetic effects. Good performance in week 1 ensures a slightly heavier weight at the end of the week, and this will then be immediately rewarded by a greater feed allowance than competitive candidates. This greater feed allowance now allows superior gains to be made. The weight-based feeding scale is therefore open to the serendipity of the early phase of the test, rather than utilizing the whole span of the test for pig differentiation. Additional problems of the weight-based scale are (i) the need to weigh all candidates at least weekly, (ii) the problem (impossibility) of interpretation of the feed intake data in order to make proper between-candidate assessment of performance and (iii) the need to individually calculate and individually allocate the feed allowance to each pig, with the result that there may be as many different daily feed allowances in the test house as there are pigs! The ultimate in apparently sensible, but in reality disastrous, test regimes is the one which is operated on a weight-based feeding scale to a fixed end weight.

4. Candidates are housed in groups and fed *ad libitum*. This reflects most closely the likely production environment, but suffers from the disadvantage that individual feed intake is not known. It is, however, a simple method to manage and, where the main selection pressure is on reducing the fatness of pigs, can be effective, providing that some degree of appetite reduction is an acceptable consequence. Although not as sophisticated as scheme 2 above, this choice of rationing may be more

accurate if the management structure cannot cope with individual candidate feed provisioning.

5. Candidates are housed in groups and each pig fed *ad libitum* through the medium of an electronically controlled feeding device which identifies each individual candidate, allowing continual access for voluntary feed consumption. The computer-controlled system logs the daily individual pig-feed intake. It is evident that this system contains within it all the advantages and none of the disadvantages of the alternatives described above and therefore has much to commend it. There are, nevertheless, provisos:

    i. the equipment is dependable;
    ii. the possibility of adverse selection for behavioural aggression is avoided.

This last point is made on evidence that aggressive individuals may deny access to the feeder of those pigs which are more placid. This results in feed being available *ad libitum* only to the more aggressive animals. Whilst aggression may be helpful in ensuring individual success in group situations, it is questionable whether the industry would be best served by the creation, through positive selection, of a whole population of pigs with raised aggression levels.

At present the most favoured regime for an index with positive pressure upon improvement of lean-tissue growth rate would appear to be either scheme 2 (the development of scheme 1 which helps to improve appetite), or scheme 5, both being operated from a fixed weight at start and conducted over a fixed time of 10–12 weeks.

So far the discussion of feed regime has rested on the assumptions that the index targets for selection will be either a reduction of backfat through appetite reduction (see Figure 6.15, B towards A), or increase in lean-tissue growth rate (Figure 6.14, (a) towards (b)). The case of fat reduction at lower feed levels has not, however, yet been dealt with. It will be noticed in Figure 6.16 that the ratio of fat to lean has been represented as not changing at feed intakes between O and Q for either pig (a) or pig (b). If this level of fatness is satisfactory for meat quality, then clearly no action is necessary. But in some traditional pig types (and certainly in most pig types before the modern era of selection against fat) even the amount of fat grown by pigs in circumstances of limited feed supply is more than that required by the customer/consumer. Reduction in fat at lower feed levels requires a change in the metabolism of the animal such that feed is diverted from the production of fat to the production of lean so that there is an effective increase in the regression slope of daily lean gains on feed intake and an effective decrease in the regression

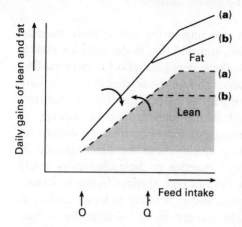

Figure 6.16 At feed intake levels which support lean-tissue growth rates which are less than the potential (feed intakes of less than Q), the fatness can only be reduced by achieving a change in animal metabolism and the favourable partitioning of feed away from fat toward lean; as shown by the arrows.

slope of daily fat gains on feed intake, with a consequential improvement of the fat:lean ratio in favour of leanness. This description is typical of the different behaviour of castrated and entire males, and also of ordinary pigs and those to which porcine somatotrophin has been administered. It appears from past experience that selection against fat on a restricted feeding regime may well bring about these changes in the population, and that such did indeed occur in some of the early central test selection programmes. For most improved contemporary pig populations, however, the problem of fatness at limited feed intakes should no longer exist.

Choice of diet requires a view on genotype:environment interaction effects as have been discussed earlier. Where there is a particular need to identify genotypes that are especially competent to deal with forages there may be merit in considering a low-density diet. But, in general, traits such as lean growth rate, feed efficiency, fatness and meat quality are likely to be best expressed on diets of high quality. The use of the average diet of the industry where the industry itself tends to be conservative with regard to diet quality may well only ensure slow genetic progress and perpetuate an unsatisfactory situation for the industry concerned.

Future objectives for within-population selection

Before 1980, primary selection objectives in northern Europe were for fat reduction and improvement in feed-conversion efficiency. Test regimes were chosen to achieve this and, by and large, were successful. Over the last decade, emphasis has changed to enhancement of lean-tissue growth rate with associated increases in appetite, and continued improvement in feed-conversion efficiency (the latter now consequent upon a reduction in maintenance requirement due to more rapid growth to slaughter, rather than as previously upon a reduction in the rate of fat deposition). There are developing contemporary interests in selecting in the female more strongly for reproductive capacity and longevity, and selecting in the male for muscle size and shape independent of the halothane gene, and also for meat quality. These characters have already been added to lean-tissue growth rate, and are now the important issues for considering in the index. However, the industry is not yet particularly well advanced in the identification of appropriate testing regimes to facilitate the maximum differences between candidates for characters such as litter size and meat quality, and the determination of appropriate test regimes for individuals superior in these characters would benefit from attention. Health, in the form of both specific disease resistance and general robustness to environmental abuse, is likely to be an important preoccupation for the future. If fat reduction *was* the initial target for within-population selection, and if lean-growth rate *is* the current target, then future selection pressures are most likely *to be* in the additional arenas of meat quality, reproductive rate and health.

Breeding plans

Logistics, not science, is the underpinning force of a successful breeding policy. Without a system for handling the details of livestock identification, classification and movement, science is of little avail.

Combining together the three breed improvement strategies, (a) importation of new breeds, (b) selection within nucleus populations and (c) utilization of hybrid vigour, and then organizing the subsequent supply of improved pigs through the system down to meat-production level, is a logistical exercise of scale and complexity which requires the most excellent of management planning.

Use of central testing facility by breeders (Figure 6.17)

Of the total population of pigs, a very small proportion is in the hands of breeders; but the influence of the breeder upon the efficiency of both an individual commercial grower and of the national herd in general, is far-reaching. A national structure for pig improvement identifies superior individuals, helps to concentrate improved characters into nucleus herds and then allows improved genes to filter down from the pure-breeding nucleus herds through the multipliers and into commercial slaughter pigs. Alongside,

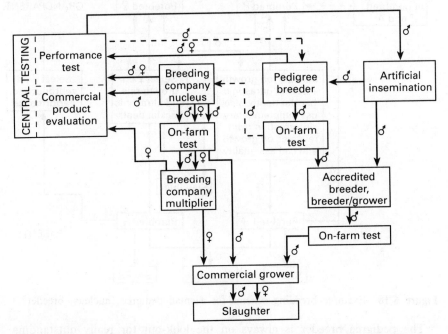

Figure 6.17 National structure for progressive pig improvement by identification, selection and dissemination of pigs of above average merit. Such structures may now be of historical interest only.

central testing facilities – usually a Government or *quasi*-Government agency – help to identify the presence of genetic merit in pure-bred stocks and can compare the results of breeders' efforts. The need for, and effectiveness of, a national test facility depends greatly on the individual country's pig industry structure and the level of central control possible. There also needs to be a generally understood and agreed need to service the pig-breeding companies, and smaller nucleus breeders aspiring to become larger.

The independent pedigree (nucleus) breeder of limited herd size (Figure 6.18)

To achieve generation-by-generation progress in a population of foundation (nucleus) stock, the breeder will usually concentrate on pure breeds and usually only on one type within that breed, taking the best individuals as parents for the next generation. Selection might be for one character, perhaps lean-growth rate, or for a complex of characters by use of an index, together with positive selection for reproductive performance.

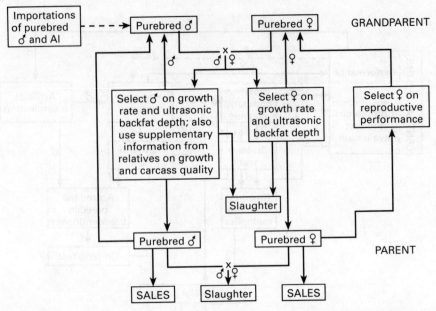

Figure 6.18 Example breeding scheme for a small pedigree (nucleus) breeder.

The pedigree breeder is always on the look-out for really outstanding individuals – males with very rapid growth, or females with extreme prolificacy – and once they have been found these individuals will tend to be held in the herd for as long as possible. If the rate of improvement slows, or a neglected character slips back, it may be opportune to import new genes into the herd in the form of stock from another breeder. The availability of semen from males of merit standing at artificial insemination centres makes this a relatively simple move.

The smaller breeder finds it difficult or impossible to make progress in reproductive performance for reasons outlined earlier. Many breeders therefore resort to a policy of ensuring merely that all candidates for selection are from reasonably large litters, which is unlikely to improve reproductive performance but may prevent its decline.

Growth characteristics are more easily tackled, and can be measured directly in both males and females. As the proportion of female candidates necessarily selected is large (about 20 times higher than the males), then accurate testing is not particularly cost beneficial. Candidates for selection as breeding mothers may be taken from group grow-out pens at 80kg or so, their growth rate determined from their age and the backfat depth estimated with the use of an ultrasonic meter. Liberal feeding of candidate pens would ensure that no animal is prevented from demonstrating its potential because of an inadequate feed supply. But where carcass quality is a significant factor in the value of slaughtered animals, a large number of fattening pens put on to an over-liberal feeding regime may be costly as those females not selected as herd replacements may be downgraded at slaughter for being over-fat.

There is more reward for increasing the accuracy of tests for male candidates. A greater differential between selected animals and herd average is possible as the numbers required are few, and the influence of the male on the herd is deeper and longer lasting. Candidate males can be penned separately or in small groups of two or three brothers, so that some estimate of intake and feed efficiency can be made. In addition to growth rate, feed efficiency and ultrasonic backfat measurement, it is helpful if information about carcass quality is brought to bear from the less-fortunate slaughtered relatives.

Central testing can offer help to pedigree breeders by providing facilities at which candidate males and/or their relatives can be grown and assessed. To test for carcass quality, slaughtered relatives can be dissected into lean, fat and bone, and the muscle quality assessed. At central testing stations the breeders' stock can be scored in comparison to contemporary stock on the same test. On the basis of the score achieved, the breeder can decide whether the animal should be returned to his herd or sold off. There can be a reciprocal arrangement whereby the central testing agency publishes the performance characteristics of animals tested. The information is of benefit to other breeders and enables pigs of merit to remain in herds of merit. Some contemporary small breeders may find benefit in the central testing of sibs, but in combination with the on-farm testing of candidate sires.

Speciality lines may also be rewarding, such as types with especially high meat quality, or shape, or ability to deal with extensive production systems. As always for the small business, a unique selling quality is essential if the large breeding companies are to be successfully competed against.

The general approach of the breeding company (Figure 6.19)

The companies have the advantage of size. With upwards of 1,000 sows in each nucleus, the benefits of picking a few animals from a large diverse population can be exploited, and there can be a wide difference between the best selected great-grandparent males and the average of the pig population upon which they are to be used. The breeding company is looking for a rapid turnover of animals of outstanding merit. Progress is made by frequent small steps forward, and the turn-round time in such a scheme is crucial. Stock sires are not kept for as long as possible, but for as short as possible, so the next generation contributes its small effort to positive progress as soon as it is able.

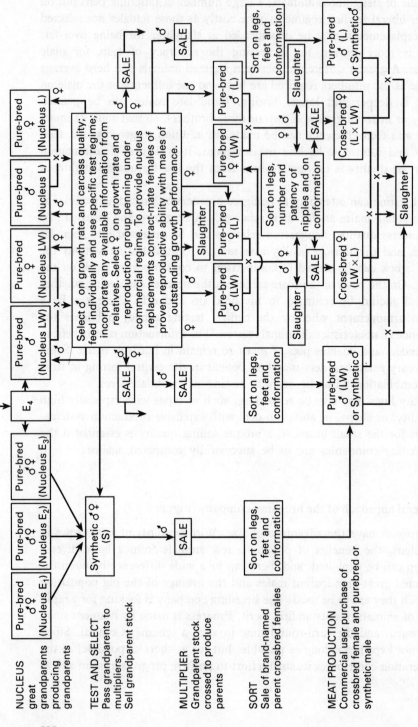

Figure 6.19 Example of a breeding scheme for a breeding company. E_1, E_2 and E_3 represent three different exotic breeds. E_4 represents the possibility of gene importations from types such as superprolific females or speciality males. S represents a synthetic 'sire line' made from three exotics and the main breed type. The sire line should have outstanding growth and carcass quality characteristics. LW and L represent Large White and Landrace types. These female lines should have outstanding reproductive (as well as growth) characteristics.

Whereas the pedigree breeder stakes his reputation upon the sale of a relatively small number of males of outstanding merit, the breeding company is in the animal-supply business. Breeding companies sell both males and females, often in the form of a stock package. The purchaser is offered brand-named parent females as herd replacements, together with parent males specifically bred as sires for use with the branded females. Breeding companies will also offer – at higher prices – grandparent stock to those wishing to breed their own replacements. Typical breeding company commercial products are shown in Figure 6.20.

The company therefore offers two services: (a) trouble-free provision of male and female herd replacements, and (b) improved pig stock. But it is only by the latter that the company will stay in business. It is fundamental to commercial success in the open market that the goods offered are demonstrably competitive in terms of quality and price, and a step ahead in terms of improved growth rate, carcass quality and reproductive characteristics. Breeding companies, because of their larger size and stronger financial base, will be able to test in their pure-bred nucleus herds to a much greater degree of rigour than smaller pedigree breeders. In addition, they will cash in on hybrid vigour by presenting a cross-bred female (and often also a cross-bred male) as the branded product.

Besides considerations of improved genetic merit, companies concerned with the sale and movement of stock are rightly preoccupied with the health status of the animals. Disease prevention and control will be a significant part of company policy. These matters are dealt with at length in Chapter 7.

The structure of the breeding company reflects its needs. Nucleus (great-grandparent) pure-breeding herds are maintained in strict isolation. Within these herds vigorous testing for growth, carcass and meat quality and reproductive performance is undertaken on both male and female lines. Selected out of these nucleus herds are the grandparent stock. A few of the best of these are used to refurbish the nucleus, whilst most others are sent to the multiplier herds which generate parent stock for sale. Because of health control, animals once shipped to multipliers cannot come back into the nucleus. To improve reproductive performance in the nucleus herds, proven

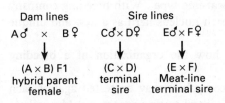

Dam lines

A♂ × B♀

↓

(A × B) F1
hybrid parent
female

Sire lines

C♂ × D♀

↓

(C × D)
terminal
sire

E♂ × F♀

↓

(E × F)
Meat-line
terminal sire

Figure 6.20 Use of nucleus 'pure breeding' lines to create genetically improved hybrid commercial breeding pigs. (Speciality parent female [A×G] may also be produced for outdoor systems, or speciality terminal sires [G×X] with particular meat characteristics.) (Cotswold Pig Development Company Limited.)

females may be contract mated to the best males and the offspring earmarked for selection back into the nucleus rather than going on to the multiplier phase.

The type of grandparent maintained in the nucleus herds has developed from three requirements. First, stock must already be of high genetic merit, but open to further improvement. It should not be assumed that nucleus grandparent stock is necessarily 'pure-bred'; it may well have synthetic lines from a variety of origins. However, the term *pure-bred* may be understood to cover breeding colonies where both the male and female are of similar genetic type, and they are bred together to yield offspring of similar type to the parents (although – if selected – hopefully also improved with regard to that type). Next, types chosen for a nucleus herd must exhibit hybrid vigour when crossed. Lastly, the breeding company may wish to maintain a nucleus herd of a variety of breeds of pigs which might impart, currently or in the future, special characteristics into the final product.

A breeding company could keep, for example, two large nucleus herds, one of Large White type and the other of Landrace type. The nucleus breeding herds may number 100–1,000 sows or more. Smaller, 50–500, breeding colonies of pigs, such as the Hampshire, Pietrain, Duroc, Belgian Landrace, Lacombe, Saddleback, or whatever, may also be maintained to produce the male side of the sire line, or in hope of their being found useful, at some future date, as providers of characteristics such as hardiness, shape, lean mass and so on. Alternatively, or in addition, a line may have been developed that contains an amalgam of various breeds and this is then treated as if it were a new breed. The result is the creation of a new *synthetic* line, particularly appropriate as a top-crossing sire for the slaughter generation level.

It is often the case with breeding companies that their pure-breeding nucleus stocks are, *de facto*, synthetics having been subjected to various gene importations. Thus, whilst a breeding company nucleus may be referred to as 'Large White type', these will not be the same as the Large White that might be found in a pedigree herd book, nor indeed the same as another Large White type to be found in another breeding company. It is part of breeding company policy that its nucleus stocks are demonstrably different from those of the competition; despite the fact that they may be classified as 'Large White type' or 'Landrace type', or 'Meat-line type'. With breeding company nucleus stocks, it is best to consider each nucleus herd as a new and unique breed.

Figure 6.19 gives an example of how the organization of a breeding company might be put together.

The products sold from breeding companies may be tested against each other in central testing facilities. The central testing agency would purchase from multipliers random samples of young females and would mate them with boars supplied from the company nucleus. The females could be tested for reproductive performance and their offspring for growth rate, feed efficiency and carcass quality. The naivety of commercial product evaluation and the possibility of misuse of results therefrom has, however, been alluded to elsewhere.

A particular breeding company

Breeding companies create a progressively improving product that will sell in competition with others. Each breeding company has different selection goals, systems of management and trading policies, but all ultimately wish to maximize sales of products to commercial pig producers (mostly F1 parent females and top-crossing sires). To do this, the companies are usually organized into four production levels. Level one comprises super nucleus herds of great-great-grandparent (GGGP) pigs, which are highly intensively selected. From these, pigs of the next generation move down to nucleus level (level two) which multiplies the improved genes whilst maintaining the pure-breeding population at great-grandparent (GGP) level. Testing and selection continues at GGP level but is less rigorous in test regime design and less intensive than that in the GGGP super nucleus. From the nucleus, the next generation progresses on to level three in the form of GP stock which moves to the multipliers. Multiplier herds organize the crossing of the grandparents and create the parent stock (level four) available for sale. These parents produce the slaughter generation.

Health status at the top of this pyramid is paramount and all herds are closed, any entry being only through hysterectomy, embryo transfer or artificial insemination. To transfer improved genes from the super nucleus where they are first identified to the parent generation where they are used may take up to 5 years. This genetic lag time can be reduced significantly by cutting out the pure-bred multiplication phase at GGP level, and creating instead a much larger pool of nucleus animals, but in which intensive selection takes place such as would be expected at GGGP level. From this GGP nucleus, improved stock can be moved directly to multiplication level. Maintaining such large nucleus herds and testing so many candidates at the maximum level of rigour and intensity is expensive, but the rate of genetic progress is much improved due to greater population size, more accurate measurement, a higher selection rate, and a lower generation interval. The customer also benefits by the genetic lag time being cut to only 2–3 years.

Breeding companies will be sure to utilize to the full all three methods for maximizing improvement of performance by genetic means:

1. Selection within populations.
2. Incorporation of new genes by importation.
3. Creation of hybrid vigour through cross-breeding.

Selection objectives:

a. Lean-tissue growth rate.
b. Economy of production and lean-tissue feed conversion.
c. Litter size.
d. Quantity, quality and distribution of carcass lean.
e. Control of the halothane gene.
f. Improvement of meat quality.

Breeding lines (A–Z):

Each breeding line is a population of a particular pig type. These populations are maintained in the form of pure breeding nucleus herds of breeding females and males. Within each nucleus breeding line, a particular programme of selection is undertaken. The selection programme will not be the same for each of the lines, the lines having different end-uses. Males and females are passed from the nucleus to multiplier herds from where cross-bred parent females and terminal sires are made available for sale to customers (see Figure 6.20).

(A) **Male line for cross-bred parent gilt production:**
Landrace type: nucleus herd size 1,000–2,500 sows
Selection emphasis on growth and litter size (a, b, c).

(B) **Female line for cross-bred parent gilt production:**
Large White type: nucleus herd size 1,000–2,500 sows
Selection emphasis on litter size and growth (c, a, b).

(C) **Male line for general-purpose terminal sire production:**
Large White type: nucleus herd size 500–1,000 sows
Selection emphasis on growth and carcass quality (a, b, d, f).

(D) **Female line for general-purpose terminal sire production:**
Large White type: nucleus herd size 500–1000 sows
Selection emphasis on growth and carcass quality (a, b, d).

(E) **Male line for cross-bred terminal meat-line sire production:**

Synthetic type originating from a mix of highly meaty breeds with high lean percentage and excellent muscle distribution: nucleus herd size 250–750 sows

Selection emphasis on growth and carcass quality (a, b, d, e, f).

(F) **Female line for cross-bred terminal meat-line sire production:**
Large White type: nucleus herd size 250–750 sows
Selection emphasis on growth and carcass quality (a, b, d, f).

(G) **Speciality line for cross-bred dam and sire production:**
Duroc type: nucleus herd size 200–600 sows
Selection emphasis on growth, carcass quality and litter size (a, b, c, d, f), and also for special qualities such as hardiness, suitability for lower-input systems, muscle quality characteristics and so on.

(Z) **Zoo:**
Small breeding populations of different pig types: nucleus herd sizes 50–250 sows
As the future need for these types is not known there is specific care for no selection objective to be pursued, but for maximum variation in the gene pool to be maintained.

Method:

Especially high health status of all herds is maintained, from nucleus, through multiplier and down to customer level.

There is a three-tier pyramid with large nucleus herds at the apex to maximize rate of genetic improvement and minimize the genetic lag between improved genes appearing in the nucleus and being passed on to the customer. Figure 6.21 shows the pyramid for lines A and B, producing the F1 hybrid parent female.

The various nucleus herds are supplied with genes through the medium of artificial insemination (Figure 6.22). Only the few best selected males from

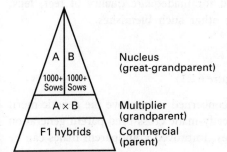

Nucleus (great-grandparent)

A × B — Multiplier (grandparent)

F1 hybrids — Commercial (parent)

Figure 6.21 Breeding company three-tier pyramid for the production of hybrid parent females (Cotswold Pig Development Company Limited).

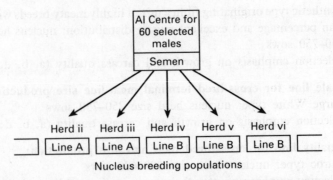

Figure 6.22 Artificial insemination centre servicing many separate nucleus herds to maximize the rate of genetic improvement (Cotswold Pig Development Company Limited).

each of the populations (A–G) are available at the AI centre. The selection intensity for males is around 1% or less. Best Linear Unbiased Prediction methods are used to compare candidates from different nucleus herds within each line. Each of the lines A–F requires about 15 males at any one time at the AI centre. The boars are used for no more than 10 weeks, and only serve 40 sows each, in order to maximize the rate of genetic advance and minimize generation interval. Sixty males per year will handle, through artificial insemination, 1,000 nucleus sows.

Female selection takes place within each of the nucleus herds on the basis of an index combining litter size and lean growth, and using maximum information from as many relatives as possible. The female selection intensity is around 40%.

Pigs are moved from the nucleus units to multiplier farms who hold grandparent stocks supplied from the nucleus. Rates of replacement in the herds should be about 100% annually for males and 50% annually for females. The grandparents are crossed to produce parent females and top-crossing sires available for sale. The latter may also be derived directly from nucleus level. Around 60% of cross-bred females born at multiplier level will be selected as appropriate for use as parents of the slaughter generation. The remaining 40% will have been culled for inadequate quality of feet, legs, conformation, insufficient nipples or other such blemishes.

The commercial breeder/grower (Figure 6.23)

Some commercial growers, although concerned to improve the genetic merit of their herds, may not wish to buy ready-made cross-bred parent generation replacements from breeding companies. Improved grandparent males can be

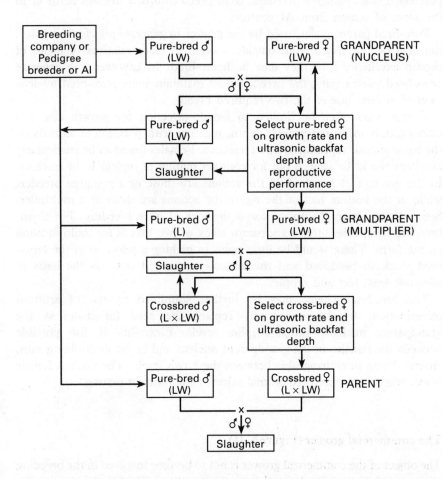

Figure 6.23 Example of breeding scheme for commercial breeder/grower (LW and L represent Large White and Landrace breeds to serve as examples). For each 100 cross-bred parent females about 30 pure-bred animals are required. Of these, about 10 regenerate the nucleus while 20 or so are mated with a male of a different breed to produce the cross-bred females from which 50 replacements are selected yearly for the main herd. The size of the pure-bred herd needed depends first upon the proportion of pigs coming from the herd but rejected because of poor legs, conformation and number of patent nipples, and second upon the proportion of pigs rejected as having failed to achieve adequate performance during the growing period. If 100 cross-bred females have four litters each in the course of 2 years of reproductive life, then 50 replacements are required yearly. Twenty pure-bred females will produce about 320 piglets per year, of which half will be female. If two-thirds to three-quarters of these animals are forwarded to test, about half those candidates need to be selected for herd replacements. To refurbish the nucleus at a rate of 15 females annually, 10 contract-mated females will generate 80 female piglets, of which 50–60 would go forward to test and the top third of those candidates would need to be selected.

purchased from pedigree breeders, from breed company nucleus herds or in the form of semen from AI centres.

Pure-bred parent males, used by the grower to generate pigs for slaughter, can also be purchased from outside – often as close relatives of males of proven excellence – or they may be home-bred. In any event, to generate cross-bred parent gilts, the breeder must maintain some pure-bred nucleus sows of at least one of the two required breeds.

Testing may often be limited to female selection for growth rate and ultrasonically measured backfat depth, together with rejection of animals on the basis of poor reproductive performance. Females found to be particularly excellent would be singled out for contract mating to replenish the nucleus. In the top half of Figure 6.23 the actions are those of a pedigree breeder, while in the bottom half of the figure the actions are those of a multiplier. Some producers might opt to forgo the joys of pure breeding. For them, breeding companies offer grandparent stock of two breeds for multiplication on the farm. There would be little point in making a selection of the cross-bred stock so produced and the animals are sorted only on the basis of adequate legs, feet and nipples.

The breeding scheme may be further simplified by use of artificial insemination, which can entirely replace the need for males at the grandparent nucleus, and multiplier levels. Flexibility is also possible between the females in the grandparent nucleus and in the multiplier group, animals being interchangeable between these two levels. The nucleus female is essentially any pure-bred animal selected for contract mating.

The commercial grower (Figure 6.24)

The object of the commercial grower is not to become involved in the breeding of improved parent pigs, but rather to procure them. The breeding companies service the grower by offering parent cross-bred females and selected parent males to go with them (Figure 6.24), or indirectly by provision of grandparent stocks for farm multiplication (Figure 6.25). Growers wishing not to depend on the breeding companies as a source of supply of breeding females may maintain a cross-bred herd by simple selection on a visual basis, and hope for some improvement in genetic merit by the purchase of males from reputable herds (Figure 6.26). To sustain the benefits of heterosis, the breed of male may be alternated every so often. Although pure-bred boars are the more likely option, some producers may go for a cross-bred boar.

The most commonly occurring breeding plan for commercial growers is to purchase directly from the breeding company hybrid F1 parent females, together with the top-crossing sire line which goes with them (Figure 6.24).

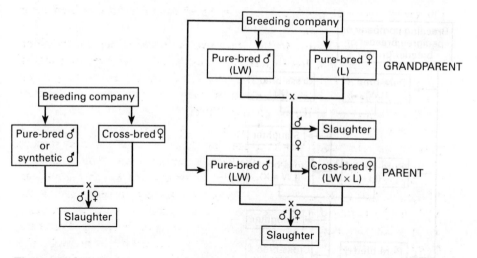

Figure 6.24 Example of breeding scheme for commercial grower purchasing parent stock from a breeding company.

Figure 6.25 Example of breeding scheme for commercial grower purchasing grandparent (pure-bred) stock from a breeding company (LW and L represent Large White and Landrace breeds to serve as examples).

Such a policy must maintain commonality of health status between company and customer, and it is therefore important that stocks from different companies are not mixed on a single customer unit. The commercial herd will usually have, per 100 breeding sows, approximately eight maiden gilts on hand at any one time. In terms of monthly purchases from the breeding company, the commercial grower will need, per 100 breeding sows in his herd, approximately four hybrid gilts and 0.3 sires monthly.

New biotechnology techniques

Genetic engineering and its associated technologies has a reputation for breed improvement well in advance of a firm foundation for the evidence! One may be forgiven for believing that genes could even now be shifted at will from one animal to another, transferring particular genetic attributes into populations where they never before existed. This is so only to a very limited extent, and not yet so for production traits associated with reproduction and growth in pigs. However, DNA recombinant technology has allowed the manufacture of hormones and drugs from simple organisms, as witness r-PST. It is likely that at some time in the future genes *will* be transferable

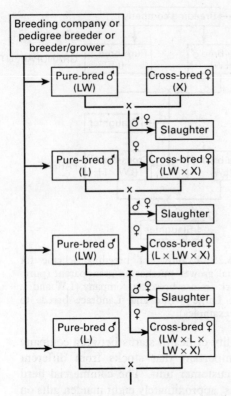

Figure 6.26 Example of breeding scheme for commercial grower maintaining the breeding herd by a home-breeding cross-bred policy.

with facility and genetic change achieved, although the difficulties in multiplying up a single successful gene transfer to an effective commercial breeding population are not to be underestimated.

Meanwhile, there are many other biotechnologies which have already added significantly to the rate at which genetic progress can be made. It has been considered by some that the likely increasing availability of hormones and drugs, through recombinant DNA technology, will obviate the need for genetic selection, characters such as lean-tissue growth rate and reproductive efficiency being open to enhancement by hormone therapy. Such activity would, however, need to deal with factors such as:

1. Public concern about repeat (at each and every generation) hormone dosing of livestock for human consumption.

2. The cost of individual treatment of each generation, as compared to genetic improvement by selection which creates a lasting effect and (being inherent) does not require any further development effort once achieved.

3. Adequate knowledge is still not available to predict the consequences of hormone or immunization therapy, in particular with regard to associated effects. Most production responses are compound, thus the animal requires a complex of appropriate responses to a metabolic change. An example would be the ability of an individual to cope with an elevated lean-tissue growth rate imposed upon a body system developed for a lesser rate of gain, or the ability of a reproductive tract to cope with an elevated ovulation rate.

4. Exogenous therapies are likely to add on to the existing genetic base for the expression of any given character; therefore the higher the base created by positive conventional genetic selection, the better will be the overall effect.

Gene transfer

The insertion of genes from other pig populations or species will greatly help in the importation of particular characters, as only the wanted gene, or gene complex, would be transferred (rather than the whole gene complement). Thus, genes for muscle size might be transferable independently of the genes for stress susceptibility. Again, the gene complexes for enhanced ovulation rate and embryo survival which are characteristic of the Meishan, might be transferred independently of the counter-productive gene complexes for slow growth and fatness also present in that breed.

Transfer of genes with particular disease resistance capabilities which are simply inherited is a particularly attractive notion, and one which is receiving active attention presently, with some degree of success. It appears that a gene offering resistance to swine influenza has been effectively transferred. Obvious other candidates are genes which may enhance resistance to the common respiratory and gastro-intestinal diseases outlined in Chapter 7.

Transgenic pigs are still best created by micro-injection of DNA into the pronucleus of the ovum where the transgene may replace or be in tandem with the resident gene at a single chromosomal site. Transgenic pigs have already been created and shown to contain genes from other species, but there remain many problems before these are fixed and express positive benefits generation after generation. Micro-injection of DNA gives about 0.5% transgenic offspring, and these usually with very low viability. Integration of genes into chromosomes still remains random, whereas an effective transfer system would require close targeting. Targeting assumes detailed knowledge of the gene map. Firing genes into an existing genome complement also raises a problem for the gene previously resident; logic would suggest that the previous resident should be evicted before the incomer is inserted; however, the means for identifying genes and gene constructs, and operating in this way, is now shown as possible, but very difficult

The most likely first contenders for effective gene transfer are characters that are simply inherited, characters that result from major genes with large single effects, or characters which achieve their expression through complexes of genes which are grouped closely together. The operational genes must be known and identifiable. They must be discrete. They must not interact with other genes in other places before they can be fully expressed. This latter can be a severe shortcoming as many genes only operate through cascade effects down-line of the initial metabolic impulsion. The genes must be free of negative associated effects such as reproductive or health failure.

Associated negative effects are a problem which has frustrated such progress as has been made thus far. Transgenic pigs containing implanted genes for growth hormone have tended to suffer from unfortunate associated disorders – sometimes terminally. But it remains likely that the most fruitful way of using genetic engineering to improve growth, lactational and reproductive traits is through transgenic manipulation of the physiological processes by targeting genes involved in the production of hormones, the sensitivity of organs to hormone action, or the control (limitation) of expression of hormones, in all cases resulting in the same effect: elevated levels of response. In the case of successfully transferred growth hormone genes, the transgenic pigs had dramatic increases in lean tissue growth and reductions in fatness, but they also suffered from sterility, lethargy, arthritis, general ill-health and early death.

Genes to be transferred must remain held, generation after generation, in the genome which receives them, and not get lost in the multiplication process. Intuitively, present methodologies for inserting genes would appear to be rather too physically disruptive for one to imagine that a stable gene complement could be happily achieved in consequence of such a shotgun approach.

Perhaps most important of all, and most difficult to ascertain, is that the level of improvement effected by gene transfer must be greater than could be achieved by conventional selection, and reach the commercial level of usage at a faster rate. Recent calculations would suggest that for most (if not all) of our production traits this might be a fond hope, and one which certainly should not be depended upon.

Gene transfer and the creation of transgenic pigs is the most exciting of the novel biotechnologies, but as far as the production traits are concerned it is not yet with us.

Exogenous hormone manipulation

The endogenous hormone system controlling growth and reproduction can be manipulated either by direct administration of exogenous hormone (as for example by r-PST administration) or, more viably, by administration of substances which act against endogenous hormone blockers and feedback mechanisms; so enhancing endogenous hormone production. Such treatments may be relevant to breed improvement strategies because:

1. they may replace conventional selection in areas where the latter is particularly difficult, slow, expensive or inconvenient; or

2. they may interact with some aspect of conventional selection and allow more effective expression of the selected character in commercial situations. Thus, selection for increased ovulation rate may be assisted in its expression in the form of increased litter size if the system is augmented by hormone therapy to increase embryo survival. Another conventional selection:hormone interaction might be that between lean tissue growth rate and cholecystokinin which is a gastro-intestinal hormone acting as a satiety controller and available in the form of an exogenously administrable material.

Embryo transfer and cloning

Embryo transfer is especially interesting in dairy cattle and sheep as a means of both promulgating improved genes and allowing a more effective progeny test of candidates. The first interest in embryo transfer in pigs was for the transfer of genes into nucleus herds of high health status by use of fertilized embryos and so avoiding importation of disease that would come with the whole live animal.

   Once removed from the tract, the fertilized embryo can be divided by biotechnological manipulation of nuclei and in this way clones created. The presently favoured cloning idea is to transfer nuclei from the parent embryonic cell into an enucleated ovum that has had its own nucleus removed. The transferred nucleus then takes over. The parent embryo can continue to create nuclei available for transfer and, if hopes are achieved, the clone size is infinite and relates only to the availability of recipient cells for enucleation. A clone of very many identical individuals offers exciting opportunities.

1. Superior genotypes could be better identified through both more efficient estimation of σ^2E (σ^2A being approximately equal to zero), and a greater

number of relatives being made available to produce information on a particular candidate.

2. The end-product could be created at source, without needing to go through multiplier levels. Thus, the best genes available from dam and sire could be combined in an embryo which is then subdivided to make many identical, highly superior, individuals which could move directly down to the user end of the breeder pyramid. In fact, genetic lag is effectively made negative by this technique; the end-user would get a better product than the average of the great-great-grandparent nucleus!

3. Sire lines which combine excellent carcass characteristics with poor reproductive capacity are expensive to breed because the dams are less efficient. With a combination of cloning and embryo transfer, sire lines can be produced from dams of high uterine capacity. This will increase selection rate within the line, reduce cost and increase the rate of proliferation of the line.

Early life markers

The identification of genotype through measurement of the phenotype is tedious and time-consuming as exemplified by the descriptions of the necessary procedures in the foregoing sections of this chapter. The possibility of identifying the presence and/or strength of a character by assessing a marker in early life has many attractions. Growth rate might well be assessable from the presence, and the strength, of a hormone or enzyme, or some other biochemical element, already present in the body of the newborn piglet. Better still, inherent characters could be tested for by seeking the presence of the required genes or gene constructs in the genetic material itself. Marker genes, those associated with the genes in question but particularly easy to identify, would prove as useful. Early life markers which could be tested for at the very early embryo stage would be the ultimate in efficiency. The embryo could be tested for its genetic excellence by assessment of its genetic complement (DNA fingerprinting) and a decision made at that stage whether or not it should be cloned and multiplied up.

The genetic basis to porcine stress syndrome (PSS) and pale, soft exudative muscle (PSE) has led to a search by molecular biologists for the particular part of the DNA coding which is responsible; thereby increasing the possibility of the separation of the positive aspects of slaughter pigs possessing the halothane gene (muscle depth, carcass yield) from the negative (predisposition to sudden death, reduced meat quality). The halothane test itself (administration of the halothane gas and observation of presence or absence of consequent stress) is not only tiresome, it is also imperfect in its rigour and, furthermore, can only find homozygote (nn) forms, failing (by and large) in its detection of the heterozygote carrier (Nn) (Table 6.6 refers). It is now clear that a biochemical blood test for the particular DNA elements responsible for PSS/PSE has been found. This will allow easier and more rigorous testing for the gene associated with PSS/PSE, and an increase in the likelihood of its elimination. By being able to identify the Nn carrier as well as the nn homozygote, the test will speed complete elimination of the n gene from female lines. The gene may, however, remain used in male lines on account of its positive benefits; it is not yet certain whether or not the positive and negative aspects of the halothane gene can be disassociated. Alongside this potential advance must also be weighed the substantial progress made by conventional selection techniques in pig populations toward the creation of a halothane-negative (NN) pig type which also possesses the positive characters of a high carcass yield of lean meat.

Best linear unbiased prediction

Selection of pigs by individual performance, or with an index including information from progeny or siblings, allows comparison of candidates, and the derivation of breeding value, usually only amongst those in a contemporary group at the same time and place, and preferably reared prior to test in similar circumstances. Increased accuracy is achieved in an individual's assessment if a much wider and catholic variety of relatives from other generations, times, places and pig herds, contribute to the judgement. This demands not only sophisticated computerized record-keeping but also appropriate statistical techniques which combine information of different types and correct for the various confounding factors. Best Linear Unbiased Prediction (BLUP) programmes are complex computer-based calculating systems which utilize various mathematical models to allow accurate evaluation of genetic effects and improve greatly the accuracy of test and the construction of indexes, especially for traits of low heritability such as those associated with reproduction. BLUP procedures are especially useful in that, together with artificial insemination, they allow a large single nucleus population to be distributed over numerous herds and geographical sites – the concept of the group nucleus.

Sex determination

The separation of male (Y) from female (X) sperm still remains extraordinarily difficult. The fertilized embryo can, of course, be sexed as it has possession of its complete gene complement immediately upon fertilization.

243

Embryo transfer and cloning techniques can therefore accommodate determination of sex. Sex determination is much less vital at commercial levels now that the entire male is as acceptable a meat producer as the female. But for the breeder there would be considerable benefit in ensuring all the piglets in a litter generating terminal sires were to be male and all the piglets in a litter generating F1 hybrids were to be female.

CHAPTER 7

DISEASE PREVENTION

Introduction

Disease organisms are present at low levels in most pig units most of the time; these do not always present a problem because immunity builds up naturally. Disease outbreaks (as opposed to the presence of disease organisms), and consequential loss of productivity, can be controlled (i) by maintaining a low level of the organism in the environment and encouraging the build-up of natural immunity, and (ii) by reducing the susceptibility of the stock to the disease with high standards of nutrition, housing and care; or, occasionally, the breeding of resistant stock. Often it is the initial outbreak of disease which is responsible for the greatest loss; after primary infection natural immunity usually builds up in the adult population and the organism can become endemic on the unit without being punitive. In other cases, however, the level of immunity is never adequate to prevent continuing or resurgent problems, especially in young growing stock. Thus a breeding herd may show full resistance to a disease which continually plagues the grow-out unit (for example, mycoplasma (enzootic) pneumonia).

There is always the danger, however, that the natural build-up of immunity may not be complete amongst all animals in the herd, or the strength of the immunity level may fade so that, when initiating factors appear in the system, the disease flares up again. In order to avoid periodic resurgence of disease, or continuing economic loss, it is often more cost-effective to maintain the herd free of the disease and to endeavour to prevent its entry. Monitoring for the absence of diseases is well-established practice in preventive medicine. It has been possible for 20 years to repopulate herds with the absence of diseases such as transmissible gastroenteritis, Aujeszky's disease, swine dysentery, atrophic rhinitis, enzootic pneumonia, mange, streptococcal meningitis, actinobacillus (haemophilus) pleuro-pneumonia and the major epizootics, such as foot-and-mouth disease, classical swine fever and African swine fever.

Some diseases can be completely eradicated from a region by a slaughter policy, as has been the case for swine fever and foot-and-mouth disease, and in some cases for Aujeszky's disease and swine vesicular disease. Equally, some disease organisms, such as those producing swine dysentery and mange, can be eradicated from a pig unit by management and medication procedures. For pig units of high health status, free of the major diseases, it is important that these diseases are permanently excluded, for the clean herd will have no immunity. If the herd does become infected the consequences may be particularly severe. Examples would be outbreaks of mycoplasma pneumonia and actinobacillus (haemophilus) pleuro-pneumonia, in herds where these organisms were previously absent.

Where an organism multiplies and symptoms appear, if non-viral, the disease must first be treated, and then preventive medication and management techniques employed. For those disease organisms which are ubiquitous, the build-up of a natural acquired immunity may be an effective long-term strategy. For *Escherichia coli* diarrhoea infections, the consequences of which are particularly severe for the young pig, the receipt of passive immunity through mothers' milk is an essential element in herd protection. At all levels, good management is paramount in disease control. For example, mycoplasma (enzootic) pneumonia need not be a problem where standards of care are high and other organisms likely to induce secondary infections, such as actinobacillus and pasteurella, are maintained at low levels.

Most pig herds are infected with respiratory coronavirus and with mycoplasma pneumonia, while streptococcal meningitis, atrophic rhinitis and mange should only be present at levels of 5–10% in herds. There is some difference of opinion amongst veterinarians as to the exact levels of subclinical disease which exist in pig herds. It is possible that up to 20–30% of all herds may have, for example, some level of mange and atrophic rhinitis infections. These levels reflect the pervasiveness of different diseases, and the potential for management to prevent clinical occurrences. Complete exclusion strategies are much more likely to be successful with streptococcal meningitis or atrophic rhinitis than with mycoplasma infections. Mycoplasma may be airborne spread for up to 5km.

Density of stocking

Even for perfectly healthy pigs, as space allowance per pig decreases so also does growth rate and efficiency of feed use (Figure 7.1). But, additionally, at higher stocking densities the likelihood of heightened aggression, cannibalism and disease outbreaks rapidly rises, and when this happens the negative relationship between space and growth becomes even worse (Figure 7.2). For pigs likely to be challenged by disease, space allowance needs to be increased if economic growth and feed usage are to be maintained. Benefits of increasing space allowance, and the penalties of decreasing space, are thus greater if certain disease organisms are present (Figure 7.1). By reducing growth rate, disease itself causes stocking densities to rise, creating a downward spiral of enhanced disease level and reduced performance.

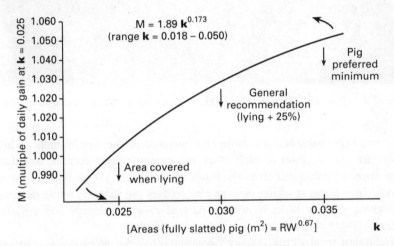

Figure 7.1 If the value given as **k** is multiplied by live weight to the two-thirds power ($W^{0.67}$), the area allowance per pig (m^2) for pigs of any given weight can be calculated. As the value for **k** is increased so that the allowance becomes greater than would merely provide the area a pig needs in which to lie down, growth rate (given as a multiple of the daily gain when **k** = 0.025) will increase according to the equation, $M = 1.89k^{0.173}$. As may be seen, increasing to the generally recommended allowance gives a 3% increase in growth rate, whilst a further increase to the pig-preferred minimum gives another 3% increase in growth rate. The curve illustrated relates to conditions in the absence of disease. If disease is present the curve tilts dramatically, as shown by the heavy arrows. This results in the negative consequences of increased stocking density and the positive consequences of decreased stocking density, both being greater in the presence of disease organisms (interpolated from Edwards *et al.* (1988) *Animal Production* **46**:453, together with other sources).

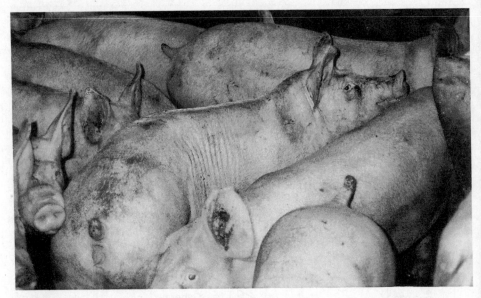

Figure 7.2 Growing/finishing pigs which have been given inadequate space allowance.

Pathogenic organisms become more concentrated in the environment, while the pigs are placed closer to each other and closer to their excreta, so aiding disease spread by droplets from the lungs or mouth, or via particles of faecal material. Increasing stocking density exacerbates pig disease, while decreasing stocking density is an important part of most preventive and curative regimes.

A reduction in stocking density necessitates special attention to ambient temperature. The ability of pigs to keep warm depends on the amount of feed they are eating. Newly weaned young pigs eat little feed and therefore are likely to get cold more easily. At around 6kg live weight a baby pig requires an ambient temperature almost as high as 30°C. This should be combined with sufficient ventilation to keep the air fresh and clean, but with no cold draughts. As the pigs begin to eat and grow, this temperature can be reduced until by 12kg live weight 25°C is usually adequate. For healthy, fast-growing pigs temperatures as low as 22°C can be satisfactory at 20kg. As room temperatures rise from being too cold toward the optimum, the incidences of pneumonias, and pen soiling, reduce. As the temperature rises above the optimum, the incidences of diarrhoeas, tail biting and meningitis increase (Figure 7.3). It is commonly understood that fluctuations in temperature may be as much involved in triggering respiratory and enteric diseases as the absolute temperature.

Lameness

Up to one-third of sows are culled from the breeding herd because of lameness, joint problems, strained tendons or infections of the toe, foot and leg. Some of these follow from abrasions, knocks and sprains from poor housing, slatted concrete floors and inadequate husbandry during the growing period (Plate 7.1a). Abrasions can become infected, while physical injuries debilitate the animals. Pigs reared intensively in high density housing situations with slatted floors may often carry some sort of blemish or injury to the legs or feet. Growth and feed efficiency can be affected and the stress caused could render the pigs prone to other infections. In the breeding herd sows can be restricted in their movement when housed in tethers, stalls and crates for most of their lives, and they may find themselves placed on to improperly prepared or inadequately maintained flooring. Such situations, if they arise, together with unhygienic conditions, provide a ready cause of internal skeletal and muscle damage, lameness, sprains, erosions and cracks of the foot (hoof, toe and heel), necrosis and various limb infections. Slatted floors may become slippery and predispose to the sow's legs sliding forward or sideways, making rising difficult. If the gap is too wide, part of the foot may enter and be damaged.

Mineral and vitamin deficiencies are usually only the cause of leg disorders when there are frank and gross inadequacies of required nutrients; however, hoof tissue can be strengthened by the addition of dietary biotin. But over-

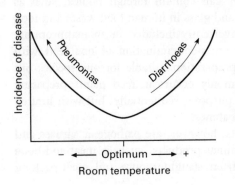

Fig 7.3 When room temperature is at the optimum, the incidence of all diseases will be minimized. As temperature falls below the optimum, the incidence of pneumonias, and also pen soiling, will increase. As temperature rises above the optimum, the incidence of diarrhoeas, meningitis and tail biting will increase.

and undernutrition in respect of the total quantity of food (animals being too fat or too thin) may predispose to lameness. Fat sows are heavy and clumsy and the legs must bear an extra burden; while thin sows are prone to abrasions, sores, abscesses and skin infections. Pigs forced to lie on cold floors suffer more frequently from arthritis and joint problems. Slippery, dirty floors in pregnant sow and lactating sow houses cause the sows to lose their grip and fall. Lameness in the rear limbs, dropped pasterns (Plate 7.1b), weakened tendons and ligaments and foot abrasions are commonly caused by excessively close confinement and slippery floors and also by over-abrasive floors, the sows having difficulty in rising and lying down without incurring injury.

The cause of splay-leg in newborn piglets is not clear, being of multifactorial origin. It may be due to developmental abnormalities in pregnancy that could occasionally be associated with infectious agents. There is also a heritable factor more common in the Landrace breed. Feed mycotoxins have been implicated. Where farrowing floor surfaces are smooth and wet, these are probably the greatest aggravating factor to a splay-leg problem. (Splay-leg may be helped if the hind limb muscles are massaged vigorously regularly during the first day of life.)

Mycoplasmal synovitis (arthritis) is the consequence of invasion of the joints by *Mycoplasma hyosynoviae* causing stiffness in growing pigs. It may be treated with specific antibiotics.

Feed pathogenic contaminants

Occasionally pig feedstuffs may be contaminated with industrial chemicals, pesticide residues (such as organo-chlorines) and the like, which can have deleterious toxic effects. Some feeds can contain foreign bodies, such as plastics in feed-grade fats, and metal and glass in human food waste and food by-products. Medicines may be inadvertently included in an inappropriate diet or, more likely, there may be cross-contamination of one feed with another which contains a medicinal preparation suitable for one species but unsuitable for pigs. This most commonly occurs in feed plants preparing compounds for a variety of different purposes sequentially, but with limited ability to completely separate feed batches.

The most serious feed contaminants, however, are pathogenic viruses and bacteria which may be present in animal products, such as meat-and-bone meal, blood meals and fats, made from slaughterhouse and meat-packing plant wastes. These by-products of human meat preparation are valuable sources of animal feed, especially high-grade proteins and lipids; and animal feed represents a highly satisfactory outlet for an otherwise troublesome – and potentially polluting – by-product. Of the total animal product entering slaughterhouses and meat processing and packing plants, some 60% leaves as appropriate for human use; the remaining 40% goes to the rendering plant and 80% of rendered material returns to animal feedstuffs, having been effectively recycled. The possibility of transfer of infectious agents (for example, BSE and *Salmonellae*) through the food chain by use of inadequately prepared animal by-products in animal feed has received much recent attention in Europe and has resulted in more stringent control of the product processes for meat-and-bone meals and fat rendering. Bacteria which are often thought to require particular vigilance are *Escherichia coli*, *Salmonellae* and *Treponema hyodysenteriae*, all of which may cause diarrhoea in pigs. It may be noted that in a recent survey of pig feedstuffs for the presence of salmonellae, more positive samples were found amongst vegetable protein concentrates than meat-and-bone meals, and in no case was *Salmonella enteritidis* found in meat-and-bone meal. Feedstuffs are only very rarely suspected of containing *Clostridium botulinum* or *Bacillus anthracis*.

Feed mycotoxins

In many parts of the world mould-contaminated feedstuffs and resultant mycotoxin levels in the diet constitute a considerable health hazard. Feed of all kinds may become infected due to faulty production, harvesting or storage of ingredients before compounding and/or faulty storage of the mixed diet on the production unit. Mycotoxins in feed cause reduced appetite, poor growth, poor efficiency of feed use, anoestrus, reduced ovulation rate, reduced fertilization rate, embryo death and resorption, foetal death and mummification, stillbirths and low birth weights. These problems are wholly avoidable if only well-stored, wholesome ingredients are purchased, and storage bins on the production unit are kept in good order and regularly cleaned. Commonly occurring toxins are aflatoxins and ochratoxins, such as from *Aspergillus* species; and zearalenone (F2), T2 toxins and vomitoxins (trichothecenes), such as from *Fusarium* species. At least 17 species of toxigenic fungi may be readily isolated from various contaminated feedstuffs, and at least 40 mycotoxins identified as having important negative consequences upon animal performance. New mycotoxic compounds continue to be isolated in this rapidly expanding area of knowledge. The seriousness of the chronic effects of mould toxins upon growth rate, feed intake and fertility of pigs is only yet partly realized.

Porcine stress syndrome (PSS)

Often, but not always, PSS is associated with the halothane gene and with rapidly growing meaty pigs. Pigs will show an acute temperature rise, muscular rigidity, sudden death and a blueing of the extremities. Muscles will be pale, soft and exudative (PSE). Symptoms will occur after mixing, fighting, transport, hot weather, absence of water-cooling facilities, bad lairage, ill-treatment or any other acutely stressful situation. Less acute stressful situations will bring about changes in hormone levels, reduce the immune response, lower productivity and induce behavioural aberrations.

Stomach ulcers, colitis, rectal prolapse, strictures, atresia ani, hernia and constipation

All of these conditions may be absent in some herds, whilst quite frequent in others. Stomach ulcers are present in most herds and associated with stressful housing conditions, poor environment, poor management, overcrowding, excessive confinement and the use of diets which are high in energy or low in fibre, or finely ground, or pelleted, or all of these. Colitis causes sloppy, but dark, faeces and dirty pigs. Colitis is becoming a frequently seen condition with chronic consequences of loss of growth and efficiency. It rarely produces acute symptoms. Its cause is unknown but was thought at one time to be associated with finely ground and pelleted feedstuffs, especially those with high levels of non-soya vegetable proteins and supplementary dietary oils. It is probable that an infective agent is also involved, proliferating in a poor gut environment where there has been a history of previous digestive disturbance. Rectal prolapse (Plate 7.1c) is more prevalent where there is constipation, straining, coughing and overcrowding. Rectal stricture may be associated with prolapses due to the build-up of scar tissue after a prolapse. Strictures may occur also in young pigs, causing them to be unthrifty and distended. Atresia ani (Plate 7.1d) is a congenital abnormality and usually lethal. Inguinal and umbilical hernias (Figure 7.4) have a variety of causes but sometimes can be linked with particular boars. Constipation is most commonly found in pregnant or newly farrowed sows, where it is usually associated with inadequate water supply, high ambient temperature, inadequate dietary fibre – especially from cereal grains (particularly wheat, which appears most beneficial), excess of dietary fibre from grassmeal and lucerne sources, or an elevated body temperature.

Low numbers of viable sperm

Usually associated with high temperature, malnutrition, ill health, boar over-use or can be characteristic of a particular individual. Boars that are tentative in mating also show reduced conception rates and litter size. This may be an individual characteristic or may be due to inadequate mating pens, especially slippery floor surfaces and insufficient space. Boars should not be reared in isolation and should start mating once a week when 7–9 months of age. At about 12 months they can manage up to five copulations/ejaculations a week

Figure 7.4 Umbilical hernia.

(Figure 7.5). The average age of all the boars on the unit should not be greater than 2 years. Normal adult boars at full work should manage three or four copulations on successive days, followed by 2–3 days' rest. For every 100 sows in the breeding herd there should be about five sows weaned (and therefore available for service) weekly. To this should be added the sows that may return to oestrus as the farrowing rate per sow mated will often not exceed 0.8. This means that for every 100 sows in the breeding herd there should be about six sows served weekly. If 15 matings are allowed for these six sows, and active boars are managing three or four copulations weekly, then five boars are required. At any one moment in time the herd will be likely to be carrying one novice male, making a total boar stock complement of six per 100 sows. Often the rate of serving will not be a smooth six sows per 100 breeding sows per week; there will be peaks and troughs. Peaks can be accommodated by carrying an extra stock boar, or by using artificial insemination. An overall sow:boar ratio of 16.6:1 is not unreasonable.

Boars giving less than 1,000 offspring born alive per 100 matings should be considered suspect. To mitigate the effects of low sperm numbers and poor viability, multiple mating of individual sows, and the culling of the inadequate male once detected can be considered. All matings should be supervised carefully.

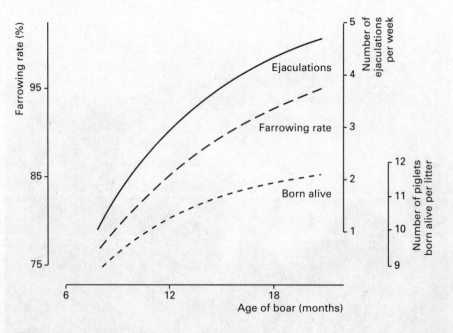

Figure 7.5 Influence of boar age upon limits to frequency of use, farrowing rate and litter size.

Boars used for artificial insemination are usually collected from once or twice weekly, each collection of some 50–100ml sperm-rich fraction of ejaculate, containing about 50×10^9 sperm at $0.5–1 \times 10^9$ sperm/ml, being used to cover 20–30 females, depending on sperm concentration and sperm quality and motility. An effective sperm dose is about $2–3 \times 10^9$ mobile sperm in about 100ml of diluent fluid, with three doses per oestrus, each set 12 or so hours apart. Double insemination routines should be spaced 24 hours apart. The insemination of a sow artificially is illustrated in Figure 7.6.

Sucking piglet mortality

Normal losses of live-born piglets between birth and weaning amount to 12% (one per litter) but can range from 5% to 25%. Common causes of variation are the presence or absence of agalactia or *E. coli*. Agalactia *may* be exacerbated by high-level late pregnancy feeding, but this is not the cause.

Figure 7.6 Artificial insemination of sow showing oestrus.

Failure to obtain colostrum, together with its milk antibodies, predisposes the piglet to early infection and death, especially from diarrhoeas. In circumstances where diarrhoea is the cause of less than 1% of the mortality of live-born piglets, the major cause of neonatal death is likely to be crushing by the sow. Litters of low birth weight (less than 1.25kg average piglet live weight) will have higher levels of mortality associated with starvation and disease. Birth weight influences vigour and survival approximately according to the relationship: percentage survival of total pigs born $= 55W_b^{1.3}$, where W_b is the weight at birth (kg). But it appears that large and small piglets are both equally prone to being crushed. While the sow is retained in a farrowing crate, crushing levels are, in general, significantly reduced. However, there are wide differences between crate designs, and other factors such as sow restlessness have a considerable bearing on the number of piglets trampled to death in early life. Piglets which have not been injected or orally dosed with iron soon after birth will suffer from piglet anaemia.

Cannibalism and aggression

Outbreaks of cannibalism amongst growing pigs can be disturbing and costly. Occasionally one rogue pig is to blame and the solution is to isolate it. More often pigs attacking each other's tails, flanks and ears do so because of management malpractice. The severity of the outbreak and the number of animals in the unit affected can be variable, and the trigger for this aberrant activity is sometimes inexplicable. Cannibalism can sometimes be seen amongst pigs even with a high level of welfare kept in good environments at proper density and well fed. It is most frequent amongst pigs which are on slatted concrete or metal floors, are in barren environments, are over-crowded, are suffering from other diseases, are too hot or too cold, have too frequent or too infrequent a change of air (faulty ventilation), have the relative humidity of the atmosphere too high or too low, are in environments where the concentration of gases (ammonia, carbon dioxide, hydrogen sulphide) is excessive, have a lighting intensity in windowless houses which is too bright and so on. Nutritional deficiencies in terms of feed composition (energy, protein, minerals, vitamins), and especially feed level, are also implicated. Increasing feed fibre and trough space can be helpful, as can docking of the tail. When faced with an outbreak there is often no other satisfactory action than to manipulate the environment on a trial and error basis to try to identify the predisposing factors.

Pigs will fight to assert dominance and determine their place in the social hierarchy. Fighting will therefore occur whenever pigs are mixed into new social groups such as at weaning, moving on to grower or finishing pens, or during transport and lairage. Aggression will cause loss of appetite, loss of growth and inefficiency of feed use; sometimes such losses can be considerable. If fighting occurs near to the time of slaughter, meat quality can be severely reduced.

Fighting amongst sows in communal groups will cause anoestrus, embryo losses, reduced litter size and poor milk yield. Pregnant sows loose-housed are especially prone to aggression and to acts of cannibalism such as vulva biting. Whilst release from sow stalls and tethers may provide greater freedom of movement, sow welfare may be compromised in many loose-housing systems.

Infertility in the sow

Infertility can occur at each stage of the reproductive process: anoestrus, poor ovulation rate, reduced conception, poor fertility, reduced embryo or foetal survival and dystocia. There is no aspect of diet nutrient content, pig-feed level, genetics, housing, management or disease which does not impinge upon the fertility of the breeding sow.

Pigs are normally lactationally anoestral when confined. But this was not so in the Edinburgh Family Pen System of Pig Production, where sows loose-housed, in contact with a boar, and in an appropriate sociological context, conceived within 42 days of parturition but did not wean themselves until 84 days or so post-partum. Nevertheless, in conventional systems weaning is prerequisite for oestrus, although it does not ensure it. An anoestrus rate of 1–5% of sows weaned not coming into oestrus within 14 days of weaning is considered acceptable. Of sows mated, some 85–90% should farrow, or if 5.2 sows per 100 in the breeding herd are mated weekly, 4.4 should farrow weekly (the farrowing rate can be calculated as the number of sows farrowing in the month divided by the number of sows mated in the same period 4 months previously). In a fertile herd about 7% of sows may repeat regularly 21 days after mating (conception failure), 3% repeat randomly (embryonic death), 0.5% abort, 0.5% have womb infections, 1% be not-in-pig and 2% be culled through injury and death. Numbers greater than this indicate an infertility problem (Table 7.1). Primiparous sows should have 10 live-born

Table 7.1 Tolerance levels for sow fertility

Repeat oestrus 21 days after mating (%)	8
Repeat oestrus at more than 21 days after mating (%)	2
Abortions (%)	1
Not-in-pig (%)	1
Culled while pregnant (%)	2
Numbers live-born per litter (primiparous sows)	10
Numbers live-born per litter (multiparous sows)	11
Numbers born dead per litter	0.5
Numbers of mummified piglets per litter	0.2
Live-born pigs not surviving to weaning per litter	1
Gilts ready for mating (per 100 breeding sows)	6
Weaning-to-mating interval (average) (days)	7
Weaning-to-conception interval (average) (days)	12
Litters per sow per year	2.3

and multiparous sows 11. Normally 0.5 will be born dead, 0.2 mummified and 9 (primiparous) or 10 (multiparous) weaned.

The causes of reduced litter size may include an inadequate ovulation rate, poor fertilization or losses of embryos or foetuses. Reduced ovulation may result from nutritional problems, disease, inadequate management or pig genotype. It may be associated with an extended weaning to oestrus interval. Poor fertilization is often associated with *regular* returns to oestrus, and may be caused by infertility of the male or faulty management at the time of mating. Embryo and foetal death is often associated with *irregular* returns to oestrus, and is usually related to disease invasions of one type or another.

The most frequent cause of oestrus failure within 7 days after weaning is poor sow body condition due to inadequate feeding and excessive weight and fat losses in lactation. All lactating sows should be fed to appetite and high feed intakes encouraged. High lactation feed intake will follow from the frequent provision of a high energy diet given wet in a cool environment.

Regular repeats, low levels of abortions, sows not in pig, foetal deaths in late pregnancy, stillbirths and low litter size are usually *not* caused through infective agents unless there is evidence of vulval discharge or mummified pigs. Regular repeats, low litter size and sows not-in-pig point to the need for better mating management and higher feeding levels in lactation and immediately after weaning. Time of mating is crucial to both conception and litter size. Three times mating at 12-hourly intervals usually gives at least one pig per litter more than once-only mating. The period over which the sow is mated should always extend to more than 24 hours. Late insemination means ageing eggs at fertilization and these are prone to death as embryos. Litter size is also related to previous lactation length; increasing the age at weaning from 3 to 4 weeks can increase litter size by one piglet or more per litter on some units (lactation length, 18, 23, 28 days; litter size 9, 10, 11 pigs born alive). After weaning, sows which have lost body condition in lactation are not helped by having their feed reduced when they are already in nutritional stress. Weaned sows should be fed to appetite twice daily until mated. When sow body condition is a cause for concern, they can also continue to be fed up to 3–3.5kg daily after mating. Poor body condition will have a more deleterious effect upon ovulation rate and embryo survival than elevated feed levels. Losses of embryos at implantation can also be caused by cold conditions and inadequate (less than 16 hours) day length at this time. Regular returns to oestrus usually mean a failed conception; irregular returns result from embryo losses such that less than four remain and the pregnancy self-terminates through lack of hormonal support from the products of conception.

Infective agents may cause uterine infections and vulval discharges (Plate 7.1e), and result in embryo losses giving irregular returns to oestrus (the latter often between 23 and 40 days), foetal death and part-resorption in mid-pregnancy resulting in more than one mummified foetus, reduced litter size, increased numbers of pigs stillborn, abortions and sows found not-in-pig. The most frequent agents to cause the above symptoms are mycotoxins, infections of the reproductive tract, parvovirus or Aujeszky's disease (pseudorabies) or *Leptospirae*. Uterine infections and infertility often associate with cystitis, the latter causing bloody urine and sow death in some afflicted units. Parvovirus is widespread throughout all pig populations in the world. Infection may take place at any time resulting in permanent immunity. However, if infection occurs in a herd for the first time during the first third of pregnancy, then there will be an outbreak of the disease with resultant foetal death. The end result is the presence of mummified pigs. Breeding animals at risk should be vaccinated as gilts.

On some units natural immunity to opportunist invaders of breeding sows is built up by challenging new entrant gilts when they come into quarantine. This can be done by exposing them to weaner excreta, or to finishing pigs on their way out of the unit. One routine is the presentation to acclimatizing gilts of piglet faeces from the farrowing rooms and weaner pens, for a period of 2 weeks. This practice is terminated 3 weeks before conception. Faeces (only) may then be offered again for a week in mid- and late pregnancy.

Vaginal discharges due to a range of ascending infective organisms (such as *Pasteurella* and *E. coli* and *Corynebacterium*) entering through the vagina and disrupting the reproductive tract will give painful urination, cystitis, regular and irregular returns to oestrus, reduced litter size, foetal death, abortion and, most confusing of all, may stop the non-pregnant animal from cycling altogether (only determined at term or by pregnancy testing). Foetal resorptions and mummified foetuses are not usually associated with ascending infections, but rather with parvovirus. The normal levels of vaginal discharge are about 1% of sows; when the level rises to 1.5% or more, action is required. Treatment demands cleanliness in the pregnant sow house, antibacterial washing and douching, antibiotic injection of all affected sows and routinely after farrowing and at weaning and a delay of one cycle before remating. The boar may be carrying the organism himself, so he also will require treatment by injection, and by systematic delivery of intramammary antibiotic into the prepuce. A more positive, if rigorous, treatment programme is to cull all discharging animals without any attempt at remating, and to increase the rate of entry of gilts into the herd to compensate.

Internal and external parasites

Pigs may be infected by a range of helminths invading the stomach, the large and small intestines, the lungs and the liver. The consequence is reduced growth and efficiency, diarrhoea and general debility. Sometimes worm infections will cause abortion in sows which are otherwise in good condition. The major external parasites are (i) lice, whose presence will also reduce growth and cause anaemia as well as irritation, and (ii) mites. Both may be dealt with by use of external anti-louse and anti-mite preparations. Mites are a serious menace causing sarcoptic mange. The animals scratch and rub and dark skin lesions form. Growth rate and efficiency can be severely affected.

Modern anthelminths are available which will be active against all internal parasites of pigs, with a possible exception of lungworm. This latter parasite, however, would be unusual in intensively reared pigs since access to the earthworm (an intermediate host) is essential to complete the life cycle. Preparations in the Ivermectin group deal comprehensively with all internal and external parasites. Perhaps the best known internal parasite is the roundworm *Ascaris suum* which infests the small intestine, causing reduced growth weight and diminished carcass quality. The parasite is evidenced by white lesions on the liver (milk spot).

It is impossible to maintain an absolutely parasite-free population due to the contamination and mechanical transmission of worm eggs by birds and other means. It *is* possible, however, by good management procedures, to break the life cycle and hence maintain minimal populations and obviate the requirement for worming. This is particularly so in indoor pig units where dung is removed every 3 days or sows have no access to dung. There is no longer any reason for internal worms, external lice or mites or mange to be a problem in pig herds.

Barrier maintenance against infective agents

The principal carriers of infective agents into the pig unit are pigs. So, first, breeding-company supply herds should demonstrate the absence of specific diseases (the presence of organisms associated with disease and the presence of the disease itself are, of course, different; the latter can be more easily and meaningfully monitored). Secondly, incoming pigs can be immunized against those disease agents which are present on the receiving unit. This latter can be achieved by vaccination (there are vaccines available for erysipelas, parvovirus, pseudorabies, foot-and-mouth disease, *Actinobacillus pleuro-pneumoniae*, atrophic rhinitis, *E. coli*, and many others) or by natural immunity (for example *E. coli*, mycoplasma pneumonia, transmissible gastro-enteritis and parvovirus). The receiving farm should always try to buy its replacement females from the same single dispatch farm of the breeding company. Pigs from different breeding companies are best not mixed together on the same production unit. If pigs must be purchased from different sources, it is important to ensure that the health status of each source is compatible.

Disease control is best achieved by keeping many infectious agents out of the unit. It is now a practical proposition to both establish and maintain herds with the absence of streptococcal meningitis, swine dysentery, atrophic rhinitis, Aujeszky's disease (pseudorabies), transmissible gastro-enteritis, lice and sarcoptic mange (and, of course, swine vesicular disease, foot-and-mouth disease, African swine fever, classical swine fever, leptospirosis, brucellosis and *Salmonella choleraesuis*). Pigs should only be purchased from sources that are able to demonstrate through regular monitoring the required absence of specific diseases. The siting of the herd, however, is important in relation to airborne diseases, such as mycoplasma pneumonia, Aujeszky's disease and foot-and-mouth disease. Naturally, mycoplasma-free herds must ensure their replacement stock comes only from mycoplasma-free sources; but the decision whether or not to purchase stock *with* mycoplasma pneumonia is one that should be made on an individual farm basis. It may well be better to introduce mycoplasma pneumonia-free pigs into herds which have the mycoplasma organism present, but show no clinical signs of pneumonia. While often, in herds where frank pneumonia is clinically present (as is so often the case), it may be better to buy in stock which has been infected with mycoplasma and thus avoid problems of introducing stock with no immunity into an infected environment.

Newly entered pigs should receive a low level challenge of organisms present within the herd to establish immunity levels before they are mated. This can be done in the quarantine quarters, either by an appropriate veterinary vaccination programme or by challenging the pigs with excreta (or other pigs). The point of entry of livestock to the unit must be separate from the point of exit, to avoid the 'walking-in' of contaminated material, such as may happen if a soiled truck carrying pigs to the slaughterhouse contaminates the ramp over which newly delivered replacement females may walk soon after. The unit should operate on a one-way system with no contraflow permitted; this may be seen in the flow (e) to (f) to (g) to (h) to (i) in Figure 7.7. Boundary fences will stop casual visitors.

Additional precautions can be taken with advantage. These might be the provision of a disinfectant spray for vehicles approaching the unit, the passing of bagged feed and equipment through a fumigation room and the provision of canteen facilities in the office (unit-side) to allow personnel to remain on the unit for the whole of the working day. The unit-side office complex could also provide sleeping accommodation for overnight staffing. Between each of the main sections of the unit (farrowing, mating, gestation, nursery (7–20kg), grower (20–60kg) and finishing (60–110kg), disinfectant footbaths may be provided (Plate 7.2b). Separate boots for the farrowing house, with a footbath for each room, can also be helpful (particularly for *E. coli* diarrhoea control).

Disease organisms come in with pigs, lorries or dirty footwear. Swine dysentery and transmissible gastro-enteritis are examples of diseases spread by faecal contamination. Respiratory diseases are spread by infected droplets from the exhaled breath of diseased pigs. Breeding companies maintaining high health status in their stocks will deliver in their own truck and maintain high standards of hygiene. They will take extra precautions to avoid contamination from the farms to which they deliver. Feed supply lorries should unload from outside the ring fence via augers. Stock lorries removing finished meat pigs should load up from a ramp at the perimeter fence. It is better that lorries visiting should do so on Monday morning, after having been cleaned and disinfected and rested over the weekend. Trucks already part-loaded with infected pigs are a serious risk if they come into the vicinity of the unit. Where the unit is integrated with the feed mill and the meat packer, a special fleet of vehicles can be identified for clean units and another restricted to visiting dirty units only.

Pigs entering the unit should be quarantined for 40 days or so in quarters specially arranged for the receipt of incoming stock, which should be as far

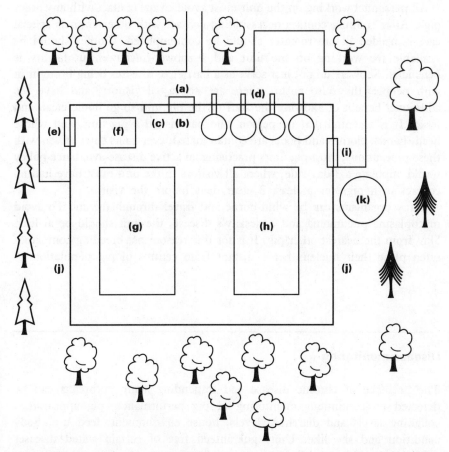

Figure 7.7 Lay-out of breeding/feeding unit for disease prevention: (**a**) and (**b**) points of entry of personnel through shower unit; (**c**) inner offices; (**d**) point of entry of feed through auger; (**e**) point of entry of replacement females; (**f**) quarantine quarters; (**g**) breeding unit; (**h**) feeding/grow-out unit; (**i**) point of exit of finished pigs; (**j**) boundary ring fence; (**k**) slurry lagoon.

distant as reasonable from the main unit, but within the perimeter fence. The more dependable the supply source, the shorter the period of quarantine needed. Before moving on from quarantine to the main herd, the source should be contacted for a report of any disease outbreak which may have occurred in the supply herd during the intervening period. Essential visitors to the unit should be asked to shower in and wear unit clothing, and there should be as little contact as possible with the outside world. There is no evidence that man himself will transmit pig diseases, but diseases can be spread by disease organisms which have contaminated the human body (head and hands) and boots and clothing.

All personnel working on the unit must avoid casual contact with any other pigs. After training courses necessitating contact with other animals special care is needed before re-entry should be contemplated. Clothes should be available for working on the unit and a shower-in/shower-out facility is beneficial. Showers do not just serve as a barrier to diseases being brought in with people, they also ensure there are no casual visitors, and have the additional benefit of allowing the staff of the pig unit to go home clean and fresh. It is essential that no person who has visited a pig unit, and whose head, hands, clothes and boots are contaminated, ever gains entry. Even with these precautions, most pig units practising an active disease-avoidance policy would stipulate a 3-day rule, where all visitors to the unit must have had no contact with pigs for at least 3 clear days before the visit.

Many organisms can be wind-borne and travel through the air. To avoid mycoplasma pneumonia and Aujeszky's disease, the unit should be at least 5km from the nearest other pig. It is for this reason that breeding companies often place their nucleus herds distant from centres of pig population.

Disease monitoring

The presence of chronic disease and impending acute outbreaks can be detected by continuous monitoring of pig performance, pig appearance, coughing levels and diarrhoea levels, house environment, feed use, body condition and the like. Units guaranteed free of certain stated diseases (minimal disease) are obliged to monitor for those conditions, and to inspect at the slaughterhouse for evidence of diseases such as atrophic rhinitis and pneumonias. It is necessary for expected farrowing rates, numbers born, stillbirths, growth rate, vaginal discharges and all other such characteristics to be noted and used for comparing records week-by-week. Interference levels can be set appropriate to each unit. If expected mortality post-weaning is 1%, then when a 3% level is reached the management and veterinarian are required to interfere, find a cause and ameliorate it. Problems for which normal and interference levels are set should relate to the particular diseases of importance on a particular unit and at a particular time. Disease monitoring schemes should be special to the problems of the unit, the requirements of the stockman and manager and prevailing circumstance. Above all, such records should be kept up to date and be seen as useful by the person who is collecting them. Usually the paper record will require identification of date and place, the animal afflicted, the symptoms, the suspected cause, associated observations such as appetite, environment, stock movement and so on, treatment (including the name of the drug and identification of its source) and any subsequent actions that may be thought likely to be required.

Diseases associated with infections of the digestive tract and with diarrhoea

Colibacillosis

Escherichia coli is a ubiquitous organism with many serotypes causing septicaemia in neo-natal piglets, diarrhoea in sucking piglets, diarrhoea in newly weaned piglets and oedema in young growing pigs. There are many pathogenic strains. Septicaemic strains invade susceptible tissue and cause lesions. The enteric strains attack the brush border of the intestine, causing reduced absorption, diarrhoea and toxicity. Young sucking pigs affected show severe loss of condition, acute septicaemia, diarrhoea and possibly death. Antibodies in both colostral and normal milk provide protection to sucking piglets from sows which have built up immunity to specific *coli* strains. This protection is lost at weaning consequent upon the withdrawal of sows' milk. Piglets weaned at 3 weeks of age are therefore particularly susceptible; a susceptibility which is exacerbated as the intestine is simultaneously required to cope with the change from a liquid to a solid diet (especially if the latter has a high concentration of vegetable proteins and carbohydrates). Delayed weaning to 4–5 weeks often reduces the seriousness of the condition. Immunity is readily acquired naturally and may be passed on from the mother to the sucking piglets. Vaccines are available against a range of serotypes, while incoming breeding stock can also be subject to a natural challenge before they are mated for the first time. Gilts can be vaccinated twice before first parturition. This will allow them to build up their own immunity and pass the appropriate protection on to the litter. Failure to encourage the build-up of natural immunity may lead to consistent outbreaks of diarrhoea in the sucking piglets of primiparous gilts brought into the breeding unit from remote locations not harbouring the same serotypes. Within a stable population of pigs, acquiring natural immunity in the normal way, there may still be outbreaks of diarrhoea, especially in the young piglets, consequent upon rapid multiplication of *coli*. *E. coli* will multiply when there is over-stocking, environmental problems or the presence of other disease organisms. Its control is primarily by antibacterials and good management and a high level of cleanliness and disinfection within the lactating sow house. Diarrhoea is seen mostly in young sucking piglets and in newly weaned piglets. In both cases animals lose appetite and weight and may die from dehydration. Animals suffering, together with others in the same pen, should be orally dosed or injected with antibiotic immediately, offered glucose and electrolytes in their water and given protective levels of antibiotic additives in their feed. Control is aided by maintaining the proper environment and ambient temperature, by hygiene and by reducing the density of stocking. For pigs challenged with *coli* organisms, there is a strong inverse relationship between the percentage of pigs showing diarrhoea and the temperature of the weaner accommodation. In the course of the infection, *E. coli* produces massive amounts of toxin under conditions of rapid multiplication and there is also loss of gut motility and destruction of the villi lining the intestinal wall. The diarrhoea can range from mild to extreme. All young animals in stress are likely to suffer; the extent of the debilitation caused by the disease will be directly proportional to the degree of immunity that has been built up in the animal by the time it was infected and the level of prevailing management.

Faulty nutrition predisposes sucking and weaned pigs to *E. coli* attack. This is particularly the case with young newly weaned animals offered diets containing high levels of uncooked starch and vegetable protein sources. Pigs weaned at less than 30 days of age do not have the digestive capacity to deal effectively with raw vegetable carbohydrates and proteins. There may also be additional problems in the gut lining of adverse antigenic reaction to the presence in the gut of particular vegetable protein moieties. Some vegetable protein sources appear to be more severe in this respect than others. Where problems are initiated by dietary factors for weaned pigs between 20 and 35 days of age, *E. coli* organisms invade opportunistically and rapidly, and severe diarrhoea, rapid loss of condition and death may result. Young pigs having suffered from *E. coli* diarrhoea during their early life will never fully recover from the experience and will always grow slower and less efficiently thereafter. The economic consequences of damage by this organism are considerable.

It is helpful if piglets have been accustomed to consuming substantial quantities of solid feed before they are weaned and, furthermore, are not removed from milk as the main nutrient supply until the digestive tract is fully competent to deal with non-milk substrates. Weaning at a greater age and heavier live weight is therefore helpful in the control of post-weaning diarrhoea. As the disease may sometimes be associated with a high stomach pH, probiotic treatments to enhance gastric acidity may be helpful.

Young piglets may also suffer coccidiosis-induced diarrhoea associated with poor hygiene and dung contamination of pen floors.

E. coli may be found in breeding sows as an opportunist invader of the reproductive tract, and injections of antibiotics at farrowing and weaning and appropriate treatment of the boar are required. Outbreaks in the lactating sow house are reduced by effective environmental control, the presence of foot dips at the entrance to each room, and a high level of cleanliness for floors, equipment and personnel.

Campylobacter coli shows similar symptoms to *E. coli* infections and is prevalent in young piglets. Treatment is also similar through administration of antibacterials.

Swine dysentery

Found in many units, it is caused by a *Treponema* spirochaete (*hyodysenteriae*) that infects the large intestine resulting in mucus and blood in a thin light-brown coloured diarrhoea. Morbidity can rise to 100%, with mortality up to 8%. The disease is transmitted by the ingestion of faeces from infected pigs. Antibiotics are valuable in combating the disease, with subsequent in-feed preventive medication. Overcrowding and a poor environment exacerbate the severity of dysentery. Possibilities for live oral vaccines are being explored. Those animals which do not die, and are not effectively treated, become thin and the economic loss is very severe. On breeder/feeder units, the disease can be eradicated by the long-term medication of sows and boars, farrowing into clean quarters, and then moving the (now clean) weaners into fully disinfected and rested growing/finishing accommodation.

Transmissible gastro-enteritis (TGE)

This contagious viral disease causes greenish diarrhoea and a high level of mortality especially in young sucking pigs. When infection enters the unit for the first time, most piglets under 3 weeks of age die. The disease may become endemic with continuing outbreaks of diarrhoea resulting in ongoing growth reductions in fattening pigs. The organism is transmitted through infected pigs, faecal material, vehicles that have been carrying infected pigs, dirty boots and sometimes birds. There is no effective treatment or vaccination programme available. The avoidance of TGE is one of the reasons for carrying out detailed procedures for the prevention of entry of pigs, people and vehicles from unknown sources into the unit. Purchased animals entering into the unit should come from sources known to be free of TGE. Testing for the presence of the TGE is currently compromised by cross-reactions that take place with a common respiratory coronavirus (which appears to cause little harm to the pig).

Epidemic diarrhoea

Rather similar to TGE and, like TGE, caused by a coronavirus. Usually characterized by green-brown diarrhoea, vomiting, dehydration but low mortality, especially in sucking and young weaned pigs. Appetite is reduced and growth and feed efficiency mildly affected. As with TGE, an immunity to the virus develops within the herd but in growing herds there is the potential for an ongoing chronic problem. It can enter the herd via contaminated feedstuffs (such as meat and bone-meal), rodents or infected pigs.

Paratyphoid

The organisms *Salmonella choleraesuis* and *S. typhimurium* cause high mortality from septicaemia and enteritis, mostly in growing pigs, and abortion in sows. It may or may not be associated with diarrhoea. A vaccine is available and *Salmonella* may be treated with antibiotics. Due to the high level of mortality, the best strategy is to ensure the organism does not enter into the unit in the first place.

Porcine intestinal adenomatosis (PIA)

PIA causes necrotic enteritis, ileitis and proliferative haemorrhagic enteropathy (intestinal haemorrhagic syndrome). Both morbidity and mortality rates are variable. Often the bowel is thickened and there is bleeding into the intestine, causing loss of weight and growth. Pigs look thin and do not thrive and the faeces are often loose and dark. When there is a large amount of blood loss the pig itself becomes white in appearance. No treatment is fully effective but the animals are helped if they are given glucose antibacterials and electrolytes. This is a strange disease about which little is known. It tends to strike worst in high health status herds with fast-growing and efficient animals. Little can be recommended with regard to its avoidance. Fortunately outbreaks appear relatively infrequent but mortality is high where there is haemorrhage into the lower part of the intestine. It is associated with *Campylobacter* organisms, which respond to most antibiotics. In-feed medication can be used as a means of prevention.

Specific diseases of the respiratory tract

Mycoplasma (enzootic) pneumonia

Caused by a ubiquitous organism, *Mycoplasma hyopneumoniae*, bringing respiratory distress, with characteristic plum-greyish lesions on the anterior lobes of the lungs, and a sharp, dry (barking) cough when disturbed, with, in the acute form, deep (thumping) breathing and chest heaving. Most pigs in intensive units are affected, but the symptoms and consequences are absent or slight if the management is good. Outbreaks in herds previously unaffected can be severe, causing debility, loss in growth and mortality. Treatment with specific antibiotics can be helpful. Vaccination may be an option. If mycoplasma does not exist in the herd, it is essential to maintain this state by purchasing incoming stock from proven pneumonia-free stock. Attention requires to be given to maintaining the defensive barriers of the pig unit to the proximity of other infected pigs (as, for example, on visiting stock lorries), bearing in mind that the organism can be wind-borne for distances of up to 5km. The organism is usually passed from the sow to the piglet and becomes active at around 8–12 weeks of age. In herds carrying mycoplasma at high levels, doses of antibiotic at this time will help control. It is vital that young growing pigs be kept warm and well-fed, in small groups (less than 12) and at low stocking density if this delicate phase is to be passed through successfully. It has been suggested that there is predisposition to disease when there are more than 200 pigs in a single air space, when less than $0.5m^2$ of *lying* space is allowed per 100kg of pig live weight, where ammonia levels are above 20 parts per million, where dust levels are above $5mg/m^3$ and where the rate of air movement is $<0.2m^3$ per hour per kg of pig live weight.

On units carrying chronic levels of pneumonia the consequences are usually not severe. On well-managed units with excellent environmental conditions there can be no detectable consequences at all. Problems usually arise when mycoplasma pneumonia is associated with secondary bacterial infection, often triggered by stress in the pigs, inadequate environmental control, the presence in the air of dust, mixing pigs, overcrowding and failure to achieve batch throughput and effective disinfection of growing pens between batches. Sometimes acute outbreaks in chronically infected herds may be triggered by the purchase of infected pigs, and/or by the mixing of sources of pigs. Where a problem is likely in the herd, pigs should be treated with in-feed antibiotics in order that the organism can be controlled to subclinical levels.

In badly infected herds for which it seems that the disease level cannot be controlled by effective management, units can be depopulated and repopulated with mycoplasma pneumonia-free stock. In this case, unless the unit is far distant from other pigs and from roads down which pigs may travel on lorries, then a breakdown can usually be expected within 5–10 years. When a herd becomes infected, the consequences may be severe unless it is controlled. In herds where there are chronic levels of the organism, severe disease outbreaks with coughing, loss of condition, and sometimes death are often associated with the introduction of other pneumonia organisms, usually *Pasteurella*, *Bordetella* or *Actinobacillus (haemophilus) pleuropneumoniae* bacteria. It is perhaps in predisposing the pigs to secondary respiratory attack that mycoplasma pneumonia is most threatening to productivity.

Atrophic rhinitis

Usually contracted early in life while sucking. The incidence can be reduced if the farrowing quarters are kept particularly clean and free from ammonia and dust. The condition is particularly exacerbated by dusty conditions and excessive density of stocking. Infected pigs show nasal discharge, sneezing and respiratory distress, reduced appetite, loss of feed-conversion efficiency and reduced growth rate. In the most acute cases the snout will become distorted (Plate 7.2a) and the turbinate bones within the snout will atrophy. Outbreaks are particularly associated with bad housing and poor environmental control. There used to be some doubt as to which organism is responsible, both *Pasteurella* and *Bordetella* organisms usually being present. But work at Compton Laboratory in England has now pointed to a necrotizing toxin from *Pasteurella* as the causal agent. As a bacterial disease, it is controlled through the use of antibiotics. Vaccination of sows is an option. The severity of atrophic rhinitis is usually proportional to the standard of husbandry and the quality of the environment. Pigs should only be purchased from an atrophic rhinitis-free source.

Actinobacillus (haemophilus) pleuropneumonia

When *Actinobacillus (haemophilus) pleuropneumoniae* enters a previously uninfected herd it is often first identified as soft coughing in the growing pigs. The disease can sometimes follow an outbreak of mycoplasma pneumonia. There are vaccines available although the organism has many serotypes which sometimes causes vaccination to be ineffective. Susceptible pigs should be medicated with antibiotics in the feed and, if necessary, in the water, on entry into the grow-out unit. The disease may appear suddenly and there can be high morbidity and acute distress with a sharp cough. Death usually occurs in young and growing pigs, and when exposed to large numbers of organisms they suffer from pleurisy, high temperature and sudden death, going blue at their extremities and having dark blue lung lesions spread throughout the upper lobes of the lungs, together with pleurisy. Because breathing is painful, pigs do not breathe deeply or thump, rather the breathing is seen to come from the stomach (as opposed to the chest). For infected animals, or animals thought likely to become infected (consequent upon an outbreak in the grow-out unit), the first line of control is an injection of antibiotics, followed by antibiotics in the drinking water and, finally, in-feed antibiotic medication. After the initial outbreak has swept through the unit, during which time it may cause a significant number of deaths and losses in growth rate and performance, there will need to be long-term medication and then general control levels in order to 'live with' the disease organism and allow animals to build up adequate immunity. In this regard it is important to reduce the concentration of the organism to as low a level as possible by appropriate cleaning and disinfection of the houses and by operating an all-in all-out batch system, cleaning the houses between batches. The animals should avoid being stressed and should be adequately fed and cared for. Stocking levels require to be reduced and a high level of environmental control exercised. Outbreaks in chronically (sub-acute) affected herds are usually associated with some sudden stressor, such as increased stocking densities or changes in environment. The organism is spread by contact in droplet form, so it is usually brought in with infected pigs. It has a short life outside a living pig. Those units failing to keep a chronic level of infection under adequate control may consider the possibility of clearing the unit and repopulating with stock free of the disease. There is some danger, however, in that an actinobacillus-free population as such would be highly susceptible to infection. The best option is exposure to non-pathogenic strains, with the consequence of subsequent immunity and the absence of disease.

Glasser's disease

The organism *Haemophilus parasuis* may produce similar symptoms to *Actinobacillus (haemophilus) pleuropneumoniae* but mostly also infects the joints and causes meningitis. It is present in many herds but strains vary in their capacity to produce disease. The complaint is open to treatment by antibacterials.

Swine influenza

A virus disease, similar to that in the human, affecting all pigs but with low mortality. There will be considerable loss of productivity during and after the attack.

General control of respiratory disease

Much can be done to ameliorate the effects of endemic respiratory disease in growing/finishing pigs. In addition to high standards of husbandry and cleanliness, the pigs should have greater than $3m^3$ of air space per pig; a total floor area of greater than $0.75m^2$ per 100kg of pig live weight; an absence of draughts (maximum wind speed across pen of less than 0.4m/s); good ventilation and clean air; the correct air temperature (normally 18–22°C depending on live weight); maximum temperature fluctuations of ±2°C; pen sizes of <15 pigs/pen and room sizes of <150 pigs/room.

Diseases associated with reproductive problems in the breeding herd

Parvovirus

Immunity to this virus is obtained either naturally or through vaccination. Infection is a once-for-all event resulting in life-time immunity. Boars and gilts should be vaccinated on entry into the herd or, if a vaccine is not available, by exposing gilts to weaner faeces before mating, or by penning new herd entrants in isolation quarantine pens directly adjacent to pens holding sows due to be culled from the herd, and allowing limited contact between the two groups. When the virus strikes pigs that are not immune for the first time, it has little effect unless those animals are about to become pregnant or have recently been made pregnant. In this case the virus will cause embryonic death, foetal death, piglets weak and small at birth and dead piglets at birth. It is a prime agent associated with infertility and reproductive losses in sows.

Most herds world-wide are infected with parvovirus. Within a herd, infection may range from a quarter of the animals to all of them, which allows periodic flare-ups of the disease in apparently infected herds due to natural immunity not necessarily being universally acquired. The virus is not dangerous in animals a week or more prior to mating, but if the infection strikes pregnant animals severe losses occur. In early pregnancy all foetuses may die and the pregnancy may be lost, the sow returning to oestrus or found not-in-pig. In later pregnancy the foetus becomes mummified, stillborn or stunted. The disease is progressive and therefore varying sizes of mummified foetuses may be seen at parturition. The symptoms of parvovirus infection are similar to those of 'SMEDI' syndrome: stillbirths, small piglets at birth, aborted or mummified piglets, embryo losses, born dead and infertility in the sow, which is usually associated with enteroviruses and for which treatment is similar.

Whilst parvovirus disease is ubiquitous, and its effects traumatic, its influence can be readily curtailed by the use of a simple vaccination programme. Young (susceptible) pigs entering an infected farm may also be protected by deliberate infection through faeces of fattening pigs, or contact with outgoing stock, provided that this exposure is completed by three weeks prior to first mating.

Mastitis

Caused by E. coli, Staphylococcus or Streptococcus and other infection of the mammae, resulting in loss of milk, elevated temperature and reduced piglet growth in early life. Responds well to antibiotics and improved farrowing house hygiene. Mastitis can also be associated with metritis (vaginal discharge) and agalactia (MMA) in the early phase of lactation, with resultant starvation of the sucking litter. Oxytocin can be helpful.

Aujeszky's disease (pseudorabies)

Caused by a herpes virus. When first introduced into the herd, there is high mortality and loss of growth in breeding sows with abortions, mummified piglets and meningitis in sucking piglets. The disease usually becomes chronic leaving animals sneezing and coughing, with reduced appetite and pneumonia complications. If there is a relatively low incidence in a particular country or area, a slaughter policy may be considered worthwhile. It is essential, however, that the borders of the slaughtered-out area are effectively maintained. Vaccinated animals do not suffer from the symptoms of the disease but they may contract a field infection, excrete the live virus, and therefore become carriers. Thus, if a vaccination policy is to be adopted, then, like the slaughter policy, it is best if total across all pigs. Breeding companies should maintain freedom from this major disease and where animals are traded to countries where it is prevalent, vaccination is advised. Vaccination recommendations currently vary between vaccines, countries and veterinary opinion.

After infection the virus hides in cells and the animal is quite healthy with no viral excretion. After stress, or another disease, the animal excretes the virus but itself shows no symptoms. The newly infected herd collapses. In piglets the nervous disease causes loss of appetite, epilepsy or paralysis. Mortality rate can be as high as 90% for pigs of less than 10 days of age, 70% for pigs between 10 and 20 days of age, and 10% for pigs between 20 and 50 days of age. After 60 days of age there is no longer such a serious mortality problem. In growing pigs there is pneumonia, weight loss and a 10% mortality at maximum. The loss of productivity, however, is catastrophic. In sows the disease causes abortion, sterility, foetal death, but no mortality. After an epidemic the situation will clear itself, but during the 2 months after the initial infection up to 75% of all sucking piglets may be lost and 15% of all pregnant sows could abort.

Aujeszky's disease is transmitted mainly by pigs but may also be carried 2km through the air. The most usual source of infection is by reactivation of the latent virus. After an epidemic animals in the unit will progressively lose their own natural immunity. But within the herd there will be latent virus carriers. If these carriers excrete the virus, all the non-immune animals go down. Slaughtering diseased pigs does not solve the problem of latent carriers who would not be destroyed by such a policy.

Vaccines must be used (a) in sows, in order to give protection to the baby pigs via the colostrum, and (b) it is also necessary to vaccinate growing pigs.

It is possible to start an eradication scheme through vaccination and such schemes are now being pursued in Europe. Blanket covering of sows and growing pigs is required in the eradication region, and genetically engineered vaccines ('marked') allow vaccinated animals to be distinguished from carriers. All animals are then blood-sampled in order to identify latent carriers of field infection, which are then slaughtered. If this scheme is adopted over large populations, it should eradicate the disease in the due course of time. It is necessary to combine the slaughter policy with the vaccination policy. Vaccination alone is unlikely to eradicate the disease due to the problem of latent carriers.

Endemic disease is a major problem in spite of vaccination, because ultimate control is through the elimination of the latent carrier. Chronic Aujeszky's disease exists by way of the herd having within it sows which have gained natural immunity and which become latent carriers. New sows can come into the herd not protected, or even young pigs bred within the herd may not develop adequate protection if there has not been an outbreak for some time. Should the latent carrier excrete, animals not naturally protected will suffer. If a commercial herd becomes infected, and is in an area where the disease is endemic, depopulation should only be considered in special circumstances, such as a need to trade with producers in areas where the disease is not endemic.

Porcine reproductive and respiratory syndrome (PRRS)

A new virus-like infection, Porcine Reproductive and Respiratory Syndrome (blue ear disease) struck a small number of European pig units in 1991, and may previously have existed as MRS (Mystery Reproductive Syndrome) elsewhere in the world. The effects were marked, as herds had no immunity. Symptoms are inappetence, lactation failure, respiratory distress, abortion, diarrhoea, poor growth and general distress. These wide-ranging symptoms are probably a result of the disease attacking the immune system of the pig. Thus units with high underlying disease levels will be harder hit, and those underlying diseases will be the ones which become evident after the first phase of the disease has struck. Upon taking its course, the disease subsides and immunity builds up in the infected herd. Piglet losses of 50% may be experienced during the active phase of the disease (around 8 weeks), which will reduce overall annual production by 5–10%. It is possible that the pigs on some units may have become infected mildly and unknowingly, and now carry antibodies to the disease. A vaccine is required.

Leptospirosis

There are various *Leptospirae*, including *pomona* and *bratislava*, which may be associated with digestive disturbances, abortions and reproductive failure. The disease may be associated with poor rodent control and poor hygiene, and may be treated with oral antibacterials.

Streptococcal meningitis

This is a bacterial disease, causing joint infections, pneumonia, brain dysfunction, nervous convulsions and death. If treated early with penicillin,

80% of affected pigs will recover. Individual animals may be seen *in extremis* whilst others show reduced growth rate. During treatment stocking density requires to be considerably reduced and ventilation increased. It is important that the proper temperature in the house is maintained and variation in temperatures is avoided. This distressing disease can be prevented from entering the unit by purchasing stock from a herd with no history of disease over a minimum 5-year period. Its presence, once in a unit, is greatly to the disadvantage of production efficiency, especially during the first 28 days after weaning; during this time in-feed medication is helpful. Streptococcal infections can also cause septicaemia, meningitis, joint-ill, arthritis, joint swellings and sore feet in baby pigs.

Exudative epidermitis (greasy pig disease)

Found in young sucking and newly weaned pigs, caused by a *Staphylococcus* infection of the skin through abrasions. The skin becomes a dirty brown colour and exudative scales and scabs form (Plate 7.2c) and the animal may die from toxaemia and liver damage. It responds in about half the infected cases to early use of antibiotic treatment together with medication of the skin surface with an antibacterial agent. It is often to be found where young pigs have the surface of their skin abraded due to fighting or roughened floor surfaces. It tends to affect individuals, and should not be seen frequently. In herds at risk, the teeth of newborn piglets should be carefully clipped down to the level of the gum, leaving a smooth surface. Disinfection of the farrowing accommodation is required frequently and this is helped by a batch-farrowing system.

Erysipelas

This is caused by a ubiquitous organism but may be readily controlled by vaccination. All breeding animals should be vaccinated twice. If this routine is not undertaken then adult animals are at high risk. Symptoms include sudden death, usually (but not always) septicaemia with raised, purplish, diamond-shaped marks on the skin, chronic joint and heart conditions and inappetence, constipation and loss of productivity.

African swine fever

Transmitted through infected feeds, parasites and carriers. The virus causes high temperature, loss of growth, abortion and general deep malaise.

Mulberry heart disease

Sudden death, with evidence of heart damage, caused by the unavailability of vitamin E and/or selenium, and thus sometimes associated with organic acid-treated cereals and/or high fat diets.

Practicalities of prevention and avoidance

The most common syndromes concerning the day-to-day management of commercial pig units relate to colibacillosis diarrhoea, pneumonias, cannibalism and, most important of all, infertility in the breeding herd.

The following diseases are invariably found on pig units and management and husbandry routines must be paramount in controlling the extent and the severity of them:

colibacillosis; lameness; parvovirus; cannibalism; mycotoxins.

The following diseases are commonly found, but their incidence may be controlled and their effects minimized by good management:

agalactia; anaemia; mastitis; coccidiosis; cystitis; leg weakness;
erysipelas; exudative epidermitis; porcine intestinal adenomatosis;
porcine stress syndrome; internal parasites; mycoplasma pneumonia;

mycoplasma arthritis; oedema disease; parvovirus; rotavirus; stomach ulcers; joint infections; vulval discharges; prolapses.

It is possible to establish and maintain freedom from the following diseases:

streptococcal meningitis; swine dysentery; atrophic rhinitis; external parasites; actinobacillus pleuropneumonia; Aujeszky's disease; transmissible gastroenteritis; African swine fever; classical swine fever; foot-and-mouth disease; mycoplasma pneumonia; swine vesicular disease; leptospirosis; swine pox; trichinella parasites.

Vaccines are being rapidly developed for an ever-widening variety of diseases. At present they may be available for parvovirus, leptospirosis, Aujeszky, erysipelas, greasy pig disease, atrophic rhinitis, *E. coli* diarrhoea, transmissible gastroenteritis, pasteurellosis, and pleuropneumonias. Antibacterials (of which there are some 40 in-feed and more than 100 injectable products listed) are effective against *Streptococcus, Staphylococcus, Treponema, Leptospira, E. coli, Campylobacter, Corynebacterium, Pasteurella, Mycoplasma, Actinobacillus, Haemophilus* and *Salmonella*.

Classical and African swine fever, foot-and-mouth disease and the clinically similar swine vesicular disease, brucellosis and Aujeszky's disease (pseudorabies) are the subject of slaughter policies in the UK.

The price paid by a pig industry for unhealthy pigs is the sum of:

the cost of veterinary visits, veterinary medicines, vaccines, etc. (a recent survey in North America estimates some 40% of growing pigs and 30% of sows are regularly vaccinated, and some 60% of growing pigs and 30% of sows receive antibiotic treatment);

the cost of in-feed medicines (in Europe, more than half the feed offered to weaner pigs contains in-feed disease preventatives or curatives; the average on-cost is conservatively an additional 5% of the feed cost; sustained use of in-feed antibiotics through the point of slaughter can accumulate on-costs equivalent to the value of 1–2kg of pig carcass meat);

loss of product quality and market sympathy (wholesomeness of meat product) post-farm gate;

loss of saleable product through meat processing condemnations;

cost of lost efficiency pre-farm gate due to reduction in sow productivity, diminished growth rate and reduced efficiency of feed use.

It is as difficult to assess the cost of a disease outbreak as to calculate the cost-benefit of its eradication. First there is variation in the severity of disease attack, next there is variation in the level of defence available in the pig and, last, there are a plethora of critical environmental factors which influence the consequences of a disease presence on a unit, from trivial to extreme. It is self-evident that it is best to create an environment that minimizes the effects of unfriendly organisms, to avoid the incursion of overt disease into the unit, and to build animal immunity defences. The investment for such actions can be high, but worthwhile. Repopulation of a disease-ridden unit is drastic financially, but may readily reduce days to 100kg from 200 to 150, reduce feed conversion ratio from 3.0 to 2.7, and increase pigs sold per sow per year from 17 to 23. The international pig health consultant, Mike Muirhead, has estimated that even merely chronic outbreaks of transmissible gastro-enteritis, mycoplasma pneumonia, actinobacillus (haemophilus) pleuro pneumonia, atrophic rhinitis, and mange may each individually add respectively to the time from birth to 90kg about 5, 20, 15, 15 and 10 days. But he also points out that poor management alone may add 30 days.

CHAPTER 8

ENERGY VALUE OF FEEDSTUFFS
FOR PIGS

Introduction

Energy is yielded from oxidation of the carbohydrate, protein and lipid components of feed. Respective gross energy (GE) values are 17.5MJ GE/kg for carbohydrate (starch and cellulose), 39.3MJ GE/kg for fat and oil and 23.6MJ GE for protein. These nutrients can be used by the body in three ways. First, in the context of their existing chemical structures: amino acids from feed to amino acids of pig muscle growth, lipids from feed to lipids of pig fat growth and simple sugars from feed to liver glycogen or to milk lactose. Secondly, nutrients may be converted from one type of structure in the feed to another type of structure in the pig; thus, the end-products of the digestion of feed carbohydrate can take part in the anabolism and retention of pig protein or pig fat. The third and major use of energy-yielding nutrients is, however, for just that: the yielding of energy that is needed to fuel the metabolic activities of the life and work of the pig, and which ultimately leaves the body as heat.

The primary carbohydrate energy source is starch, yielding up its energy after enzymic digestion in the small intestine and absorption in the form of simple sugars. More complex non-starch polysaccharide also has some part to play, especially if the feedstuff is high in fibre; but in this case much of the digestion is by way of bacteria in the caecum and large intestine, and the end-products are volatile fatty acids which are less efficiently used. Protein yields energy after absorption as amino acids and then through the (rather inefficient) process of gluconeogenesis. Dietary fats may be absorbed in their constituent parts of glycerol and free fatty acids.

Feed energy is used by pigs (i) for the basic maintenance of normal body processes such as muscle movement, digestion, respiration, blood circulation, the renewal of worn-out body tissues and the recycling of existing body tissues, (ii) for driving the manufacturing activities of milk synthesis, reproductive effort and protein and lipid growth, (iii) for maintaining body temperature in the face of a cold ambient environment and (iv) energy is retained in the body products of secreted milk, the foetal load and lean and fatty tissue growth. All the energy but that which is retained (iv) leaves the body as heat. The more heat created in the work functions of maintenance and manufacture, the less likely will be the need for energy to keep warm; energy for cold thermogenesis is therefore only needed for that specific purpose when 'by-product' heat from other purposes is not available to do the job.

Not all the energy in a feed is available to be metabolized for purposes of maintenance and production. Some energy-yielding ingredients are not digested and pass out in the faeces, while other energy may leave the body as gases or in the urine. The total energy-yielding potential of the feedstuff may be determined by its complete oxidation and by measuring the heat directly generated in a calorimeter (usually an adiabatic bomb calorimeter). This is the gross energy in feed (GE_i). After subtraction of the undigested energy associated with the faeces (GE_f), the digestible energy (DE) can be determined:

$$DE_i = GE_i - GE_f$$

(Equation 8.1)

Metabolizable energy (ME) is estimated as the DE less the energy found in the urine (GE_u), and also less the energy that may be found, mostly as CH_4, as gaseous escapes – mainly from the large intestine via the anus:

$$ME_i = DE_i - (GE_u + GE_{gas})$$

(Equation 8.2)

The metabolizable energy is now available to be metabolized and would usually be handled by the animal in the order (i) maintenance, (ii) cold thermogenesis (if needed), (iii) driving the manufacture of products and deposition within products. When feed intake is particularly low a reverse order can apply; previously deposited body products – especially fatty tissue but also protein tissue – can be catabolized to yield energy essential for maintenance of the basic life processes.

A schematic representation for energy use in the body is shown in Figure 8.1. Table 8.1 shows some example values which might pertain for each of the various aspects of energy use.

The ultimate energy value of a feedstuff will depend upon both the proportional contents of the various energy yielding nutrients and the end-product destinies. The latter are dependent (as may be apparent from Table

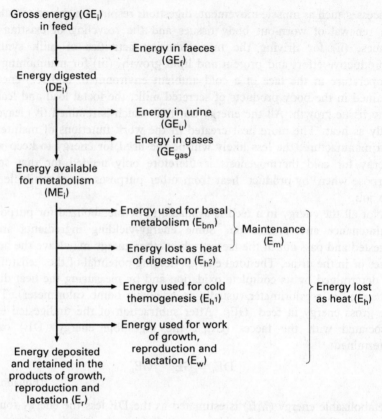

Figure 8.1 A schematic representation for energy use in the body of the pig.

8.1) not only upon the feedstuff itself but also upon the balance of metabolic activities pertaining in the pig at the time of the feedstuff's utilization.

Whilst it will be argued that the most useful pig-independent descriptor of a feedstuff is likely to be the digestible energy (DE), it is germane at this point to consider more deeply the way a pig can handle its energy resources, and how these can be measured.

Measurement of the energy economy of pigs

Heat (E_h) output can be measured directly but is best measured indirectly as the difference between the metabolizable energy and the energy stored in products (E_r):

Table 8.1 Example values for various aspects of the energy economy of a 60kg growing pig gaining about 850g live body weight daily. All values are given in terms of MJ per day[1]

$$GE_i = 34$$
$$GE_f = 8$$
$$DE = 26$$
$$GE_u = 0.8$$
$$GE_{gas} = 0.2$$
$$ME = 25$$
$$E_{bm} = 8.5$$
$$E_{h2} = 1.5$$
$$E_m = 10$$
$$E_{h1} = 0.4$$
$$E_w = 6.7$$
$$E_r = 7.9$$

[1] E_r and E_w have been calculated as 160g protein retention at 3.8MJ and 105g lipid retention at 4.1MJ, with the associated work energies being 5 and 1.7MJ respectively.

$$E_h = ME_i - E_r$$

(Equation 8.3)

ME in turn is measured as the difference between the energy intake and the energy losses in faeces and urine (the energy losses in gas are usually taken to be trivial):

$$ME_i = GE_i - (GE_f + GE_u + GE_{gas})$$

(Equation 8.4)

The energy retention (E_r) is measurable by comparative slaughter; the total body content of lipid and protein at the end of a period over which the metabolizable energy intake has been measured is compared to the protein and lipid content at the beginning of that period. The difference between these two masses is the retention of protein and lipid pertaining to the period for which the metabolizable energy intake had been measured. Knowledge that protein contains 23.6MJ per kg and lipid contains 39.3MJ per kg allows simple calculation of the energy retained. Of course, a relatively long period of time must elapse if this difference measurement is to be accurate, and this means painstaking measurement of the ME intake in addition to the non-trivial exercise of analysing the pigs' bodies at the beginning and end of the experiment.

In order to overcome the difficulties of the comparative slaughter technique, heat output may also be estimated from a knowledge of the oxygen consumed and the carbon dioxide, methane and urinary nitrogen emitted. It is, of course, in the oxidization of feed nutrients that heat energy is created, with the consumption of oxygen and the release of carbon dioxide in the process of respiration. Measurement of heat production in this indirect way requires quantitative collection of oxygen and carbon dioxide. Accurate measurement of gaseous exchange is sufficiently complex itself for the comparative slaughter method to recommend itself to many research workers. It does appear reasonable to assume that energy ingested (GE_i), but not subsequently found in faeces (GE_f), urine (GE_u) or animal products (E_r), must have left the body as heat (E_h), as Figure 8.1 confirms.

Measurements of heat production do not, of course, distinguish between energy that has been used for maintenance (E_m), for cold thermogenesis (E_{h^1}) or for work (E_w). Experimentation at a thermoneutral temperature may assume cold thermogenesis to be zero, while for animals which are creating no products heat losses may be assumed to measure maintenance. Zero product formation may be considered to occur when no feed is ingested; in this case maintenance energy is derived from the catabolism of body tissues (mostly lipid). A more rational interest might be in the amount of energy being used for maintenance (E_m) when the body is neither anabolizing nor catabolizing body tissues; that is, when it is in stable energy balance. This will occur at some given (unknown at the time of measurement) nutrient intake, and may be approached by the use of graded levels of nutrient intake in the experimental protocol. The efficiency of use of energy for maintenance is higher than the efficiency of use of energy for growth. Thus, the point of zero energy retention and of the true measurement for maintenance will coincide with a change in the slope for the regression of energy retention on metabolizable energy intake (Figure 8.2).

While for a mature and non-productive animal the notion of maintenance is credible, the idea that maintenance can be a process which looks after the basal metabolic activities independently of all the other energetic processes going on inside the body of a productive pig may be seriously challenged.

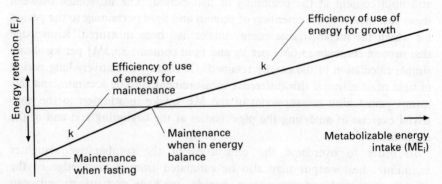

Figure 8.2 Showing the relationship between energy intake and energy retention, and the measurement of maintenance (E_m).

The whole concept of maintenance at the point of energy balance is difficult to accept as a rational way of estimating non-productive energy expenditure in a rapidly growing pig on full feeding or in a lactating sow. For purposes of quantifying the energy economy of productive pigs, there is a view that maintenance should be considered not so much as an indicator of a basal metabolic requirement, but more simply as a residual cost accruing to the system as an unavoidable fixed loading. The comparative slaughter technique is particularly adept at measuring maintenance with this latter idea in mind. Groups of animals are each given graded levels of nutrient intake. Some animals on the experiment are therefore slow-growing, and others progressively more rapidly growing. The analysis of the carcasses will yield graded levels of protein retention (Pr) and lipid (Lr) retention from which the energy deposition in products can be calculated (using the values 23.6 and 39.3MJ per kg as the energy contents of protein and lipid respectively). Expression of available energy as metabolizable energy from knowledge of faecal and urinary losses now allows the construction of a regression relationship, the constant for which must be maintenance. The regression technique is used here to divide up the metabolizable energy intake into its constituent parts:

$$ME_i = E_m + k(E_r)$$

(Equation 8.5)

The system may be expressed graphically (Figure 8.3).

The slope of the line gives the efficiency of the use of energy for retention of protein + lipid; that is, for growth. The same set of data may be used in a multiple regression analysis, and this may differentiate between the efficiency with which energy is used for protein deposition and the efficiency with which energy is used for fat deposition – it being implicit that the graded levels of energy intake will also have resulted in different levels and proportions of protein and lipid retention:

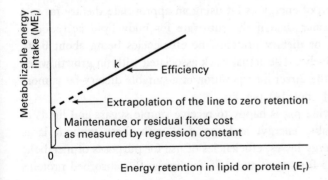

Figure 8.3 Determnation of 'maintenance' requirement by regression of ME_i on E_r.

$$ME_i = E_m + k(Pr) + k(Lr)$$

(Equation 8.6)

Energy costs of maintenance are invariably related to the live body mass. This is logical because maintenance costs include the work of movement (about 20%), pulmonary and cardiac activity, and all the vital processes of tissue turnover and refurbishment. The relationship, however, is such that per unit of mass a lighter animal will have a higher requirement than a heavier animal, possibly it used to be thought because of the changing relationship between body surface area and body weight; thus the frequent use of an exponent of less than 1 (often in the range 0.6–0.8). For example:

$$E_m \text{ (MJ/day)} = 0.440 \text{ LW}^{0.75}$$

(Equation 8.7)

where LW is the live weight.

The amount of heat lost from the body is also influenced by the digestive processes themselves; that is, the work of obtaining the metabolizable energy from the dietary gross energy. The heat of digestion (E_{h^2}) results from energy costs of mastication, its propulsion through the tract, heat losses from enzymic digestion and also from microbial fermentation; E_{h^2} usually amounts to about 5% of the total GE and, of course, is highly dependent upon feedstuff type. The more complex and demanding the digestion for the particular nutrient concerned, the more energy that will be lost in consequence of making it available. For example, simple sugars and lipids are much less energy demanding in their digestion than proteins and complex structural carbohydrates. The influence of feed type upon heat loss does not, however, stop there. Work heat tends to be described by single values either for protein accretion or for fat accretion. Thus, 31MJ of work is required for each kilogram of protein deposited ($k_{Pr} = 0.44$), and 14MJ for each kilogram of lipid deposited ($k_{Lr} = 0.75$). However, this cannot be realistic as the efficiency of deposition of energy as fat using an appropriate dietary fat as a substrate is much greater than if the substrate for body lipid accretion is dietary carbohydrate, or dietary protein (the efficiencies being about 0.90, 0.80 and 0.70 respectively). The actual work associated with fat growth when this is achieved from the direct incorporation of a suitable dietary fat is more probably around 5MJ (not 14) per kg.

The body of a growing pig is happy to receive simple sugars and lipids as sources of metabolizable energy, and can deal with these without large consequential heat energy losses, efficiencies of use for purposes of metabolic work being about 0.85 for sugars and 0.75 for lipids. But absorbed proteins and volatile fatty acids (the latter from fibre digestion) incur greater costs in their metabolism (efficiencies here being around 0.65 and 0.50). To yield energy from protein deamination requires a considerable input of work energy itself (about 5MJ/kg protein deaminated), and there are further losses of energy in the urine consequent upon deamination and the formation of urea (about 7MJ/kg protein deaminated). So, of the gross 24MJ/kg in the absorbed protein, only some 12MJ are realistically available for use as energy. The heat losses occurring from the yield of the given amount of available energy are therefore much greater from a feed whose energy is mainly in the form of fibrous carbohydrates and protein, than with a feed whose energy is mainly in the form of simple starches, sugars and (especially) lipids. The carbohydrate fraction of wheat, for example, is about 85% starch and 13% non-starch polysaccharide; respective values for wheatfeed are 32% and 59%. Lupin meal has negligible starch, while more than 80% of the carbohydrate is as non-starch polysaccharide. Differences in the effective energy values of these three feeds would only be evident consequent upon a study of the complete energy economy of pigs being given them.

The possibilities for generalization

It can be taken that as the urinary losses of energy in the form of nitrogenous compounds (urea) comprise normally about 3% of the digestible energy, and gaseous losses about 1%, ME will then be around 0.96 of DE. High rates of tissue turnover and greatly enhanced protein intakes, as is often now usual, would lead to the expectation of urinary losses being around 4% (rather than 3%) of the DE; and higher if the protein is not well balanced, or is fed in excess. High-fibre diets, being degraded by microbial action in the hind gut, will also enhance gaseous losses. It is unusual, however, for the ME of pig diets to be much different from 0.95 of the DE, and as gaseous losses are rarely measured, the standard value of 0.96 is commonly used.

The transfer of ME to the absolute amounts of net energy (NE) available for maintenance and production is more difficult to handle than with a commonly agreed multiplier. The efficiency of transfer of ME to NE is dependent upon the metabolic activity of the animal; for example, the ratio of maintenance to growth, and the ratio of lipid to protein accretion in that growth. This is why often NE is constrained to be measured as 'NE for fattening', and estimated in mature animals whose product deposition is in the form of the single entity, lipid. The efficiencies of use of ME for fattening are in the region of 0.75 for starch, 0.90 for oil and 0.65 for protein. That part of diet crude protein not used for accretion in lean tissue, but rather for the creation of work energy, incurs a high work cost of deamination itself, with the result that the NE yield is only about half of the ME apparently available. On the other hand, the direct transfer of lipid from diet to tissue can result in the net energy yield being close to 100% of the metabolizable energy. Net energy yield from carbohydrate ME is highly variable, the efficiency of transfer ranging from 0.5 to 0.75, depending principally upon the proportions of fibrous to starchy carbohydrates. This is because of the dramatic difference in the efficiency of utilization between volatile fatty acids and glucose for the creation of work energy. Overall, the NE of a conventional pig diet given to a conventional growing pig is around 0.65 of the ME. However, there is sufficient variation in the value for any generalization of this factor to be of limited worth.

Energy classification of pig feedstuffs

Expression of energy value in feed by measurement of digestible energy (DE) takes the most simple line that the estimate does not concern itself with efficiency of absorbed energy utilization, but only with the difference between GE ingested (GE_i) and GE excreted from the body in the faeces (GE_f). Thus the DE measurement cannot give a totally accurate picture of energy value as perceived by the animal. Especially, DE fails to account for feed-dependent energy losses in urine and gases which occur. DE further fails to register the differences in the heats of digestion of various feedstuffs. Lastly, DE takes no account of the important distinctions between the efficiencies of utilization of digested sugars, amino acids and volatile fatty acids as energy generators.

The DE measurement does give, however, a straightforward view of the level of the apparently digested energy in a feed, and does so in a way that is independent of animal effects. It is therefore realistic to use DE as a descriptor of the energy value of a feed in a range of different circumstances and at different times. For example, the DE of a diet, in contrast to the ME, is not confused by the quantity and quality of dietary protein supply in relation to the ability of the pig to accrete products in the form of lean tissue or milk protein.

It may also be readily accepted from the foregoing sections that net energy (NE) is influenced to such a great extent by the activities of the animal itself that a single NE value cannot be accurately or properly ascribed to a feedstuff.

The determination of energy value of feeds by DE measurement (rather than GE, ME or NE) has been found particularly useful because it is quite readily undertaken with the use of simple experimental techniques (in contrast to live animal calorimetry), and it has also tended to yield adequately consistent relationships with achieved animal performance. This latter has much to do with pig diets usually comprising a conservative and restricted range of ingredients, so the coefficients relating DE to ME, and ME to NE are relatively stable. But these relationships are different for pigs in different physiological states, and will not hold over a more liberal range of diet ingredients. The digestibility and use of complex carbohydrate components differs between young and adult, when fibrous non-starch polysaccharide replaces dietary starch, DE values being some 10% higher for sows.

The major practical limitation of DE is indeed to do with its inability to cope with the consequences of elevated levels of high grade dietary lipid on the one hand (whose value it underestimates for direct lipid accretion), and low grade dietary fibre on the other hand (whose value it overestimates). The former effect may be limited for pigs of modern type which are not rapidly fattening whilst growing, but might be expected to be of interest to the lactating sow whose milk contains 8% of lipid. The latter effect most definitely requires some accommodation if the dietary crude fibre level rises above 6%, as may well be the case when pigs are fed by-product or forage-based diets. Apparent digestibility merely relates to disappearance, saying nothing of the nature of the end-product disappearing nor its usefulness. Thus similar DE values would be erroneously given to simple sugar energy disappearing from the ileum as glucose as to energy derived from complex carbohydrate energy leaving the caecum as volatile fatty acids, or even the anus as gaseous methane escapes. The change in the site of digestion of dietary components from the small intestine to the large intestine will diminish the relationship between dietary DE and animal response. Workers at the Rowett Research Institute in Scotland have proposed the use of correction factors for fibrous feeds, bulky feeds and otherwise indigestible feeds. This correction is based upon the proportion of energy disappearing from the large, rather than the small intestine. Thus, whilst 15MJ/kg of dry matter (DM) of both barley and grass are digested from the gut, the ascribed DE value for grass is 3MJ less than barley at only 11MJ, because nearly half of the digested energy is absorbed from the large intestine whereas the equivalent value for barley is less than one-fifth. A table of some suggested corrections is given (Table 8.2).

Table 8.2 DE (MJ/kg DM) and corrected DE to allow for reduced utilization of energy from hind gut digestion of feedstuffs (Rowett Research Institute)

	GE	DE	Proportion of DE arising from microbial digestion in the hind gut	DE (corrected)	DE/DE$_c$
Barley	18.6	14.9	0.16	13.8	0.93
Cooked potato	17.0	15.5	0.23	14.1	0.91
Raw potato	17.0	13.8	0.48	10.8	0.78
Fodder beet	15.5	13.2	0.21	11.4	0.86
Raw cabbage	18.0	11.9	0.33	10.1	0.85
Young grass	21.2	14.8	0.40	10.9	0.74

These caveats should not unduly detract, however, from the use of digestible energy (DE), which remains the most effective means of classifying individual feedstuff ingredients and mixed feeds in order to allow predictable pig performance, monetary evaluation of diets and diet ingredients, proper incorporation of ingredients into mixed diets at the correct levels to give a balanced feed and the ultimate rationing of compounded feeds to pigs in the form of the final total dietary allowance.

While DE is the chosen level for *feed description*, it is self-evident that when it comes to determining the ultimate utilization of energy by the animal, and predicting the animal's performance, it is the NE that gives the only final and definitive statement of energy balance. NE is dependent upon the nature and activities of both the animal and its environment, as well as upon the intrinsic energy-yielding characteristics of the food. The calculation of NE yield from DE supply cannot be satisfactorily achieved through a series of fixed or semi-fixed conversion factors. The only effective route for the determination of NE is by dynamic estimation of nutrient utilization through the medium of a simulation model for prediction of nutrient flow and production response within the body. Such models will be discussed at length in Chapter 19. But the first requirement for any model is a (reasonably) universal description of the feed energy content which will apply to a range of conditions and pigs. This is provided by the measurement of DE.

Determination of digestible energy by use of live animal studies

The determination of DE is achieved by feeding a measured amount of the given feed material to a pig, crated in such a way that the faeces can be separated from the urine, quantitatively collected, and then preserved to avoid energy losses after excretion (Figure 8.4).

Sample collection may be undertaken with the help of a marker, the concentration of marker giving a means of determining the proportion of the whole represented by the sample. Markers themselves have problems: indigestibility alone does not ensure exactly proportional excretion due, among other things, to differential flow rates. In any event, the need is not great as crating is no particular problem, and total faecal collections for 1 week or so present no particular hardship for either pig or scientist. It is essential, of course, that animals are adequately habituated to the material being given (habituation can take from 3 days to more than 3 weeks) and for the collection to last long enough to avoid the problems of end-effects. At Edinburgh, a continuous period of 10 days is usually employed for total faecal collection under acid. (In the case of specific work on fibre or lipid digestibility, which would be disrupted by the acid medium, the material is collected fresh daily and frozen.)

An average level of feeding is usually chosen. Too low a level will cause an underestimate of digestibility due to the endogenous faecal losses, but this effect is small if the feeding level is at or above maintenance. Usually 1.5–2.0 times the maintenance level is chosen, as this may be expected to be readily eaten.

High levels of feeding (at least more than three times maintenance) – an important consideration in ruminant balance trials where rate of passage may significantly affect digestibility – does not appear greatly to affect digestibility of energy determination in growing pigs. However, whilst this latter may have been exhaustively studied and well proven with young pigs given concentrate diets, the matter is inadequately researched for adult pigs, for pigs given diets high in roughage or for pigs given diets of low DM or based on liquid by-products.

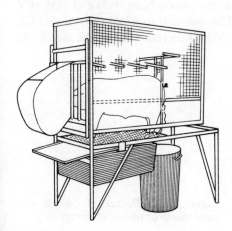

Figure 8.4 Metabolism crate for balance studies. Faeces fall into the bucket and urine into the box. Each container holds dilute acid. The feed trough reduces spillage but there is also a tray below. The floor of the cage on which the pig stands is of mesh. The crate is suitable for pigs of 20–100kg and of male sex (from Whittemore, C T and Elsley, F W H (1976) *Practical pig nutrition*, Farming Press, Ipswich).

As particle size has some effect upon digestibility, most probably through the simple mechanism of surface area exposed to enzyme attack, it is essential that an adequate, but not excessive level of comminution of the feedstuff is achieved (Figure 8.5).

For a 40–50kg crated pig, 1,500g of feed given daily in two separate meals might be appropriate. GE determinations are carried out on both ingested (i) and faecal (f) material. Where GE_i is the MJ of GE ingested daily, and GE_f is the daily faecal excretion, then $(GE_i - GE_f)/GE_i$ gives the digestibility coefficient which, when used with the GE concentration of the ingested material, will give the MJ DE per kg (fresh or dry) of the diet in question (see Table 8.3).

Increased accuracy may be gained from replicating the treatment four or more times and/or by feeding three or more levels. This latter would also allow an additional check against a level of feeding effect (Figure 8.6).

When determining the digestibility of the energy of a diet ingredient, additional complications arise in inverse proportion to the normally expected level of inclusion of the ingredient in a balanced diet. The DE of a cereal of reasonable protein level could be determined by feeding the material alone, or with a small supplement of minerals and vitamins. However, it is inconceivable that physiological normality would pertain if fish meal or rice bran was given alone – let alone soya oil or tallow. Usually, then, the DE of individual feed ingredients is determined within the environment of a diet including other ingredients. The procedure in this event is to manufacture for the test ingredient a balancer meal. The DE of the balancer meal is determined as a first step, and the DE of the balancer plus test ingredient as a second step. The difference between the two estimates gives the contribution of the test ingredient. Suppose the GE of the test (T) ingredient per kg of dry matter (DM) is 22.0MJ/kg DM, and the GE of the basal balancer (B) is 18.0MJ/kg DM. First, 2kg DM of B is given, providing 36MJ of GE per day. Of this, 7.2MJ GE daily is recovered in the faeces. Next, 0.5kg of DM of T is given, in addition to the 2kg of B. This amounts to 47MJ of total diet GE (36 from B + 11 from T). Of this, 8.3MJ GE is recovered in the faeces. As

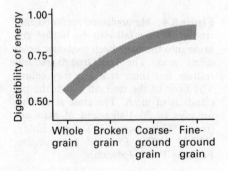

Figure 8.5 Influence of particle size upon the digestibility of energy.

Table 8.3 Example energy balance for determination of digestible energy (DE)

Intake (kg/daily)	= 1.500kg fresh feed at	
	0.870 dry matter (DM)	= 1.305kg DM
Gross energy of feed	= 17.65MJ/kg DM; GE_i	= 23.03MJ
Faeces (plus dilute	= 1.179kg slurry at	
preservative)	0.212DM	= 0.250kg DM
Gross energy of		
faeces	= 15.56MJ/kg DM; GE_f	= 3.889MJ
$(GE_i - GE_f)/GE_i$	= 0.831	
DE value of feed	= 0.831 (17.65)	= 14.67MJ DE per kg DM
	or	= 12.76MJ DE per kg of fresh feed

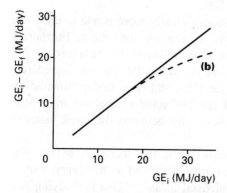

Figure 8.6 Estimation of digestibility using three levels of intake. A negative effect of intake upon digestibility would be shown by curvilinearity (**b**). The slope gives the digestibility coefficient (Y/X = 0.80). The constant term should be equal or close to zero.

7.2MJ was recovered from B, then the energy recovered from T was (8.3 − 7.2) = 1.1MJ. It can now be concluded that, as 11MJ of T were given, the digestibility of T is $(GE_i - GE_f)/GE_i$; (11 − 1.1)/11 = 0.90. The DE of T is therefore (0.90 × 22.00) = 19.8MJ/kg DM.

When the test ingredient is added into, rather than on top of, the basal meal, then the calculation must, of course, be weighted by the relative contributions of the basal balancer and the test ingredient to the final diet; bearing in mind that it is not the proportional contribution to the DM that is of concern, but the proportional contribution to the gross energy:

$$x(DE_T) + y(DE_B) = DE_D$$

(Equation 8.8)

Suppose the GE of the test (T) ingredient is determined as 24.0MJ/kg DM and the GE of the basal balancer (B) as 18.0MJ/kg DM, and the two materials are mixed together in the DM proportions of 0.33 and 0.67: the GE of the mixed diet (D) is 7.92 + 12.06 = 19.98. Of this, 0.396 is from the test ingredient and 0.604 from the basal balancer. On the first step, 36MJ GE of

B is given daily and 7.2MJ GE recovered in the faeces. $(GE_i - GE_f)/GE_i = 0.80$; the digestibility coefficient (DE_B). On the second step, 40MJ GE of the mixed diet is given daily and 6MJ GE recovered in the faeces. $(GE_i - GE_f)/GE_i = 0.85$; the digestibility coefficient (DE_D). Where $x = 0.396$ and $y = 0.604$; $DE_B = 0.80$ and $DE_D = 0.85$; and DE_T is unknown; then the equation above solves to:

$$0.396 (DE_T) = 0.85 - 0.604 (0.80)$$

(Equation 8.9)

$$DE_T = 0.926$$

(Equation 8.10)

The digestibility of the test ingredient is therefore 0.93 and the DE value $(0.93 \times 24.00) = 22.2$MJ DE per kg DM.

This method – the difference method – is clearly more prone to error the lower the proportion of GE contributed by the test ingredient. Further, a small error in the determination of the digestibility of the balancer, or the mixed diet, can have a large effect upon the result for the test ingredient. Say, for example, that the coefficient for DE_D had been underestimated by merely 2%. DE_T would be 0.88 and the DE value of the test ingredient underestimated by fully 5%. Lastly, additivity between the basal balancer and the test ingredient is explicit.

It is difficult to interpret a response when the inclusion of the test ingredient has been achieved simultaneously with, and at the expense of, a withdrawal of the equivalent amount of basal balancer. One is studying two events and not one. This becomes especially important if the type of energy in the balancer (say, starch) is suspected of being physiologically different from the type of energy in the test ingredient (say, protein), or if a chemical component of the energy fraction is being studied, such as would be the case for an examination of the digestibility of neutral-detergent fibre (NDF).

These problems can be greatly diminished by the procedure used for many years at Edinburgh. To a single level of basal diet are added incremental levels of test ingredient. In one experiment, to 1kg of basal balancer (B) given daily was added 0, 0.2, 0.4 and 0.6kg of test ingredient (T_1), and again 0, 0.2, 0.4 and 0.6kg of a second test ingredient (T_2). Three animals were used at each level. The type of plots obtained were as shown in Figure 8.7. The slope of the line indicates the DE value of the added test ingredient (T), and the constant indicates the DE value of the basal diet (B). Similar values for B in the cases of both experimental procedures for T_1 and T_2 give reassurance as to the correctness of that value in different dietary combinations, whilst the lack of curvilinearity demonstrates additivity. The use of the regression slope rather than the single difference estimate gives much greater accuracy of determination of the DE value of T.

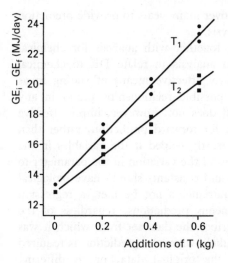

Figure 8.7 Determination of digestible energy (DE) of test ingredients T by means of regression. For T_1: $Y = 13.1 + 16.5X$ (●). For T_2: $Y = 12.9 + 11.8X$ (■). Where 1kg of basal balancer (B) is given, $DE_B = 13.0$MJ/kg, $DE_{T_1} = 16.5$MJ/kg and $DE_{T_2} = 11.8$MJ/kg.

Estimation of digestible energy from chemical composition

It is feasible for most common ingredients to be run through live animal digestibility trials to provide updated DE values; and feed evaluation units may be routinely used for such purposes. It is not feasible, however, to undertake live animal trials on individual batches for individual feedstuff importations, which may differ one from another in important ways. Common cereals and protein-rich residues from the oil extraction business often display a range in their fibre, lipid and protein contents, with consequent effect upon DE.

Equally in need of solution is the problem of compounded feeds, the nutrient specifications and ingredient compositions of which frequently change, so altering the final DE value.

It is also the case that live animal feed evaluation units are invariably to be found only at research institutes, university departments and some research facilities of larger feed compounders. Very many other organizations without live-animal metabolism facilities nevertheless wish to monitor the energy content of pig feeds and pig-feed ingredients. Not least among these are pig producers, agricultural businesses, feed-compounding mills and commodity buyers and importers.

For continuous monitoring of the nutritional content of individual feed ingredients and of complete compounded pig feeds, a scheme based upon simple laboratory chemical analysis would be beneficial. It has been towards this end that attempts have been made over many years to provide prediction equations for DE from chemical analysis.

Live animal determinations of DE, together with analysis for chemical components, allow multiple regression analysis to relate DE to chemical composition. Regression is a particularly effective means of stating in as closely exact a mathematical way as possible relationships found in any particular set of data. Regression itself does not, however, imply effective prediction. If regression is to be used for forward prediction, rather than historic description, then accuracy is greatly helped if the variables in the equation are themselves the causal forces of the variation in the parameter to be predicted. Further, the coefficients and constants should have biological logic and relevance. Should these requirements not be met, a regression equation is in danger of giving erroneous predictions, regardless of the numerical accuracy with which it describes the data set from which it was constructed. Errors become the more likely when the prediction is required to relate over a wider range than the original data, or to differing combinations of ingredients.

Regression of the data set in Figure 8.8 may, for example, give a line of best fit (a) with a constant term of around 20 and a slope of 0.010, indicating each additional gram of oil to increase DE by 0.010MJ. Such a line states: (i) oil has a DE value of about 30 and (ii) diets with no oil will have a DE of about 20. These propositions require to be checked against the common sense that (i) the DE of a diet without oil should approximate to the expected DE of the remaining components (fibre, starch and protein), and (ii) that oil is highly digestible to the pig and will have a DE value not dissimilar from

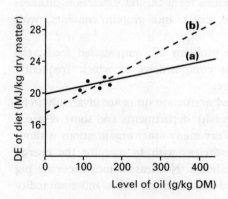

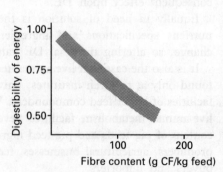

Figure 8.8 Choice of regression line for accurate prediction of DE of edible grade oil: (a) slope = 0.0100; k = 20.0; (b) slope = 0.0215; k = 17.5.

Figure 8.9 Influence of fibre content upon the digestibility of energy.

the GE. Such ideas as these latter would accord much more closely with a line (b) of slope 0.022 and constant term 17.5.

The major causal forces of variation in DE concentration in the DM of feedstuffs may be put forward as: a negative contribution from the presence of fibre (Figure 8.9); a positive contribution from the presence of oil; no contribution from the presence of carbohydrate (starch contributing primarily to the average DE of the diet, rather than to deviations from that average) and a small positive effect of protein. Active variables in useful regression equations should bear some relationship to the above suppositions.

Expectations for the form of an effective regression might be something as follows:

1. A constant term close to zero if either gross energy or carbohydrate is included in the equation.
2. A constant term close to the average DE of the diet if carbohydrate or gross energy are not included in the equation.
3. Fat and fibre should have large effects within the equation, with protein (and ash) contributing less.
4. Coefficients should bear some relationship to the logic set out in Table 8.4.

It may be postulated, therefore, that prediction equations for DE may approximate to the examples presented in Table 8.5. Examination of the research literature yields a range of published equations relating DE to chemical composition. Some examples are given in Table 8.6. The first equation has a negative multiplier for crude fibre that is higher than might be expected. This suggests either that crude fibre does not measure all the indigestible fibrous components negatively active in the diet, or that the presence of crude fibre disrupts the digestibility of other diet components. The constant term in this equation is also a little low, implying that a diet with no ether extract and no crude fibre (i.e. made of carbohydrate and protein) has a DE value of 16.0MJ/kg DM. The second equation would please phlogistonists as it appears that, upon digestion, more GE is yielded than was ingested. This is then corrected by a negative constant term. The third equation was, as it happens, a remarkably accurate description of the data set in question, but has unexpectedly high multipliers on crude protein, ether extract and 'starch' (estimated as nitrogen-free extractives – NFE). Fibre is absent from the equation, but is included by implication via NFE. These issues are resolved in statistical terms by a weighty negative constant.

Table 8.4 Expected contributions of chemical components to the concentration of diet DE

Chemical component	Approximate GE (MJ/kg)	Assumed digestibility	Contribution made by 1g (MJ DE per kg DM)
Starch	17.5	0.8–1.0	0.016
Oil	39.0	0.8–1.0	0.035
Protein	24.0	0.7–0.9	0.019
Fibre	17.5	0–0.1	0.001

Table 8.5 Postulated forms for equations to predict DE from chemical components

1. DE (MJ/kg DM) = 0.016 starch (g/kg) + 0.035 oil (g/kg) + 0.019 protein (g/kg) + 0.001 fibre (g/kg).

2. DE (MJ/kg DM) = 17.0[1] + 0.018 oil[2] (g/kg) + 0.002 protein (g/kg) − 0.016 fibre[3] (g/kg) − 0.001 starch (g/kg)

[1] 70% starch, 20% protein, 5% oil, 5% fibre. This term may also be expressed by using GE (MJ/kg) multiplied by the expected overall digestibility coefficient for GE, e.g. 0.84 GE.
[2] Removal of 1g of average DE and substitution of 1g of oil DE: − 0.017 + 0.035 = 0.018.
[3] Removal of 1g of average DE and substitution of 1g of fibre DE: − 0.017 + 0.001 = − 0.016.

Table 8.6 Three example prediction equations drawn from the literature (all elements per kg DM)

1. DE (MJ/kg DM) = 16.0 − 0.045 CF(g) + 0.025 EE(g)
2. DE (MJ/kg DM) = − 4.4 + 1.10 GE (MJ) − 0.024 CF(g)
3. DE (MJ/kg DM) = − 21.2 + 0.048 CP(g) + 0.047 EE(g) + 0.038 NFE(g)

CF = crude fibre; EE = ether extractives; CP = crude protein; NFE = nitrogen-free extractives.

The roles of fibre and fat

To quantify by experiment the exact roles of fibre and fat in influencing the digestibility of feedstuffs, a series of investigations were completed at Edinburgh through the eighties.

Acting on the proposition that fibre was the most important component reducing diet digestibility, the relationship between analysed chemical fibre content and digestibility in the live animal was examined using three different fibre types. These were those found in oat feed, rice bran and beet pulp. Oat feed contains about 250g crude fibre per kg DM, rice bran about 160g, and sugar beet pulp about 190g. The basal balancer used was wheat/soya, and to 1kg/day of this were added graded levels of the test materials. By the regression method of plotting DE intake ($GE_i - GE_f$) against level of addition of test material (X, kg), the following equations were derived:

$$GE_i - GE_f \text{ (oat feed)} = 13.53 + 8.56X$$

(Equation 8.11)

$$GE_i - GE_f \text{ (rice bran)} = 13.76 + 10.25X$$

(Equation 8.12)

$$GE_i - GE_f \text{ (beet pulp)} = 13.86 + 11.16X$$

(Equation 8.13)

The regressions suggest that the DE of the basal balancer is relatively unaffected by the source of added fibre, but that the DE value for the three materials (MJ DE per kg, as given) were 8.56, 10.25 and 11.16 respectively.

Crude fibre (CF) is a term used to describe an analytical procedure and it has little meaning in terms of structural plant elements. The Van Soest procedures attempt to differentiate between various fibre fractions in a sequential analysis. Lignin is the most complex fibre material, followed by cellulose and hemicellulose. Lignin is a totally indigestible polyphenolic polymer. It becomes associated with the structural carbohydrates cellulose and hemicellulose. Cellulose is a glucan, like starch, but with beta links, and therefore digestible only by micro-organisms. Hemicellulose, unlignified, would be supposed to be more digestible, being of pectins and gums derived from xylans and mannans. Both cellulose and hemicellulose may be lignified to varying degrees. There is agreement that lignified polysaccharide is indigestible to the small intestinal digestive enzymes of the pig, and also indigestible to all other enzymes, even those of bacterial origin found in the large intestine and caecum. It may be supposed that cellulose is indigestible in the small intestine, but potentially digested posterior to the ileum. Argument rages as to the potential and site for the digestibility of hemicellulose. But the weight of current opinion is that hemicellulose may in fact be rather less digestible than cellulose, being resistant to degradation in the large intestine and caecum. In the analytical series, neutral detergent extraction separates out the cell content of lipids, sugars, starches and protein, leaving the neutral detergent fibre (NDF) comprising the cell wall hemicelluloses, celluloses, lignins and ash. Acid detergent extraction of this residue separates the hemicellulose from the acid detergent fibre (ADF) comprising the cellulose, lignin and ash. The final separation of cellulose from the acid detergent lignin residue (ADL) is achieved by acid hydrolysis.

Fibrous components of four example fibrous feeds are shown in Table 8.7. Straw is rich in cellulose, oat feed and rice bran have similar structural carbohydrate compositions, while sugar beet pulp has relatively more cellulose and relatively less lignin. The digestibility of the feed material will be consequent upon the absolute amount of total fibrous components and the balance of hemicellulose, cellulose and lignin within the fibre. Thus, the relatively higher total fibre content of sugar beet pulp as compared with rice bran appears to have been more than offset by sugar beet pulp having a much lower proportion of lignin within the fibre. The digestibility of oat feed as compared with rice bran reflects the greater absolute quantities of structural carbohydrate components in oat feed as compared with rice bran.

When the various ways of analysing for fibrous components were examined with a view to their efficacy in predicting DE, it was evident from both the logical and the statistical point of view that NDF was much more effective than the others, whilst crude fibre (CF) was especially poor. This might be considered as a predictable outcome, as CF is an empirical analysis not based on any particular fibre source (see Table 8.7), whereas NDF includes all three fibre fractions considered to be more or less indigestible in the pig.

The statistically determined best-fit prediction equation based on the combined data for all three fibrous sources was:

$$DE \ (MJ/kg \ DM) = 17.4 - 0.016 \ NDF$$

(Equation 8.14)

Table 8.7 Fibrous components of some fibrous feeds (g/kg DM)[1]

	Straw	Oat feed	Rice bran	Sugar beet pulp
Chemical composition				
Crude fibre	478	252	155	192
Neutral-detergent fibre	775	579	353	349
Acid-detergent fibre	544	309	188	241
Lignin	152	136	76	22
Structural carbohydrate composition				
Hemicellulose	231 (0.30)[2]	270 (0.47)	165 (0.47)	108 (0.31)
Cellulose	392 (0.51)	173 (0.30)	112 (0.32)	219 (0.63)
Lignin	152 (0.20)	136 (0.23)	76 (0.22)	22 (0.06)

[1] In comparison, the respective values of CF, NDF and ADF are: for maize 20, 80 and 22; for barley 42, 152 and 50.
[2] Values in parentheses give the relative proportions.

which accords remarkably well with the propositions laid out in Table 8.5.

Acting next on the proposition that oil was the most important component enhancing diet digestible energy value, the relationship between oil content (EE) and digestibility was examined using three different sources of human-edible grade oil: tallow, palm oil and soya oil, added at progressively increasing levels (X, kg) on to a basal balancer of 1kg of a barley/soya diet:

$$GE_i - GE_f \text{ (tallow)} = 12.90 + 38.82X$$

(Equation 8.15)

$$GE_i - GE_f \text{ (palm oil)} = 12.60 + 37.54X$$

(Equation 8.16)

$$GE_i - GE_f \text{ (soya oil)} = 12.92 + 40.83X$$

(Equation 8.17)

In the case of the added oils, the test materials were pure, whereas with the fibrous feeds the test materials were high in fibre, but not solely comprising fibre. As the GE of oil is 39.6MJ/kg, the regressions indicate high digestibilities. The expectation that the more highly hydrogenated tallow was not as digestible as the unsaturated palm and soya oils was not realized.

The best-fit regression over all oils produced a positive coefficient of +0.025 for ether extractives (EE) (g/kg), which is higher than the 0.018 proposed in Table 8.5, but not too far distant from the 0.023 that logic would have proposed for a very high-grade oil, which might be completely digestible.

These results gave the Edinburgh workers sufficient confidence to set up a full-scale experiment involving the use of 36 compounded diets covering most, if not all, of the ingredients to be found in pig feeds (33 in total), and giving a composition range of 23–123g crude fibre per kg, 149–258g crude protein per kg, and 21–116g EE (oil) per kg. These diets were each given to four growing pigs housed in metabolism crates and the complete data set used for multiple regression analysis to derive equations for the prediction of DE from the chemical analysis of compounded diets.

A very large number of equation forms were derived. Of the 62, of particular interest are the following:

1. Using crude fibre, ash and oil:

$$DE \text{ (MJ/kg DM)} = 18.3 - 0.037CF - 0.019ASH + 0.011OIL$$

(Equation 8.18)

This is not particularly accurate (residual standard deviation = 0.65), neither are the coefficients particularly logical. The high multiplier for crude fibre (CF) may indicate either that CF estimates only a part of the indigestible diet constituents or that the presence of CF disrupts digestion of other nonfibrous diet components. If either of these possibilities pertains, the naive use of an elevated multiplier is unlikely to be sufficient means of ensuring the accuracy of a prediction equation. In any event, it was perfectly clear from the whole exercise that, of all the analyses for fibre, for the purpose of DE prediction, NDF was considerably superior to any other measure, and CF was particularly inferior.

2. Including gross energy:

$$DE \ (MJ/kg \ DM) = 3.77 - 0.019NDF + 0.758GE$$

(Equation 8.19)

This equation is rather accurate, with a residual standard deviation of only 0.38. The coefficient for GE reflects the digestibility coefficient itself, although it is less than the mean digestibility, due to the presence of a positive constant term of 3.77MJ. The coefficient for NDF is within expectation. Unfortunately, many analytical laboratories are not well enough equipped to be sure of an accurate measurement of GE.

3. Using NDF and oil together with a constant term:

$$DE \ (MJ/kg \ DM) = 17.0 - 0.018NDF + 0.016OIL$$

(Equation 8.20)

This equation is quite accurate, having a residual standard deviation of 0.44, and has the benefit of restricting itself to the two major causal forces of digestibility: the negative force of fibre and the positive force of oil. It conforms with the propositions put forward in Table 8.5. Further, the coefficient for NDF accords with the expectations derived from the first Edinburgh fibre experiment. The coefficient for oil is lower than that in the first Edinburgh oil experiment, where the three human-grade oils were used. In that previous case the coefficient was 0.025, as compared with 0.016 here. The value of 0.016 is, however, more in line with expectations for feed-grade added oils and oils naturally occurring in feed ingredients, these latter being the effective contributors to oil in the commercial-type compounded diets used in the 36-diet matrix experiment. The coefficient of 0.016 for oil indicates an achieved DE of dietary fats of around 33MJ/kg. For simplicity and logic, this equation has much to commend it.

4. Using NDF, oil, ash, protein and a constant term:

$$DE \ (MJ/kg) = 17.5 - 0.015NDF + 0.016OIL + 0.008CP - 0.033ASH$$

(Equation 8.21)

This equation is the most accurate of all, having a residual standard deviation of only 0.32. The inclusion of ash and crude protein (CP) helps to make the equation more robust over a wide range of circumstances.

Following a review of digestibility studies completed in Edinburgh, Nottingham, Australia, Denmark and elsewhere, a recent European working party has put forward an equation which includes nitrogen-free extractives (NFE) as a measure of starch, avoids the use of a constant term, and uses crude fibre (CF) as the measure of structural carbohydrate on account of the readiness with which CF may be analysed. The resulting equation is used to predict ME (which will be some 0.96 of predicted DE):

$$ME \text{ (MJ/kg DM)} = 0.016NFE + 0.032OIL + 0.018CP - 0.015CF$$

$$\text{(Equation 8.22)}$$

This equation may be (favourably) compared with the form originally postulated from Edinburgh some 10 years earlier on theoretical grounds (Table 8.5). But it will be noticed that while the Edinburgh equation takes the contribution to DE from fibre (as NDF) to be low, the European equation uses CF and proposes a negative contribution. The logic here is difficult to follow as it is known that some CF does indeed disappear in the gut, and it will be noted that this equation does not have a constant term (only equations with constants requiring a negative fibre term as also shown in Table 8.5). Perhaps the negative sign is to allow for the disruption which fibre may cause to protein digestion. More practically, it may be noted that the multipliers used are more close to DE expectations than ME expectations, and the CF term is about the right magnitude to effect the (necessary) 4% reduction on the predicted value.

Guide values for energy content of pig feedstuffs

Reports of *in vivo* digestibility experiments and *in vitro* chemical analysis of pig feedstuffs from many countries have been collated into a set of guide values and feed composition tables now given in Appendix 1. Feed composition tables are invaluable for the proper formulation of diets and for the calculation of monetary worth of individual feedstuffs. There is no substitute, however, for regular reappraisal of both chemical composition and *in vivo* digestibility of conventional feedstuffs. Nor is there any substitution for live animal digestibility studies and performance response trials for new feedstuffs, feedstuffs of questionable quality or feedstuffs of novel origin.

CHAPTER 9

NUTRITIONAL VALUE OF PROTEINS AND AMINO ACIDS IN FEEDSTUFFS FOR PIGS

Introduction

In addition to carbon, hydrogen and oxygen, the major functional element in protein, and the amino acids which link together to make up protein, is nitrogen (N).

The protein content of the whole live pig will vary between 15 and 18%, depending on fatness. On average, 21% of the fat-free carcass is protein, the protein content of lean meat varying between 20 and 25%, increasing with age. Sows' milk contains about 6% protein. Because the construction of absorbed diet amino acids into pig protein is an energy-consuming process in itself, some 30–50% of the energy used by growing pigs may well be involved in the anabolism of lean tissue. The substantial involvement of protein synthesis in the overall energy economy of the pig gives some measure of its importance in proper nutrient provisioning.

Feedstuff protein is the essential source for amino acids. Amino acids are required: (i) to build protein in the body of the pig, mostly in muscle and (ii) to replace proteins lost in the course of protein tissue turnover (maintenance) and voided both in the form of cells, amino acids, other nitrogenous compounds and unabsorbed enzyme secretions from the intestine (termed metabolic faecal losses or endogenous faecal losses), and also in the form of urea from the kidney (termed urinary losses). Some nine amino acids (lysine, methionine and cystine, threonine, tryptophan, histidine, isoleucine, leucine, phenylalanine and tyrosine, valine) cannot be made by the pig itself from more simple compounds, nor interconverted from other amino acids (although about half of the methionine and tyrosine requirements can be met from cystine and phenylalanine respectively). These nine are the essential amino acids.

On the other hand, protein can be used to yield energy. Because the process of making pig protein from feed protein is usually only about 30–60% efficient, the energy yield from the deamination of amino acids not deposited in pigs' body or pigs' milk can be quite significant. Amino acids not destined for pig protein may also, upon their deamination, yield useful precursors for other body activities, for example the formation of fatty tissues. Mainly, however, feed proteins are seen as providing the amino acid elements for the deposition of pig protein in growth or milk, and for the maintenance of proteinaceous pig tissues.

Feed proteins are not constructed in the same way as pig proteins. The necessary restructuring from the one to the other involves costs. In the digestion, breaking-down, absorption and building-up processes: (i) energy for biochemical work is used up; (ii) amino acids unwanted for deposition and retention are turned to other uses; (iii) they are finally excreted in faeces, urine, sweat or (iv) oxidized to be lost as heat. This apparent inefficiency is, in reality, a highly valuable and effective way of turning vegetable proteins (such as from cereals) which are not of high value to humans, and animal proteins (such as fish by-products) which humans may not wish to eat, into lean pig meat which is highly nutritious and readily eaten.

Although most diet protein for pigs comes from cereals, the concentration of crude protein (CP) is usually low at around 8–12%. This means that the diet requires to be supplemented with higher-protein seeds (such as peas or beans) with 20–30% protein, or with plant products in which the protein has been concentrated to 30–60% due to the extraction of other components such as oils (such as extracted soya bean and extracted rape seed meals). Vegetable proteins are less easily handled by the pig (especially the young pig) because they can be more difficult to digest, the concentration of essential amino acids is relatively low, and there is much restructuring and reorganization of the amino acids needed in order to make pig protein out of vegetable protein. For this reason, protein of animal origin, such as from milk, meat-and-bone meal or fish meal, may have an important part to play in the compounding of high quality pig diets. Not only are animal proteins easily digestible, but their amino acid balance is, self-evidently, already much closer to that required for the construction of pig protein in lean meat muscle tissues.

As has been described, the unavoidable inefficiencies of the conversion of diet protein to pig protein in the form of growth, products of conception or sows' milk is acceptable to the extent of creating useful products from less useful ones. However, within any given system efficiency must be optimized by reconciling the biological cost and benefits with the economic ones. This requires information on the net yield to the pig of usable amino acids obtainable from feedstuff proteins. The yield of amino acids from a feedstuff is a combination of amino acid content, amino acid digestibility and amino acid utilizability.

The final amounts of amino acids deposited and retained by the pig are also dependent upon the balance of amino acids available at the time. This quality is therefore usually a function not of any individual feedstuff, but of the combination of feedstuffs in the diet as a whole. The influence of the balance of amino acid supply will be described in the chapter dealing with nutrient requirement (as opposed to the present preoccupation which is with nutrient provision).

There is interest from countries with a dense population of intensive pig units in the minimization of the amounts of nitrogen excreted both in faeces and in urine. Nitrogenous output from the pig in the form of slurry is considerable, and this may contribute to environmental contamination. The higher the digestibility of the diet protein, and the better the amino acid balance, the less will be the level of faecal nitrogen and the lower will be the level of urea excretion. This improvement in biological efficiency, although probably only attainable by increasing the cost of the diet, may well have more than proportional benefits in terms of pollution control.

Digestibility of protein

Meat and vegetable protein contains about 16% nitrogen. Total protein, or crude protein, is therefore usually measured as $N \times 6.25$. (To be exact the N multipliers for estimation of protein should be 6.25 for animal tissue, 5.8 for cereals and soya bean meal and 5.5 for lower quality proteins, the differences being the relative presence of non-protein and nucleic acid nitrogen. CP is really, therefore, more a measure of N than of protein.) The linkages between the amino acids which form them into proteins are broken down in the hydrolysing processes of digestion.

After some hydrolysis in the stomach to peptides, the major enzyme attack is by proteases and peptidases reducing the proteins to short chains of two or three amino acids which are either absorbed as such or further hydrolysed by the intestinal wall to free amino acids. There is some utilization of amino acids by the gut wall itself, but most amino acids pass through into the body proper and are taken by the bloodstream to the liver and body cells. In the large intestine there is microbial degradation of the residues from digestion and of the endogenous secretions. This releases ammonia and amines which, while disappearing from the gut and being apparently digested, are of no nutritional worth.

Digestion is invariably incomplete, the extent of digestion (digestibility) depending upon circumstances of both animal and feedstuff, but primarily the latter. Protein which has been ingested, but which has not been subsequently excreted in the faeces, has, by definition, disappeared from the tract and is assumed to have been digested. Thus if 20g of protein (as N × 6.25) appear in the faeces for every 100g protein (N × 6.25) ingested, then digestibility is 0.80. This value usually ranges between 0.75 and 0.9, with an average for high quality pig feeds of 0.8–0.85 (Table 9.1). Digestibility, as measured by the difference between ingestion and faecal excretion, is termed *apparent* digestibility because it measures that part of what is eaten which disappears from the intestine, appears within the body-proper of the animal, and is available for metabolism. The apparent digestibility is less than the *true* digestibility of feedstuff protein. This is because some feed protein has indeed been digested, but then returned back into the tract in the form of digestive juices and cells lost from the intestinal wall lining as the feed passes down the tract. In a growing pig of 50–80kg there can be around 23g nitrogen or 140g endogenous (metabolic) protein secreted into the tract daily in this way (Table 9.2). Endogenous N can be a significant proportion of the digesta, especially at lower feed intake levels. True digestibility can be 5–20% higher than apparent digestibility, the extreme often being consequent upon a high level of fibre in a diet fed at low level. Most, but not all, of the endogenous protein secreted into the tract is

Table 9.1 Digestibility coefficients of protein for some feedstuffs

Feedstuff	Apparent faecal digestibility[1]	Apparent ileal digestibility[2]
Fish meal	0.90–0.95	0.80–0.90
Meat-and-bone meals	0.50–0.70	0.40–0.70
Maize meal	0.80–0.85	0.70–0.80
Wheat meal	0.80–0.85	0.70–0.80
Barley meal	0.75–0.80	0.70–0.80
Wheat offals	0.50–0.75	0.30–0.60
Extracted soya bean meal	0.85–0.90	0.75–0.85
Extracted cotton, coconut and rape-seed meals	0.50–0.70	0.40–0.60

[1] These values are the difference between ingested and faecally excreted proteins (N × 6.25) divided by the amount ingested (i−f)/i. Artificial amino acids may be assumed to be completely digestible.

[2] These values are the difference between ingested proteins and those found passing the terminal ileum divided by the amount ingested (as N × 6.25). Artificial amino acids are completely digestible.

Table 9.2 Endogenous secretions into the intestines of a growing pig eating around 2kg of feed and 400g diet crude protein daily

	Protein[1] secreted daily (g)
Pancreas	20
Bile	15
Saliva and stomach	15
Cells	10
Intestinal secretions	80
Total[2]	140

[1] The threonine and tryptophan content, relative to other amino acids, is higher in the digestive secretions than in the digesta. As a result, more of these amino acids will pass the terminal ileum and their ileal digestibilities may be expected to be relatively lower than the other amino acids.

[2] About 80% of the total secretions are reabsorbed and available for reutilization, some 20% pass unabsorbed beyond the terminal ileum and will be degraded by microbes in the hind gut or lost in faeces as metabolic faecal protein.

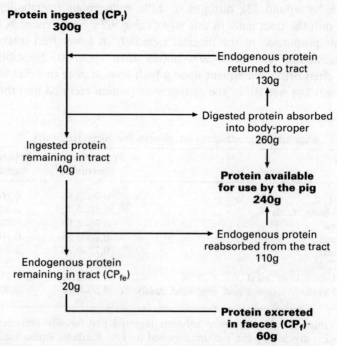

Figure 9.1 Scheme of movement of protein in and out of the intestine of the pig. Digestibility measured as apparent digestibility is determined as $(CP_i - CP_f)/CP_i$ or $(300 - 60)/300 = 0.80$. Digestibility measured as true digestibility is determined as $CP_i - (CP_f - CP_{fe})/CP_i$ or $300 - (60 - 20)/300 = 0.87$.

reabsorbed, the residue being passed out in the faeces as the metabolic faecal protein component, joining the indigestible dietary protein in the faecal residues. This is shown schematically in Figure 9.1.

The presence of the endogenous component in faecal protein means that apparent digestibility is somewhat intake-dependent and measurements of digestibility must be made at realistic intake levels. If endogenous faecal protein (CP_{fe}) amounts to 20g, and 150g of protein are ingested with a true digestibility of 0.87, then 40g of faecal protein will be lost; comprising 20g of metabolic faecal protein and 20g of indigestible diet protein, giving an apparent digestibility of 0.73. If, however, 300g of protein are ingested, and metabolic faecal protein remains at 20g, then 60g of faecal protein will be lost, comprising 20g of metabolic faecal protein and 40g of indigestible feed protein. Apparent digestibility for the feed protein is now 0.80.

The measurement of endogenous loss is difficult as it is so dependent upon the animal and dynamic state of the gut, so is usually taken as the level of appearance in the faeces when the level of ingested protein is low or absent. As animals grow bigger and eat more, the absolute levels of metabolic losses tend to increase, which causes apparent digestibility to slightly decrease. More importantly, as the amount of feed eaten increases, both the rate of enzyme secretion and the loss of cells from the gut lining are elevated. Fibrous and abrasive feedstuffs are associated with greater levels of metabolic faecal losses, and thus the inclusion of straws, cereal husks or sand and like materials into pig diets will induce a decrease in the apparent digestibility of the protein and the net yield of feed protein to the pig. Feeds of low protein content and high in fibre may, through their influence upon metabolic faecal losses, contribute little, or negative, amounts of protein to the nutritional economy of the animal.

Usually metabolic faecal protein amounts to 6–12g for each kg of compound feedstuff if this has around a 5% level of fibre. For a pig consuming 150g of protein in each kilogram of feed, the metabolic faecal losses will amount to around 30% of total faecal protein.

The true digestibility is more interesting than it is practical. The animal can only benefit from the digestive processes to the extent of the apparently digested protein moieties; only that which has disappeared between ingestion and faecal excretion being available for productive uses (Figure 9.1). For practical purposes, therefore, interest is usually confined to the determination and expression of the apparent digestibility of feedstuff nitrogen, protein and amino acids. The term 'digestible' may therefore be taken as synonymous with 'apparently digestible'. It is important, however, when placing values on feedstuffs as to their content of digestible protein – or digestible amino acid – that the determination has been undertaken at reasonable levels of intake, and in conjunction with balancer meals of reasonable fibre level.

Measurement of digestibility and retention of crude protein (N × 6.25)

Digestibility is determined by balance using the same methodology as described in the previous chapter for energy. Crude protein (CP) ingested is determined as the product of the amount of feed eaten and the concentration of N × 6.25. Apparent digestibility is calculated as the difference between crude protein consumed (CP_i) and crude protein (N × 6.25) excreted in the faeces (CP_f):

$$(CP_i - CP_f)/CP_i$$

(Equation 9.1)

It is important that no ammonia leaves the excreta between the time of being voided and the time of analysis; this can be prevented by collection under dilute sulphuric acid.

While the protein value of a feedstuff is considerably influenced by the digestibility of the crude protein, there is also great interest in the part of the gastro-intestinal tract where the CP_i moieties were digested and disappeared. Digestion and disappearance from the small intestine yields usable end-products, while disappearance from the caecum or large intestine yields useless products. It is evident that the greater the proportion of a feedstuff's protein content that is digested in the small intestine, the greater is the value of that feedstuff. To differentiate between small intestinal and large intestinal digestion, a cannula can be inserted at the end of the ileum (Figure 9.2). Withdrawal of the digesta from the terminal ileum allows a direct measure of the level of digestion which has occurred in the small intestine – the ileal

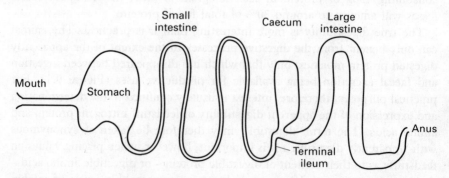

Figure 9.2 Diagrammatic representation of the digestive system of the pig to show the position of the terminal ileum.

digestibility. This is considered to be a truer reflection of realized protein value than the faecal digestibility.

Estimation of the retention of energy by difference – that is, heat losses – is not without its complications, as has been described. For CP, however, such complications do not arise to the same extent, although there may be gaseous losses of N and also losses of N in skin and hair. CP retention (CP_r), however, can be adequately measured by determining the urinary nitrogen losses (CP_u):

$$CP_r = CP_i - (CP_f + CP_u)$$

(Equation 9.2)

Urine can be readily collected, under acid, in the course of a standard digestibility procedure as described in the previous chapter (Figure 8.4).

CP retention can also be determined by comparative slaughter techniques, the difference between the CP mass in the pig at the end of the trial period and at the beginning of the trial period being – evidently – a direct measure of the crude protein (N × 6.25) retention. The efficiency of the retention is determined with knowledge of the intake of crude protein over the period:

Daily feed intake × number of days (day_1 to day_n) × CP concentration of feed = CP_i

Whole body mass × CP content of body at day_n − Whole body mass × CP content of body at day_1 = CP_r

CP_r/CP_i = Efficiency of CP retention

The efficiency of CP retention may not, however, be a particularly fruitful measurement for comparing the relative value of the protein content of feedstuffs because it is pig-dependent as well as feed-dependent. The amount of CP retained in relation to the amount ingested is: (i) a function of the digestibility and the quality of the protein in terms of the amino acids absorbed; (ii) a function of the amount given in relation to the amount needed and (iii) a function of the other diet constituents. Thus a good protein may show a poor efficiency of retention if supply greatly exceeds requirement, because the unwanted excess has overflowed out via the urine. Equally a protein may have a low efficiency through a shortage of one amino acid which prevents effective construction of pig protein, but which may be simply and cheaply redressed by balancing with another protein high in the missing amino acid, or even with an artificial amino acid source.

The digestibility of protein in feedstuffs is only rarely predicted from laboratory analyses, involving digestion with pepsin and enzyme mixtures. But these methods are general for nitrogen (N), and do not address the organic nature of the protein moiety itself. Damage to protein, which reduces digestibility, can be assessed through analysis of the extent to which the amino acid lysine is bound up and rendered unavailable. In general, however, laboratory methods for estimating the amount of amino acid absorbed through the small intestinal wall are much less useful than those available for the equivalent measure of digestible energy. Protein quality, however, as perceived from the balance of total amino acids, is readily measured *in vitro*.

Because protein quality is as much related to its amino acid content as to its digestibility, chemical analysis for amino acids, especially the nine essential amino acids, allows an effective evaluation of the protein worth of a feedstuff. Knowledge of the amino acid make-up enables ready calculation of the cost that will be incurred in balancing the feed to create an appropriate amino acid mix for the needs of the pig. For example, fish meal and soya bean proteins have more lysine than cereal proteins; thus the former may effectively complement the latter when placed together in a diet mix.

At present the best estimate of protein value in a feedstuff would appear to arise from information on the amino acid composition and on disappearance proximal to the terminal ileum.

Factors influencing digestibility of protein

Although differences exist in the apparent digestibility of crude protein (N × 6.25) of pig-feed ingredients, the concept of digestible crude protein (DCP) has not been readily taken up by feed formulators. Variation in digestibility has – wrongly – been considered as not as important for protein as for energy. The digestibility of most protein of animal origin (milk, fish and meat) is about 0.90, and most of the proteins of vegetable origin (cereals, legumes, roots and young leaves) about 0.80. However, there are many important exceptions to these generalities which are deeply significant to cost-effective feed formulation. Despite the reluctance of some nutritionists to accept the importance of DCP as a feedstuff descriptor, protein digestibility is a most useful indicator of feed value and is profoundly affected by both extrinsic and intrinsic factors.

The basic amino acids that make up protein are highly digestible. Nevertheless, in the small intestine there is more protein passing down the gut than was eaten. This is because the gut contents are inundated with proteinaceous secretions (mostly enzyme-containing fluids), and also the cells lining the intestine have a high turnover rate and are lost in quantity into the gut lumen. Fortunately, reabsorption of these secretions is just as highly efficient as the primary absorption of the amino acid products of enzyme degradation of diet proteins.

Heat damage

Digestibility falls dramatically if the protein is heat damaged; such damage may occur when either animal or vegetable proteins are overcooked as part of the production process. The consequences of such damage are difficult to predict due to their variable nature, and therefore almost impossible to incorporate into a quantitative nutritional evaluation of feedstuffs. Suffice it to say that heavily damaged materials – once identified – should be treated circumspectly as sources for amino acids for pigs. If used, on the grounds of economic expediency, the effects upon growth and reproduction are difficult to elucidate other than by empirical trial.

Heat damage changes the protein structure, which reduces its digestibility. The greater the degree of heating, the higher the loss in digestibility. To some extent, all cooked proteins are less digestible than raw ones (about 5% less). But in cases of overcooking, when actual damage occurs, the matter becomes serious and digestibility may be reduced to 50% or less as, for example, in overheated soya bean meal, fish meal, meat meal and blood meal. Specific amino acids may be bound on heating. The classic case is the binding of lysine to sugar compounds; this reduces the digestibility and utilizability of lysine. 'Available lysine' is particularly important because: (a) it is a general indicator of heat damage and (b) cereals are lysine deficient; lysine is the most valuable amino acid in most supplementary protein sources. If the lysine is unavailable, then all the other amino acids in the protein, even though utilizable, cannot be utilized.

Fur, feather and hide

The protein may be bound in the feed in a form resistant to enzyme attack as, for example, the proteins in leather and feather meal.

Abrasion

Certain foodstuffs may increase the rate of cell wall loss from the gut lining and/or prevent efficient reabsorption of this protein of body origin. This will increase the protein in faeces and decrease perceived digestibility. Straw has been implicated in such activity, and also other highly fibrous and abrasive feeds. The evidence is, however, not straightforward.

Rate of passage

Digestive enzymes need time to work. Digestibility can therefore be reduced by anything that increases the rate of passage of digesta through the intestine. High feeding levels of liquid diets of low DM based on by-products or liquid wastes may produce just such an effect.

Protection and comminution

Enzymes also work most efficiently if they have a large surface area to act upon. Digestibility will be reduced if feed material is only coarsely ground and in large fragments, or if the diet contains large masses of material, which may surround the protein moieties and prevent enzyme penetration. Proteins may be found mixed up with cell-wall and other poorly digestible carbohydrate fractions. Until these resistant carbohydrates have been broken down, the proteins cannot be released and will remain undigested. Similarly, protein constituents of cells will only be available for enzyme attack and absorption after such time as the cell wall has been digested away, or ruptured mechanically (by milling, grinding, biting or chewing) prior to reaching the small intestine. Given that a diet contains no protease inhibitor activity, is not heat-damaged and is pulverized to a reasonable extent, protein digestibility appears to be related to some extent to the amount of structural or unusual carbohydrates present. Both heat damage and structural carbohydrates are potent forces in shifting the site of digestion for some diet protein away from the small intestine to the large intestine, with important negative repercussions for the efficiency of use of ingested amino acids, which are, of course, best absorbed from the small intestine.

Anti-nutritional factors

Most, if not all, plant proteins are associated with anti-nutritional factors. These are present to very different extents in different plant species and even in different varieties within species. Thus cereal protein is low and pea protein is quite low in toxic factors, whilst raw soya bean is rich in protease inhibitor activity, and raw potato rich in inhibitors and also in alkaloids, if green. Winter sown rape seeds can be goitrogenic, containing generous levels of glucosinolates, as well as harbouring erucic acid; while in some spring-sown varieties these poisons are at much lower levels. Cyanogenic glycosides, such as linamarin, may be found in inadequately prepared cassava meals, and occasionally in sorghum. Tannins are usually associated with brown sorghum, but are also present in rape seed, sunflower and some bean varieties; they disrupt digestion, reduce digestibility and are difficult to destroy. Alkaloids can be found in lupin (but not sweet lupin), potatoes and ergot sclerotia. Lectins may be found in most legumes to varying degree; they disrupt the intestinal epithelium and impair nutrient absorption. Phaseolus beans are especially rich in lectins, and when given raw to pigs are toxic; even after cooking by toasting there remains serious reduction in performance when phaseolus beans are fed. Although phytates (found in legumes and many other plants) are usually associated with the tying-up of phosphorus and the chelating of calcium, magnesium, zinc and iron (and thereby render all these less available), they also have the capacity of reducing the effectiveness of protease (and amylase) enzymes.

Many cruciferous plants, not just oil-seed rape, can be goitrogenic. The plant glucosinolates are hydrolysed by enzymes to yield the goitrogens which depress growth.

Goitrogens, tannins, saponins, gossypols and alkaloids are heat stable, but the lectins, protease inhibitors and other poisonous amino acids are destroyed by heat. Heat treatment remains the most effective way of counteracting heat-labile, anti-nutritive factors. Such processing requires careful control in order to achieve adequate cooking to detoxify but to avoid overcooking and heat damage; this balance is crucial to the effective production of soya bean meal following the extraction of soya oil. Heat-stable poisons can only be extracted by complex exchange technology and/or washing (as for oil-seed rape meals) or, more commonly to date, by breeding improved strains of plants not carrying the offensive material. The most notable recent case of breeding toxins out of plants has been that of the spring-sown 'Canadian' rapes.

Protease, or specifically trypsin and chymotrypsin, inhibitors act as anti-enzyme factors, reducing protein digestibility in the gut. Protease inhibitor activity also increases pancreatic enzyme secretion and consequential endogenous protein losses. In some foodstuffs, levels of trypsin inhibitor activity are extremely high, particularly in the bean families (field beans, soya beans, etc.) and potatoes (Table 9.3).

Uncooked soya and potato can have protein digestibilities lower than 30%. These protease inhibitors act not just specifically in relation to the feeds that contain them, but generally within the whole of the gut environment. Thus *all* the protein in a diet containing raw soya will have reduced digestibility. In general, it has to be stated that anti-nutritional factors in plants adversely affecting protein absorption remain mysterious and poorly understood.

Table 9.3 Presence of anti-nutritional trypsin and chymotrypsin inhibitor factors in some feedstuffs

Present in sufficient quantity to influence pig performance greatly	Present in sufficient quantity to influence pig performance significantly	Present in sufficient quantity to influence pig performance slightly	Not present in sufficient quantity to influence pig performance at all
Raw soya bean	Chick-peas	Rye	Maize
Adzuki bean	Pigeon peas	Triticale	Oats
Raw potato	Cow-peas	Lupin	Wheat
	Field peas	Barley	
	Mung beans		
	Field beans		

It is because protease inhibitors are themselves made of protein that their activity is destroyable by heat treatment. Partial cooking causes partial destruction, and it is this gradation of treatment effect that makes quality control so important for soya beans (and for potato) intended for pig diets. One way of counteracting a low level of inhibitor activity is to give luxury amounts of the protein, and it has been accepted for many years that diets containing field beans should be compounded to a more generous protein content. Recent varieties of low trypsin inhibitor soya beans have been developed. However, when fed raw, protein digestibility is still significantly depressed and full heat treatment remains necessary.

Site of digestion

Usually the digestibility of protein determined as $(CP_i - CP_f)/CP_i$ is some 10% (5–20%) higher than the digestibility determined at the point of exit of digesta from the terminal ileum. Much of the reason for this is the disappearance of ammonia and amines from the large intestine. Thus while 'digestion' may mostly occur in the small intestine proximal to the terminal ileum, 'disappearance' may occur anywhere in the tract including the large intestine. Nitrogenous materials disappearing from the tract count towards crude protein absorbed but not towards protein moieties realistically available to the pig, as the nitrogenous products of fermentation in the large intestine tend to go directly to urinary excretion. Where the measured DCP value may therefore often be in the region of $0.85 \times CP$, the effective DCP value may well be nearer $0.70 \times CP$. Because of this, interest in the *ileal* digestibility of protein is now rather greater than interest in *faecal* digestibility.

Diet protein is, of course, no better or worse than its constituent amino acids; it is the amino acids that the animal is seeking to be provided from feed proteins. Therefore protein content is often best expressed in terms of the amino acid constituents, and a measure of the effective utilizability of the protein expressed in terms of the digestible amino acids. Because of the confusing role of the hind gut, digestibility at the point of the terminal ileum for individual essential amino acids appears presently to be a most useful way of expressing the protein value of either a feedstuff or a mixed diet.

Ileal digestible amino acids

Ingested protein is attacked in the stomach by pepsin and hydrochloric acid, and hydrolysed by enzymes in the small intestine secreted from the pancreas and the intestinal wall. The amino acids thus yielded can be absorbed from the small intestine and are available for use by the pig. The effectiveness of the small intestine in digesting some of the amino acids appears rather superior to its effectiveness in digesting others. Thus the balance of amino acids, one to the other, in the feed will not be perfectly reflected in the balance of amino acids found at the terminal ileum (see Table 9.4).

The differential absorption of amino acids from the small intestine is variable amongst feedstuffs and therefore an important factor for consideration in feed evaluation.

There is some bacterial degradation of amino acids in the small intestine, but it is slight. It is only in passing from the ileum to the caecum and large

Table 9.4 The balance of amino acids at the point of ingestion (in comparison to lysine = 100) compared to the balance at the terminal ileum. Notice how the balance has changed, the digestive processes dealing unequally with the various amino acids. These different rates of absorption are important to the correct evaluation of a feedstuff and add weight to the need for knowledge of digestion in the small intestine

	At point of ingestion	At the terminal ileum
Lysine	100	100
Methionine + cystine	60	50
Threonine	65	70
Tryptophan	20	25
Leucine	100	130

intestine that the protein residues are subjected to a significant level of microbial attack, while enzymic yield of amino acids is now low. Hind gut microflora, present at 10^{10} bacteria per gram of digesta, may well deal with a further 15–20% of proteinaceous material in the digesta distal to the terminal ileum; but the disappearance of nitrogen from the hind gut does not reflect a yield of material useful to the pig. The nitrogen is mainly in the form of unusable non-amino moieties and is passed out directly via the urine. As no useful amino acids are absorbed from the hind gut, digestibility measured at the terminal ileum is likely to give a better estimate of utilizability than digestibility measured from the rate of faecal excretion. The faecal excretion of amino acids is, in fact, pig-dependent to quite some degree, being in part of endogenous cells and secretions, and this element of total faecal loss is therefore relatively less affected by feed types within the conventional range.

Values for digestibility of crude protein at the terminal ileum – ileal digestibilities – are usually about 8% lower than faecal digestibilities (over 43 feedstuffs, ileal digestibility of crude protein = 0.92 faecal digestibility of crude protein). If this relation were to be constant, no greater precision would be achieved by using ileal rather than faecal digestibility values. But it is not constant and varies between amino acids and between feed ingredients. One reason for this is that for some feeds (such as meat and bone, meat offal and some more difficult to digest plant protein sources) a larger proportion tends to escape digestion in the small intestine and passes into the large intestine, where protein is more likely to be degraded by the microbial gut flora to non-amino nitrogen than be absorbed as amino acids. In the important case of soya bean meal, whereas the ileal digestibility of amino acids is normally some 8% less than faecal digestibility, for under- or overheated beans the difference increases to almost 20%.

While the difference between faecal and ileal digestibility of protein is evidently feed-dependent, it is also the case that this difference is variable amongst amino acids. Thus, while the difference between faecal and ileal protein digestibility of cereals is about 10%, the difference between faecal and ileal lysine digestibility of cereals is about 5%. Threonine and tryptophan are particularly liable to disappear by deamination or decarboxylation from the large intestine, whilst for methionine net bacterial synthesis may even occur. The rate and type of degradation occurring in the large intestine is highly dependent upon the ingesta environment, especially the level and type of carbohydrate present and the presence or absence of toxins.

These differences in the behaviour of amino acids in the small and large intestine are exemplified by a general tendency for the ileal digestibilities for threonine and tryptophan to be relatively lower than the general digestibility of protein, whilst the ileal digestibility of methionine is generally higher.

The ileal digestibility of protein and four of the most interesting amino acids in some feedstuffs is given in Table 9.5, whilst Table 9.6 shows the ratio of ileal digestible amino acid to ileal digestible crude protein averaged over 43 feedstuffs. Although the ileal digestibility for lysine is shown in Table 9.5 as being quite similar to the ileal digestibility for crude protein as a whole, this relationship is feed-dependent. Thus, while the ileal digestibility of lysine in fish meal is somewhat higher than the ileal digestibility of crude protein, the ileal digestibility of lysine in feather meal is considerably lower than the ileal digestibility of feather-meal crude protein.

Table 9.5 Ileal digestibility of some amino acids in some feedstuffs (values mostly from few measurements only)

	Protein[1]	Lysine	Methionine	Threonine	Tryptophan
Barley	0.70–0.80	0.70–0.80	0.75–0.85	0.65–0.75	0.70–0.80
Wheat	0.70–0.80	0.70–0.80	0.75–0.85	0.65–0.75	0.70–0.80
Maize	0.70–0.80	0.70–0.80	0.75–0.85	0.65–0.75	0.70–0.80
Wheat offals	0.30–0.60	0.30–0.70	0.30–0.50	0.40–0.50	0.50–0.60
Maize gluten feed	0.40–0.60	0.50–0.60	0.57–0.70	0.40–0.55	0.30–0.40
Extracted soya bean meal	0.75–0.85	0.80–0.90	0.80–0.90	0.75–0.80	0.75–0.80
Rape-seed meal[2]	0.40–0.60	0.50–0.70	0.60–0.80	0.50–0.70	0.50–0.70
Meat-and-bone meal	0.40–0.70	0.50–0.70	0.60–0.80	0.50–0.65	0.40–0.60
Fish meal	0.80–0.90	0.85–0.95	0.85–0.95	0.75–0.85	0.70–0.75

[1] In general, ileal digestibility of crude protein is about 8% lower than faecal digestibility of crude protein. Interesting exceptions are rape-seed meal (12% lower), corn gluten feed and corn gluten meal (16% lower), wheat middlings (17% lower), and meat-and-bone meal (14% lower).

[2] Often typical of cotton-seed meal, coconut meal and lupin meal, some other by-products, and some secondary ingredients.

Table 9.6 Ratio of ileal digestible amino acid:ileal digestible crude protein (average of 43 feedstuffs)

	Ileal digestible amino acid/ Ileal digestible crude protein
Isoleucine	1.06
Lysine	1.02
Methionine	1.20
Threonine	0.95
Tryptophan	0.95

[1] Methionine is almost twice as variable for this character as the other amino acids. In the case of maize, the ratios for lysine, threonine and tryptophan are 0.88, 0.94 and 0.92. For feather meal the ratio for lysine is 0.72.

The use of ileal digestible essential amino acids appears to be an appropriate way of assessing the potential of feed ingredients to satisfy the amino acid specifications for a pig diet. This supposition has been borne out by feeding trials in which growth rate was much closer correlated to the ileal digestibility than to the faecal digestibility.

Effective multiple measurements of the apparent ileal digestibilities of essential amino acids in feedstuffs for pigs are still inadequate in number. There remains a lack of accurate documented measurements for many feedstuffs, and the variation between measurements made, with no explanation for that variation, remains a worrying aspect of feed definition in this way. Nevertheless, the concept is so clearly an improvement upon the simple measurement of crude protein that nutritionists and feed compounders are rapidly developing the database of ileal digestibilities for the essential amino acids. New methodologies for estimating ileal digestible amino acids for pigs are under current development. Often a good appreciation of these values can be gained from making the measurement in other species, such as chickens or rats. *In vitro* enzymic techniques, together with a knowledge of the carbohydrate components of the diet, are clearly attractive. Modified nylon bag techniques have been tried and show promise. Whilst *in vivo* studies with live pigs cannulated at the terminal ileum will remain the datum for the measurement of digestible amino acids in feedstuffs, more rapid predictors will certainly come forward. It is also reasonable to assume that variation between batches of feedstuffs in their digestible amino acid content will become more readily interpreted.

Available amino acids

It might be assumed that, once absorbed, amino acids are available for use by the pig for purposes of maintenance, growth or lactation. Level of ultimate use is, of course, modified by the level of requirement (excess over requirement not being used), and by the appropriate balance, one to the other, of amino acids absorbed in relation to the balance of amino acids needed for anabolic processes (amino acids supplied, but not in balance with others, not being used).

However, even after these predictable losses of absorbed amino acids are accounted for, ileal digestibility values still overestimate net nutritional value. The rate of utilization of balanced ileal digestible amino acids not provided surplus to requirement is seen to be less than completely efficient. This phenomenon has been the long-time study of Dr Ted Batterham at Wollongbar in New South Wales, Australia. The ileal digestibility of lysine, threonine, methionine and tryptophan amino acids in both cotton-seed and soya bean meal may be measured as 0.7–0.8. But the amount of lysine retained from cotton-seed protein is only about 40% of that ileally digested; while some 80% of the lysine is retained from ileal digested soya bean protein. Equally, while some 45% of ileal digestible threonine in cotton-seed meal may be retained, the equivalent value for soya bean threonine is some 65% of that digested. For methionine, equivalent percentages for retention of ileally digested amino acids are 60% and 70%. This problem of utilizability of digested amino acids seems to pertain particularly to protein meals which have been processed, or which require some degree of heat treatment.

It may be surmised that no biological process can possibly be 100% efficient. A loss due to mechanical efficiency should be justifiably allocated. It appears that an appropriate value for the efficiency of transfer of absorbed balanced amino acids to tissue protein (v) may usually range from 0.80 to 0.95 for conventional high quality feedstuffs. Such a value appears to be supported by achieved retentions of ileally digested amino acids from fish meal and soya bean. However, as already shown, the information from other feedstuffs such as cotton seed and lupin shows this mechanical efficiency value to be feed-dependent; it may in practice vary from 0.5 to 0.95. The cause of this failure of use of absorbed amino acids is yet unknown, but may be associated with relative rates of absorption, the presence of toxins, the binding of absorbed amino acids at gut level or beyond, or some other undiagnosed biochemical phenomenon.

The concept of v, and its possible variability amongst feedstuffs, raises the importance of considering the ultimate availability of amino acids as the true measure of nutritional worth, availability adding to the coefficient of ileal digestibility a further coefficient for utilization. Due to the mysterious nature of v, overall available amino acids may often be best measured by biological assay such as growth-response studies. One such method is the slope ratio assay which uses graded levels of the test ingredient against graded levels of standard feed, or artificial amino acid (Figure 9.3). The relative responses (as N retention, growth, carcass gain, efficiency or whatever) are measured. Where the comparison is against an artificial amino acid, the ileal digestibility coefficient and the v value for the artificial amino acid can be taken as unity. The ratio of the slopes of the performances will give a coefficient of the relative utilization of the tested feed amino acid.

It is sufficiently evident that feed description requires some approach toward a definition of available amino acids, as given by the product of ileal digestibility and v. The measurement of v and/or the determination of overall availability by growth trial assay requires continuing scientific attention.

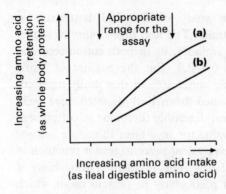

Figure 9.3 Slope ratio assay comparing the efficiency of use of ileal-digested amino acid for growth in pigs: (**a**) from a standard feedstuff and (**b**) from a test feedstuff. Note that the assay must be completed over the linear response range. Amino acid retention may be measured by comparative slaughter or by prediction. The slope for the test feedstuff is less than that for the control, indicating relatively more interference in the availability of amino acid post-absorption in that case.

Where conventional feed ingredients (such as high-quality animal products, cereals, and high-quality vegetable proteins from oil extraction processes) are readily available, these issues may seem matters of scientific rather than practical interest. But where less conventional feedstuffs must be used, or a wider range of feedstuffs due to inconsistent product quality, or variable product availability, or when new feedstuffs require to be evaluated, the matter of the efficiency of utilization of absorbed amino acids (v) is paramount. It is the latter position which pertains in many pig-producing countries world-wide, and should be considered as the norm rather than the exception.

Guide values for protein and amino acid content of pig feedstuffs

Guide values to the levels of crude protein, digestible protein, faecal and ileal digestible amino acids, together with notes referring to likely final availabilities, are to be found in Appendix 1. These values may be used to assess the monetary value of feedstuffs and to formulate them correctly into diets. Book values for nutritional content are never as precise as those determined directly for the sample feedstuffs concerned. However, it is often neither cost-effective nor necessary always to determine nutritive values directly. Given a carefully constructed set of guide values, and a mechanism for monitoring performance response, formulation of diets on the strength of information such as in Appendix 1 can be quite adequate for most purposes. This comment particularly pertains where well-known and conventional ingredients are used.

CHAPTER 10

VALUE OF FATS IN PIG DIETS

Introduction

The available energy in utilizable fats is about 2.25 times that in utilizable carbohydrates.

The major fatty acids in feed fats are palmitic (16:0), stearic (18:0), oleic (18:1) and linoleic (18:2); but lauric (12:0), myristic (14:0), palmitoleic (16:1), linolenic (18:3), arachidonic (20:4), eicosapentaenoic (20:5), erucic (22:1) and docosahexaenoic (22:6) may also be found. (In '18:2', etc., the number before the colon refers to the number of carbon atoms in the chain, and the number after the colon refers to the number of double bonds.) Those fatty acids with no double bonds (for example, 18:0) are termed saturated and have high melting points. One double bond (for example 18:1) gives a mono-unsaturated fatty acid, and when there are two or more double bonds (for example 18:2) the fatty acid is polyunsaturated (PUFA). The ratio of unsaturated to saturated fatty acids in feed has important consequences for the dietary value of the feedstuff and for the human food value of the pig meat which results.

The major sources of vegetable oil, in order of importance, are soya bean (17.5% oil) [rich in 18:2 fatty acids], palm (50% oil) [rich in 16:0, 18:1 mesocarp; 12:0, 14:0 kernel], sunflower seed (40% oil) [rich in 18:2], rape seed (35% oil) [rich in 22:1 {erucic} or 18:1 {zero varieties}], groundnut (45% oil) [rich in 18:2], coconut (65% oil) [rich in 12:0], cotton (20% oil) [rich in 18:2, 18:1]. Erucic acid is toxic, and cotton-seed oil may contain toxic elements. The major sources of animal fats are cattle tallow [18:1, 16:0, 18:0], sheep fat [18:1, 16:0, 18:0], pig lard [18:1, 16:0, 18:2] and fish oils [18:1, 16:0, 20:1, 22:1]. Fatty acids of chain length greater than 18 are noticeable in fish oil, and a change in balance of saturated [18:0] to unsaturated [18:2] fatty acids noticeable when ruminants are compared to pigs.

Commercially available soya oil and soya-acid oil contain about 10% 16:0 and 5% 18:0 (20% saturated); 30% 18:1 and 50% 18:2 (80% unsaturated). Palm-acid oil contains about 45% 16:0 and 5% 18:0 (50% saturated); 40% 18:1 and 10% 18:2 (50% unsaturated). Tallow contains about 25% 16:0 and 20% 18:0 (50% saturated); 40% 18:1 and 5% 18:2 (50% unsaturated). Fish oils are interesting for their highly unsaturated 20 and 22 fatty acids. This makes them prone to rancidity and off-flavours, but some (like C20:5) are presumed beneficial to human health. Other waste fats are available, of various composition. A major waste source is recovered vegetable oil (RVO). This may have up to 45% 16:0 and as little as 1% 18:2, but is highly variable. RVO is less digestible than other fat sources and is of dubious value in a blend. Table 10.1 shows the fatty acid compositions of some common feed-fat sources.

Tallow is available in a wide range of qualities and requires to be closely examined for fatty acid spectrum, peroxide levels and general impurities. The level of free fatty acids is used as a quality control, and less than 15% free fatty acids is desired. A high quality tallow (FFA <5%, peroxide 2.5m.eq (kg)) may be fully digestible by adult pigs (DE = 39 MJ/kg); while badly stored tallow (FFA >40%, peroxide 4.0), or a lower feed-grade tallow, may be 75% digestible (DE = 30) or less.

Acid oils, such as soya-acid oil and palm-acid oil, have high levels of free fatty acids (about 80%) and are available from industrial fat hydrolysis processes. Often a blend containing soya-acid oil or palm-acid oil, together with, for example, tallow and soya oil, may readily contain 40–50% of free fatty acids. Acid oils now comprise a significant proportion of non-tallow feed grade fats. There is less problem perceived with a high level of free fatty acids in vegetable oil products, and, once stabilized, these are considered by some to be almost as valuable as tallow and soya oil triglycerides. Fat products with high levels of free fatty acids, particularly above 30%, are, however, less well digested, and are invariably of more variable digestibility. The process of industrial hydrolysis is not identical with that taking place in the pig's intestine, and this is assumed to be the reason for the decline in DE value with increasing FFA content – which appears more marked with younger than with older animals. Varying levels of free fatty acids from varying sources can also affect the palatability of diets. This is one of the reasons why a high level of free fatty acids may not be volunteered for by a feed manufacturer, but only accepted on the basis of favourable pricing.

Short and medium chain fatty acids are softer but less palatable than longer chain fatty acids, while those above 20 may be prone to off (fishy) flavours.

Table 10.1 Approximate fatty acid compositions (% of total lipid)

	Soya oil	Rape oil	Palm acid oil	Soya acid oil	Coconut[1]	Beef tallow	Pig lard	Fish oil	Mixed soft acid oils
C16:0	10	5	45	10	35	25	20	20	10
C16:1	–	–	–	–	–	5	–	10	–
C18:0	5	2	5	5	15	20	15	5	5
C18:1	25	55	40	30	–	40	45	15	40
C18:2	50	25	10	50	10	5	15[2]	5	30
C18:3	5	10	–	5	–	1	–	3	5
C20:3	5	–	–	–	–	–	–	–	–
C20:4	–	–	–	–	–	–	–	3	–
C20:5	–	–	–	–	–	–	–	20	–
C22:5	–	–	–	–	–	–	–	2	–
C22:6	5	–	–	–	–	–	–	15	–
FFA[3]	1	1	80	65	–	10	–	1	65

[1] Also rich in C12.

[2] Typical value when the diet contains 15g linoleic acid/kg. If dietary linoleic acid rises to >5% linoleic acid in the diet, the level of linoleic acid in pig fat can rise to levels as high as 35% of linoleic acid in pig lard. Pig fatty tissues are broken down and reassembled as the pig grows. Withdrawal of high dietary levels of linoleic acid will cause a decline in the linoleic acid content of pig fat which is related to the amount of pig growth which has taken place following the reduction in dietary linoleic acid. Where Y is the linoleic acid in pig fatty tissue (%) and X is the amount of live-weight gain (kg) which has occurred between withdrawal of a diet high in linoleic acid and the point of measurement of the linoleic acid content of pig fat, then $Y = 34 - 0.3X$.

[3] Free fatty acids.

Note that maize seed has 4–5% of oil which contains over 50% 18:2 + 18:3 which will impart improved digestibility of saturated fatty acids added to maize-based diets. Most other cereals have 2–3% fat with about 50% 18:2. Meat-and-bone meal is low in linoleic acid (about 6% of the fat only).

Fatty acids below 16 and above 20 should therefore each be minimized to below 2.5%.

Alpha-tocopherol (vitamin E) inhibits oxidation of unsaturated fatty acids, so if the fat source is high in polyunsaturates, then vitamin E is needed at the rate of 0.6mg/g linoleic (18:2) acid. Vegetable oils are usually already high in vitamin E. Because of the double bonds, polyunsaturates are prone to oxidize to peroxides, which even at low level dramatically reduce palatability. Oxidative rancidity is a chain reaction, so once started proceeds ever faster; fats high in linoleic are particularly susceptible. The antioxidants, butylated hydroxytoluene or toco-pherol (vitamin E), are often used in fat-supplemented animal diets.

Animals can anabolize, although at considerable metabolic cost, 18:0 and

16:0 from carbohydrate, then (if necessary) introduce double bonds into saturated fatty acids (i.e. desaturate) at position 9 (thus to create oleic from stearic), but not at position 12. So, 9,12-linoleic (18:2) is essential (i.e. cannot be manufactured within the animal body). Arachidonic is needed for prostaglandins and is obtained from linoleic through gamma-linolenic (18:3). The linolenic series is essential to normal heart and blood supply functions, and to the endocrine and immune systems. Ideally, linoleic acid should provide a minimum of about 4% of total dietary energy; for the lactating animal, 8% of total dietary energy. This is not only because of its value as a percursor for essential body components, but also because dietary polyunsaturated fatty acid is some 20 times more efficiently incorporated into animal body and milk fat than if these latter fatty acids had to be made by the body from other dietary nutrients.

Fats high in saturated fatty acids, or of medium chain length (12:0, 14:0, 16:0, 18:0), are generally more poorly absorbed, but vegetable oils high in unsaturated fatty acids, of longer chain length (18:1, 18:2, 20:1, 22:1), are readily absorbed and highly digestible. The DE value for a high-grade soya oil is likely to be around 37MJ/kg, while that for a high-grade tallow is more likely to be around 33MJ/kg. Oils that have been hydrogenated have reduced digestibility (unhydrogenated fish oils are 90% digestible, while partially hydrogenated fish oils and hydrogenated fish acid oils are only 60–80% digestible). Digestible energy values for unhydrogenated fish oils range between 34 and 37MJ DE/kg, while for partly hydrogenated fish oils the estimate is 28–34MJ DE/kg.

To be absorbed, fatty acids from hydrolysed fat can be bound to a protein which prefers unsaturated to saturated fatty acids, and long-chain to short- or medium-chain fatty acids. Digestibility thus improves with chain length and degree of unsaturation. Unsaturated fatty acids also help to increase the solubility (and absorbability) of non-polar lipids, such as stearic acid, into the absorptive micellar phase; thus the addition of linoleic may enhance by synergism the digestibility and absorption of the less well digested saturated fatty acids. Synergism is highest at lower levels of fat inclusion.

Young pigs, although using the dietary fat from milk efficiently (obviously), do not digest feed fats well, but unsaturated long chain fatty acids are more readily digested than saturated and/or medium chain fatty acids. It is not clear why vegetable fats are less well digested than sow milk fats, but this is evidently the case for newly weaned pigs who are not as effective at digesting fatty acids as older pigs.

It does not appear that levels of inclusion of fats up to 10% will significantly decrease the efficiency of digestion (or utilization) in pigs, although opinion on this is hotly contested and some authorities suggest that DE value is likely to fall off at levels of above 7% of inclusion of fat in the diet. Thus the effective digestible energy value of fats may fall with level of inclusion, giving a quadratic rather than a linear change in digestible energy value of diets with increasing levels of added fat. In comparison to the DE of the highly digestible unsaturated fats, the less digestible tallows (16:0, 18:0), and the less well used free fatty acids, are the most likely to show the phenomenon of the DE value being reduced as level of dietary fat inclusion increases.

Progressive increments of usable fats into pig diets up to levels of 8%, and possibly more, may be assumed to give *pro rata* increments in energy density of the diet. Increasing the energy density of baby-pig and growing-pig diets appears to be associated with linear corresponding improvements in growth rate of animals restricted by limited appetite. The economics of this growth depends upon the cost of the fat additions, but not only will the faster growth improve efficiency in modern strains of pigs, but the reduction in days to slaughter increases throughput, defrays fixed costs and allows lower stocking rates (with consequent further improvements in growth rate). There is a proposition that monogastrics will 'eat to energy'; that is, will voluntarily eat less of a diet rich in energy. Whilst this is the case to some extent, the effect is not counterbalancing, so energy intake rises on energy-rich diets. The addition of fats to diets of animals already given a restricted allowance is, of course, not logical (unless cost per unit of available energy is less for fat than for carbohydrate), and so long as carbohydrate energy is cheaper than fat energy per megajoule of net energy yielded, more of a less-dense feed is a better production tactic than the same amount of a more-dense feed.

Feed fats are often competitive with cereal energy sources on the basis of cost per unit of utilizable energy. But there are other advantages that accrue from the inclusion of fat into the diet.

In monogastrics, diet fats can slow the rate of passage and improve the digestibility of other feed components – especially carbohydrates – thus enhancing the assumed metabolizable energy value of the carbohydrate fraction of the diet and attributing to the fat a remarkable energy value. Because the presence of diet fat can improve the efficiency of use of the assumed total diet metabolizable energy, it is technically possible for the metabolizable energy values for fats in monogastric diets to be equal to or above the gross energy value.

At equal feed intake, a diet containing added fat will provide more energy; this is particularly important where physical capacity limits appetite, as in young pigs and lactating sows. In addition, diet fats can increase appetite (feed consumption) in pigs through: (a) increasing palatability and (b) decreasing the need to dissipate body heat (being efficiently absorbed and then efficiently converted from diet constituent into adipose tissue, energy provided in the form of fat has the consequence of a low heat burden in comparison to carbohydrate energy). This phenomenon is particularly important in animals which can have appetite depressed due to a high level of heat output (such as lactating sows), and in pigs kept in hot climates where the level of voluntary feed intake is seen to be unacceptably low on account of a high environmental temperature.

Fat coated pellets are of enhanced quality and reduce dust. Not only does this reduce wastage and improve the environmental atmosphere, it will also reduce respiratory disease and thereby improve growth rate and feed conversion.

Value of fats in pig diets

The ME value of acceptable quality feed fats with an unsaturated:saturated ratio of more than 1.5:1 should be about 32MJ ME/kg. The Edinburgh School of Agriculture prediction equation indicates 33MJ DE/kg. Metabolized energy from fat sources is less well used than energy from carbohydrate sources for maintenance, but better utilized for growth (body fat synthesis from diet fat is about 90% efficient, whilst body fat synthesis from carbohydrate is only 70% efficient). The fact that less heat requires to be dissipated not only means improved efficiency, but also enhances appetite in a warm environment. However, in a cold environment the same phenomenon means that high-fat diets give pigs a reduced resistance to cold. Thus, in comparison to carbohydrate, it is possible for fat to have an effective ME value equal to (or higher than) the gross energy value. Because of the metabolic consequences of fat consumption, effective net energy in a hot environment could well be around 40MJ/kg, while in a cold environment the failure to dissipate heat, and the consequent need to burn off nutrients specifically for cold thermogenesis, could make the effective net energy value of dietary fat to be as low as 25MJ/kg.

Fat additions to lactating sow diets can increase milk production by up to 25%. The use of high-energy density diets for lactating sows whose total energy intake is limited by appetite is bound to be helpful in reducing the rate of body fat losses. As sow fatness is an important contributor to reproductive efficiency, where rapid fat losses or low absolute levels of body fat are perceived as a problem in lactating sows, enhancing diet density by the addition of fat is likely to be an effective nutritional tactic.

Neonate pigs of less than 1kg at birth are likely to have a greater mortality rate. Feeding of 200g/day of fat in the last 2 weeks of pregnancy can improve postnatal survivability of pigs newborn to sows that are thin.

Fat reduces the digestibility of fibre at low-fibre levels. Dietary calcium and magnesium levels above 1% may significantly reduce the DE value (from 35 to 30) due to the formation of soaps. The soaps of the saturated fatty acids are the least digestible. Young pigs utilize tallows poorly and fully hydrogenated tallows hardly at all. Digestibility of fats (both vegetable and animal) improves with pigs up to 10 weeks of age.

One of the most immediate causes of lowered energy value from reduced digestibility is the presence in a fat blend of useless fatty materials which analyse as fat but are not available to the pig. Typical of these are oxidized fats and unsaponifiable fats which may comprise more than 10% of the total analysed lipid in a poor sample, but normally should be less than 3.5%. Fat sources may also contain valueless non-fatty impurities, such as water and polythene (maximum total, 1.0). Glycerol may be found in some fat blends at low levels; although not a fat it is, of course, nutritionally useful.

In pigs, saturated fatty acids have a reduced potential for micellar formation, and are therefore absorbed less well than unsaturated fatty acids, but this potential is itself improved by the addition of unsaturated fats. As the ratio of unsaturated:saturated fatty acids rises up to 2:1, fat digestibility of the mix continues to improve (from about 40% for a totally saturated fat to 60% for a 1:1 mix and to 80% for a 2:1 mix).

The addition of fats to the diet will enhance appetite, but diets higher in free fatty acids, and therefore proportionately lower in triglycerides, are less palatable to pigs. For pigs, a free fatty acid level that is high (above 30%) can cause digestibility to fall to 70%.

Incorporation of diet fats into animal carcass fat

Monogastric adipose tissue reflects diet fatty acid composition as fatty acids are absorbed largely unchanged. Use of unsaturated vegetable fats in pig diets will result in enhanced levels of linoleic acid in the pig fat. It is because long-chain fatty acids can be incorporated directly, or with little change, into animal adipose tissue that they have such high net energy values. A high linoleic acid supplement will produce high linoleic adipose tissue in pigs and poultry; the tissue can increase from a normal 10–15% linoleic acid in adipose tissue up to 30–40% linoleic with a consequent dramatic softening of the carcass fat. (This, of course, does not happen in ruminants, as the linoleic is hydrogenated in the rumen.) Linoleic acid is, of course, essential for young pigs up to 30kg live weight at a minimum of 1.5% of the diet, and for older pigs at a minimum of 0.75%. Levels above 1.5% may cause soft fat, but not invariably so. Soft fat can be avoided by the addition of saturated fatty acids (tallow) into the blend; but there is a penalty of reduced digestibility. Happily, the digestibility of tallow is better for the older animals in the weeks immediately prior to slaughter when it is more important that the fatty tissue being grown at that time does not contain too high a level of unsaturated long-chain fatty acids.

In pig diets fat increases palatability and intake as well as nutrient density, so the addition of fat to the diet can result in a frank excess of energy intake. This will cause an enhanced predisposition by the animal to lay down extra carcass fat. It is therefore reasonable to presume that adding fat into the diet will militate towards the possibility of an increase of fat in the body and in the carcass. Certainly, the body fat formed in the case of pigs will reflect the direct transfer of fatty acids in the diet to fatty acids in the body. The absolute rate of fatty tissue growth will, however, only increase if there is excess total energy supply over the needs for protein growth and minimum fat growth. Thus there will be a tendency for highly nutrient-dense diets, enhanced with added fats, to produce increased levels of carcass fat only if (a) the amount of energy relative to protein in the diet is imbalanced in favour of excess energy or (b) the total energy intake (balanced or otherwise) is excessive due to an increase in energy ingestion because the animal's appetite has increased as a result of reduced heat embarrassment or increased diet palatability (or any of the other related causes of diets with added fat giving enhanced intake).

Diet compounding using fats and oils

● Pure oils can be of reasonably certain parentage, acid oils are more variable, while tallows are extremely variable in quality – from high-grade tallows, which are excellent, to fully hydrogenated low-grade materials, which can be effectively useless in pig diets. Recovered vegetable oil and other diverse sources may be of low value. Variation in nutritional value of fat supplements can therefore range from almost zero to even above 40MJ effective ME/kg, depending entirely upon the sources chosen and the integrity of the blender. It is not unreasonable to suggest that, at an assumed given value of 32MJ ME/kg, available commercial materials may show effective values ranging between 24 for some commercial blends and 40 for high-grade pure oils. Some commercial blends may have effective DE values little greater than 25MJ DE/kg, particularly if used at very high levels of inclusion.

- High-quality materials may be of considerable benefit to a diet above the given ME value. Thus assumed multipliers for chemically analysed oil, often around 30–35MJ ME or DE/kg, may underestimate as well as overestimate the effective value of fat. Such added benefits would be improved digestibility of other diet components, improved pellet quality and increased feed intake. Even with a favourable tallow price in comparison to vegetable oils, it is always worth blending at least 20% vegetable oil into the tallow in order to obtain synergism and improve tallow digestibility.

- Young pigs digest fat rather poorly, but soon improve. Monogastrics always do better on vegetable oils than tallows but, apart from very high-quality starter diets, high-quality tallow can be used in the blend at levels increasing with age.

- Pigs can benefit, in the right conditions, from up to 8–10% fat in the diet. This can be added into the mixer, at the die, coated after the pellet has left the cuber, added through the medium of full-fat soya, or any combination of these.

- The monogastrics are best suited to unsaturated fatty acids and blends for pigs therefore contain little or no palm oils, and relatively higher levels of soya together with the tallow.

- Given the current improvements in the general level of fat quality entering the trade, and discrimination against the use of low-grade materials, there is every reason to suppose that, if the price is competitive with other energy sources, the amounts of fat used in animal diets will increase. Recently many benefits of dietary fat in addition to the value per MJ ME have been identified. Care must be taken, however, to balance the extra energy from fat with adequate high-quality protein, or the resultant imbalance will cause excess energy supply and fat carcasses.

- Specifications for the various fat blends required by feed manufacturers are best obtained from the feed manufacturers themselves. Broadly, the manufacturer may use a general-purpose blend for all species, or choose to use a special blend for pigs. Three blends allow for the possibility of two types for the monogastrics (a higher tallow for the older animals). Sometimes, a general-purpose mix of about 75% tallow and 25% soya can be used together with full-fat soya for pigs to redress the shortage of unsaturated fatty acids.

● Sources of fats and oils for pigs are usually blends of tallow, the vegetable oils from soya bean, maize, sunflower, rape and palm, together with acid oils (free fatty acids) and by-product and recovered oils. The best quality blend is likely to be a soya bean oil and high-quality tallow mix. A specification appropriate for young, growing pigs may set a maximum of 30% free fatty acids, a maximum of 30% for saturated fatty acids, a minimum of 20% linoleic (18:2) and 4% linolenic (18:3), with about 40% oleic (18:1) acids (the balance containing, amongst others, around 5–10% of stearic and 15–20% palmitic) and a 'tallow:soya' ratio of 50:50. The ratio of unsaturated to saturated fatty acids in soya oil is around 7:1, while the ratio of unsaturated to saturated fatty acids in tallow is around 1:1. The ratio of unsaturated to saturated fatty acids in a 50:50 mixture will be around 2:1. For older, growing pigs the free fatty acid maximum may be lifted to 40%, the maximum for saturated fatty acids to 40%, and the 'tallow:soya' ratio to about 60:40. Often, these standards for pigs will move towards more tallow in the blend if the price differential between tallow and soya oil widens, or a danger of soft carcass fat is perceived, and a general-purpose mix for pigs of 70:30 would not be unusual.

CHAPTER 11

ENERGY AND PROTEIN REQUIREMENTS FOR MAINTENANCE, GROWTH AND REPRODUCTION

Introduction

Nutrient requirements are met by nutrients in feedstuffs. Requirement must therefore be expressed in terms wholly compatible, and open to 1:1 linkage, with descriptors for feedstuffs.

Many thousands of experiments have been completed measuring the response of pigs to graded levels of nutrient input. Figure 11.1 shows the results of one such experiment to identify the requirement for an amino acid. The requirement is taken to be around the point of the diminishment of response, in the region of 5.0g per day.

The problems of this approach are many, including:

1. a different requirement for each different criterion of assessment (feed efficiency, carcass quality, economic margin and so on);

2. a complex multigraded experiment being needed for each nutrient, each environment, each moment in time and place and for each type of pig (weight, sex, genotype), and for all the interactions between nutrients, pig types, times, environments and places;

3. the limiting factor often not being the achievement of the requirement, but rather a shortage of another nutrient (Figure 11.2).

The more appropriate approach, with general applicability, is to identify requirement in terms of constants and efficiency factors for each of the specific products required. The nutrient supply necessary to provide for any given level of product or function can then be calculated. For present

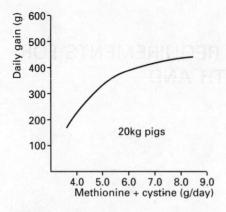

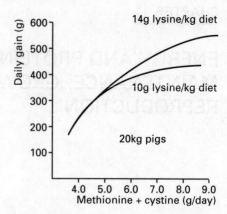

Figure 11.1 Growth rate response to graded levels of methionine + cystine identifying requirements around 5g amino acid daily.

Figure 11.2 Growth rate response to graded levels of methionine + cystine in two diets, one with 10g lysine per kilogram of diet and the other with 14g lysine per kilogram of diet, identifying requirement at around 8g amino acid daily. Even with adequate lysine, response may be limited by some other nutrient, perhaps energy.

purposes these are taken to be some combination of protein retention, lipid retention, the foetal load, lactation, cold thermogenesis and maintenance.

Energy

The energy requirements of pigs are most conveniently stated in terms of the metabolizable energy (ME). Energy requirement is then calculated from the net energy of a particular function, and the associated efficiency of utilization of the ME.

Not all energy is equally well used for all purposes. In order to obtain energy from protein it must first be deaminated, which itself costs energy. While protein contains some 23.6MJ/kg, ME yield from digested protein which has been deaminated is about 11.5MJ/kg (deamination costing about 5MJ/kg protein and 7MJ/kg protein being voided in the urine in association with nitrogen moieties). Energy from fibre is yielded through the end-products of digestion in the form of volatile fatty acids, which are only about 50% as efficiently used as energy from glucose; while energy from lipid is especially well used if it is reincorporated into body fat. The effective ME of digested fibre may be taken as around 0.5 of the DE value. The ME of carbohydrate and lipid may be taken as 1.0 of the DE. Because fibre levels are relatively low in most conventional pig diets (but not all), the major controller on this system is in fact the amount of energy yielded from protein. ME is usually about 0.96 of DE, for a mixed diet varying from 0.94 to 0.97; the lower value applies when there is a high rate of deamination and urinary N excretion due to either oversupply of diet protein or the supply of poorly balanced dietary amino acids.

Maintenance

Energy for maintenance (E_m, MJ ME per day) is usually expressed as a function of $W^{0.75}$ because larger animals have a smaller maintenance requirement in relation to body weight.

$$E_m = 0.440W^{0.75}$$

(Equation 11.1)

remains the best-used of all estimates of the energy requirement for maintenance. The three-quarter exponent has no special biological meaning, and the United Kingdom Agricultural and Food Research Council Working Party found $0.72W^{0.63}$ to give the line of best fit for growing pigs. There is a suspicion that the exponent for live weight is helpful merely in adjusting for the pig carrying more fat as it grows heavier and, because much of the energy expenditure for maintenance is concerned with the turnover of body protein tissues, there may be some logic also in using a function based on protein mass (Pt), such as:

$$\text{Growing pigs: } E_m = 1.75Pt^{0.75}$$

(Equation 11.2)

the exponent now accounting for larger pigs turning over their protein mass at a slower rate than smaller pigs. Equation 11.2 is suitable for growing pigs. For breeding sows the exponent in Equation 11.2 is also satisfactory at higher values for Pt, and:

$$\text{Breeding sows: } E_m = 0.44W^{0.75} = 1.75Pt^{0.75}$$

(Equation 11.3)

Ambient temperature (T) below the critical or comfort temperature (Tc) induces cold thermogenesis. The energy cost of cold thermogenesis (E_{h^1}) may be estimated as 0.018MJ ME for each kg of metabolic body weight ($W^{0.75}$) for each degree (C) of cold (Tc $-$ T) for pigs housed singly, and 0.012MJ for pigs housed in groups:

$$E_{h^1} = 0.012W^{0.75}(Tc - T)$$

(Equation 11.4)

Tc is itself modified by the level of metabolic activity of the animal. Higher feed intakes and higher rates of growth or milk production will give a lower value for effective critical temperature and therefore a greater resistance to cold by creating an elevated rate of heat output. Critical temperature (°C) may be estimated as:

$$Tc = 27 - 0.6H$$

(Equation 11.5)

where H is total body heat output, which is the sum of all the bodily work functions for maintenance, growth of protein and fatty tissues, lactation, foetal growth and protein deamination.

The quality of the environment modifies the effective environmental temperature, thus:

$$T = T(Ve)(Vl)$$

(Equation 11.6)

where Ve is an estimate of the modifying effect of the rate of air movement and insulation, and Vl is an estimate of the modifying effect of floor type. Values for Ve and Vl are given in Table 11.1.

Production

One kilogram of retained protein (Pr) contains 23.6MJ. On average, some 31MJ per kg of Pr is required in addition to cover for the energetic work of protein tissue turnover associated with protein anabolism, and the energetic cost of tissue synthesis itself. This gives a total energy cost for protein retention (E_{Pr}) of 55MJ per kg:

$$E_{Pr} = 23.6Pr + 31Pr$$

(Equation 11.7)

the efficiency of protein retention (k_{Pr}) being 0.44. This, however, is an inadequate descriptor. Although the cost of tissue synthesis – the joining

Table 11.1 Scores for Ve and Vl for use in calculating the effective environmental temperature (T) in the equation $T = T(Ve)(Vl)$

Rate of air movement and degree of insulation	Ve
Insulated, not draughty	1.0
Not insulated, not draughty	0.9
Insulated, slightly draughty	0.8
Insulated, draughty	0.7
Not insulated, draughty	0.6

Floor type in lying area	Vl
Deep straw bed	1.4
Shallow straw bed	1.2
No bedding on insulated floor	1.0
Slatted floor with no draughts	1.0
No bedding on uninsulated floor	0.9
Slatted floor with draughts under	0.8
No bedding on wet, uninsulated floor	0.7

together of the amino acids into body protein – is probably independent both of the rate of synthesis and of the size of the pig and is around 5MJ per kg of protein synthesized (estimates range from 4 to 8), the energetic cost of tissue turnover is likely to be positively related to both the rate of turnover and the mass of protein tissue (Pt). In younger pigs of about 20kg live weight, the ratio of new protein retained to total protein turned over is about 1:5, while for older pigs at 100kg live weight the ratio is about 1:8. The major part of the energy expenditure for protein synthesis may therefore be variable rather than fixed, and positively related to pig size. Where Px is the total mass of protein turned over in a day (including new protein deposited):

$$E_{Pr} = 5Px + 23.6Pr$$

(Equation 11.8)

The definitive estimation for the daily rate of body protein turnover, and the factors affecting that rate, are not yet well elucidated. It may, however, be assumed that there is a minimal value, probably in the region of 5% of the total body protein mass (Pt):

$$Px_{min} = 0.05Pt$$

(Equation 11.9)

It may further be assumed that turnover is some function of the degree of maturity, and the following has been estimated:

337

$$Px = Pr/(0.23[Pt - Pt]/Pt)$$

(Equation 11.10)

where Pt is the mature total protein mass. It remains impossible to quantify the possibility that Px is also related in some proportional way to Pr, such that higher daily rates of protein retention may be associated with higher rates of daily protein turnover.

Values require to be assigned to Pt and Pt. Pt is accumulated as the pig grows and is therefore self-generated; usually it approximates to 0.16W. Pt is strongly influenced by pig type and may vary between 25 and 50kg, but for modern genotypes is now usually accepted to be some 40–50kg. As the daily rate of protein growth is strongly related to mature size, it may be possible to calculate Pt from knowledge of the maximum rate of protein deposition (Pt) associated with any particular sex or genotype (the Gompertz growth coefficient, B, is described in Chapter 3 and Table 3.4):

$$Pt = e(Pt)/B$$

(Equation 11.11)

As at least some of the above remains conjectural, and not well quantified, it may be forgivable to settle for a single fixed value for k_{Pr} of 0.44. However, many nutritionists believe k_{Pr} to have higher values of up to 0.54 for younger pigs, and to have lower values down to 0.38 for heavier pigs of 100kg or more.

For sows, the calculation of E_{Pr}, with the involvement of estimates of turnover, has the unfortunate consequence that as maturity is approached the cost of the small remaining increments of Pr approaches infinity – which is unacceptable – and a fixed value for k_{Pr} of, say, 0.44 has much to commend it.

One kilogram of retained lipid (Lr) will contain 39.3MJ. Most of the work energy associated with lipid retention is to do with the creation and linking together of appropriate fatty acids, and there is little energy cost of lipid turnover. Only some 14MJ per kg Lr is therefore suggested to be required to cover for the energetic work of fatty-tissue turnover and tissue synthesis, giving a total energy cost for lipid retention of 53MJ per kg:

$$E_{Lr} = 39.3Lr + 14Lr$$

(Equation 11.12)

the efficiency of lipid retention (k_{Lr}) being 0.75. In contrast to the situation for protein synthesis, this is likely to be an entirely adequate descriptor because turnover of fatty tissue is much slower and less important. In the event of a significant proportion of the substrate for lipid synthesis being dietary lipid, k_{Lr} may be greater than 0.75, the efficiency of direct transfer being high at up to 0.90.

Reproduction

Pregnancy involves the progressive development of the foetal load, the placenta and foetal membranes, foetal fluids, the uterus and the initial development of the mammary glands. The final total gravid uterus at term weighs around 25kg and contains almost 3kg of protein and 85MJ of energy. The rate of energy deposition in the uterus (E_u) is slight in early pregnancy but rises exponentially. French and UK workers at Shinfield determined:

$$E_u(MJ/day) = 0.107e^{0.027t}$$

(Equation 11.13)

where t is the day of gestation. The efficiency of use of ME for energy deposition in the uterus (k_u) is probably around 0.5. The energy cost of deposition in the gravid uterus therefore appears to be for days 20, 80 and 110 of pregnancy, 0.4, 1.9 and 4.2MJ ME respectively (if t = 110, then E_u/ 0.5 = 4.2).

The rate of energy deposition in mammary tissue (E_{mam}) has been handled similarly to E_u by the French/UK team:

$$E_{mam}(MJ/day) = 0.115e^{0.016t}$$

(Equation 11.14)

Efficiency of deposition (k_{mam}) is probably similar to k_u at 0.5. Thus on days 80 and 110 of pregnancy some 0.8 and 1.3MJ ME respectively are required.

Lactation level is highly dependent upon litter size (Chapter 4). Because the energy cost of sucking piglet live-weight gain is approximately 22MJ ME/ kg and the energy value of milk is 5.4MJ ME/kg (being 55g protein, 50g lactose and 80g fat per kg), milk yield may be estimated as:

Daily piglet gains × number of piglets × 4

(Equation 11.15)

Individual sucking pigs may gain 200–400g daily depending upon their age and size.

The efficiency of use of ME derived from dietary sources for milk production (k_l) may be taken as 0.70. The energy requirement for the formation of 1kg of milk, inclusive of the energy retained in milk, is therefore 7.7MJ ME per kg. During lactation, sows will lose significant quantities of lipid from the maternal body and this is used to support milk synthesis – especially milk fat synthesis – with a high efficiency of 0.85 or more. Fatty tissue loss of 0.5kg would therefore contribute 16.7MJ (39.3 × 0.5 × 0.85), which at 5.4MJ ME per kg milk is enough to satisfy the requirements for 3.1kg of milk (the same 3.1kg of milk would require 23.9MJ from dietary sources; in terms of diet equivalent, therefore, 1kg of body lipid is worth 48 dietary MJ).

In the event of body protein being catabolized for purposes of providing substrate for milk lactose or milk fat (rather than milk protein), the efficiency of conversion will be much lower at 0.5.

Protein

Diet proteins are digested and absorbed as amino acids. Ileal digestibility values are an appropriate guide to the likely rate of appearance of amino acid in the body, although the efficiency factor v (Chapter 9), the proportion of absorbed amino acids available for metabolism, brings about a writing-down of the ileal digestible amino acids in order to achieve a true estimate of available amino acids.

The proteinaceous content of feedstuffs comprises amino acids and non-protein nitrogen. Of the 22 commonly found amino acids, nine are essential to the pig. These cannot be manufactured in the body. Some non-essential amino acids can be synthesized from non-protein nitrogen, and the non-essential amino acids are interconvertible. Protein requirement is therefore usually expressed in terms of (a) the total protein supply as crude protein (CP) or digestible crude protein (DCP) and (b) the essential amino acids which must be supplied to the pig in the diet.

Both these fractions must be supplied adequately for the pig to have an effective protein economy. It is taken for granted that if the essential amino acid provision is adequate, then so also will be the non-essential balance. This will be true for pigs fed conventional feedstuffs, but the possibility of purchasing pure amino acids, and the fashion of presenting diet protein content in terms of the single amino acid lysine, may mislead. It is possible – but unlikely – for the key amino acids lysine, threonine, methionine and tryptophan to be supplied industrially from artificial sources well up to the level of dietary requirement, but for the pig to have a frank undersupply of digestible crude protein.

Protein is required in the body (i) to replace losses consequent upon body protein tissue turnover which is highly active but not perfectly efficient, (ii) for the manufacture of body enzymes, replacement of intestinal epithelial cells and synthesis of various gut secretions and (iii) for deposition and retention in lean tissue growth, the foetal load and milk.

The utilization of digested protein: ideal protein

The amount of diet protein which is available for metabolism is dependent upon the level of essential amino acid supply. Special interest is often centred on those amino acids required at high levels by pigs but which are comparatively lower in feedstuffs: lysine, histidine, threonine, methionine and tryptophan.

Because the body does not use the amino acids supplied to it as individual entities, but rather for purposes of synthesizing proteins from mixtures of essential amino acids and non-essential amino acids, the balance of the essential amino acids is crucial to optimum protein utilization. Protein metabolism requires a supply of 'ideal protein'; that is, essential amino acids balanced correctly for the various purposes of maintenance and production. Amino acids supplied outwith the necessary balance cannot be used for protein synthesis and must therefore be deaminated.

The value of a diet protein depends upon the relationship of the balance of essential amino acids required to make up ideal protein on the one hand, and the balance of amino acids supplied from the diet on the other hand. This is not to suggest that pig diets should be purposely compounded in order to supply protein exactly in its ideal form – that would be not only extremely expensive but also would waste the biological potential of the pig to turn lower quality cereal proteins into higher-quality pig-meat protein. The proposition, however, does allow knowledge of the proportion of the total diet protein that will be used by the pig for productive purposes. This latter is necessary both for the effective provision of requirement and for the proper economic evaluation of the worth of pig feedstuffs and mixed diets.

The ideal protein content of a diet is not the sum of the contributions of ideal protein from the ingredients. This is because the amino acids in the various feedstuffs can be complementary. A judicious mixture of feedstuffs, each lacking in a different amino acid, can produce a diet of a quality far in excess of its individual components. In short, the content of ideal protein in a mixed diet is not the weighted mean of the ideal protein contents of the ingredients.

The 50/50 combination of one foodstuff of which the protein is only 0.5 ideal because of shortage of lysine, with another which is only 0.4 ideal because of a shortage of threonine, is not a diet of which the protein is 0.45

ideal but one which is 0.6 ideal, because the essential amino acid falling short in the first feedstuff is complemented by excesses in the second (Figure 11.3).

Calculation of protein value requires a strict definition of the essential amino acid balance in ideal protein for pigs. Much painstaking work has gone into the determination of amino acid requirements for pigs, especially in the US, Poland, France, England and Scotland. The spectrum of amino acids in ideal protein is largely that pertaining in the product for which the amino acids are required; namely, pig-meat protein, pig milk or maintenance. If there are relatively small differences in the efficiency of utilization (v) of individual absorbed amino acids, then ideal protein should look rather like pig-meat protein, or sow-milk protein, there being similarity between the two as shown in Table 11.2. Perhaps, because there is some inefficiency in collecting up sulphur amino acids during protein turnover, methionine + cystine levels in ideal protein might be a little higher than in pig tissue. The essential amino acid composition of ideal protein is suggested in Table 11.3. The total of all the essential amino acids in ideal protein seems to be just less

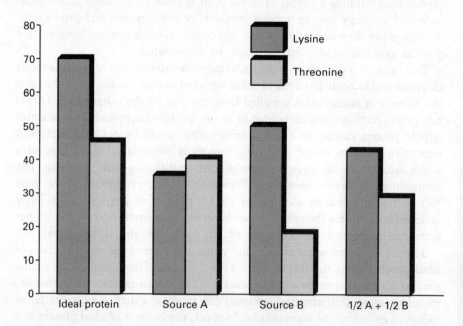

Figure 11.3 Compared to ideal protein with 70g lysine/kg, feed source A, having only 35g lysine/kg, is limited by inadequate lysine to being used with an efficiency of 0.5. Compared to ideal protein with 45g threonine/kg, feed source B, having only 18g threonine/kg, is limited by inadequate threonine to being used with an efficiency of 0.4. As is shown, a 50/50 mixture of A + B is used with an efficiency not of 0.45 (the mean of the separate efficiencies of the two sources) but of 0.60.

Table 11.2 Amino acid composition of pig-meat protein and sow-milk protein (g/kg)

Amino acid	Tissue protein	Milk protein
Histidine	30	25
Isoleucine	35	40
Leucine	75	85
Lysine	70	75
Methionine + cystine	35	30
Phenylalanine + tyrosine	70	80
Threonine	40	40
Tryptophan	15	15
Valine	50	55
Other non-essential[1]	500	

[1] Alanine 70; glutamine 140; glycine 100; proline 65; serine 40; asparagine 80.

Table 11.3 Ideal protein for growing pigs and pregnant and lactating sows. The required balance of amino acids is the balance of the utilizable ileal digested amino acids

Amino acid	Amino acid (g per kg ideal protein)
Histidine	25
Isoleucine	40
Leucine	75
Lysine	70
Methionine + cystine	40[1]
Phenylalanine + tyrosine	75[1]
Threonine	45
Tryptophan	15
Valine	50
Total essential amino acids	435
Total non-essential amino acids	565

[1] At least half as the first-named.

than half the total protein; the other half represents the non-essential amino acids.

Having provided the criterion for ideal protein, the amino acid spectrum of the protein in a mixed diet can now be compared with it and the value of the diet (V) derived (Table 11.4). The effective value of the protein in the diet is that proportion of the dietary protein judged utilizable on the strength of the most-limiting amino acid. In the case shown, lysine is the most limiting, and the maximum efficiency of use of the diet protein (V) for growing pigs is 0.64 or 64%. 0.64 of the ileally digested and subsequently utilizable protein will be available (although not necessarily all used) for maintenance and for protein tissue synthesis, while 0.36 will be deaminated and excreted via the urine. If there is oversupply, the rate of deamination will be proportionately higher. There is possibly further reason to believe that, in the case of some amino acids, oversupply may have a negative influence upon protein value.

The addition of artificial lysine to the diet shown in Table 11.4 could raise the lysine concentration in the protein to, say, 55g/kg, in which event threonine and methionine + cystine would become the limiting amino acids and the protein value (V) would become 0.75. However, there may be some doubt as to the efficacy of large additions of synthetic lysine to boost protein value. There is a view that the efficiency of its use may be lower than natural lysine if the total artificial lysine contributes more than about 2.0g/kg of total feed lysine. But recent work from Eire appears to suggest that up to 4kg lysine, 1kg threonine and 0.5kg methionine per tonne of feed may be added efficaciously in the form of artificial amino acids to diets limiting in these amino acids from natural sources. In some circumstances nutritionists may force in an extra 1kg of artificial lysine per tonne of feed where dubious or unusual feed ingredients are used. (Artificial amino acids may be assumed to be 100% digestible. L-lysine HCl contains 780g lysine/kg, L-threonine contains 980g threonine/kg, DL-methionine contains 980g methionine/kg, and DL-tryptophan contains 800g tryptophan/kg.)

The proportion of ideal protein in cereal protein is usually about 0.45. For pig diets, protein supplements are added to raise the value of cereal-based diets (see Table 11.5), and these supplements are usually particularly rich in lysine (cereal proteins being particularly poor). Although lysine is often the first limiting amino acid in pig diets, and is usually the controller of protein value, this is not invariably the case.

Table 11.4 Derivation of the protein value (V) of a mixed diet

	Ileal-digested and utilizable amino acid (g per kg diet protein) (A)	Amino acid (g per kg ideal protein) (B)	V (A/B)
Histidine	20	25	0.80
Isoleucine	40	40	1.00
Leucine	70	75	0.93
Lysine	45	70	0.64
Methionine + cystine	30	40	0.75
Phenylalanine + tyrosine	65	75	0.87
Threonine	30	45	0.67
Tryptophan	10	15	0.67
Valine	40	50	0.80

The protein value (V) is taken as the lowest score in the column A/B.

Table 11.5 Some protein sources used in pig diets grouped according to their ability to impart high protein value to diets in which they are included[1]

Class 1 (good) protein sources	Class 2 (mediocre) protein sources	Class 3 (poor) protein sources
Fish meals	Meat, and meat and bone	Cereals
Milk products	meals	Cereal by-products
Single-cell proteins	Rape seed	
Soya bean	Groundnut	
	Field beans	

Type of diet	Approximate protein value (V)[2]
Class 3 sources alone	0.45–0.55
Class 3 + class 2	0.55–0.70
Class 3 + class 1 + class 2	0.55–0.80
Class 3 + class 1	0.65–0.85

[1] This table refers to the quality of the protein in feeds. This is independent of the level of protein, although many feeds high in protein are in the class 1 category and many feeds low in protein in classes 2 or 3.

[2] The higher the inclusion level of any protein source, the greater its effect on dietary protein value. When lysine is first limiting, V can be calculated as V = (lysine in diet (g/kg)/CP (g/kg))/0.07.

The balance of essential amino acids in ideal protein and in pig tissue and pig-milk protein is represented in Figure 11.4, while Figure 11.5 shows how only around half of unsupplemented cereal protein can be utilized by the pig because the proportion of the total cereal protein which is ideal is limited by lysine.

In pig diets, dietary protein values (V) usually vary between 0.65 and 0.85. Even when the calculated number for diet protein value is in excess of 0.85, it is often best to take 0.85 as an assumed pragmatic upper limit for commercial diets.

Deamination of a proportion of the absorbed diet protein is a proper part of efficient pig production. Pigs convert lower-quality plant proteins into higher-quality pig proteins, the unbalanced amino acids being excreted. To feed ideal protein in a diet would be grossly uneconomic, and it is appropriate to offer more of a lower quality. As the pig requires its protein in terms of daily mass, not protein proportion, one diet protein with half the biological value of another can provide the same requirement if given in twice the amount. Energy considerations aside, it only needs the second protein to be less than half the price of the first for the poorer quality to be worthy of consideration for inclusion into the diet.

With the capability of calculating V, the proportion of diet protein that is ideal, protein supply can be presented in terms of ileal digested and available essential amino acids supplied in ideal balance; namely, ideal protein. Crude

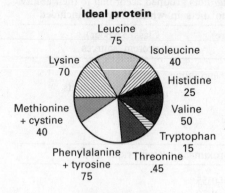

Ideal protein

Leucine 75
Isoleucine 40
Histidine 25
Valine 50
Tryptophan 15
Threonine .45
Phenylalanine + tyrosine 75
Methionine + cystine 40
Lysine 70

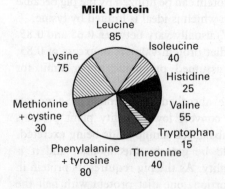

Tissue protein

Leucine 70
Isoleucine 35
Histidine 30
Valine 50
Tryptophan 15
Threonine 40
Phenylalanine + tyrosine 70
Methionine + cystine 30
Lysine 70

Milk protein

Leucine 85
Isoleucine 40
Histidine 25
Valine 55
Tryptophan 15
Threonine 40
Phenylalanine + tyrosine 80
Methionine + cystine 30
Lysine 75

Figure 11.4 Showing the similarity of balance of essential amino acids in ideal protein, pig tissue protein and pig milk protein.

protein supplied in pig diets is usually in the range 140–240g CP/kg diet. Ileal digestibility of protein (D_{il}) may vary between 0.60 and 0.90 with a reasonably expected average of about 0.75, depending upon source and circumstance. The protein value V for conventional pig diets may range from 0.65 to 0.85, being around 0.75–0.85 for young pigs, 0.70–0.80 for growing pigs and 0.65–0.75 for finishing pigs and for sows. The efficiency of use of ileally digested ideal protein (v) may vary from 0.70 to 0.95.

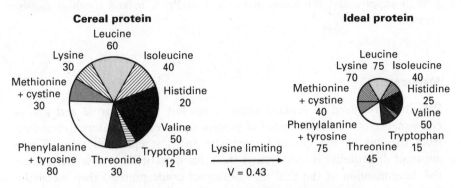

Figure 11.5 Only 0.43 of unsupplemented cereal protein can be utilized because of the particular imbalance and shortage of the amino acid lysine.

Where the total available ideal protein required daily to support the protein need for maintenance (IP_m), production (IP_{Pr}) or reproduction (IP_u and IP_l), is represented as IP_t, then:

$$IP_t = (F)(CP)(D_{il})(V)(v)$$

(Equation 11.16)

where F is the feed intake, CP is the crude protein concentration, D_{il} is the ileal digestibility of protein, V is the protein value in comparison to ideal protein in terms of the amino acid balance and v is the efficiency of use of absorbed ideal protein.

IP_t is the source of supply of ideal protein for maintenance and production functions, thus:

$$IP_t = IP_m + IP_{Pr} + IP_u + IP_l$$

(Equation 11.17)

The above model is effective in estimating the supply of balanced amino acids (ideal protein) to satisfy the needs of maintenance and production, but falls short in desiring only a single value for D_{il} and v to represent all the amino acids, because the algorithm works on the base of protein (IP) and not the various constituent amino acids. Given that D_{il} and v both differ amongst the amino acids, an alternative approach is to describe protein supply in terms of each of the 9 (+2) essential amino acids, $aa_{(i)}(i = 1–9)$:

$$aa_{(i)} \times F \times D_{il(i)} \times v_{(i)}$$

(Equation 11.18)

to give the supply of utilizable $aa_{(i)}$ ($uaa_{(i)}$) at tissue level. (v) ($uaa_{(i)}$) is then compared to the amino acid spectrum of ideal protein for maintenance (m), body tissues (Pr), milk (l) or conceptus (u). The amino acid in shortest supply on the strength of this comparison gives the value for V for the diet protein supply, and the levels of product of Pr, l, m and u which can be created.

Maintenance

Protein is required to replace amino acids lost via the urine and gut in consequence of the inefficiencies of protein turnover, and failure to reabsorb from the gut all the enzymic secretions and lost epithelial cells. Where apparent digestibility is used (rather than true digestibility), as is the case for the determination of the ileal digestibility of crude protein, then metabolic faecal losses are set against the diet and should not be double-counted. Maintenance, therefore, will accord with the obligatory losses in the urine following deamination. Ideal protein requirements for maintenance (IP_m) may be expressed on the basis of metabolic body weight:

$$IP_m = 0.0013W^{0.75}$$

(Equation 11.19)

or, more logically, as a function of the protein mass:

$$IP_m = 0.0040Pt$$

(Equation 11.20)

There is no reason to doubt that this estimate is as valid for the breeding sow as for the growing pig.

Production

Given that the supply is already calculated in terms of available absorbed ideal protein, all inefficiencies have been included in the estimate of IP_t and it follows that the requirement for production (IP_{Pr}) exactly equals the daily rate of protein retention:

$$IP_{Pr} = Pr$$

(Equation 11.21)

$$Pr = IP_t - IP_m$$

(Equation 11.22)

This relationship will hold until $Pr = P\hat{r}$, the potential maximum daily rate (see Chapter 3). Beyond this position there will be excess ideal protein (IP_{xs}), and this adds to the pool of protein destined for deamination (Pm).

Reproduction

Protein deposition in the uterus over the course of the pregnancy (t, days) has been estimated from the data of the combined French and British research teams of Noblet and his co-workers in the mid-eighties:

$$Pr_u = 0.0036e^{0.026t}$$

(Equation 11.23)

This equation gives a gravid uterus at term with some 3kg of protein therein, and daily deposition rates at days 20, 80 and 110 of pregnancy of 6g, 29g and 63g respectively. As for Pr, IP_{Pr_u} requirements equal Pr_u.

The rate of protein deposition in the mammary tissue, Pr_{mam}, has been described as:

$$Pr_{mam} = 0.000038e^{0.059t}$$

(Equation 11.24)

The product Pr_u is, of course, lost at parturition. There is also a loss of nitrogen in urine after parturition associated with regression of the maternal uterine tissues. At weaning there is similarly regression of Pr_{mam} and consequential losses of nitrogen in the urine.

The protein content of milk is around 55g per kg, and daily milk yield can readily rise to 12kg. The requirement for Pr_l may be expressed as:

$$IP_{Pr_l} = Pr_l$$

(Equation 11.25)

Where Pr_l exceeds the availability of IP_{Pr_l}, then maternal body-protein catabolism will supplement the dietary supply. It may be assumed that efficiency of transfer of body-tissue protein to milk protein would be in the region of 0.8–0.9, given that the amino acid profiles and the likely values for v for muscle protein and milk protein are so similar.

Example calculations for the energy and protein requirements of growing and breeding pigs

As may be gathered from the foregoing, a sophisticated calculation of requirement necessitates the use of a complex model to determine net energy and protein usages, and to deal with all the various interactions. Such

349

techniques are dealt with later. For the present, a more simple approach with generalized factors is utilized.

Naive calculation of the energy and protein requirements of a young pig

Energy:

$W(kg) = 10$; $Pt(kg) = 1.6$; $E_m(MJ\ ME/day) = 1.75Pt^{0.75} = 2.5$
$Bp = 0.018$; $Pt = 45$; $Pr(kg/day) = 0.096$; $E_{Pr}(MJ\ ME/kg) = 55$; $E_{Pr}(MJ\ ME/day) = 5.3$
$Lr(kg/day) = 0.7Pr = 0.067$; $E_{Lr}(MJ\ ME/kg) = 53$; $E_{Lr}(MJ\ ME/day) = 3.6$
$H = E_m + 31Pr + 14Lr = 2.7 + 3.0 + 0.9 = 6.6$; $Tc = 27 - 0.6H = 23.0°C$
if
$T(Ve)(Vl) = 25(0.8)(1.0) = 20°C$
then
$E_{hi}(MJ\ ME/day) = 0.012W^{0.75}(Tc-T) = 0.2$

Total energy requirement = 11.6MJ ME = *12.1MJ DE per day*

Protein:

$W(kg) = 10$; $Pt(kg) = 1.6$; $IP_m(kg/day) = 0.0040Pt = 0.0064$
$IP_{Pr}(kg/day) = Pr = 0.096$
$IP_t(kg/day) = IP_{Pr} + IP_m = 0.102$
if
$v = 0.90$, $V = 0.80(0.056kg\ lysine/kg\ protein)$, $D_{il} = 0.80$
then
the efficiency of use of protein = 0.58

Total crude protein requirement = 0.102/0.58 = *0.193kg CP per day*

Summary of daily requirements:
   Energy: 12.1MJ DE
   Protein: 0.193kg CP
   Lysine: 0.056 × 0.193 = 0.0109kg lysine
   g CP/MJ DE: 16.0
   g lysine/MJ DE: 0.90

Where the pig eats 0.75kg feed daily:
   Dietary concentration of CP = 260g CP/kg diet
   Dietary concentration of DE = 16.1MJ DE/kg diet

Dietary concentration of lysine = 14.5g lysine/kg diet

Naive calculation of the energy and protein requirements of a growing pig

Energy:

$W(kg) = 60$; $Pt(kg) = 10.2$; $E_m(MJ\ ME/day) = 10.0$
$Bp = 0.0125$; $P\dot{t} = 47.5$; $Pr(kg/day) = 0.196$; $E_{Pr}(MJ\ ME/day) = 10.8$
$Lr(kg/day) = 0.7Pr = 0.137$; $E_{Lr}(MJ\ ME/day) = 7.3$

Total energy requirement = 28.1MJ ME = *29.3MJ DE per day*

Protein:

$IP_m(kg/day) = 0.041$
$IP_{Pr}(kg/day) = 0.196$
$IP_t(kg/day) = 0.237$
if
$v = 0.90$, $V = 0.80(0.056kg\ lysine/kg\ protein)$, $D_{il} = 0.78$
then
the efficiency of use of protein = 0.56

Total crude protein requirement = 0.237/0.56 = *0.423kg CP per day*

Summary of daily requirements:
  Energy: 29.3MJ DE
  Protein: 0.423kg CP
  Lysine: 0.0237kg lysine
  g CP/MJ DE: 14.4
  g lysine/MJ DE: 0.81

Where the pig eats 2.2kg feed daily:
  Dietary concentration of CP = 192g CP/kg diet
  Dietary concentration of DE = 13.3MJ DE/kg diet
  Dietary concentration of lysine = 11.6g lysine/kg diet

Naive calculation of the energy and protein requirements of a finishing pig

Energy:

$W(kg) = 100$; $Pt(kg) = 17.0$; $E_m(MJ\ ME/day) = 14.7$
$Bp = 0.011$; $P\dot{t} = 40$; $Pr(kg/day) = 0.160$; $E_{Pr}(MJ\ ME/day) = 8.8$
$Lr(kg/day) = 0.9Pr = 0.144$; $E_{Lr}(MJ\ ME/day) = 7.6$

Total energy requirement = 31.1MJ ME = *32.4MJ DE per day*

Protein:

IP_m(kg/day) = 0.068
IP_{Pr}(kg/day) = 0.160
IP_t(kg/day) = 0.228
if
v = 0.88, V = 0.78(0.055kg lysine/kg protein), D_{il} = 0.76
then
the efficiency of use of protein = 0.52

Total crude protein requirement = 0.228/0.52 = *0.437kg CP per day*

Summary of daily requirements:
Energy: 32.4MJ DE
Protein: 0.437kg CP
Lysine: 0.0240kg lysine
g CP/MJ DE: 13.5
g lysine/MJ DE: 0.74

Where the pig eats 2.5kg of feed daily:
Dietary concentration of CP = 175g CP/kg diet
Dietary concentration of DE = 13.0MJ DE/kg diet
Dietary concentration of lysine = 9.6g lysine/kg diet

Naive calculation of the energy and protein requirements of a pregnant sow

Energy:

W(kg) = 200; Pt(kg) = 31; E_m(MJ ME/day) = $1.75Pt^{0.75}$ = *22.9*
Day of gestation (t) = 80; E_u(MJ ME/day) = $0.107e^{0.027t}/0.5$ = *1.9*
Day of gestation (t) = 80; E_{mam}(MJ ME/day) = $0.115e^{0.016t}/0.5$ = *0.8*
Parity = 3; Pr(kg/day) = 5/116 = 0.043; E_{Pr}(MJ ME/kg) = 55; E_{Pr}(MJ ME/day) = *2.4*
Parity = 3; Lr(kg/day) = 7/116 = 0.060; E_{Lr}(MJ ME/kg) = 53; E_{Lr}(MJ ME/day) = *3.2*
Replacement of 4kg of lactation lipid loss = 4/116 = 0.034; E_{Lr}(MJ ME/day) = *1.8*
H = E_m + 0.5(E_u + E_{mam}) + 31Pr + 14Lr = 26.8;   Tc = 27 − 0.6H = 11°C
if
T(Ve)(Vl) = 15(0.8)(0.8) = 9.6°C
then

$E_{hi}(MJ \ ME/day) = 0.012W^{0.75}(Tc - T) = 1.0$

Total energy requirement $= 34MJ \ ME = 35.4MJ \ DE \ per \ day$

Protein:

$W(kg) = 200; \ Pt(kg) = 30; \ IP_m(kg/day) = 0.0040Pt = 0.120$
$IP_{Pr_u} \ (kg/day) = 0.0036e^{0.026t} = 0.029$
$IP_{Pr_{mam}} \ (kg/day) = 0.000038e^{0.059t} = 0.004$
$IP_{Pr}(kg/day) = 0.043$
$IP_t(kg/day) = 0.196$
if
$v = 0.87, \ V = 0.75(0.052kg \ lysine/kg \ protein), \ D_{il} = 0.75$
then
the efficiency of use of protein $= 0.489$

Total crude protein requirement $= 0.196/0.49 = 0.400kg \ CP \ per \ day$

Summary of daily requirements:
Energy: 35.4MJ DE
Protein: 0.400kg CP
Lysine: $0.052 \times 0.400 = 0.0208kg$ lysine
g CP/MJ DE: 11.3
g lysine/MJ DE: 0.59

Where the pig is given 2.6kg of feed daily:
Dietary concentration of CP = 154g CP/kg diet
Dietary concentration of DE = 13.6MJ DE/kg diet
Dietary concentration of lysine = 8.0g lysine/kg diet

Naive calculation of the energy and protein requirements of a lactating sow

Energy:

$W(kg) = 240; \ Pt(kg) = 36; \ E_m(MJ \ ME/day) = 25.8$
Piglets $= 10;$ milk yield $= 12kg$
if
0.25kg of maternal fatty tissue is lost daily
then
$0.25 \times 39.3 \times 0.85 = 8.4MJ$ available for milk synthesis $= 8.4/5.4 = 1.55kg$ milk;
leaving
$12 - 1.55kg = 10.45kg$ milk to be provided from dietary resources

$E_l(MJ\ ME/kg) = 7.7$; $E_l(MJ\ ME/day) = 7.7 \times 10.45 = 80.5$

Total energy requirement $= 106.3MJ\ ME = 110.8MJ\ DE\ per\ day$

Protein:

$IP_m(kg/day) = 0.144$
Protein content of milk $= 0.055kg\ CP/kg$
if
maternal body protein tissue contributes toward the protein of 0.4kg of milk
then
$0.022/0.85 = 0.026kg$ body tissue protein daily will be lost
and
this leaves 11.6kg milk to be provided from dietary resources
$IP_{Pr_1}(kg/day) = 0.055 \times 11.6 = 0.638$
$IP_t(kg/day) = 0.782$
if
$v = 0.90$, $V = 0.77(0.054kg\ lysine/kg\ protein)$, $D_{il} = 0.78$
then
the efficiency of use of protein $= 0.54$

Total crude protein requirement $= 0.782/0.54 = 1.45kg\ CP\ per\ day$

Summary of daily requirements:
Energy: 110.8MJ DE
Protein: 1.45kg CP
Lysine: 0.0783kg lysine
g CP/MJ DE: 13.1
g lysine/MJ DE $= 0.71$

Where the pig eats 8kg of feed daily:
Dietary concentration of CP $= 181g$ of CP/kg diet
Dietary concentration of DE $= 13.9MJ$ DE/kg diet
Dietary concentration of lysine $= 9.8g$ lysine/kg diet

It is evident that maternal tissues may well contribute in excess of that estimated above, or that the milk yield may well be curtailed.

Recommended energy and protein concentrations in diets for different classes of pigs

Recommended nutrient requirements expressed in terms of nutrient concentrations clearly have shortcomings in their generality of application

354

(and hence a likelihood of being wrong) and their dependence upon the amount of the diet the pig may be consuming. Nevertheless, such guides to diet specifications are used world-wide as the basis of diet formulation, and their derivation remains the main preoccupation of both research and industrial nutritional scientists. Diet specifications and the recommendation of nutrient concentrations represent the culmination of the sciences of growth, reproduction and nutritional biochemistry, as has been demonstrated in this chapter. Recommended guide diet specifications are to be found in Appendix 2.

Energy and protein requirements of the breeding boar

The common factor associated with experimentation to determine the nutrient needs of working boars is the equivocal nature of the response. Results are often contradictory, some experiments showing positive response to the nutritional treatment and some negative, but usually there is no response at all. Even where response might be expected, there is bound to be a considerable lag-time due to spermatogenesis itself taking some 4 weeks to complete, and passage through the male tract to storage awaiting ejaculation taking a further 2 weeks more, giving a lag of almost 2 months for any nutritional effects on sperm formation to be seen through effects on fertility.

In general it may be assumed that there is little effect of energy or protein nutrition on the efficiency of the breeding boar, provided that diet protein levels are maintained at a reasonable level and amino acid quality. In this latter regard, one of the more consistent findings has been that long-term protein inadequacy will moderate sperm production. Enough feed should be given to maintain fitness while avoiding excessive leanness, or – and possibly most important – excessive fatness. Libido and fertility are likely to be reduced should the boar become over-fat and over-heavy.

Maintenance requirements (E_m and IP_m) for breeding boars may be taken as the same as for breeding sows. There may frequently be a requirement for cold thermogenesis (Tc $\simeq$ 20°C, and often boars may find themselves in a draughty environment with an inadequately insulated floor), the extent of the energy need to support an effective reaction to a cold environment clearly depending much upon individual housing circumstances.

Post-pubertal growth could be assumed empirically to proceed at around 500g daily to 150kg live weight, 250g daily to 250kg live weight, and 100g daily thereafter until mature size of around 350kg or more is attained.

While factorial calculations can be made for the growth of protein and fat, maintenance and cold thermogenesis, and indeed the work of mating itself and the production of sperm, such would be to imply a precision of nutritional knowledge which does not exist. Common practice would suggest that the energy requirements of working boars should be met by the provision of 30–40MJ DE per day. This would normally be between 2.5 and 3.25kg of feed, although lower-density diets are likely to be the more preferable due to the benefits of greater bulk density for the satiation of physical appetite. Dietary energy is normally balanced with protein, amino acid, minerals and vitamins in a manner similar to that for the breeding sow (for example, 12g CP/MJ DE; 0.6g lysine/MJ DE).

CHAPTER 12

REQUIREMENTS FOR WATER, MINERALS AND VITAMINS

Introduction

The body of the pig contains on average 16% protein, 16% lipid, 3% mineral ash and 65% water.

Water is a major structural element giving form to the body through cell turgidity. Water also acts as a transport medium, both in the intestines and in the blood and tissues of the body-proper. It is the medium at metabolic level for enzyme-aided biochemical reactions. Because of the crucial necessity for maintenance of body-water balance, the system is set up on the assumption that water consumption will always be in excess of requirement.

The body ash is held in fairly strict ratio to protein; ash = 0.20Pt, which is not surprising in view of the support role which the skeleton has for the lean tissue mass. Of the mineral matter the vast majority (three-quarters) is calcium and phosphorus as bone hydroxyapatite, and the remainder mostly sodium, potassium and chlorine in body fluids. Magnesium, iron, zinc and copper are measurable, while the other minerals and all the other vitamins are to be found only in traces (Tables 12.1 and 12.2).

Proportionality of body content is not indicative of importance. Vitally important vitamins and minerals are used at trace levels acting as catalysts in enzymic reactions and recycled. The low levels of daily needs, the presence of internal buffer and storage mechanisms in the body, together with the often adequate background levels occurring naturally in feedstuffs in any event, result in a diversity of understanding of requirement for trace mineral elements and vitamins, and a breadth of opinion as to the exact quantities pigs may actually need. It is a noticeable feature of nutritional standard-setting for trace elements and vitamins that National Research Councils and

Table 12.1 Mineral composition of pigs (g/kg body weight)

(Protein	170)
Calcium	15
Phosphorus	10
Potassium	2
Sodium	1.6
Chlorine	1.1
Sulphur	1.5
Magnesium	0.4
Iron	0.060
Zinc	0.030
Copper	0.003
Molybdenum	0.003
Selenium	0.002
Manganese	0.0003
Iodine	0.0004
Cobalt	0.00006

Table 12.2 Mineral composition of sows' milk (g/kg milk)

(Protein	55)
Calcium	2.1
Phosphorus	1.5
Potassium	0.80
Sodium	0.35
Chlorine	0.20
Magnesium	0.10
Iron	0.0018
Copper	0.0008

public bodies often set requirements markedly lower than levels routinely added to pig diets by feed manufacturers (Table 12.3).

Fear of inadequate dietary supply of vitamins and trace minerals, together with their expense being a small proportion of total feed costs, leads to vitamin and mineral requirements being 'talked-up' to ever-greater levels of dietary inclusion. Progressive increases in vitamin and trace mineral allowances are the more suspect because in many cases naturally occurring levels may be adequate, and considerable quantities of many of the vitamins can be biosynthesized in the pig's gut. On the other hand, the discovery of the essential nature of dietary minerals and vitamins, and the supplementation of diets accordingly, has been one of the most exciting advances of biological science in the twentieth century, and has led to the considerable enhancement of productivity. Further, nutritionists must allow for significant increases (of up to 25%) in diet energy and protein density, which must be matched by equivalent increases in mineral and vitamin density. But, most

Table 12.3 Comparison of supplementary levels of trace minerals and vitamins generally added to diets for growing pigs by UK feed manufacturers, with UK Agricultural Research Council recommendations for total requirements published 4 years previously in 1981. The levels are expressed in mg/kg of air-dried diet

	ARC	Feed manufacturers
Manganese	4–14	40
Iron	55	150
Zinc	45	100
Copper	4	125 (growth promoter)
Cobalt	–	0.7
Iodine	0.14	1.2
Selenium	0.14	0.17
Vitamin A (i.u.)	2,000	12,000
Vitamin D_3 (i.u.)	120	2,000
Vitamin E	5.0	30
Vitamin K	0.3	2.0
Thiamine	1.4	1.5
Riboflavine	2.3	4.0
Pyridoxine	2.3	2.3
Nicotinic acid	13	15
Pantothenic acid	9	10
Cyanocobalamin	0.010	0.015
Biotin	–	0.030
Folic acid	–	0.20
Choline	800	500

important of all, nutritionists must aim for dramatic increases in rates of growth and reproduction over recent years. These increases have often not been matched by equivalent (or any) improvements in appetite. All these forces conspire to demand increased concentration of dietary vitamins and minerals. A requirement of n mg of a vitamin per kg of lean-tissue formation would calculate to a diet concentration of 0.1n mg/kg diet for a pig eating 2kg and gaining 0.4kg daily, of which half is fat; but 0.5n mg/kg diet for a pig eating 1.5kg and gaining 0.8kg daily, of which one-eighth is fat. As this example is in no way an exaggerated claim for the change of fortune of growing pigs over the last decade, increases in vitamin requirements should not necessarily be considered out of order. Last, vitamin potency may be lost with storage, both before inclusion into the diet and between diet manufacture and diet use. The threat of loss of potency may be realistically countered by an increase in level of diet supplementation. Clearly, where experimental trial evidence is used to set vitamin and mineral standards rather than practical and field experience, it is essential that feed intake, vitamin potency and pig performance are properly accounted.

It should come as no surprise that factorial calculation to estimate requirements for trace elements and vitamins used in tiny amounts, primarily as catalysts in the body, is doomed to failure. The empirical approach, primarily through (a) deprivation and (b) dose–response experiments, is the only effective means of establishing level of need. However, for the major minerals – calcium, phosphorus, magnesium, potassium and sodium – estimates of net deposition rates and obligatory losses (maintenance, digestibility and subsequent utilizability) should provide factors from which requirements can be deduced. Many have been the attempts, and the results remain regrettably unconvincing. There are a diversity of problems with the factorial approach, even although the major elements are clearly readily determinable as predictable proportions of the growth of the skeleton (calcium, phosphorus, magnesium), the growth of the soft tissue (potassium, sodium) and the secretion of milk (see Tables 12.1, 12.2 and 12.4).

From Table 12.4 the expectation may be derived that a 60kg pig retaining 150g protein daily would require some 14g calcium and 9g phosphorus for purposes of deposition alone. With a daily consumption of 2kg of diet, and assumed combined coefficients for digestibility and utilizability of 0.7 for calcium and 0.6 for phosphorus, the diet supply would calculate to 10g/kg and 8g/kg diet respectively.

It is the obligatory losses, digestibility and utilizability which have proved so difficult to predict, due mostly to the intrinsic control mechanisms for minerals which are not present for either energy or protein. Thus the digestibility of minerals, measured as their disappearance from the digestive tract, can fluctuate according to the relationship between supply and requirement. For calcium the intestine is a major route for endogenous loss and homeostatic control. Equally, the phosphorus content of urine will vary widely according to the discretion of the body; thus may phosphorus be absorbed in excess of need, only to be excreted with minimal metabolic cost. Sodium and potassium are readily digested, and as readily excreted – such a system is essential if ionic balance is to be maintained in the body on the prerequisite short-term basis. Magnesium is poorly absorbed, but little is needed, so a status of luxury consumption is normal.

Table 12.4 Deposition of minerals in pig gain

Mineral element	Mineral (g)/kg protein retained
Calcium	90
Phosphorus	60
Potassium	15
Sodium	10
Magnesium	2.5

Water

The water requirement for pigs is free availability at all times. The calculation of requirement is therefore one of expected rate of intake.

Water may be derived from the natural moisture of feedstuffs and from oxidative metabolism, which yields metabolic water. These sources, however, are trivial in comparison to the water supplied through drinking, or added to the feed.

The water content of the body of a pig can vary from 80% in the newborn to 50% in a fat adult. At any one time, however, the water content of soft body tissue and blood is quite stable, while water intake may fluctuate wildly. The kidneys excrete water via the urine in direct proportion to the intake. It is a confusing characteristic of the study of water requirement that liquid intake varies so widely amongst individual pigs, and across different times. Much of the variation in water consumption can be adjusted at the level of the digestive tract, but the main control mechanism is through the continuous excretion of water into the bladder for temporary storage prior to frequent voiding.

Water is needed by pigs as a means of maintaining a rigid but malleable body form (by giving rigidity to the muscle cells of the lean mass). Water upholds the ionic balance, forms the products of new growth, foetal tissue and milk, acts as a hydraulic fluid for conveying metabolites, endogenous biochemical moieties and waste through the body, comprises part of the thermoregulatory system by its evaporation from the lung surface, its specific heat maintains body temperature and prevents fluctuations, lubricates joints and is the essential medium for the conduct of the biochemical reactions of the digestive and metabolic processes.

The initial stages of water depletion will increase withdrawal rate from the large intestine, and predispose to constipation. Loss of body fluids in the form of secreted milk and urine will diminish, reducing productivity and piglet growth in the former case, and in the latter reducing the ability of the body to clear toxins while also disrupting ionic balance and water-sensitive metabolic reactions. The consequences of diarrhoea are those of water shortage due to uncontrolled faecal losses, and dehydration of the essential tissues, with disruption of ionic balance and failure of the endogenous biochemistry to operate in the absence of an adequately fluid medium. It is self-evident that water requirements will be related to the level of lactation yield, the amount of feed eaten, ambient temperature, the need to evaporate water from the lungs and the amount of toxic product to be cleared from the system via the urine. Water output is increased in proportion to the level of dietary unideal protein and dietary protein excess consequent upon the need to dilute the deamination product (urea) to a level of concentration appropriate for urinary excretion.

There is little logic in sophisticated calculation of water requirements, as any attempt to control supply to provide for a requirement would be counter-productive. In general, however, young pigs will need more water per kilogram of body weight than older ones, due to their greater surface area of body and lung in comparison to weight and the tendency for the urine of younger animals to be more dilute. Using sows' milk as a guide, piglets will prefer a ratio of water to dry matter of about 5:1, and early-weaned pigs given a dry diet may readily exceed this. Growing pigs are conventionally (and erroneously) understood to desire a water to dry-feed ratio of 2:1. Recent European data suggest that for growing pigs:

$$\text{minimum water need (l)} = 0.03 + 3.6I(kg)$$

(Equation 12.1)

where I is the feed intake. But even this is marginal and, given free access, a water to dry-feed ratio of 5:1 is common. Pregnant sows given water at any ratio of less than 3:1 are likely to be in deficit even at low environmental temperatures. As pregnant sows are often given their feed dry, and access to water is restricted in order to prevent wastage, these animals – especially in hot climates – are prone to suffer chronic water shortage and this may often be evidenced by hot-weather constipation. Pregnant sows given restricted levels of feed may show a desire to compensate for inadequate gut-fill by an enhanced water intake. Current interest in increasing the fibre content of pig diets, especially those for pregnant sows, is also likely to increase the required ratio of water to food. Lactating sows will require not merely the replacement of some 8–16kg of secreted milk, but also enough water to void in the urine huge quantities of metabolic by-products; this, in the lactating sow, is in addition to a need to create an adequate fluid environment in the intestinal tract if effective digestion of up to 8kg of food is to take place. A water to feed ratio of 5:1 is a reasonably likely absolute minimum. It is self-evident that water needs will rise with increasing ambient temperatures.

Water usage for pigs is almost best dealt with independent of aspects of pig metabolism, which the pig can take care of more than adequately, provided that water is freely available at all times. The means of providing free availability for water is a more appropriate subject for systematic study than any factorial estimation of requirements. The presence in the pen of a water point, in the form of a trough, bowl or nipple-drinker, may not be taken to imply free availability, and careful siting of water points, and maintenance of adequate flow-rate, is necessary. It is estimated that one water point is needed for every eight pigs in a pen, this ratio covering all pig types from young growers to adult sows. Lactating sows should have individual access to high-quality water flowing freely but at low pressure.

Water is the most important of appetite stimulators. Young pigs given dry feed may be seen to alternate between feed hopper and water point during the course of a meal, and feed intake will be directly proportional to the availability of the water. The simple expedient of adding water to feed can increase the voluntary intake of pigs by at least 5–10%, and occasionally by up to 30%. Pigs with appetite-limited productivity, such as young pigs, growers and lactating sows, will benefit from being given their feed wet rather than dry, and it is well appreciated that three times daily provision of a wet meal will elicit markedly greater feed intake than the *ad libitum* provision of dry feed from a self-feed hopper. Additional benefits of wet feeding are the reduction of dust, and also of feed wastage. Dry feeding systems may often waste 5–10% of the feed supplied, whilst wet feeding waste is normally 2–4%.

Automatic wet-feeding systems are one of the most effective means of both maximizing intake and controlling nutrient supply to individual pig rooms and pens. The hydraulic medium is, of course, water and in this capacity water fulfils one of its important roles on the pig unit. Wet-mix pumping can be readily achieved at 15–18% dry matter: a water to feed ratio of 5:1. This may be considered overly dilute to maximize feed intake due to the limited volume capacity of both the intestine of the pig and the feed trough. Optimum feed and water intake is facilitated by part of the water being ingested with the feed, and part drunk subsequently as pure water. Lesser ratios for the pumped feed, with top-up provision of water from independent water points, is therefore preferred. Few pumping capacities, however, can achieve a dry matter percentage exceeding 25% (a water to feed ratio of 3:1).

Given the modifying effects of feed intake and environmental temperature on water requirements, generalized approximations of water usages are given in Table 12.5. This level of provision will result in the consumption of 7,000–10,000 litres of potable water daily for every 100 sows and their progeny.

Table 12.5 Water intake allowances for pigs[1]

Type of pig	Water intake daily (kg)
Lactating sows	25–40
Pregnant sows	10–20
Growing pigs 20–160kg	Dry feed intake × 5
Young pigs	Dry feed intake × 6

[1] Water should be freely available to all classes of pigs at all times, even when feed is given wet.

Minerals

Calcium and phosphorus

The skeleton holds 98% of total body calcium and 80% of total body phosphorus in a ratio of just greater than 2:1 in the form of hydroxyapatite. Bone contains 45% mineral, 35% protein and 20% fat, while the bone ash comprises 40% calcium, 20% phosphorus and 1% magnesium. Bone is a dynamic and labile tissue, continuously in a state of balanced accretion and destruction. Calcium in soft tissue is found in muscle and plasma (blood contains around 10mg calcium/100ml), and soft-tissue phosphorus is also mostly in the muscle mass and in blood. In addition to bone formation, ionic calcium is involved in cell permeability, muscle contraction, blood clotting, excitation of nerves and the activation of enzyme systems. Phosphorus is heavily involved in a wide range of metabolic processes, especially the release of energy and the formation of amino acids. Requirement, however, is mostly perceived in terms of that for skeletal deposition in the course of growth, and mammary secretion in the course of lactation.

The digestibility of calcium is influenced by (a) the source of the element, (b) other diet components and (c) the physiological state and the physiological needs of the animal, as both endogenous calcium and phosphorus may be voided via the intestines (especially calcium, but also phosphorus which uses both intestinal and urinary routes). The ability of the gut to control rate of entry of intestinal calcium to the skeletal and plasma pool renders calcium digestibility a primary function of animal needs, and the digestibility coefficient will fall with both increase in supply and decrease in demand. This system uses both active transport and diffusion processes, and is somewhat dependent upon the presence of ample vitamin D_3. Vitamin D also interacts with parathyroid hormone to encourage skeletal mobilization of calcium and the maintenance of calcium homeostasis. The presence in the gut of dietary sugars, such as lactose, may enhance calcium digestibility, while oxalates will decrease it. Most important, however, is the consequence of dietary fats, which at high levels can mutually inhibit absorption of both the fat and the calcium by the formation of insoluble calcium soaps. Further interactions in the tract between dietary calcium and dietary zinc may result in the depression of either if one is consumed at excessive levels. Most common and conventional sources of calcium will freely yield the element for absorption from the intestine; the order of relative digestibility is bone meal, calcium carbonate, mono-, di- and tricalcium phosphate, the potential rate of digestibility in all cases being above 70%. But it is the level of diet calcium itself, together with the perceived calcium need, which is the primary control of both the rate of inflow (digestibility) and outflow (endogenous losses) fluxes across the gut wall.

The digestibility of phosphorus is reduced if diet calcium is in excess due to the formation of insoluble calcium phosphates. The ratio of calcium to phosphorus should be maintained in the diet at 2:1 or less, the ratio being more critical in the absence of adequate supplies of vitamin D_3. Organic phosphorus forms, especially the phytates, are less readily digested than inorganic forms, which is important in terms of interpreting the general usefulness of total diet phosphorus, much of which (often half or more) will be in organic form from plant sources. Cereals will contain around 3–4g P/kg, whilst cereal by-product phosphorus can rise to 5–10g P/kg (cereal calcium, in contrast, is trivial at 0.5g Ca/kg).

The digestibility of phosphorus from di- and tricalcium phosphate and bone meal is greater than that from alumino ferric phosphate, fluorine rock phosphates or phytate phosphorus; the latter group are likely to be lower than 40% digestible, whilst the former group may be digested at levels better than 70%. While luxury absorption of calcium results in it being cycled back into the gut lumen for endogenous excretion, and thus causes a decrease in apparent digestibility, the major mechanism for endogenous loss of phosphorus is via the urine, so the effects of level of diet phosphorus intake in comparison to bodily phosphorus need are noted not at the level of digestibility, as for calcium, but rather more at the level of urinary phosphorus outflow; the diet supply and rate of urinary loss of phosphorus being directly and closely related.

Inadequate calcium and phosphorus supply will depress growth and milk yield and result in poorly mineralized bones. The pig's requirement – as expressed in terms of dietary concentration – for fully mineralized bones of maximum strength is about 1g per kg higher than normally recommended levels, but there is little reason to wish to achieve such a target. There is scant evidence to support any relationship between joint, leg or foot weakness and the diet supply of calcium and phosphorus provided these are within reasonable range. Serious deficiencies do, of course, result in osteomalacia, osteoporosis and rickets.

The recommendations given in Table 12.6 assume that most of the diet calcium is in the form of either bone meal, calcium carbonate or dicalcium phosphate, and that about half of the diet phosphorus is from bone meal or dicalcium phosphate, while of the other half some three-quarters is organic phytate.

Table 12.6 Recommended calcium, phosphorus and sodium levels in pig diets

	Pigs up to 15kg live weight		Pigs of 15–150kg live weight		Breeding sows	
	g/kg diet	g/MJ DE	g/kg diet	g/MJ DE	g/kg diet	g/MJ DE
Calcium	9–15	0.80	8–10	0.65	8–12	0.80
Phosphorus	7–11	0.60	6–8	0.50	6–8	0.55
Sodium[1]	1.0–2.5	0.14	1.0–2.5	0.14	1.0–2.5	0.14

[1] Common salt contains 390g sodium/kg.

Sodium, chlorine and potassium

Sodium, chlorine and potassium are major controllers of the delicate ionic balance in body tissues. Sodium and potassium salts are well digested, normally absorbed in excess of requirement, and excreted by the urine whose content of these elements is closely related to the rate of dietary intake. Sodium is supplied as salt, which is widely used as a diet appetizer and is usually included at supplementary levels of 1–3g salt/kg diet, while ample potassium is available from normal diet ingredients. Dietary levels of common salt above 10g/kg will predispose to salt toxicity if water supply is at all limiting. Definitive requirement appears to be about 1g/kg diet for sodium and about the same for chlorine. This would be supplied in total by the addition of 2.5g salt/kg diet. The sodium content of cereals is around 0.3g Na/kg, while fish meal contains about or above 6g/kg. The definitive requirement for potassium seems around 3g/kg diet. Cereals contain natural levels of greater than 4g K/kg, whilst soya bean has 20g K/kg. The sodium requirement is given in Table 12.6, the chlorine requirement will be satisfied along with that of sodium, and potassium will normally be provided in excess from all but the most unusual of diets.

Magnesium

Magnesium is associated with calcium and phosphorus in the formation of bone, and is also an essential part of body enzyme systems. The requirement is around 0.4g Mg/kg diet, and this is readily met from dietary sources. Cereals contain somewhat above 1g Mg/kg, although the digestibility is probably at or below 50%.

Trace mineral elements

Zinc, manganese, iron, cobalt, iodine, selenium and copper are all required for normal metabolism as a part of the biochemical reactions catalysed by enzymes or as a part of hormone systems. Their requirement is best expressed in terms of the amounts which might be expected to be needed as a dietary supplement. The supplement is usually pre-compounded to contain a given balance of all the trace minerals, and this supplement is then added to the diet at a single recommended level to provide for the whole of the estimated diet requirement. Recommendations vary hugely, due at least in part to a difference of opinion as to the extent to which trace elements contained naturally in the diet ingredients can contribute to the perceived requirement. Values given in Table 12.7 are formulated after reference to the United States National Research Council, the United Kingdom Agricultural Research Council and to common practice.

Vitamins

The gross effects of vitamin deficiencies are spectacular, but rather difficult to reproduce with natural diets. Vitamins are involved in the widest possible range of normal body functions, especially the biochemical pathways of the metabolic processes, the immune response, cofactors in enzyme function, cell membrane activity, inheritance and tissue differentiation and so on. Shortfalls in vitamin supply, such as may occur on rare occasions in the course of conventional pig production activities, will therefore bring about chronic and insidious consequences for growth rate, disease resistance and reproductive ability, all of which tend to be of a general rather than a

Table 12.7 Trace element requirements for growing and breeding pigs (mg/kg diet)

Mineral element	Recommended requirement (mg) to add per kg of diet
Zinc	50[1]
Manganese	20
Iron	80[1]
Cobalt	1
Iodine	0.3[2]
Selenium	0.2
Copper	5[3]

[1] Zinc, 150; iron, 200; if copper is included at growth-promoter level.
[2] Two or even three times higher if dietary goitrogens are present.
[3] For growth promoter levels, up to 175 for young growers and up to 100 for pigs above 60kg.

particular nature. It is essential, therefore, to be satisfied that dietary levels of vitamins are entirely adequate.

For vitamins, the determination of adequacy is fraught. Even the criteria of assessment for the adequacy of vitamin supply are difficult to determine. It may be instructive to take the case of vitamin E as an example. The requirement for alpha tocopherol is around 10mg/kg to avoid muscular dystrophy, heart and liver disease. It has been shown, however, that up to 100mg/kg may enhance the immune system, whilst up to double that amount again can help to protect the fat of pork post-slaughter from fatty acid oxidation and instability. Alpha tocopherol allowance must also accommodate for loss of potency, the influences of dietary polyunsaturated fatty acids, the deleterious effects of the presence of dietary mycotoxins and the possible antagonistic effects of vitamin A.

The level of need for diet supplementation of vitamins is further complicated by the extent to which allowances should be made for the vitamins naturally occurring in feedstuffs. Again using vitamin E as the example, barley and maize corn, which have been well stored, appear to contain all of the necessary levels. These two cereals also contain half the requirement for pantothenic acid and all of the thiamin; the question here is doubt as to the bio-availability and suitability of form of the naturally occurring vitamins which may be variable and low. Many vitamins are also biosynthesized in the body of the pig, from other nutrients, from metabolites, or through the medium of microbial synthesis in the intestine. Vitamin C (ascorbic acid), for example, may be synthesized by the pig from glucose, whilst both biotin and B_{12} are open to synthesis with the help of gut flora in the lumen of the intestine. With regard to biosynthesis, the question is always one of the adequacy of synthetic rate possible within the body of the pig in relation to the level of demand for highly productive animals.

Recommendations for vitamins are therefore presented on the same basis as for the trace elements. The whole of the requirement is proposed as being supplied in the form of a supplement. The values in Table 12.8 are derived from a compendium of many sources, including US NRC, UK ARC, and feed manufacturing practice. There is some temptation to present the full range of recommended and used levels of vitamin inclusions into diets obtainable from diverse literature sources. However, to do so would be to present ranges often spanning tenfold differences or more. Such a wide range in estimates for vitamin requirements makes recommendation most difficult, and the values given in Table 12.8 can only be taken as a guide based on best available current knowledge. As already pointed out for the trace mineral elements, diets of particularly high nutrient density should have vitamin inclusion levels raised pro rata. Concentrations given in terms of mg/kg diet in Table 12.8 assume a dietary DE content of 13MJ/kg.

Table 12.8 Vitamin requirements for growing and breeding pigs. To add, per kilogram of final diet of assumed energy content 13MJ DE/kg, and assuming no loss of potency

	Growing pigs	Breeding sows
Retinol (A)[1] (mg)	3–6[4]	3
Cholecalciferol (D$_3$)[2] (mg)	0.025–0.05[4]	0.05
D-alpha tocopherol (E)[3] (mg)	30–60[5]	30–60[5]
Menaphthone (K$_3$) (mg)	3	3
Thiamin (B$_1$) (mg)	2	2
Riboflavin (B$_2$) (mg)	6	6
Nicotinic acid (mg)	20	20
Pantothenic acid (mg)	15	15
Pyridoxine (B$_6$) (mg)	3	3
Cyanocobalamin (B$_{12}$) (mg)	0.02	0.02
Biotin (H) (mg)	0.20	0.30–1.00[6]
Folic acid (mg)	1.0	3.0
Choline (mg)	750–1000[7]	750–1000[7]
Ascorbic acid (C)[8]	–	–

[1] 1mg retinol = 3,333i.u. vitamin A.
[2] 1mg cholecalciferol = 40,000i.u. vitamin D$_3$.
[3] 1mg D-alpha tocopherol = 1.70i.u. vitamin E; 1mg DL-alpha tocopherol = 1.14i.u. vitamin E.
[4] The higher values in the range for younger pigs.
[5] Higher levels than the range given if fat is added to the diet, especially unsaturated fats; 3mg per gram of linoleic acid has been proposed. Breeding sows may merit higher levels.
[6] The higher level for sows with hoof problems.
[7] The total requirement is about 1,500mg/kg diet. About half is usually added.
[8] There may be a requirement for vitamin C, but the level is unknown.

CHAPTER 13

APPETITE AND VOLUNTARY FEED INTAKE

Introduction

Animal appetite – the desire for nutrients expressed in terms of feed consumption – is, by definition, a function of nutrient requirement. But appetite must ultimately be limited by the physical capacity of the gut; that is, gut size together with the rate of throughput of feed along it. In fact, gut size is a more important variable than rate of throughput. Feed flows out of the stomach and down the duodenum of the pig at the rate of about 10g dry matter per minute, beginning immediately after feeding. The rate of outflow is linear until the stomach is about half-cleared of the original feed inflow. After this, the outflow rate declines exponentially. Outflow rate is slightly lower for lighter pigs than for adults, but it is relatively constant for much of the total dry matter moved and is not greatly affected by the absolute amount of feed eaten. So total stomach and intestinal volume will be the most important factor controlling the physical limits of feed intake in the pig, and it may be confidently supposed that the physical limits to feed intake will be related to some function of body size.

The quantitative determination of actual voluntary feed intake achieved by a given pig in a given day is complex, and neither deduction nor empiricism have yet arrived at a satisfactory means for precise prediction. Nevertheless, because nutrient intake is the driving force for the rate and efficiency of production of meat and milk, the closest possible estimation of pig appetite is essential.

Appetite is moderately heritable (0.2–0.4) and highly correlated with both growth rate overall (+) and percentage lean (−), and somewhat correlated (+) with lean tissue growth rate. The relationship with feed-conversion efficiency is positive so long as maintenance costs are effectively offset by increasing lean-tissue growth rate, but this relationship tends towards the negative when fatty tissue deposition becomes dominant.

Pig responses to increasing daily feed intake are shown in Figure 13.1. The positive relationship between feed eaten and daily fatty tissue growth accelerates at higher feed intake, whilst the even more positive relationship with daily lean-tissue growth is shown to plateau at higher feed intake. Unsurprisingly, daily live-weight gain is the sum of these two responses. Feed efficiency increases at first, and then has a tendency to stabilize or even decline. Optimum feed intake is that which maximizes lean-tissue growth rate and provides additionally the level of fat desired; Figure 13.1 shows that this will occur at point X or beyond.

The central nervous system (CNS) is involved in appetite control through a series of sophisticated neural and endocrine interactions. In particular, the controlling roles of beta-adrenergic agonists, serotonin, cholecystokinin,

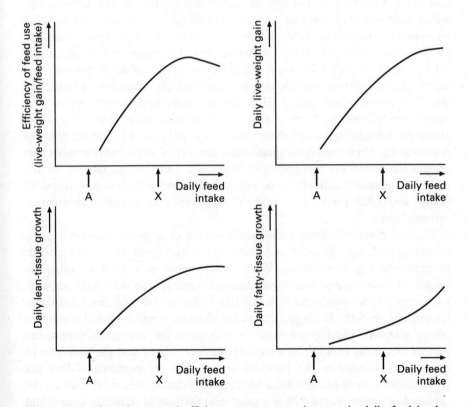

Figure 13.1 Growth rate and efficiency responses to increase in daily feed intake. Point A is an indicator of maintenance requirement. Between A and X responses tend to be linear. Optimum feed intake is at X or beyond.

opioids and neuropeptides have been the subject of study but without much elucidation. That the CNS may be open to direct communication with blood metabolites, and control feed intake through that mechanism, has been proposed for many years through the glucostatic and lipostatic theories, high blood levels of these metabolites giving negative feedback to appetite. It may readily be shown that glucose or lipid loading of the gastro-intestinal tract will reduce the size of an individual meal, but a systematic control mechanism has not been identified. The hormones insulin, glucagon and somatotrophin are clearly involved in nutritional response, but any linkage there may be between feed intake and nutrient need mediated through such hormone controls is as yet obscure. Somatotrophin has been implicated as often in appetite depression as in appetite enhancement; the illogicality of such a finding suggests that understanding of the system is unlikely in the near future.

Taste and smell are important aspects of CNS-mediated feeding behaviour and the predilection of the pig for sugar and truffles is well known. But whilst a flavour may mask an appetite-depressing unpalatability, or enhance an appetite stimulator, these phenomena also remain mysterious. Pigs will prefer to eat a feed rich in sucrose when offered a choice, but often will not eat any more of such feed when offered no choice. Flavour technology is a novel science and it is possible that genuine appetite enhancers, acting upon the CNS, may be forthcoming. However, normally such flavours will at best merely ameliorate a problem of appetite depression, rather than lift appetite above the fundamental controlling forces of (a) gut capacity, and (b) nutrient requirement. High ingredient quality and absence of unpalatable components remain the best route to appetite optimization. Self-inflicted obesity (or the converse anorexia) is likely to be consequent upon a bizarre interaction of bodily and CNS forces and unlikely to be open to simple manipulation through taste.

To date there has been some manipulation of appetite through genetic selection, both high-appetite and low-appetite genotypes now being evident in improved pig populations. Whilst feed intake may itself be relatively simple to measure and manœuvre through genetic selection, high appetite genotypes may be associated with fast lean-tissue growth rate, fast lipid-tissue growth rate or both. In former times, breeding for obesity created genotypes whose gluttony outstripped their requirement for maximum lean-tissue growth rate, thus creating an excess of nutrient supply and the accretion of depot fats. Reversal of the trend to fatness, and the breeding of lean pig types, initially went hand in hand with appetite diminishment. It remains the case that reduction in appetite is a most effective way of reducing fatness and improving percentage lean content. Between 1960 and 1990, in the UK, there was a reduction in backfat depth of about 0.5mm P2 annually (from 28 to

14mm P2 at 100kg). Over the same time period, there is evidence that feed intake at 100kg may also have been dramatically reduced, from about 4kg to around 2.5kg. Increased diet nutrient concentration can only account for some of this latter diminishment. On the other hand, given a population of adequately lean genotypes, there is a need to breed for increased appetite because only by elevating nutrient intake can genetic increases in potential lean-tissue growth rate be realized. The link between feed intake and obesity can be severed, but only while increased lean-tissue growth rate is selected for simultaneously with fatness control.

There is a hormone, cholecystokinin (CCK), locally produced in the intestine, which is secreted in response to the presence of feed there and through its action limits feed intake. If the influence of cholecystokinin can be suppressed by creating an immune response to it, then it appears that feed intake may be enhanced by 5–10%. Immunosuppression of CCK is not generally available, and it might be fair to suggest that there are other ways of improving feed intake which should perhaps be more fully exploited first.

Appetite as a consequence of nutrient demand

The demands of a pig for energy may be itemized in terms of the digestible energy requirements for maintenance, protein deposition, lipid deposition, lactation etc., as has been described earlier. Where values for the optimum rates and compositions of growth and lactation are known, then:

$$\text{DE intake (MJ/day)} = E_H + E_U + E_R$$

(Equation 13.1)

where E_H is the energy lost as heat through maintenance, metabolic and physical work and cold thermogenesis, E_U is the energy lost in urine and E_R is the energy retained in protein growth, lipid growth or milk.

Sources of nutrients are multidimensional, comprising at least energy (such as starches, fats and sugars), amino acids and a range of minerals and vitamins. In domesticated circumstances pigs are usually disallowed any choice amongst these dimensions, being given a fully compounded diet. Where the ratio of nutrients in the diet is in accord with the ratio of nutrients demanded, then appetite will reflect a rational choice comprehensive in the extent of its satisfaction of requirement. If, however, diet nutrients are not balanced in relation to demand, then appetite is likely to become irrational:

1. The imbalanced diet may be avoided to the extent which follows from the perception of the level of its inappropriateness; that is, the diet will be discriminated against as a result of a nutrient being consumed above the level of need.

2. The imbalanced diet may be more avidly consumed to an extent which follows from the pig perceiving a limiting nutrient, and, through a desire to satisfy a need for that nutrient, eating others to excess.

In the first case a pig may, for example, eat inadequate levels of energy because the diet supplies in excess a mineral which the pig wishes to avoid, whereas in the second case a pig may eat excess levels of energy because the diet supplies an insufficiency of an amino acid which it specifically desires.

Negative feedback: heat losses

Negative feedback restraining appetite may come from elevated levels of blood glucose, blood lipids and plasma amino acids; all of which may be used to monitor adequacy of supply in relation to need. Equation 13.1 implies no limit to the capacity of the pig to lose heat (E_H). However, this is often not the case and the avoidance of heat stress will limit energy intake where the environment fails to allow for adequate heat losses from the body. Consumed nutrients that have been used for metabolic work, together with energy not deposited in products, are actively lost as heat. This heat must be dissipated if stress and a body temperature rise are to be avoided. The more productive the pig, the greater is its nutrient need, and it follows that the greater also is the need to lose heat and therefore to find an environment of appropriate (reduced) temperature. Consumption of excess of any of the energy-containing lipids, proteins, or carbohydrates due to dietary imbalance will further exacerbate the need to dissipate heat.

In the absence of ability to lose adequate heat, then reverse pressure is placed upon Equation 13.1 and feed intake must fall. It may be estimated that feed intake will be reduced by 1g for every 1°C of heat above comfort level for every 1kg of body weight of the pig:

$$\text{Feed reduction (g/day)} = W(T - Tc)$$

(Equation 13.2)

where W is the live weight (kg), T is the effective ambient temperature and Tc is temperature associated with optimum pig comfort.

The influence of rate of air movement (which is important in the avoidance of heat stress) may be accommodated somewhat on the assumption that a 10cm/s increase in air movement is approximately equivalent to a 1°C reduction in air temperature.

Positive feedback: relative productivity above maintenance

Positive feedback (enhancing appetite) will come from demands for maintenance, cold thermogenesis and production exceeding the current rate of ingestion (Equation 13.1). If the most steadfast demand is for maintenance, then the basal energy intake, scaled according to live weight, would need to be amended upwards according to the rate of growth or level of lactation. Using a diet of 14MJ DE/kg feed, maintenance intake measured in terms of kilograms of feed eaten daily would be around $0.03W^{0.75}$. During rapid growth of lean and fatty tissue, intakes may rise to $0.12W^{0.75}$ (4 times maintenance) or above, while at peak lactation values of $0.15W^{0.75}$ (5 times maintenance) may be achievable.

Entire males usually have appetites 10% lower than castrated males, with the female intermediate.

Pigs of improved genotype, due to selection against percentage fat in the carcass, have lower appetites than unimproved pigs; although there is some dichotomy in the relationship as further genetic gains in lean-tissue growth rate are dependent upon appetite increase – not diminution.

Influence of diet nutrient density

A diet relatively low in energy presented to a pig with a nutrient demand that is relatively high in energy, will induce a positive feedback and enhanced appetite. In theory this will prevail until either the energy demand is satisfied, or negative feedback occurs due to the retrograde effects of all the other nutrients which would now be consumed to excess. Pigs offered diets of increasing energy concentration will be seen initially to consume more and then, once a balance has been achieved between intake and demand, to consume progressively less. This downward adjustment of feed intake with increasing diet energy concentration, and its converse upward adjustment with decreasing diet energy concentration, is rather precise and delicate, especially in pigs, such that equivalence of digestible energy intake is almost exactly attained. Feed intake and digestible energy responses to increasing concentrations of dietary energy are shown in Figure 13.2.

There is some evidence, although weak, that appetite may, in addition to

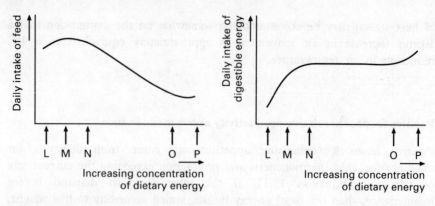

Figure 13.2 Feed intake control in pigs. L–N, range of physical control by maximum limits of gut capacity. N–O, range of physiological control to provide balance between intake and demand. O–P, range of physical control by minimum limits of gut capacity. L–M, increase in diet concentration encourages increased intake, bulk restraints in force, maximum physical capacity of the gut is utilized, energy intake improves but needs are not yet met. M–N, daily intake of feed stabilizes, bulk restraints in force, maximum physical capacity of the gut is utilized, energy intake improves but needs are still not met. N–O, energy needs are met, increasing dietary energy concentration is compensated by decreased voluntary intake of feed, balance of nutrient intake and nutrient need is achieved, gut volume changes accommodate the early part of this response but later the gut is inadequately filled. O–P, minimum requirements for gutfill are reached and compensation ceases for increasing diet energy concentration by downward adjustment of appetite, energy needs are exceeded.

the influence of energy, also be affected by the dietary amino acid level and amino acid balance. Negative feedback can be given by diets with imbalanced levels of amino acids, these being consumed less avidly than those whose amino acids are balanced more closely to the ideal, thus avoiding excess intake of unwanted nitrogen. Figure 13.3 indicates that feed intake will increase in response to increasing lysine to the point of achievement of ideal amino acid balance, whilst excess of lysine supply will bring about subsequent decrease in feed intake.

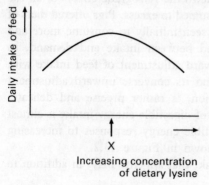

Figure 13.3 Quadratic response of feed intake to the concentration of lysine in the diet. The point X is indicative of a lysine concentration commensurate with an ideal balance of amino acids in protein.

Positive feedback, on the other hand, may arise from diets with a frank shortage of (balanced) protein in comparison with protein requirements such that, as was the case for pigs 'eating to energy' above, they would seek an increased rate of ingestion of protein, and in this event 'eat to protein' with the possibility of consuming unwanted energy. This response tends towards attempts by the pig to satisfy the demands for lean-tissue growth rate, whilst unavoidably promoting the deposition of depot fatty tissue.

Appetite as a consequence of gut capacity

The classic ability of the pig to adjust appetite in response to progressive decreases in diet nutrient density (especially energy) has its limits when ultimate gut capacity (in terms of volume and throughput rate) is reached. This leads to the remaining elements of the scheme represented in Figure 13.2.

At diet concentrations below N (Figure 13.2), gut capacity limits feed intake and the rate of digestible energy ingestion must fall. In practical terms, pigs maximizing their gut fill but failing to maximize their productivity due to inadequate intake of energy may have their performance increased by the addition to the diet of fats and oils which provide maximum energy for minimum diet space. Artificial (synthetic) amino acids will serve a similar purpose for similar reasons. The exact position of N in Figure 13.2 – the point below which required energy intake is curtailed by diet bulk – is highly dependent upon pig gut capacity in relation to the rate of pig productivity. In young pigs, potential productivity is especially high but gut capacity especially low due to the small size of the animal. Bulk effects may begin to exert some limit on performance through appetite restraints at DE values below 20MJ/kg for baby pigs (sows' milk has around 30MJ DE/kg DM), 14MJ/kg for young growers and lactating sows and 10MJ/kg for other adults. The potential productivity of lactating sows and young growers is greater in relation to their gut capacity than that of older growing pigs and pregnant sows and, because of this, the former require diets of higher nutrient density.

Although in wet-feed systems water may constitute an element of 'bulk', the bulking characteristics of feedstuffs are normally measured in terms of their fibre content; that is, their indigestibility. The relationship between fibre and DE concentration is, of course, highly negative. Fibre not only occupies much space in the gut in its own right as an entity of low (or zero) energy value, but it also attracts and holds water (which takes up further gut space). It may therefore be helpful to measure gut capacity in terms of potential daily output of undigested food material in the faeces. For example, if it is surmised that the faecal organic dry matter output of a pig approximately equals 0.013W, then values for 20, 100 and 200kg pigs would be 0.26, 1.3 and 2.6kg. Assuming dry matter digestibilities for pigs of these weights of respectively 0.80, 0.70 and 0.65, then these values equate to physical intake limits of 1.3, 4.3 and 7.4kg. These limits are similar to those that would be estimated by the expression $0.14W^{0.75}$, a function mentioned earlier and equivalent to rather more than 4 times the maintenance requirement. If this argument is credible, then:

Physical limits to feed intake (kg/day) $\simeq$ 0.013W/1 $-$ Digestibility coefficient

(Equation 13.3)

The coefficients for W may be assumed to be relatively lower for pigs of a genotype selected for lower appetite, and relatively higher for pigs with a particularly strong propensity to consume fibrous feedstuffs.

Estimation of feed intake in young and growing pigs

It is conventionally assumed that until 20kg live-weight growth rate and feed intake of young pigs is limited by the physical capacity of the gut. Equation 13.3 would suggest physical limits for pigs of 5, 10 and 15kg live weight, of 0.65, 0.65 and 0.98kg (if digestibilities of 0.90 (milk), 0.80 and 0.80 are used). These values are realistic for the weaned pig, but for the 5kg suckler consuming about 300g dry matter daily there is either a limit due to the liquid intake, or space available for the consumption of some 300g of supplementary feed. It is interesting to note that at 3kg live weight gut limits and likely milk dry matter intakes are broadly similar.

Weaned pigs do not consume to their potential appetite until around 20kg live weight, and sometimes not even then. Were this to be so, they would be more likely to achieve their growth potential which can be demonstrated in optimum conditions (see Chapter 3). Some of the reasons for young pigs failing to eat enough to satisfy their nutrient demands have been alluded to earlier. Specifically, ill health, stress, unfavourable ambient temperatures, unpalatable feed, feed of inappropriate ingredient composition or nutritional balance, inadequate water supply, insufficient space to live and eat and general mismanagement all conspire to ensure, for example, that the 400g of high-quality feed required to be eaten by the 6kg newly weaned pig to attain growth rates equivalent to those already shown pre-weaning is comprehensively not attained, nor will be until 2 weeks later. Attempts to estimate pig feed intakes between weaning and 15kg have therefore largely failed due to their overwhelming dependence upon individual management circumstance. Figure 13.4 shows estimates for daily feed intake in relation to the physical limits, the nutrient needs and the likely achievements at production level, the latter falling far short of the former.

From 15kg onwards it is likely, all other aspects of production being satisfactory, that the appetite of growing pigs may be best estimated from their nutrient demand (Equation 13.1), within the contraints of gut capacity (Equation 13.3). Empirical estimates have been determined by many workers in various countries. The UK ARC working party reported in 1981:

Digestible energy intake = 4 × maintenance requirement

(Equation 13.4)

or, where maintenance is estimated as 0.72MJ ME per $kgW^{0.63}$:

$$\text{DE intake (MJ/day)} = 4(0.72W^{0.63})/0.96 = 3.0W^{0.63}$$

(Equation 13.5)

However, this is inadequately asymptotic at higher values of W and the following was forwarded:

$$\text{DE intake (MJ/day)} = 55(1 - e^{-0.0204W})$$

(Equation 13.6)

a relationship supported by the US NAS-NRC working party reporting in 1988.

Table 13.1 presents values from this equation together with a calculation of the intake of a diet of 14MJ DE/kg and the prediction from the equation:

$$\text{Feed intake (kg/day)} = 0.13W^{0.75}$$

(Equation 13.7)

which, as mentioned earlier, is a naive expression approximating to just

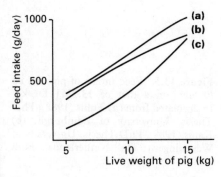

Figure 13.4 Feed intakes of weaned pigs given solid feed from 21 to 28 days onwards: (a) estimate from nutrient need assuming a diet of 16MJ DE/kg, (b) estimate from physical limits to gut capacity, (c) estimate from production achievement in good commercial practice.

Table 13.1 Limits to feed intake estimated in different ways

Live weight of pig (kg)	Equation 13.6[1] (MJ DE/day)	Equation 13.6[2] (kg diet/day)	Equation 13.7[3] (kg diet/day)	Equation 13.3[4] (kg diet/day)
20	18.4	1.31	1.23	1.30
40	30.6	2.19	2.07	2.08
60	38.8	2.77	2.80	3.12
80	44.2	3.16	3.48	3.47
100	47.8	3.41	4.11	4.33
120	50.2	3.59	4.71	4.46
140	51.8	3.70	5.29	5.20
160	52.9	3.78	5.85	5.94

[1] DE intake (MJ/day) = $55(1-e^{-0.0204W})$.
[2] Equation 13.6, using a diet with 14MJ DE/kg.
[3] $0.13W^{0.75}$.
[4] $0.013W/(1-\text{Digestibility coefficient})$; Digestibility coefficient = 0.8, 0.75, 0.75, 0.70, 0.70, 0.65, 0.65 and 0.65 at each live weight respectively.

above 4 times maintenance, and therefore likely to approach the limits to appetite.

Values in Table 13.1 show how maximum feed intake is similarly estimated by Equations 13.3 and 13.7, the former based on faecal excretion of indigestible material and the latter on metabolic body weight. Both of these estimates are in respectable agreement with the AFRC working party Equation 13.6 until 80kg live weight, when divergence occurs. The idea that feed intake limits are asymptotic, and that for growing pigs that asymptote appears before 100kg live weight, is well accepted and confirmed by many data sets (see, for example, Figure 13.5). However, known feed intakes achieved by lactating sows when 160kg support the likelihood of gut capacities of 6kg, which is closely in accord with Equations 13.3 and 13.7. Perhaps Equation 13.6 and the responses in Figure 13.5 are reflective not of maximum gut capacity but rather of a lesser nutritional need, and a following diminishment of growth rate subsequent to 100kg live weight. But caution

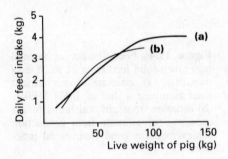

Figure 13.5 Feed intake of pigs fed *ad libitum* on a diet of 13.2MJ DE/kg. Interpolated from (a) Tullis (1982), PhD Thesis, University of Edinburgh, (b) Kanis (1988), PhD Thesis, University of Wageningen, The Netherlands. Both curves are asymptotic.

must be exercised as to assumptions made for the asymptote to growth, which should occur at much higher live weights for improved than for unimproved genotypes. The causes of the apparent asymptote to appetite which seems to occur around 100kg are possibly in large part a function of systems of production and management and deserve greater investigation.

Whereas the expected appetite of growing pigs may be described by one or other of the alternatives in Table 13.1, or by a description such as in Figure 13.5, it remains the case that practical feed intakes achieved under *ad libitum* conditions on farms more nearly approximate to values predicted by:

$$\text{Feed intake (kg/day)} = 0.10W^{0.75}$$

(Equation 13.8)

or:

$$\text{DE intake (MJ/day)} = 2.4W^{0.63}$$

(Equation 13.9)

this latter being some 3.2 times maintenance. These estimates are shown in Table 13.2.

Practical feed intakes may be less than the potential because the nutrient requirement is satisfied well within the limits of gut capacity. More realistic, however, is that production management circumscribes intake even below optimum nutrient needs. Negative factors may include feed palatability, inadequate trough space, between-pig aggression, ill-health, nutrient imbalance, ambient temperature, stocking density and so on. Some attempt can be made at quantification of the last two factors.

Temperature (T) above the comfort level (Tc) will diminish feed intake at the rate of 1g per °C (T − Tc) per kg of pig body weight, and Tc is dependent upon heat output − that is, the metabolic activity of the pig.

Table 13.2 Achieved feed intakes by pigs fed *ad libitum* under practical conditions

Live weight of pig (kg)	Equation 13.8[1] (kg diet/day)	Equation 13.9[2] (kg diet/day)
20	0.95	1.13
40	1.59	1.75
60	2.16	2.26
80	2.67	2.71
100	3.16	3.12
120	3.63	3.50
140	4.07	3.86
160	4.50	4.19

[1] Feed intake (kg/day) $= 0.10W^{0.75}$.
[2] DE intake (MJ/day) $= 2.4W^{0.63}$, using a diet of 14MJ DE/kg.

Figure 13.6 (relating to a pig of around 80kg live weight) shows how the daily intake of feed is directly and negatively related to temperature. At low temperatures there is an energy demand for cold thermogenesis. As this demand diminishes with temperature rise, intake will fall pro rata (as indicated by Equation 13.1). In the comfort zone, daily feed intake is likely to stabilize. At temperatures above Tc there is again seen to be a steep negative relationship as a progressive temperature rise induces intake depression according to Equation 13.2.

The influence of stocking density also merits further elaboration. It has been shown in Figure 7.1 that daily gain is adversely affected by increasing stocking density. Most of this response is due to a reduction in feed intake. Where the area occupied by the pig (m^2) is k$W^{0.67}$, a change in k value of 0.005 below optimum may be taken to be associated with a 4% change in feed intake (when k $\simeq$ 0.025 there is just sufficient space for it to lie down – usually k $\simeq$ 0.040 to 0.050 in acceptable intensive housing systems). The relationship between change in k and feed intake is likely to be universal for growing pigs. For a 60kg pig eating 2.0kg of feed, a change in k value of 0.005 would give a reduction of 80g daily, or about 0.1kg of potential feed intake per pig per day lost for each 0.1m^2 loss of floor space.

Estimation of feed intake in sows

From Equation 13.3 the physical limit to feed intake for a sow eating a diet which is 0.70 digestible is probably around 5kg at 120kg live weight, and 11kg at 250kg; which is, of course, much greater than the asymptote identified for growing pigs. Nutrient demand, given a reasonably balanced

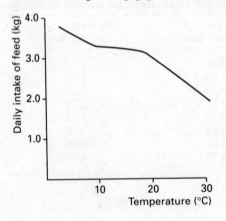

Figure 13.6 Influence of temperature upon appetite. Increase in temperature first reduces intake due to reducing demands for cold thermogenesis; next, higher temperatures depress appetite as the pig responds to the need to avoid heat stress.

diet, during pregnancy and between weaning and conception will be within this limit. Indeed, sows given concentrated diets will readily consume a further 0.5kg or more of straw daily, either to fill vacant space in the gut or perhaps to satisfy the motivation to forage. Appetite in the non-lactating adult sow is therefore likely to be a function of the various demands of maintenance, the foetal load, cold thermogenesis, growth and (equally important) the replacement of fatty tissue catabolized during lactation; that is to say, as indicated by Equation 13.1.

The lactating sow is usually unable to satisfy nutrient demand within the limits of appetite, and will lose body fat to bridge the deficit. This problem is particularly acute in the first lactation where voluntary feed intake in gilts is invariably at least 1kg per day lower than that for sows. This is probably due to a combination of lower body size and increased stress. Increasing sow appetite with parity number and live weight is frequently observed, and in general sow appetite is consistent with the '4 times maintenance' proposition, which can also be expressed as $0.12W^{0.75}$. Many types of sow will indeed well exceed this intake and manage 10kg of food daily in lactation. The propensity of many of today's breed types, however, is to consume much less than that, and lactation feed intakes of 5–6kg are the conventional expectations. These reduced feed intakes are probably not solely the result of selection against fat, and therefore against appetite. Amongst other things, modern farrowing and lactating rooms frequently place the sow into heat stress. With a comfort temperature (Tc) of around 15°C a room temperature (T) of 25°C will reduce the appetite of a 200kg lactating sow by 2kg (Equation 13.2). It may therefore be estimated that:

$$\text{Feed intake (kg/day)} = [0.013W/(1 - \text{Digestibility coefficient})] \\ - [W(T - Tc)/1000]$$

(Equation 13.10)

which is an elaboration on Equation 13.3 and likely to be general.

Low lactation feed intakes may be ameliorated by:

- feeding more frequently (maximum intake at a single meal is probably 3–4kg, indicating benefit from 3 times daily feeding);
- feeding wet rather than dry (the simple expedient of providing a water:food ratio of 3:1 will increase feed intake by 10–15%);
- allowing additional ample clean water separate from the food;
- reducing temperature to 10–14°C;
- using high concentration diets (greater than 14MJ DE/kg – increasing energy concentration by use of fats and oils rather than starches is most helpful because fats are used more efficiently, and therefore create less heat in their metabolism);

- feeding at cooler times of day, and using water drips and snout coolers (which may be worth 1–2kg of extra feed intake daily in hot climates);
- provision of a balanced diet with adequate protein.

There is a well-known negative relationship between the level of pregnancy feed intake and appetite in lactation, but this is not especially strong and is probably mediated through sow body fatness at parturition as indicated by P2 backfat depth (mm):

Lactation feed intake (kg in 28 days) = 240 − 0.20 (feed intake in 115-day pregnancy)

(Equation 13.11)

or:

Lactation feed intake (kg in 28 days) = 212 − 3.6P2

(Equation 13.12)

In addition to being negatively related to body fat, lactation feed intake is positively related to lactation yield; that is, the number of piglets sucking. Where capacity limits due to the physical size of the gut are not in play, it may be calculated that each piglet will demand about 1.2kg of fresh milk daily, and this will require 8MJ of dietary ME per kg of milk produced, or about 0.7kg of feed per sucking piglet per day. So the feed intake of lactating sows could be described as:

Feed intake (kg per day) = A + 0.7x

(Equation 13.13)

where x is the litter size and A is the maintenance requirement (say, in kg of feed daily, $0.033W^{0.75}$). The consequence of the use of these propositions is shown in Table 13.3. As discussed elsewhere, it has been estimated empirically that lactating sows of average weight, with acceptable litter size, and consuming less than 8kg of an average diet, are likely to lose body fat, a proposition in accord with Equation 13.13. Usually, the sow never achieves a

Table 13.3 Influence of litter size on theoretical feed intake of lactating sows

Litter size	Live weight of sow (kg)		
	140	180	220
8	6.9	7.2	7.5
10	8.3	8.6	8.9
12	9.5	10.0	10.3

rate of increase in feed intake of 0.7kg for each extra sucking piglet; more often a response in the region of 0.4kg of extra feed intake per extra sucking piglet can be expected. Because this response is inadequate to maintain nutrient balance, there is an inevitable loss of maternal fatty (and lean) tissues which are catabolized and go towards supplying the nutrient needs of lactation. Lactating sows appear well aware of forthcoming difficulties of nutritional supply in lactation and, like most other mammals, are happy to deposit fat during pregnancy, in order to lose it during the subsequent lactation.

Appetite as a consequence of feed characteristics

Specific dietary ingredients

There is ample evidence that pigs will eat more of feedstuffs which are sweet. Sugar itself is highly palatable, but it must not be assumed that sugar inclusion will necessarily overcome unpalatable aspects of other diet ingredients. Cooked and flaked cereals, oils and fresh and dried milk are amongst the range of feedstuffs appreciated by pigs, while appetite depression will result from dietary inclusion of ingredients such as meat-and-bone meal, some fish meals, rape-seed meal (both high and low glucosinolate, but the former is rejected more vigorously than the latter), cotton-seed meal and, to a certain extent, also some bean-meals. Pigs are also averse to lectins, mycotoxins, glucosinolates, tannins, saponins, sinapines and suchlike toxic and anti-nutritional factors. Correct choice of ingredients is fundamental to appetite maximization.

Feed enzymes

Feed carbohydrates are made up of sugars (few), starch and the structural carbohydrates or non-starch polysaccharides. The non-starch polysaccharides are mainly celluloses, beta-glucans and pentosans. The relative proportions of these latter differ between feedstuffs (barley endosperm is richer in beta-glucans, and wheat in pentosans). Carbohydrates are digested in the stomach, small intestine and large intestine.

In the pig, microbial fermentation is the only adequate system available for the hydrolysis of the non-starch polysaccharide cellulose. Microbial fermentation will take place throughout the gut, but mostly in the large intestine and caecum. Bacterial carbohydrases will yield mostly the fatty acids lactate, acetate, butyrate and propionate which are utilized about half as well as glucose.

The addition of enzymes to the feed is purported to break down carbohydrates and improve digestibility. If this is so, then, according to Equation 13.3, at any given live weight, feed intake will increase. The yield of the nutrient from each unit of feed ingested will also increase. If the enzyme attacks the fibre components of the carbohydrate fraction and reduces the fibre loading in the gut, then there is potential for a third positive factor influencing nutrient intake. It is reasonable, therefore, that enzyme preparations should be considered for use in diets for young pigs; but their efficacy should not be taken as automatic.

The enzyme system of the young pig is highly milk-orientated until around 4 weeks of age when the need to digest solid food becomes dominant (Figure 13.7). In-feed enzymes may therefore act beneficially in diets for young pigs whose own limited enzyme systems might be supplemented, or whose natural enzyme secretions might be complemented by enzymes not normally occurring in the intestine – such as fibre-digesting cellulases.

Gastric pH in the empty stomach is around 2, below the working tolerance of most carbohydrases. However, the oesophageal part into which food falls is maintained above that level, and after a meal the gastric pH rises significantly in any event. The possibility of exogenous enzyme action taking place in the stomach is supported by the finding that salivary carbohydrase itself (active at pH >4.0) probably has 2–3 hours action time there. Bacterial fermentation also occurs in the stomach producing lactic and organic acids; lactobacilli are mainly responsible in the sucking pig for lactic acid production.

The capacity of the small intestine to digest starch increases in direct response to the presence of starch in the gut. Within about 2 weeks of a significant challenge, ample carbohydrase is available for any conceivable

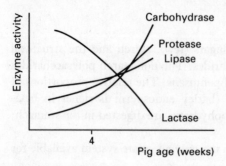

Figure 13.7 Influence of pig age upon endogenous enzyme activity.

need for simple starch hydrolysis; but this is to assume the absence of the protection of feed carbohydrate by non-starch polysaccharide cell wall materials not attacked by intestinal carbohydrases. It is also apparent that whereas most of the cereal starches are relatively simple, those in non-cereal vegetable feedstuffs such as bean-meals may be more complex and less readily digested. Perhaps most important of all, however, has been the observation that not only is weaning associated with a massive change in nutrient substrates (from milk to cereals, and animal to vegetable proteins), but the trauma itself appears to compromise the effective workings of the small intestinal and pancreatic enzyme systems.

On the face of it, therefore, it is reasonable to propose, especially for newly weaned and young growing pigs:

1. the use of in-feed enzymes may lend support to conventional starch digestion by their action in the stomach;

2. supplementation of the endogenous system by exogenous in-feed cellulase enzymes which may digest non-starch structural carbohydrates, and which may be active both before ingestion and in the gut, will not only yield useful nutrients from otherwise unavailable substrate, but will also release starches and proteins previously protected in plant material by indigestible cell walls.

In the case of the non-starch polysaccharidases, however, it is well to remember that there are between-ingredient differences in the type of non-starch polysaccharide present, and the enzymes used require to be specific.

The digestion of vegetable protein is known to be poor in young pigs immediately post-weaning, and the milk-based endogenous protease system takes some time to build up to adequate levels to provide the daily nutrient needs from vegetable protein sources. In the frank absence of adequate protein digestion capability, support for exogenous enzymes is a valid concept.

Commercial products containing cocktails of amylases, betaglucanases, pentosanases, cellulases, (lipases) and proteases are available for incorporation into dry pig diets. It is clear that these will result in active hydrolysis of starches, non-starch polysaccharides and proteins, in both cereals and vegetable protein concentrates, upon addition of water and warmth in a laboratory. Additions of a comprehensive enzyme cocktail to cereal-based diets appear invariably to increase marginally the digestibility of starch and non-starch polysaccharides in young piglets, but there is not always any noticeable improvement in performance. While not all experiments are reported, and there may be some bias toward positive results being reported more often, there is a commercial understanding that up to 2–3% improvement in daily gain may be achieved. This presumably comes from a combination of improved digestibility and enhanced appetite. Similar or slightly greater claims are made for the cooking, flaking and extrusion of cereals.

Most of the readily digested mineral phosphorus in pig diets arises from readily soluble dicalcium phosphate, but the vegetable components of the diet also contain significant amounts of phosphorus. In the latter case, however, availability is low as the mineral is bound with phytic acid to form phytate phosphorus. This is only about 0.3 digestible. The phytate is also disruptive to the digestion of calcium and some other mineral elements. The use of an exterior source of phytase enzyme (not manufactured in the body of the pig) to improve the digestibility of plant phosphorus therefore has many attractions. Such enzymes do exist, and they have been found to work successfully, improving the digestibility of plant phosphorus and reducing the need for supplemental mineral phosphate. Unfortunately the cost of sufficient quantities of enzyme to produce the desired effect is presently prohibitive.

Environmental gases

Ammonia and hydrogen sulphide are both toxic and low levels of around 10ppm or more of either will reduce appetite. The effects increase with gas concentration. Ammonia levels in intensive pig grower houses may rise above 50ppm, while levels of above 5ppm of hydrogen sulphide may occur following slurry agitation. Environmental gases are clearly best controlled by effective ventilation, but difficulty may arise when outside air temperatures are low. Recent commercial interest has been shown in the possibility of placing into the pig's feed additives which, upon passing into the slurry, appear to reduce ammonia levels. This may be achieved through binding the ammonia and buffering the system to prevent an over-rapid rise in pH. Encapsulated microbial and enzyme products may also be added into the feed, on the supposition that these may act on the slurry aiding odour-free slurry digestion. It appears that use of this material may, by decreasing gaseous ammonia levels when they exceed 10ppm, improve feed intake and thereby pig performance. But as with some other such new biotechnology developments in pig nutrition, the forces of commerce are somewhat ahead of the validations and explanations of science. Neither should it be assumed that other methods of reducing gaseous ammonia, which would clearly be beneficial, might not be equally effective. Deep-litter (sawdust) systems for providing bedding for pigs, and using in situ composting techniques, may be associated with special substances which, when added to the bed, are claimed to reduce gaseous emissions and aid fermentation. Such systems in any event can only tolerate stocking densities 30–50% more generous than conventional systems; but even then, the extent of the benefit of the additive has yet to be quantified.

Addition of dietary organic acids

Gut acidity levels are normally between pH 2–4 in the stomach, and rise to pH 5–6 in the small intestine, and again up to 7 in the large intestine. At pH levels above 6, there is increasing likelihood of reduced efficiency of enzyme action and increased proliferation of pathogenic bacteria in the gut. For both these reasons there is a close relationship between maintenance of optimum gut pH and pig appetite. Organic acids such as propionic, formic, citric, lactic and fumaric may readily be added to the diet, and these may help to counter the likelihood of pH rise in the stomach which can occur when young pigs are stressed.

It may also be important that individual feed ingredients have markedly different acid-binding capacities. The acid-binding capacity of liquid milk and cereals is low, but that for dried milk and for animal and vegetable proteins (and also limestone) is high. In the immediate post-weaning period there is particularly likely to be a predisposition to a rise in intestinal pH. There is little experimental evidence giving clear information about the mode of action or efficacy of in-feed organic acids in addressing the problem of elevated intestinal pH.

Addition of micro-organisms

Various micro-organisms have been put forward as having beneficial influence within the environment of the gut, to ameliorate appetite depression after weaning, and later to enhance appetite in young growing pigs. The objective of adding micro-organisms to the diet is to direct the gut flora away from abnormal or unhealthy species. Pathogenic organisms attach to the gut mucosa and from there produce toxins. Competitive attachment of non-pathogenic species at binding sites will therefore prevent toxin production. Cultures presumed beneficial, and often used, are special strains of *Lactobacillus* or *Streptococcus* (*Enterococcus*). Lactobacillae would, in the normal course of events, be encouraged by mother's milk for the fermentation of lactose and the production of lactic acid. But this natural population of Lactobacillae would rapidly diminish upon weaning. Loss of

Lactobacillae may have negative effects for gut acidity and also, following their absence, they may be replaced in the flora with pathogenic species. The two major claims made for in-feed additions of micro-organisms are therefore:

1. that an optimum gut pH is maintained;

2. that, by competitive and antagonistic effects, the growth of pathogens such as *Escherichia coli*, *Salmonella*, and *Staphylococcus* is curtailed.

Lactobacillus and *Streptococcus* cultures may be combined with an appropriate yeast (*Saccharomyces*) culture, which is also purported to stimulate feed intake.

Improvements in feed intake and performance as a result of dietary micro-organism additions are variable. It is pre-requisite that ingested cultures are first efficacious in maintaining pH and competing against pathogens and, secondly, are included at sufficient levels to multiply rapidly into viable gut populations. There is inadequate experimental evidence to date to show clearly that the theory is translated into effective practice.

Addition of antibiotics

In-feed antibiotics at 'growth-promoter levels' enhance appetite and growth by combating the negative effects of gut pathogens (through suppression of the pathogenic microbial flora), and also by enhancing the absorptive capacity of the gut. Positive responses in healthy animals are more difficult to show than the clear benefits of antibiotics in the treatment of intestinal disorders. Antibiotic addition to the diets of newly weaned pigs especially has been common practice for many years, and the benefits to feed intake and growth are unequivocal and may be as great as 10%. Should additions also be able to prevent a frank outbreak of disease, the benefits are, of course, considerably greater. Benefits to young growers may be somewhat less (around 5–10%), whilst the response appears to reduce to something a little less than a 5% improvement when antibiotics are included in the diets of finishing pigs. These positive responses have been confirmed over a great many individual experiments and through years of production experience. The positive responses of antibiotic additions to the diet show relatively little variation, particularly when compared to the wide variation that is presently being found for response to dietary organic acids, in-feed enzymes and 'probiotic' microbial cultures. Some in-feed antibiotic/antimicrobial compounds derived from benign micro-organisms may also be employed not so much with a view to inhibiting pathogens as to enhancing feed intake directly – especially in younger pigs.

Flavours, flavour-enhancers and masking agents

Positive flavour effects are assumed to increase feed intake either (i) by heightening the pig's appreciation of an acceptable taste in the diet, (ii) by adding an acceptable taste or (iii) by masking an unacceptable taste. Prevention of mould growth in feeds by proper harvesting and storage is basic to appetite maximization. The use of mould inhibitors, such as propionates, could be interpreted, with regard to pig appetite, as a negative activity. Pigs are susceptible to diet fat oxidation and rancidity; appetite depression through loss of palatability is marked. Antioxidants include both synthetic (such as BHA and BHT) and natural (such as tocopherols) products.

Given the absence of a need to mask off-flavours, the concept of a dietary appetite enhancer is most attractive and has raised commercial expectation and speculation over many years. However, the idea that a pig might be encouraged to eat more poses the question of why that pig is presently eating less, and the search for a reason as to how a flavour compound might address that cause. It may be surmised that in the majority of cases the positive effects of feed flavours – where they are present – are more often due to the masking of negative palatability rather than the creation of positive palatability. Exceptions may perhaps be specific attractants that relate to the provision of positive cues for pig foraging and eating behaviour which would have had helpful evolutionary benefits.

Spices have been usually associated with the masking of off-flavours, but this may be an inappropriate generality. Amongst other things, mixed spices may contain attractive aromatic oil essences which may be positive to palatability and also may stimulate enzyme production. Other – unsubstantiated – claims involve suppression of gut pathogens and feed mycotoxins.

While all possibilities remain of interest, feed palatability is likely to be ensured first and foremost by provision of fresh and well-stored ingredients of high quality which are themselves found by the pig to be pleasant to eat.

Form of feed and method of presentation

There remains debate as to whether maximum feed intake is more likely to be achieved on pellets, crumbs or with meal. There is considerable interaction with feeding method. In general, medium-ground cereals are more readily eaten than those either coarse- or fine-ground. Small pellets or crumbs are preferred to large pellets or a fine meal. Fine meals may become pasty in the mouth and stomach and reduce both palatability and rate of outflow from the gut; they are also implicated with gastric ulceration. Meals create dust in buildings which enhances the likelihood of pneumonia and reduces environmental acceptability. Meals are more costly to transport (being less dense) and are prone to being wasted at the point of consumption (comparative wastage rates for meal and pellets being around 6% and 2% respectively). Pellets are, however, more costly to produce. All forms of feed are eaten in greater quantity if presented wet rather than dry.

It is an essential part of appetite enhancement that fresh water is near at hand, trough space is ample, and feeding opportunities sufficiently frequent. *Ad libitum* feeding from self-feed hoppers rarely, if ever, maximizes feed intake. Often the flow is unduly restrictive (a necessity for wastage avoidance); but that notwithstanding, the feed is often presented dry and stale while access is often curtailed for pigs lower in the social dominance hierarchy.

CHAPTER 14

DIET FORMULATION

Introduction

More expenditure is incurred in the purchase of pig feed than in all the other costs of a pig unit put together. Pig production is an exercise in turning animal feedstuffs into high-quality pig meat at the most beneficial cost:value ratio.

The pig producer may buy diets from feed compounders, or compound them himself. Any pig feeding programme will require at least four pieces of nutritional information about the diets.

- The energy concentration.
- The protein concentration or protein:energy ratio.
- The quality of the protein used (or protein value).
- The adequacy of vitamin and mineral levels.

Diets are made by mixing feed ingredients to achieve a given nutrient specification; that is, a predetermined requirement for the concentration in the diet of certain named nutrients (for example, 14MJ DE/kg, 200g CP/kg). Pigs will eat a wide range of feedstuffs and those listed in Appendix 1 are merely representative. Diets high in fibre may, however, limit intake to the detriment of performance. Most conventional diets are formulated from grains (wheat, barley, oats, rye, sorghum, but especially maize), supplemented with higher quality protein sources (beans, meat meals, fish meals, but especially soya bean). Other energy and protein contributions may come from roots (brassicas, potatoes, manioc) and by-products (fats and oils, human food industry by-products, milk products). Diets will often also contain, where appropriate, medicines and growth-promoting agents.

Energy concentration of the diet

The energy concentration of pig diets may range from as low as 11MJ DE/kg for diets containing grass meal and other roughages, to more than 15MJ DE/ kg for diets formulated from high-energy ingredients such as maize, full-fat soya bean and added feed fats. Dietary energy concentration is negatively related to fibre levels and positively related to oil levels. Where NDF is the neutral detergent fibre (g/kg diet dry matter (DM)) and OIL is the content of total oils and fats (g/kg diet DM), then

$$DE(MJ/kg \ DM) = 17.0 - 0.018NDF + 0.016OIL$$

(Equation 14.1)

The nutrient concentration of a particular weight of diet depends, of course, upon its water content. Diets are usually prepared and offered to pigs in air-dry form; where water is added it being incorporated at feeding time. Some diet ingredients may however be available wet, such as root crops (10–25% DM), fish silage (15–25% DM) and liquid milk by-products (4–13% DM). There are many industrial wastes from the human food and drinks industries of high quality, and these may range from 2 to 30% DM.

Pig diets formulated from mixtures of cereals and soya bean are usually 85–92% DM, depending upon the moisture content of the cereals. The convention is to assume about 87% DM (13% water) in a pig diet unless otherwise stated, and nutrient concentrations for pig diets are usually, but not invariably, expressed on an air-dry basis assuming 87% DM (NRC use 90%).

Proper comparison of diets differing in water content requires correction to equal water content (Table 14.1). To provide a given quantity of energy, more feed needs to be offered if the diet is of low energy concentration, and

Table 14.1 Comparison of diets of differing water content

	'Wet' diet	'Dry' diet
Dry matter of diet (%)	32.8	87.0
DE content of diet (MJ DE/kg)	5.0	13.5
CP content of diet (g CP/kg)	75	180
DE content of diet corrected to 87% dry matter	13.3[1]	13.5
CP content of diet corrected to 87% dry matter	199[2]	180

[1] $(5.0/0.328) \times 0.87 = 13.3$.
[2] $(75/0.328) \times 0.87 = 199$.

less if the diet is of high energy concentration. To provide 30MJ DE, 2.6kg is needed of a diet with 11.5MJ DE/kg while only 2.2kg is needed of a diet with 13.5MJ DE/kg. For rapid growth or high production where substantial nutrient intakes are required, more concentrated diets are needed if the capacity of the gut for feed is inadequate in relation to nutrient requirements – as is often the case with young growing pigs and lactating sows. Pigs of lower productivity with big appetites, such as growers above 80kg and pregnant sows, are able to use diets of lower nutrient density (Table 14.2). Any pig given a feed allowance which is less than its appetite could be given a diet of lower nutrient density; but it may or may not be economic to do this.

Protein:energy ratio (g CP/MJ DE)

Dietary protein is usually discussed in terms of crude protein (CP). This is less useful than digestible crude protein (DCP) because there may be a range of 0.4–0.9 in the digestibility coefficient of different dietary protein sources. However, it appears likely that differences in protein digestibility may be accommodated best by recognition in feed compounding of the importance of the level of the ileally digested amino acids, which have been discussed earlier, and of which there will be more later.

The concentration of protein required in the diet depends upon the total amino acid needs for lean-tissue growth or for milk production in comparison to energy needs for body fuelling. Larger animals have higher energy demands for body maintenance and therefore need diets with a narrower protein:energy ratio. Pigs whose growth is destined to be low in lean content and high in fat have a lower protein requirement and therefore do not need diets of high protein concentration. Smaller pigs with lower maintenance

Table 14.2 Lower limits for energy density range appropriate to various types of pig[1]

	MJ DE/kg
Baby pig (5–15kg)	14
Grower (15–30kg)	14
Grower (30–100kg)	13
Pregnant breeders	11
Lactating breeders	12

[1] The upper limit for the energy density range is not dependent on the type of pig, but on the feasibility of the diet. Thus all pigs could be given diets with 14MJ DE if this was the most economic concentration (perhaps if maize was inexpensive compared to barley). Pigs offered higher density diets should have lower feed allowances pro rata.

needs, pigs with high lean tissue and low fatty tissue growth rates, and lactating sows need more protein and therefore diets with wider protein:energy ratios (Table 14.3).

Pig diets may be divided into three broad classes according to their protein:energy ratio. Diets with 13g CP/MJ DE or less are appropriate to pregnant adult females and growers above 80kg live weight. Diets with 13–14g CP/MJ DE are appropriate to lactating females and growing pigs between 30 and 80kg. Diets with more than 14g CP/MJ DE are appropriate to pigs of less than 30kg live weight, and to pigs of greater than that weight which have particularly high potentials for daily lean-tissue growth rate.

Although diets may be usefully described by their protein concentration alone (g CP/kg diet), this value is of little use unless the DE concentration (MJ/kg diet) is also known – that is to say unless the protein concentration is expressed in terms of the protein:energy ratio. Table 14.3 shows that to provide for the pig's protein requirements a diet of higher energy density may also need to be of higher protein density. The range of possible energy densities for pig diets is wide, the upper limit being about 17MJ DE/kg and controlled by the availability of high energy ingredients. The lower limit depends on pig type and may go below 11MJ DE/kg depending primarily upon animal appetite (Table 14.2). A concentration of 165g CP/kg diet represents a high level of protein if the associated energy concentration is only 12MJ DE/kg, but would represent a low level of protein if the associated dietary energy concentration was 15MJ DE/kg.

The number of different diets found on a single pig unit depends much upon the number of differing classes of pig requiring either different dietary energy concentrations or different dietary protein:energy ratios. Sometimes management simplicity outweighs nutritional efficacy, and at the (absurd) extreme a single diet of about 14MJ DE and with 13g CP/MJ DE may be used for all pigs. Such an arrangement is straightforward, but not optimum in terms of efficiency of feed use and will incur costs of lost production and wasted nutrients. The other (equally absurd) extreme is to use as many diets as there are pig classes, and to define the pig class as narrowly as possible (ultimately one animal at one moment in time). This will maximize the efficiency of feed use, but programmes involving a substantial number of different diets are only likely to be feasible in bigger production operations. Larger units are usually able to handle a greater number of different diet specifications than smaller ones. A frequent compromise to be found in practice is to have a special diet for baby pigs up to 10kg or so, a second diet for growers up to 30kg or so, a third diet for finishing pigs, a fourth diet for pregnant sows and a fifth for lactating sows. Many units are content to use a sixth diet which is placed into the growing pig programme, and is especially appropriate if the pigs are finished at weights greater than 100kg.

Table 14.3 Example nutrient concentrations of diets for different classes of pig

	DE density (MJ/kg)	CP density (g/kg)	CP (g)/MJ DE	Lysine (g/kg)	Lysine (g/MJ DE)	Protein value (V)
Starter (up to 15kg)	15.5	250	16	14.8	0.95	0.85
Young grower (up to 30kg)	15.0	225	15	12.8	0.85	0.80
Finisher (up to 100kg)	14.0	200	14	10.5	0.75	0.75
Finisher (up to 160kg)	14.0	170	12	8.4	0.60	0.70
Pregnant breeder	12.5	150	12	6.9	0.55	0.65
Lactating breeder	13.5	165	12.5	8.1	0.60	0.70
Improved entire male grower (40kg)	15.0	225	15	12.8	0.85	0.80
Unimproved castrated male grower (40kg)	13.0	160	12	7.8	0.60	0.70

Protein value (V)

The value of feed proteins for pigs relates to their amino acid composition in comparison to the balance required by the pig. A perfect match between the amino acid balance in feed protein and the amino acid balance required may be expressed as a protein value of unity (V = 1.0). The proteins of some feed ingredients such as soya bean and fish meal impart high protein values to the diet, while others such as the protein of cereals (which is deficient in lysine) impart a lower protein value to the diet. Reduced protein value may result from a relative deficiency in any one of the essential amino acids, but in pig diets it is usually found that lysine, threonine, methionine or tryptophan are (in that order of importance) implicated. So often is lysine the first limiting amino acid that compounders may routinely add 1kg/t, or more, of artificial lysine if it is of reasonable price, and especially if feedstuffs of unusual type or dubious quality are being used.

A high protein value (V), like a high energy concentration (DE) is not an objective necessarily to be sought after, but good information is needed about protein value (just as about DE) so that appropriate adjustments can be made to feeding levels and appropriate judgements made as to monetary value. A given protein requirement can be satisfied by provision of a higher quantity of protein of lower value, or a lower quantity of protein of higher value. Some guide as to the lower limits which may be expected for dietary protein value for various classes of pig is shown in Table 14.4. Protein of high value will be utilized more efficiently by the pig, but it may not be cost effective always to use less of a higher quality protein as compared to using more of a lower quality – but cheaper – protein. Using a protein of lower value requires the diet to have a wider protein:energy ratio, and thus a higher crude protein level.

Where information on the value of diet protein is not available then a statement of the dietary level of the amino acid lysine can be of help, as this is often the first limiting amino acid and the major controller of diet protein

Table 14.4 Lower limits for value of dietary protein appropriate to various classes of pig[1]

	Protein value (V)
Baby pig	0.80
Grower	0.70
Finisher	0.65
Pregnant breeders	0.60
Lactating breeders	0.65

[1] The upper limit for value of dietary protein is dependent on the feasibility of the diet.

value. With the knowledge that the pig's requirement, expressed in terms of ideal protein, is for 0.07g lysine per 1g of protein, then protein value can be calculated from the dietary lysine level as follows:

$$V = (lysine\ in\ diet\ [g/kg]/CP\ in\ diet\ [g/kg])/0.07$$

(Equation 14.2)

Example protein values, calculated according to the concentration of lysine in the protein, are shown in Table 14.3.

Response to dietary concentration of DE

Higher concentrations of DE in the diet are appropriate where pig productivity is below potential but when the animal is already eating to maximum gut capacity; this is often the case for young growers and lactating sows. If the more-concentrated diet ingredients are more expensive, then the extra productivity in terms of growth or milk production must be weighed against this extra cost. For pigs whose appetites are greater than the productivity required of them, a decrease in DE concentration can lead to cost-saving, provided the extra quantities that are needed of the cheaper diet do not cancel out the benefit.

If the protein:energy ratio remains unchanged then an increase in DE concentration will be exactly matched by an equivalent increase in protein concentration. The effect upon the pig will therefore be similar to it having been given (or having voluntarily consumed) more feed. To maintain constant performance while changing to a diet of lower or higher energy concentration, the feed allowance must be increased or decreased appropriately (Figure 14.1). If there is no change in feed allowance, then an increase in energy concentration will increase energy supply (Figure 14.2).

Response to energy supply made by a 60kg growing pig is depicted in Figure 14.3. The rate of lean-tissue growth will increase until maximum daily lean growth is achieved. The absolute level of this maximum is dependent, amongst other things, upon sex and genotype. During this phase of linear lean-tissue growth response to increasing energy supply, the pig is unlikely to fatten. After lean growth is maximized, however, further increases in energy supply will go to the growth of fat which from this point onwards now escalates. Pigs with higher lean tissue growth potentials will use more dietary energy before becoming fat. Figure 14.3 also shows how the response to energy will relate to diets of differing energy concentrations. 2.25kg of a diet with 11.5MJ DE/kg would fail to maximize lean-tissue growth, whilst the same level of feed supply of a diet with a concentration of 14.5MJ DE/kg would actually exceed the requirements for lean-tissue growth rate, and the pig would fatten.

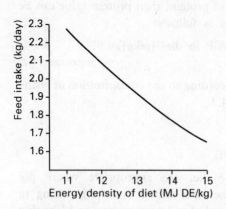

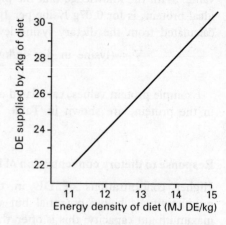

Figure 14.1 Intakes required for diets of different energy concentrations to give the same response of 625g tissue gain on a 60kg pig (500g lean and 125g fat).

Figure 14.2 Energy supplied by diets of different energy concentration fed at the same intake allowance of 2kg daily.

Responses of breeding sows to changes in the concentration of dietary energy are as readily predicted as in the case for the growing pig. If the change in DE concentration is accompanied by an equivalent and contrary alteration to the amount of feed supplied, then the nutrient intake effectively remains unaltered and there will be no change in productivity, although there may be a change in sow behaviour! If the intake level (or appetite) remains stable while the DE concentration of the diet alters, then the sow will respond to a decrease in energy supply by becoming thinner or lactating less, and to an increase in energy supply by becoming fatter or lactating more. In the case of breeding sows already thin, pregnant animals given more energy will produce heavier young, lactating sows will produce more milk, and weaned sows will have improved conception rates. Over-fat sows given more food will both waste it and become even less productive.

Response to protein:energy ratio

Given sufficient dietary energy, an increase in the supply of diet protein will increase lean growth linearly until the maximum potential for lean-tissue growth is attained. Additional increments of dietary protein cannot raise lean growth above that maximum. It should not be assumed that pigs which are not maximizing their lean-tissue growth potential would do so if they were given more protein. This would be to assume that lean growth was limited by

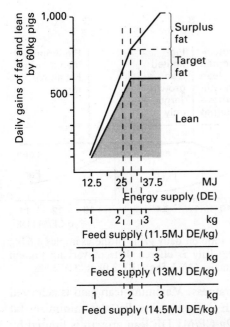

Figure 14.3 Daily gains of fat and lean for a pig given diets of differing energy density. The broken lines compare responses to an intake of 2.25kg.

a dietary protein deficit, and this is by no means always the case. Lean growth may be just as readily limited by an energy deficit.

Figure 14.4 illustrates the type of growth responses that might pertain to a 60kg pig of relatively poor genetic merit and relatively low lean tissue growth potential. The example shows the pig to have been given one or other of three intake allowances of a diet with 12.5MJ DE/kg, but with protein:energy ratios varying between <10 and >14g CP/MJ DE. An intake of 1.5kg daily (A) of a diet with only 10g CP/MJ DE does not supply sufficient protein for lean growth. Increasing the protein concentration to 12g CP/MJ DE corrects the situation and the lean-growth rate improves. Dietary energy and protein are in balance at this point, and a further increment of protein to 14g CP/MJ DE does nothing to alter the position, and although there is some 200g of potential lean tissue growth yet to be made, no enhancement of lean tissue deposition occurs because lean growth is limited by a deficit of energy (an inadequate level of food supply). The excess protein is now an embarrassment, because its disposal uses up precious energy.

A protein concentration of 12g CP/MJ DE maximizes lean growth and minimizes fat growth only within the confines of the overall feed intake allowance, which in case (A) is meagre. Lean growth is shown to be well below potential. If the feed allowance is now increased to 2.0kg (B) then there is a general improvement in performance level. But with less than 12g CP/MJ DE there is again insufficient protein to maximize lean growth, and

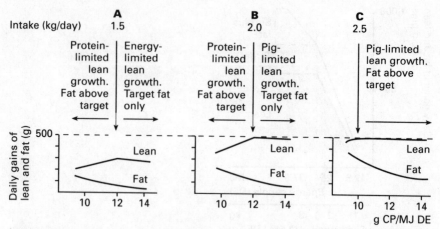

Figure 14.4 Influence of protein:energy ratio on daily gain. In the examples a 60kg pig with a lean-growth potential of 500g daily is offered diets of varying protein concentration at three levels of daily intake: (A) 1.5kg; (B) 2.0kg; (C) 2.5kg.

there is excess energy which induces fatness. Maximum lean gain is achieved when 2kg of diet with 12g CP/MJ DE is given. This diet also minimizes fat deposition. At protein levels above 12g CP/MJ DE lean growth is limited by the pig's potential (500g daily). The situation in case (B) may be contrasted to that in case (A) where the limitation was due to energy shortage. Increasing protein concentration continues to decrease the amount of fat in the pig; so although at levels above 12g CP/MJ DE there may be no further increase in daily lean-tissue growth rate, there *is* continuing increase in the carcass lean *percentage*.

In case (B) the pig has not fattened because there is no excess energy. If, however, the feed intake allowance is raised to 2.5kg of feed daily (C) then not only is maximum lean-tissue growth reached at a lower level of protein concentration (10g CP/MJ DE), but the excess energy now present causes the accumulation of surplus fat in the carcass at all protein concentrations.

Figure 14.4 has demonstrated that if the protein:energy ratio is too narrow then lean growth may be retarded, particularly at lower levels of feed intake. Widening the ratio has a positive effect upon the rate of lean growth, and only when the correct ratio is attained can lean growth be maximized. Fatness is readily induced by either (a) inadequate protein supply or (b) excess feed allowance. Fatness also may be reduced by either (a) widening the protein:energy ratio or (b) reducing food supply. Figure 14.4 shows unequivocally that pigs can become fat at any protein:energy ratio if the feed intake level is greater than the pig requires. Excess protein cannot enhance lean growth above either the limit set by available energy or that set by the inherent potential of the pig.

The classical interaction between protein supply, the supply of energy and pig potential may also be described as in Figure 14.5. This often elegantly demonstrated phenomenon shows how the slope of response to diet protein is dependent upon the utilizability of the protein, while the asymptote is dependent upon a variety of possible factors.

The lactating sow can produce 400–700g of protein in her milk daily. Milk production in the sow is a similar body process to growth and requires copious amounts of both protein and energy. A reduction in protein:energy ratio below 12g CP/MJ DE is likely to cause a reduction in milk yield and an increase in the rate of lean-tissue loss from the maternal body (Figure 14.6). Widening the ratio will correct any protein deficiency, but above 13g CP/MJ DE there is not likely to be much further positive effect upon milk yield.

The pregnant animal is relatively frugal in its protein needs. Increasing diet protein concentration may slightly increase live-weight gain during pregnancy, but within the normally feasible range of 10–14g CP/MJ DE there will be little effect on number of pigs born or upon their birth weight. The needs of the foetus are modest until the last 20–30 days of pregnancy. It has been proposed that 12g CP/MJ DE is adequate for the pregnant female. There is evidence that low protein levels may reduce conception rate and litter size, particularly if the protein is of low value. With this possibility in

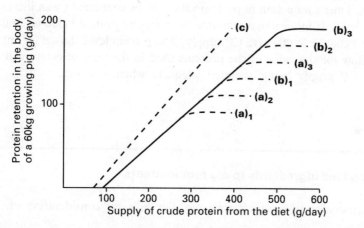

Figure 14.5 Response to increase in supply of dietary crude protein measured in terms of protein retention by a 60kg growing pig. The general slope of the solid line results from values for digestibility (D) of 0.75, for protein value (V) of 0.70, for the efficiency of use of digested ideal protein (v) of 0.85, and for maintenance requirement of 40g ideal protein daily. Lines (a)$_{1-3}$ represent possible limits to protein retention such as may occur by shortages of energy (three different levels). Lines (b)$_{1-3}$ represent possible limits to protein retention such as may occur by differing genetic potentials (three different genotypes). Line (c) represents the consequence of improvements in protein digestibility (D = 0.80) and protein value (V = 0.85).

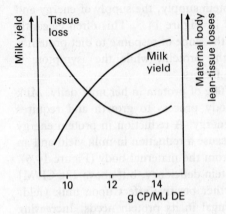

Figure 14.6 Improving the concentration of diet protein will increase milk yield and decrease the rate of body protein catabolism in lactating sows up to around 13g CP/MJ DE.

mind the diet prepared for lactating sows can be used for weaned females until they are confirmed pregnant.

Response to protein value (V)

Protein value governs the effective utilizability of the dietary crude protein (CP). A change in protein value therefore changes the effective CP content of the diet. Thus a reduction in protein value can be countered by an increase in protein level. It follows that response to change in protein value is similar to that for a change in effective CP supply. The protein level chosen for any diet must allow for the value of the proteins used in the mix, or errors will arise from a CP supply being assumed adequate when it is not.

Relating feed ingredients to pig requirements

The formulation of a satisfactory pig diet calls for consideration of:

1. The absolute levels of nutrients required daily by the pig (Requirements).
2. The daily feed intake of the pig (Intake).
3. The concentration of the nutrients in the diet (Concentration).
4. The nutrient content of feedstuffs; that is, the diet ingredients.
5. The suitability of available feedstuffs for diets destined to be fed to different pig classes.

The first three are related by the equation,

$$\text{Concentration} = \text{Requirement/Intake}$$

(Equation 14.3)

Requirements for energy, protein, vitamins and minerals are determined as the needs for maintenance, production (growth of lean and fat) and reproduction (pregnancy and lactation), and have been described in earlier chapters. Energy requirement is best expressed in terms of DE and protein requirement in terms of balanced and available ileally digested amino acids. Lysine, often being the first limiting amino acid, can be used as an indicator of the level of balanced amino acids in the mix. But where a computer is used in diet formulation there is benefit from compounding each and all the essential amino acids to the required level of provision, and as a minimum expectation the requirements for lysine, methionine, threonine, tryptophan and possibly histidine should each be considered in their own right as likely to be potentially limiting in conventional pig diets.

Example nutrient requirements are given in the upper sector of Table 14.5. The means by which a requirement may be derived has been described earlier.

To create a nutrient specification for a diet in terms of nutrient concentration to satisfy a given requirement, the feed intake must be known. The assumed feed intakes for the five example classes of pig are given in the middle sector of Table 14.5 and will result in the nutrient specifications shown in the lower section of that table.

The example diet specifications for nutrient content in Table 14.5 may be

Table 14.5 Example nutritional requirements, feed intakes and resultant nutrient specifications

	Young pigs (10kg)	Growing pigs (60kg)	Finishing Pigs (100kg)	Pregnant sows	Lactating sows
Example daily nutritional requirements					
Digestible energy (MJ DE)	12.1	29.3	32.4	35.4	110.8
Crude protein (g CP)	193	423	437	400	1450
Lysine (g)	10.9	23.7	24.0	20.8	78.3
g CP/MJ DE	16.0	14.4	13.5	11.3	13.1
g lysine/MJ DE	0.90	0.81	0.74	0.59	0.71
Example daily intakes of air-dry feed					
Feed intake (kg)	0.75	2.2	2.5	2.6	8.0
Nutrient specifications: concentrations per kg diet					
Digestible energy (MJ DE/kg diet)	16.1	13.3	13.0	13.6	13.9
Crude protein (g CP/kg diet)	260	192	175	154	181
Lysine (g/kg diet)	14.5	11.6	9.6	8.0	9.8

expanded to include all the amino acids, minerals and vitamins. For computer formulation such expansion is expected. The precise nature of the values shown accord with those calculated in the earlier chapter dealing with nutrient requirement, but their precision would not be a realistic basis for practical diet formulation because they relate to a particular individual pig at a particular moment in time. There is variation amongst individual pigs in terms of their performance (and therefore requirements), and also variation amongst pigs in terms of their feed intake. Guide nutrient specifications for practical usage are given in Appendix 2, which also identifies appropriate concentrations for all the essential amino acids, vitamins and minerals.

Given a target nutrient specification, a diet can be formulated by the appropriate combination of feedstuffs (diet ingredients). Ways of determining the nutritional value of feedstuffs have been discussed earlier and guide values for some example ingredients are given in Appendix 1. The mechanics of constructing a diet of given specification from combinations of ingredients of given value are shown in Table 14.6. In columns 1, 3, 5 and 7 of Sector A of the table the available ingredients and their nutrient contents are specified. The target nutrient specification for the final diet is shown in columns 4, 6 and 8 of Sector C as 14.0MJ DE/kg diet, 180g CP/kg diet and 9.0g lysine/kg diet. In Sector A of Table 14.6 a combination of 200kg/t of wheat, 550kg/t of barley and 200kg/t of soya bean (leaving 50kg/t of space for binders, vitamins, minerals and so on) is found to provide in the diet 12.8MJ DE/kg, 168g CP/kg and 8.3g lysine/kg. The contribution to the diet DE from wheat is calculated as $14.0 \times 0.2 = 2.8$MJ, that from barley is calculated as $12.9 \times 0.55 = 7.1$MJ, and that from soya bean is calculated as $14.5 \times 0.2 = 2.9$MJ. These sum to a total of 12.8MJ DE/kg diet. The same mechanism is used to calculate the contributions of these feedstuffs to the crude protein and lysine in the diet. It may be seen from Table 14.6 that the initial calculations shown in Sector A produce a diet mix which is not adequate to satisfy the targets given in Sector C of the table, falling short in DE by 1.2MJ/kg, in CP by 12g/kg, and in lysine by 0.7g/kg. Sector B of Table 14.6 adjusts the levels of the various diet ingredients, and uses in addition fish meal and feed-grade fat. The combination of available feedstuffs offered in Sector B now more nearly meets the target nutrient specification. Perhaps a little more fish, or wheat + soya, would improve this further, but an exact matching with all aspects of the target specification is highly unlikely. The degree of unlikelihood increases with the number of different elements in the target specification, and with the narrowness of the limitations to the range of ingredients available.

Variety and number of feed ingredients

The greater the range of available feed ingredients the more likely is the target nutrient specification for a diet to be met for the full range of energy

Table 14.6 Diet construction

1	2	3	4	5	6	7	8
		DE (MJ/kg)		CP (g/kg)		Lysine (g/kg)	
Ingredient	Inclusion rate (kg/t)	In ingredient	In diet	In ingredient	In diet	In ingredient	In diet
A							
Wheat	200	14.0	2.8	110	22.0	3.0	0.6
Barley	550	12.9	7.1	105	57.8	3.5	1.9
Fish meal		15.0		650		47	
Soya bean meal	200	14.5	2.9	440	88.0	29	5.8
Feed-grade fat		32.5					
	950		12.8		167.8		8.3
B							
Wheat	250						
Barley	425						
Fish meal	25						
Soya bean meal	200						
Feed-grade fat	50						
	950		13.9		176.4		9.2
C	950		14.0		180.0		9.0

and amino acid demands such as are shown minimally in the guide nutrient specifications of Appendix 2. However, the greater the range of ingredients used in diet formulation the more complex is the compounding operation at the mill and mix unit, and the more difficult is quality control and assessment of the veracity of the assumed nutritive content and physical quality of each of the feedstuffs. And so it is that many smaller operations will create their diets from perhaps only five or six main ingredients (or two in the case of a basic maize + soya diet!), and a mineral + vitamin supplement. Substantial feed compounding mills on the other hand may routinely have 20 or 30 ingredients available, some of which may readily be included in diets at quite trivial levels but which help to meet the target specification and optimize the diet cost. Fat, for example, is quite difficult to store and handle in the context of an on-farm mill and mix operation but is used widely by feed compounders to achieve adequate DE concentration in diets when cereals of lower energy concentration (such as barley), or by-products, are inexpensive and desirable to use.

Small-scale mixing facilities often have advantages of intimate local knowledge of a few ingredients which tend to be used in long-standing tried and tested fixed formulae. But the larger compounding mills achieve a greater precision in meeting exact nutrient specifications and are also able to minimize diet price as a result of extensive (world-wide) buying and storage opportunities when particular ingredients are cheap.

Farmers who also grow cereal grains may compromise by the purchase of a concentrate mix from a large compounder which is high in protein and includes a comprehensive mineral and vitamin supplement. This concentrate is then added to the milled cereal grains.

The range of pig-feed ingredients whose nutrient contents are given in Appendix 1 is by no means comprehensive, but most pig diets world-wide are made out of a relatively restricted range of preferred cereals, by-products and protein concentrates.

Choice of feedstuff ingredients for particular diets

Some animal feedstuffs may not be considered fit for use in pig diets because they:

- are unpalatable to pigs;
- contain factors toxic to pigs or substances which may otherwise cause ill-health or discomfort;
- do not show a cost-effective response in relation to their price;
- restrict appetite by undue bulk;
- fail to satisfy appetite by undue concentration;
- cause taints or off-flavours in the meat;
- reduce the physical quality of the meat (such as by creating soft fat);
- are likely to cause health difficulties in staff operating mill and mixing units;
- are difficult to handle through milling, mixing and pelleting machinery;
- are too costly or difficult to transport;
- are variable in nature and require excessive quality control systems in order to ensure the correct identification of the true nutrient content of various feedstuff batches.

Many of these criteria are not absolute and some ingredients may not be disallowed altogether, but restrained to be within certain maximum limits. These limits will naturally be more severe in the case of diets for young pigs than for finishing pigs. Suggested inclusion restraints for starter, grower, finisher and sow diets for a range of ingredients may be found in the lower sections of the tables given in Appendix 1. It is not possible to identify appropriate restraints on the basis of the physical quality and wholesomeness of ingredients, even though this is a most vital aspect of feedstuff assessment. In this case the appropriate restraint is dependent upon the degree of competence of feedstuff processing and the efficacy of storage. For example, soya bean meal and potatoes are prime feed ingredients for pigs, but are both highly toxic if not adequately heat treated; animal protein sources can be rendered useless by overheating; cereals may become toxic and unpalatable as a result of improper storage; feedstuffs containing high levels of sugars and lipids may be highly suitable and palatable for young pigs if of high quality, but when poorly handled can be the worst of ingredients in a starter diet.

The importance of ingredient quality and wholesomeness as factors separate from, and additional to, nutrient content cannot be overstressed in the creation of a successful diet formulation for pigs. Individual nutritionists are likely to place restraints upon the inclusion level of various feedstuffs into diets for each of the various classes of pig, according to knowledge of the quality and source of the feedstuff. This knowledge cannot be identified in the same way as can chemical analysis; more usually these decisions depend upon local knowledge and individual experience.

Sometimes doubtful feed ingredients are restrained in diet formulations as a result of feeding experiments which have shown no deleterious effects below a certain inclusion level, but some evidence of reduced performance above that level. It is conceivable that in the case of some feedstuffs there is indeed an inclusion rate below which the ingredient is satisfactory, but above which it becomes problematic due to toxic or other factors. But it is more likely that the deleterious effects of the dietary ingredient only appear to become less as the inclusion rate falls because the experimental procedures are unable to measure with statistical certainty progressively smaller effects. This does not mean that the effect is not present.

Limitations to inclusion rate can also offset the tendency for a particular feedstuff to be variable in its nutritional value. A poor batch of an ingredient will have a relatively trivial impact on a diet when included at levels of less than 5%, but would have a most serious impact if the rate of inclusion was above 20%. There is some degree of safety in a large number of ingredients being included in diets at low rates.

Too high an inclusion level of finely ground cereals may reduce digestibility and predispose the pig to gastric ulcers. High-fibre feedstuffs may need to be limited because of potential disruptions to protein digestion, because of difficulties in accurately predicting response and because of problems of low bulk density.

There is frequently a strong case for the inclusion of some feedstuff ingredients into diets at a stated minimum level. This would ensure continuity of diet palatability and familiarity. Some compounders may feel that some ingredients such as fish meal may have attributes which impart special qualities to the diet. This gives rise to minimum inclusion rates regardless of price. Similarly, a given cereal may be included in all diets, again regardless of price, at a level of 25–35%, in order to maintain continuity between formulations.

There is some evidence that, especially for younger growing pigs and lactating sows, dramatic changes in diet formulations can affect feed intake due to an alteration in diet, taste and texture. Large changes in formulation are therefore not advised within any particular diet specification, and changes in diet formulation from one specification to another should be as gradual as possible. This is especially important for young pigs changing from creep to starter diets and from starter to grower diets.

Although some ingredient restraints for feedstuffs are given in Appendix 1, these are forwarded as guides and not recommendations. Where diet ingredients are cheap, the more courageous feed compounder may be well rewarded. Equally, in some countries individual feedstuffs may be considered especially good, whilst in others the same feedstuff may be especially poor. What constitutes an appropriate ingredient restraint is often as much opinion as biological reasoning.

Example ingredient compositions of diets for various classes of pig are given in Appendix 3. The propositions contained therein are compatible with those already forwarded in Appendices 1 and 2. Clearly, however, a multiplicity of other ingredient compositions would achieve the same compatibilities. Diets for younger pigs contain more higher quality products and the range of ingredients used is often more restricted in number in order to maximize knowledge of ingredient quality and source. No 'doubtful' ingredient is included as the cost-benefit would be likely to be negative. Finishing pigs, however, may be offered a much wider ingredient choice, and some move towards the use of less expensive ingredients and by-products is to be expected.

Nutrient density is an important component of ingredient choice. The needs of young pigs and lactating sows for diets of greater energy concentration on the grounds of productive potential being high in relation to gut capacity, and the similarly important need for high ratios of lysine and protein to energy, result in nutrient specifications (Appendix 2) which can only be met from the use of certain ingredients such as wheat, maize, full-fat soya bean, milk and animal proteins, and feed oils and fats (Appendix 3). For finishing pigs and pregnant sows, however, these considerations are less important as gut capacity may usually accommodate the nutrient requirement when the latter is provided – at least in part – from less-dense feedstuffs and by-products. Provided that the appropriate ratios of nutrients one to the other are observed, and that the feed allowance or voluntary intake can accommodate a lower nutrient density, then overall cost benefit may accrue by creating diets of lower nutrient concentration from a wider range of ingredients.

The nutritional specification for dietary concentration of energy is often given great weight in the formulation procedure, the other nutrients following according to the appropriate ratio. It will have been apparent from the values given in Appendix 2 that whilst recommended ratios of essential amino acids to lysine, lysine to energy and crude protein to energy are indeed closely specified, the energy density itself is described in terms of a DE range. For pigs with appetite limitations it is of vital importance to formulate diets of high energy concentration (and of high concentration of other nutrients also). But it is of equal importance that, when appetite limitations do not apply, the DE concentration of the diet is allowed to vary within as wide limits as possible. This will encourage maximum opportunity to be taken of feedstuffs of either particularly high energy density, or particularly low energy density, when one or the other is especially beneficially priced. It is necessary only to know with reasonable precision the energy concentration of the diet (as described in Equation 14.1), and to allow for more of a lower density diet to be consumed by the pig (Table 14.7). The two pricing situations shown in the Table may readily occur over a 6-month period; in the first the high density diet is most cost-beneficial, whilst in the second case the low density diet is the better option.

The density restraint is often used in practice, and a definitive target (rather than a wide range) for MJ DE/kg is given for the feed compounder to achieve. This facilitates exact feeding recommendations along with the particular diet to be fed, and it avoids the confusing issue of needing to change feeding level every time diet concentration is changed. Nevertheless, however valid the reasons for strict adherence to a stipulated energy concentration for a given diet, real savings in feed cost per unit pig growth may be lost by unnecessary rigour.

On some pig units the policy for the amount of feed to be given to pigs is

Table 14.7 Relative values of three diets of differing density

Diet	Cost (p/kg)	Intake allowance (kg) to provide 28MJ DE and 420g CP	Feed conversion ratio (feed:gain)	Cost (p) to provide 1 day's nutrient requirements
A. High density	15.0	1.8	2.4	27.0
B. Medium density	14.5	2.0	2.7	29.0
C. Low density	13.0	2.2	2.9	28.6
A. High density	16.9	1.8	2.4	29.7
B. Medium density	14.7	2.0	2.7	29.4
C. Low density	13.0	2.2	2.9	28.6

rigidly adhered to. Management rules are expressed in terms of daily allowances for growing pigs of given weight (or in given pens), and for pregnant and lactating sows. Such values are of little use when quoted independently of the nutrient concentrations of the diets which are to be offered, because pigs require quantities of this or that *nutrient* daily, not quantities of feed. There is therefore a certain safety in having a pig diet always of a known standard energy density. This obviates problems of feeding management. Particular feeding allowances for particular classes of pigs become familiar to the feeder, and these may even be an integral part of a recording system. There is plenty of scope for confusion where a change in nutrient density could even require a *reduction* in the daily feed allowance in order to effect an *increase* in nutrient supply.

Over recent years the nutrient concentration of pig diets has tended to rise. This has been for a number of good reasons:

● the unit cost of energy in energy-dense feedstuff such as maize, wheat, full-fat soya and feed-grade fats has lowered;

● the appetite of many hybrid pigs has been reduced and become a factor limiting the achievement of potential growth up to 80kg live weight and of potential lactation yield in milking sows;

● the industry is more aware of the growth possibilities of young pigs and the benefits of giving young animals diets of the highest nutrient density and palatability. If a growing pig is not yet up to the maximum rate of protein deposition then any further increment of nutrient will be used primarily for lean-tissue growth. A premium can therefore be justified for a quality, energy-dense, protein-dense diet specially formulated for this class of pig;

● the type of feedstuffs used in the more concentrated diets are of more consistent quality, and the response is more predictable. Conversely, feedstuffs of low nutrient density tend to be more unpredictable. High-density diets can therefore be formulated more precisely;

● the more concentrated the diet, the more likely there is to be saving in transport and handling costs. These savings are achieved as a result of the same weight of diet carrying more nutrients within it. Cost savings start at the mill, continue through storage and haulage, and are also achieved by the producer who has less material to feed to the pig and less slurry to remove from the pen.

There is a direct relationship between the concentration of nutrients in a diet formulation and the efficiency of conversion of food to live-weight gain.

Containing more nutrients per unit weight of food, a diet of higher concentration will have a better feed conversion efficiency. A high efficiency of conversion (gain per unit of feed), or a low feed conversion ratio (feed per unit of gain) is intrinsically attractive, but it may not necessarily be most cost effective (see Table 14.7).

Unit value of nutrients from feedstuffs

It is evident that:

- feed ingredients with higher concentrations of the required nutrients are more valuable than feed ingredients of lower nutrient concentration;

- some feed ingredients have a worth far above that which may appear from their own nutrient content and price alone; this is because they may provide (albeit at high price) a specific nutrient lacking in an ingredient which otherwise provides other nutrients at a specially low price, the overall mix being more cost-effective;

- where there is no disadvantage in meeting the nutritional specification of a diet by use of low-priced ingredients rather than higher-priced ones, then this should be done. Conditions of no disadvantage should prevail if (i) feedstuffs are correctly evaluated according to quality, wholesomeness and nutrient content and (ii) maximum and minimum limits (restraints) to the inclusion of particular feedstuffs in particular diets are correctly set. Doubts about (i) and (ii) often, correctly, constrain feed compounders into not necessarily minimizing diet costs at all times. Thus each new diet formulated in consequence of a change in ingredient price (and therefore value per unit nutrient) will be considered by the nutritionist on a basis of informed, but qualitative, judgment as to whether or not that formulation should go forward for manufacture and feeding to pigs. This last step, where the nutritionist's knowledge, experience and caution can override the mathematical and mechanical processes of diet formulation is most important. It arises not because subjectivity and prejudice should be allowed to supersede objectivity and science, but rather because the science is often not adequate to cover all eventualities at a quantitative level.

Table 14.8 shows three cereals at three prices and compares the unit values of

Table 14.8 Unit values of nutrients in three feedstuffs

	Barley	Wheat	Maize
Price (p/kg)	11.0	12.0	12.2
DE (MJ/kg)	12.9	14.0	14.5
CP (g/kg)	105.0	110.0	90.0
Lysine (g/kg)	3.5	3.0	2.6
Cost (p)			
1 MJ DE	0.853	0.857	0.841
1g CP	0.105	0.117	0.136
1g lysine	3.14	4.00	4.69

the energy and protein which they contain. In this example the cheapest source of energy is maize, whilst the cheapest source of protein and lysine is barley. The cost for 1MJ DE, 1g CP and 1g lysine is calculated on the basis of the whole of the feedstuff price being apportioned to each nutrient. This serves for purposes of comparison and to help identify the most cost-effective ingredient, but it is not a true reflection of reality because for a single price each feed contains all three nutrients, albeit at different levels of importance. These three cereals are usually considered primarily as energy sources, so the cost per unit energy would tend to be the most important. Maize appears most opportune even though the cost per unit lysine is high. But again, such simplicity is spurious as in most pig diets at least half of the protein comes from the cereal fraction (see Table 14.6), and the cost per gram of protein and per gram of lysine *is* important. Whether maize or barley is the more cost-effective cereal is not clear until the unit values for all nutrients are compared simultaneously. To provide both 15MJ DE and 3g lysine from barley alone would cost 12.8p (energy being limiting), whilst to provide the same from maize alone would cost much more at 14.1p (lysine being limiting). Neither strategy is correct; buying both and mixing them together at a rate of 0.25 units barley and 0.75 units maize will provide both 15MJ DE and 3g lysine for the price of 12.6p. This logic can be developed further to the realization that the relative value which should be ascribed to the different grain proteins depends not only on comparison between the two grains, but rather upon the relative cost of protein which might be available from alternative protein sources. The calculation comparing barley, wheat and maize does not therefore approach realism until a feedstuff such as soya bean meal is added into the set of equations for simultaneous solution. Say the price of soya bean meal was 15.0p/kg. Containing 14.5MJ DE, 440g CP and 29g lysine/kg the unit costs of 1MJ DE, 1g CP and 1g lysine in soya bean meal are respectively 1.03, 0.034 and 0.52p. The unit cost of energy is therefore higher than that for cereals, but the unit cost of protein and lysine is very much less. The fact that barley lysine was cheaper than maize lysine now becomes irrelevant, and a maize/soya mix becomes the preferred option. This is not to say that barley, even at the example prices quoted, could never be competitive with maize; that would depend upon the relative costs of other energy sources such as feed fats.

In order to judge the relative economic worth of different feedstuffs, it is true that the unit values for energy, in ingredients included primarily for their energy content, are useful comparators. Equally useful are the unit values for protein and amino acids in feedstuff ingredients included primarily for their protein content. But it will now also be evident that full optimization of a diet formulation requires many simultaneous comparisons and a complexity of calculation that can be handled only by a computer.

Number of different diets required

The nutrient requirement, the balance of protein to energy, and the appetite of the pig all change daily. Ideally, therefore, a different diet formulation should be offered day by day. Individual pigs also vary in their requirements and appetites, which should be accommodated by diet formulae customized to each pig. With the possible exception of self-choice feeding arrangements these ideals are, of course, impractical. Different diet formulations will tend to be required in circumstances when:

- the specification becomes sufficiently over-generous that the excess of nutrients supplied above need is wasteful in economic terms;
- the specification is sufficiently under-generous that there is economic loss of potential production;
- a particular class of pig has particular ingredient restraints not common to the next class;
- one class of pig requires a diet of different nutrient concentration because of appetite constraints;
- a diet requires fortification with growth promoters or medicines appropriate only for a particular group or class of pigs.

Table 14.9 presents good reasons for five diets being used in a pig-feeding programme, but contraction to three would be feasible if a single diet were used for both lactating and pregnant sows, and two diets were used to cover the young, growing and finishing stages. Equally appropriate might be a more sophisticated – and potentially cost-beneficial – programme of an even

Table 14.9 Some diet characteristics for different classes of pigs

	Young pigs (up to 15kg live weight)	Growing pigs (up to 50kg live weight)	Finishing pigs (up to 100kg live weight)	Lactating sows	Pregnant sows
Energy concentration (MJ DE/kg)	14–17	14–16	13–15	12–15	11–14
Protein:energy ratio (g CP/MJ DE)	16	15	14	12.5	12
Lysine:energy ratio (g lysine/MJ DE)	1.0	0.90	0.75	0.60	0.55
Appetite factors	+++	++		++	
Special feedstuffs	+++	++		+	
Special additives	+	+			

greater number of diets. A starter creep of high palatability and quality with especially chosen ingredients could be used from 14 days of age through weaning and up to 10kg or so; a reduced specification would follow on until the pigs were 20kg live weight, after which would follow first a grower, then a finisher diet. For pigs growing on to weights above 100kg a final (fifth) diet may be cost-effective.

The final number of diets chosen will depend greatly on the size of the plant mixing the diets, the potential for storage of different feedstuffs, and the ability of the machinery and management on the pig unit itself to cope with a wide variety of different diets. In dry-feeding systems, often a particular feed may be associated with a particular bulk feed bin which is used to hold the mixed feed formulation prior to delivery to the pigs. These may relate to one, or a number, of houses – but never to individual pens of pigs. The maximum flexibility for diet number in this case is therefore controlled by the number of holding bins available. In this latter respect the use in diets of feed additives, medicines and growth promoters which require withdrawal periods of weeks or days before dispatch for slaughter causes considerable problems in the finisher stages by creating the need for an additional diet from which these additives are excluded, and which is used only for the latter part of the finishing period.

Wet feeding systems allow the use of a wide range of by-products, such as those from the human food manufacturing industries, whey, skimmed milk, root crops, distillery wastes, liquified fish, starch and so on. Further advantage accrues if the dry- and wet-diet components are mixed, as may readily be done to maximize the number of different dietary options. Through the wet feeding system individual diet formulations may be readily created to match more closely the nutritional requirements of pigs in particular parts of the production unit. Because each diet is mixed within a few hours of feeding, automatic wet-feeding systems can cope with a substantial number of different diets being offered through the growing phase.

Enhancement of the precision of formulation

The better the closeness of fit between animal nutrient requirement and diet nutrient provision, the more likely will optimum performance be achieved and the greater will be the efficiency of conversion of feed into pig products.

But the statement of diet nutrient specification is not an end in itself;

rather it is only an interim point towards the ultimate objective of animal requirement and production response. It is implicit that the defined requirement should bear a reasonably close relationship with growth and reproductive response. Nevertheless, the nutrient specification is not itself the response, even though it may be confused as such by some feed compounders on account of satisfaction of the nutrient specification being the primary objective of the diet-compounding, mill-and-mix unit. Only in a fully integrated operation is the response to the diet seen as the ultimate target of the diet formulator. It is for this reason that some feed compounders were in the past (wrongly) more preoccupied with provision of a recommended nutrient requirement given in terms of diet concentration than with nutrient requirement expressed in terms of achieved production response.

Better definition of dietary utilizable energy

The major variation amongst feedstuffs with regard to their content of utilizable energy is a result of variation in digestibility. Definition of energy content of feedstuffs – and the complementary definition of energy requirement – in terms of digestible energy (DE) rather than gross energy (GE) clearly enhances accuracy of diet formulation. The next step in precision improvement is to use metabolizable energy (ME) which allows for the energy losses in urine associated with protein deamination. Some diet formulation procedures do indeed express pig requirement and diet nutrient specification for energy in terms of ME, and the various available feedstuffs are described in terms of their ME rather than DE content. The improvement in accuracy of formulation is, however, trivial. First, because most ME values for feedstuffs have been assessed simply as a fixed proportion (0.96, usually) of the DE, and there is therefore no effect upon the relative energy values of the various ingredients. Secondly, variation in the relationship between DE and ME is as much a function of the animal and its circumstance (for example potential rate of protein accretion in relation to protein supply), as of the feedstuff itself. Use of ME thus gives an appearance of increased formulation precision, but one which is largely spurious. Diet formulation is indifferent to the use of either DE or ME because constant factors require to be assumed both in the animal and in the feed elements of the formulation activity.

The ultimate requirement of the animal is for net energy (NE) and this would effectively account for differences in the efficiency of utilization of diet energy for various different productive functions. Some of these have been discussed earlier and are of considerable significance. Thus the contribution of total ME requirement made by diet protein which is accreted is twice that made by protein which is deaminated; while diet lipid is used much more efficiently for growth of body lipid than for energy fuelling. The ME of feedstuffs is contained in feedstuff protein, feedstuff carbohydrate and feedstuff lipid; the carbohydrate can be in the form of sugars, or non-starch polysaccharides. All these various sources of energy may be used with different efficiencies for different purposes. It is not true, therefore, to assume that all the energy measured as ME from one feed source has the same value as another. This being the case it follows that feedstuffs could be better assessed in terms of their comparative values (cost per unit of utilizable nutrients) if NE values rather than DE or ME values were used. Diet formulation on the basis of NE is the norm in some countries and requires only that the energy value of feedstuffs (Appendix 1) and the nutrient specification of the diet (Appendix 2) be given in terms of NE. The calculation of NE requirement is not an issue because the proper expression of requirement in terms of DE or ME requires first an understanding of body energy metabolism in *net* terms. Equally pertinent is that the description of the value of a diet in terms of its response (rather than in meeting a given nutrient specification) explicitly involves a measurement of net energy response. It is not such issues as these which create difficulties in diet formulation. Rather the problems of using NE for practical diet formulation in the conventional sense of compounding to a given nutrient specification, as is being addressed here, are twofold:

1. It is necessary to define clearly and closely the way in which the pig which is eating the diet will use it. This is often not possible if a single diet is to be used over a range of appetites, genotypes, environments and pig weights.

2. A single NE value cannot be ascribed to a single feedstuff. The value will change according to the way it is used by the pig. This is difficult to accommodate in tables of nutritive value and, of course, in conventional analytical procedures, which are usually made in the laboratory and not with optimally performing animals behaving in a similar way to those about to be given the compounded diet.

In the event, there has been no evidence whatsoever to show that any greater precision of energy formulation or predictability of response will follow from the use of an NE system rather than a DE system. This is largely because of the need for fixed factors and fixed assumptions which, being fixed, render the system indifferent to the definitions used.

This is not, however, a complete picture because in many cases an NE system is used within the context of a conventional DE descriptor at both feedstuff and nutrient specification level. It is known, for example, that the efficiency of utilization of DE from fibrous foods is less than that from starch. This is because of differences in relative difficulty of metabolism of VFAs as the end products of fibre digestion as against glucose as the end product of starch digestion (see also Table 8.2). DE values of fibrous feeds containing complex non-starch polysaccharides (and feeds whose starch is fermented in the large intestine rather than digested in the small intestine) *may* therefore be written down to accommodate these phenomena. It is probably reasonable to surmise that most of the benefits of using an NE system come from:

1. Writing-up determined DE values for dietary lipids.
2. Writing-down determined DE values for fibrous foods.

These adjustments can, of course, be made within a DE formulation system, thus avoiding the difficulties of attempting a true net energy basis for diet formulation.

Given the ease of routine determination of DE in feedstuffs, and the possibility of making realistic adjustments for lipids and fibres, it is doubtful if any greater precision would result from formulation on any other than a DE basis.

Better definition of dietary utilizable protein

Individual feedstuffs differ quite widely in the digestibility of their protein (see Table 9.1). Fish meals have an apparent faecal digestibility coefficient for protein up to 0.95; that for soya bean meal is around 0.85, while that for other vegetable proteins such as wheat feed and cotton seed may be as low as 0.60. To ascribe the same value to protein emanating from wheat feed as to protein emanating from soya bean is therefore to greatly over-value the former in comparison to the latter. It is remarkable therefore that for so many years pig diets have traditionally been compounded on the basis of the crude protein (CP) requirement, that the nutrient specification for diets has been given in terms of CP, and that a CP ($N \times 6.25$) analysis of feedstuffs has stood as being an adequate descriptor of protein content (in passing, this also is in error; the multiplier for N to derive CP should vary according to the amounts of NPN in feedstuffs, but convention has never accommodated this despite true multipliers ranging from 6.25 for animal tissue to between 5.75 and 6.25 for most cereals and soya bean meal and down as low as 5.5 for some low-quality protein sources).

There would be considerable benefit to expressing protein needs in terms of digestible protein and the protein content of feedstuffs similarly. This may be readily achieved by using the digestibility coefficients given in Appendix 1 and setting up feed descriptor files with the more accurate protein definition of (ileal) digestible crude protein (DCP). Complementary expression of requirement and nutrient specification in DCP rather than CP terms equally presents no problem of either concept or mechanism, and existing knowledge is more than adequate to allow this simple process of improvement in compounding precision.

It remains mysterious how nutritionists may insist upon the expression of the energy system minimally after accounting for at least variation in digestibility between feedstuffs, but be willing to ignore the same logic for protein. That, however, has been the case and the present discussion, and the tables and appendices which relate, have bowed to convention and dealt largely in terms of CP. Nevertheless, information on digestibility value is also presented in the hope that a move to DCP may be forthcoming.

Many authorities would now consider the issue of formulation to protein requirement to have been largely superseded by the issue of formulation to essential amino acid requirement. And this is indeed so. It is necessary for a diet both to provide for the protein requirement overall and to provide for the requirement of the individual essential amino acids in addition. The nutrient specification for minimum levels of dietary essential amino acids has been set up in the present case in terms of balance to the amino acid lysine; all the other essential amino acids given are required in the diet at least at the level of the given ratio (that is the ratio required for ideal protein; Table 11.3 and Appendix 2) if a further reduction in the protein value (V) is to be avoided.

It is therefore simple to formulate diets to the provision of nutrient specifications for the nine essential amino acids. It is often the case that, with conventional ingredients, once adequate lysine is provided in a diet then the other amino acids are adequately provided for as well; this is because lysine is usually the first limiting amino acid in pig feedstuffs. In order of likelihood of limitation, next comes threonine, and then methionine, and then tryptophan and histidine. The liberal use of artificial lysine in the diet will, of course, cause some other amino acid to be first-limiting. Artificial sources of threonine, methionine and tryptophan are now also available as dietary ingredients, and they should be evaluated in the same way as any other potential feed ingredient; in this case, probably by comparison with the unit cost of amino acids in soya bean protein.

Satisfactory formulation of pig diets requires, of necessity, provision of the nutrient requirement of all the essential amino acids. To fail to provide for all the essential amino acids in the diet specification, and to fail to ensure that the final ingredient mix meets the minimum specification in terms of all the

421

essential amino acids, is seriously to fail to achieve adequate accuracy of formulation. Information for compounding to amino acid requirements is to be found in Appendices 1 and 2.

Formulating diets to a nutrient specification detailing the essential amino acids in addition to total protein avoids the possibility of a shortfall in the expected efficiency of utilization that might happen as a result of the inadequate supply of one or other amino acid. However, this does not deal with differences in *digestibility* of amino acids which are quite evident amongst feedstuffs as has been witnessed in Table 9.5, and detailed in Appendix 1. It is clear that the expression of the diet nutrient specification in terms of ileally digested essential amino acids and the description of feedstuff ingredients in terms of the same ileal digestible essential amino acids is a significant step forward in increasing the precision of diet formulation and allowing greater flexibility of ingredient choice. Only when compared on the basis of the ileal digested amino acids can, for example, the cost benefits of the use of cotton seed protein, wheat offal protein, lupin meal protein and soya bean protein be properly assessed (refer to Table 9.5).

There is no good reason why feedstuffs should not be defined in terms of DCP and ileal-digestible essential amino acids, nor any good reason why the nutrient requirement, expressed in the same terms, should not be best matched through these means in the routine formulation of diets. The determination of nutrient requirement for ileal-digestible amino acids is the simple sum of the amino acids deposited in lean tissue growth and the amino acids used for maintenance, divided by the factor v (the utilizability of the ileally digested amino acids), $(IP_m + IP_{Pr})/v$, as has been shown in Chapter 11. The necessary dietary concentrations of digestible amino acids may be determined (just as for the dietary concentration of digestible energy) with knowledge of expected feed intake. Formulation of pig diets on the basis of ileal digested amino acids is now becoming common practice amongst many progressive feed compounders. There remain two outstanding problems which detract from this apparently major step forward.

1. The use of ileal-digestible essential amino acids improves upon total amino acids in the same way as DE improves upon GE; but equally, optimum precision is not achieved until a view of net utilization is obtained. Although it is now appreciated that absorbed amino acids may be used with different efficiencies for body processes (v, Chapters 9 and 11), effective quantification of this character has not yet been achieved.

2. The determination of ileal digestibilities is arduous, and there is still a great range of values being proposed for individual feedstuffs. It is therefore difficult to state with any degree of certainty exactly what the

422

ileal digestibilities of the amino acids in the protein of a particular feedstuff may actually be. This means that the apparent increase in precision of formulation by use of ileal-digestible amino acid values can be somewhat exaggerated. Nevertheless, work on ileal digestibility has shown that particular feedstuffs can have especially poor digestibility values for their amino acids (see Table 9.5 and Appendix 1)´and this information is vital if feedstuff protein is to be credited with its true worth, and feedstuffs are to be objectively and correctly compared with each other.

Recent attempts to obtain more information on both the ileal digestibility and the subsequent utilization of absorbed amino acids (v) have led to some feeding trial work which, through measurement of the ultimately achieved pig-growth response, can try to rank the net values of proteins from various feedstuffs.

In many instances feed compounders will allow for reduced values for ileal digestibility of amino acids in some feedstuffs by only permitting their inclusion in diets at strictly limited levels. This tendency is also evident in the recommendations found in the lower sections of Appendix 1. Conventional pig diets tend to try to avoid dubious feedstuffs, but in so doing might be missing opportunities for cost-beneficial ingredients of lower nutritive value, but also of lower price.

Conclusion to possibilities for enhancement of precision of formulation

The present state of affairs allows formulation according to DCP and ileal digestibility of essential amino acids, and to do so would enhance, but not make perfect, the precision of diet formulation. Ileal digestible amino acid and protein levels in feedstuffs can be determined for individual feedstuffs and such values are given in Appendix 1 and Tables 9.1 and 9.5. The dietary nutrient specification for ileal-digestible amino acid and digestible crude protein requirement may be determined from the dataset in Appendix 2 on the assumption that the ileal-digestible lysine and digestible protein requirements were determined with diets that were around 0.8 digestible and digestible requirements are therefore some 0.8 of the total lysine and total protein requirements. Or, better, the requirement for digestible amino acids can be determined from first principles (Chapter 11) and the nutrient specification presented in terms of diet specification for the concentration of nine ileal-digested amino acids. The relative proportions of the other essential amino acids to lysine should, of course, remain the same as given in Appendix 2.

There is little benefit in attempting energy definition beyond DE.

Benefits from enhanced precision of formulation

The objective of the progression:

- identification of nutrient content of feedstuffs;
- formulation of feedstuff ingredients into a compounded diet to satisfy a nutrient specification;
- consumption by the pig of its diet at an adequate level of intake to provide for its daily nutrient requirement;
- achievement of required daily production response;
- obtaining adequate accuracy at each point in this programme

is to enable flexibility (a) in the face of changing feedstuff availability, price and nutrient content according to time and source and (b) in the face of changing market demand for product type and the changing abilities (through genetic improvement) of pigs to grow and reproduce.

In the absence of the need for flexibility the whole procedure is rendered unnecessary. Where a given feedstuff combination may be depended upon (when eaten at a given rate), to result in a predictable growth or reproductive response, and where feedstuff prices and availabilities are not capricious, and where that response is considered entirely satisfactory by both producer and ultimate consumer, then a fixed diet formulation is all that is required. Even knowledge of the nutrient concentration of that formulation is an uncalled-for sophistication. It is adequate to know that two or three feedstuffs of repeatable quality, when mixed together in a certain ratio, result in an output of acceptable and repeatable quality. Such was the basis of pig-feeding regimes over many years in the US using only maize-corn and soya bean meal, and in parts of northern Europe using only barley, wheat feed and fish meal (or barley, wheat feed and skimmed milk). However, fixed formulae for pig diets were unable to allow minimization of cost because cheaper sources of energy and protein were precluded (such as the use of maize and soya bean in European pig diets), and changes in pig genotype or customer requirements (such as greater lean tissue growth rate, and a demand for leaner meat) could not be accommodated. Fixed ingredient formulations however remain potentially beneficial where:

1. A limited range of ingredients are available, and these are of known quality. This case would apply for diets based on home-grown ingredients provided from mixed and integrated farming operations with both livestock and arable interests.

2. Ingredient quality, and therefore knowledge of ingredient source, is of paramount importance. This would apply in the case of diets formulated for baby and young growing pigs.

3. The definition of value of production response is not likely to change, and is expressed in terms of performance rather than cost-benefit. This is invariably the situation in the case of post-weaning diets where the primary requirement is for a high level of feed intake and the avoidance of diarrhoea.

In most other circumstances, cost-benefit will result from the widest possible range of ingredients being available for formulation in as wide as possible a range of combinations. If such flexibility is to be combined with effective precision of prediction of the pigs' response to nutrient supply it is essential that:

1. the relationship between nutrient supply and animal response be understood and precisely quantified at the level of the animal;

2. there is precise identification of the nutritional contents of feedstuffs in terms that relate to the final utilizability at animal level.

These aspects of diet formulation demand the further attention of nutritionists, and enhancement of diet formulation precision will result from:

a better definition of utilizable energy;
a better definition of utilizable protein;
a better definition of utilizable vitamins, minerals and trace elements.

But before these are considered it is proper to reiterate that all attempts at greater precision in the definition of animal requirements and in the analysable nutrient content of feedstuffs is quite fruitless – indeed ludicrous – if either:

1. the feed intake of the pig is compromised or significantly different to that assumed;

2. the assumed nutrient contents of feedstuffs are invalidated by poor storage, inadequate heat treatment, excessive heat treatment, the presence of toxins, and the like.

It is a self-evident truth, which is all too often ignored, that it is irrelevant to try to increase the accuracy of nutrient provision in a diet if the feed intake is compromised and equally irrelevant to reduce the price of a diet if the response is reduced by a greater financial margin.

It is frequently the case, world-wide, that further improvements in diet formulation by better nutrient specification and better feedstuff definition, are unlikely to be worthwhile unless and until optimum pig-feed intakes can be attained and the feedstuffs can be certified as free from negative factors not readily identified by simple chemical analytical procedures.

Least-cost diet formulation by computer

Although the same rules apply as for the hand calculation of a diet formulation, the computer has no effective limit to the number of ingredients which can be examined for dietary inclusion. Neither is there any limit to the nutritional characteristics which may be taken into account, so it is possible to adjust ingredient inclusions to get close to the nutrient concentrations specified for a diet.

The linear program for least cost demands not only that the available ingredients are mixed to provide the diet nutrient specification requested, but that this is done for the minimum cost. A particular balance of energy, protein, amino acids and minerals can be reached by a number of different ingredient combinations. But only one particular ingredient recipe will achieve the nutrient specification at the least possible cost. This least-cost formulation will change as frequently as the relative prices of feedstuffs, just as has been shown earlier.

The greater the freedom of choice given to the computer the better will be the chance of reducing the price of the diet as far as possible. The efficiency of the programme will be improved if the limits to the lower and upper inclusion levels allowed for the various feedstuffs are wide apart. Minimum inclusion rates for a particular ingredient can prove restricting on a computer formulation. It is not just that this ingredient has to be included at a particular stated minimum level, but with that ingredient goes a particular balance of energy, amino acids and minerals, and this balance influences the other ingredients which can be placed into the mix. Again, a low maximum inclusion rate for a feedstuff which is currently advantageously priced can be expensive. The fewer the restraints, the more efficiently the programme will run and the cheaper will be the final diet. The great beauty of the linear programming methodology for least-cost diet formulation is that it is able to look at the cost-benefit of all the various nutrients in all the various feedstuffs simultaneously and can therefore calculate in the optimum way alluded to earlier but found to be too complex to handle with a hand calculation.

Appendix 3 shows examples of minimum and maximum inclusion restraints, and further guidance is given in Appendix 1.

Nutrient specification targets, as has been seen from Table 14.6, are also difficult to meet with exactitude, particularly when there may be many of them. Again, a target with some width is helpful to the optimization process. Thus targets for a DE concentration of 13.0–13.3MJ/kg and for a lysine concentration of 11–12g/kg will result in a lower-cost diet than would exact specifications of 13.2 and 11.5. It follows, however, that whereas in the latter case there will be 0.88g lysine/MJ DE, in a former case the lysine:energy ratio can range from 0.92 to 0.83. It is important the ratio does not stray outside realistic maximum and minimum values.

In practice it is remarkably easy to restrain a least-cost computer program to producing diets little different from what may be calculated simply by hand, and no cheaper. It is quite possible to produce an apparently innocent set of restraints which are mutually incompatible and for which no solution is obtainable. Occasionally an unusual ingredient may be forced in at a very high price simply to adjust for some trivial imbalance of a minor amino acid or mineral. As a general rule therefore all restraints (of both ingredients and nutrient specification) should be kept to as few as possible and based on hard evidence.

Description of raw material feedstuff ingredients

The computer files require precise descriptions of the nutrient contents of available feed ingredients. These nutrients will be called up by the computer from the ingredients in order to satisfy the various aspects of the diet nutrient specification. There must be a complete match between the elements specified as required within the diet and these elements described for each of the individual feedstuffs. The matrix of nutrient content descriptors for feedstuffs, together with their recommended maximum and minimum inclusion levels for the various diets to be formulated, with a statement of the current price of each feedstuff, comprise the necessary information files for the least-cost diet formulation package to draw upon in the course of the formulation process. Appendix 1 is an example of an appropriate source for some of the information needed in an effective feed descriptor file. Example files for a particular source of barley and a particular source of soya bean meal are to be found in Appendix 4 which, in this case, shows an adequate, but not comprehensive, analysis. The same information file also, of course, gives the price for each potential diet ingredient.

Description of the nutrient specification for the diet

Appropriate nutrient specifications for various pig diets may be found from sources such as Appendix 2 and it is at this point that maximum and minimum levels should be set for the concentrations of the various nutrients required. Maximum (and minimum) inclusion limits for the various feedstuffs (which naturally differ for different diets) have already been suggested in the lower parts of the tables in Appendix 1.

Computerized least-cost diet formulation

The simple example described here used a computer program prepared by Format International (Singlemix).

The output shown in Table 14.10 describes in columns 3 and 4 the target nutrient specification to be achieved in the diet. The formulation specification is not comprehensive for vitamins and minerals as it is assumed that the vitamin and trace element supplements will provide fully for the dietary requirement (Appendix 2 refers). It may be seen from Table 14.10 that the minimum and maximum levels for DE requirements are quite narrow

Table 14.10 Nutrient specification: target (columns 3 and 4) and achieved (column 1)

Column	1 Analysis achieved	2 Limit	3 Minimum	4 Maximum
(Volume)	100.0	(max)	100.0	100.0
DE	14.0	(min)	14.0	14.5
CP	208.5		200.0	240.0
DCP	179.6		150.0	200.0
Lysine	11.3		11.0	13.0
Methionine + cystine	6.8		5.5	8.0
Methionine	3.2		3.0	4.0
Threonine	7.0		6.6	10.0
Tryptophan	2.7		1.6	5.0
Linoleic acid	10.6		10.0	20.0
Calcium	10.2		9.8	11.5
Phosphorus	6.5	(min)	6.5	9.0
Sodium	1.5		1.3	3.0
Oil	40.3		40.0	56.0
Fibre	45.0	(max)	25.0	45.0
Ash	65.4		49.0	70.0
Neutral detergent fibre	142.8		100.0	250.0
Copper[1]	93.9		1.6	150.0
Selenium	0.3		0.1	0.5

[1] Included as growth promoter.

(14.0–14.5), but those for protein quite wide. The lysine minimum is set to give at least 0.75g lysine per MJ DE (11.0/14.5), and the other essential amino acids (three others are identified) are balanced to lysine in the appropriate ratio as described earlier. Such a set-up might be expected to formulate on basis of energy, and lysine, possibly being modified by phosphorus, fibre, oil and even protein. In the event, as seen in column 1, the final analysis of the mixed diet gives the minimum value for DE and phosphorus, and the maximum for fibre, showing that energy and phosphorus were relatively expensive (and therefore to be minimized). The cheapest ingredient mix was also associated with an analysis at the maximum level of fibre. Releasing the restraints of minimum energy and phosphorus and of maximum fibre would allow a cheaper diet, but being of lower energy and mineral concentration, more of it would have to be eaten per day.

Frequently lysine is the most expensive nutrient to find from conventional feedstuffs and least-cost diet programs often formulate to the minimum limit for lysine. This is not the case here because of the presence in the diet of a significant quantity of fish meal which is likely to be a generous supplier of both phosphorus and lysine. This particular program was not set up to offer any artificial amino acids as possible feed ingredients. If artificial amino acids are available it is usually beneficial at least to offer them as available feedstuffs. Their inclusion would, of course, depend upon the relative prices of amino acids found within conventional feedstuffs such as soya bean meal and fish meal.

Table 14.11 shows the least-cost mix of feedstuff ingredients that will provide a nutrient specification within the limits detailed in columns 3 and 4 of Table 14.10. When formulated, the feedstuff combination in columns 1 and 2 of Table 14.11 will provide a compounded diet of the nutritional analysis shown in column 1 of Table 14.10. From Table 14.11 it is seen that the optimal currency cost is 136.17, achieved only by including the ingredients as shown in columns 1 and 2 of the Table. The major cereal used is wheat, but there is also some barley and wheat feed. Column 2 gives the exact percentage of each ingredient and column 3 the amounts to make a 2500kg mix. Column 5 shows which limits were operational in this particular formulation, while columns 6 and 7 show respectively the minimum and maximum inclusion levels to be allowed. Wheat is limited to a maximum of 40% and at a currency cost of 120 per tonne is clearly favourably priced as it is included up to the maximum limit allowed. Similar comments are appropriate for the fat supplement. Additives, vitamins and minerals are forced in at exact quantities regardless of price. Fish meal appears expensive, but is forced in at 4% on the basis of the judgement of the feed compounder. A cheaper diet would result if more wheat and fat were allowed in, and fish meal was less strongly insisted upon. There may be an (understandable) fear,

Table 14.11 Output from a least-cost linear programming formulation package (Format International – Singlemix)

Name: Pork Diet
Currency cost: 136.17

Column 1		2 %	3 kg	4 Currency cost	5 Limit	6 Minimum %	7 Maximum %
Included							
1	Barley	12.7	316	110.0		0	100.0
2	Wheat	40.0	1,000	120.0	(max)	10.0	40.0
3	Wheat feed	5.5	137	96.5		5.0	10.0
6	Soya 44	25.2	631	148.0		10.0	40.0
12	Dicalcium phosphate	0.6	16	215.0		0	5.0
19	Fat supplement	5.0	125	182.0	(max)	0	5.0
27	Additives	5.0	125	76.0	(min)	5.0	5.0
37	Fish meal 66	4.0	100	286.0	(min)	4.0	7.5
59	Vitamin/mineral mix	2.0	50	295.5	(min)	2.0	2.0
		100.0	2,500				
Rejected							
4	Maize			194.0		0	30.0
5	Flake maize			217.0		0	30.0
7	Full fat soya			246.0		0	10.0
8	Meat-and-bone meal			165.0		0	2.5
10	Grass meal			125.0		0	2.5
11	Limestone			35.5		0	5.0
13	Salt			91.0		0	1.0

however, that there would be a consequential uneconomic fall in performance.

Many ingredients available for inclusion have been rejected on the grounds of price per unit cost of their nutrients. These are listed in the lower section of column 1 of Table 14.11. Reductions in the costs of these latter feedstuffs, or increases in the costs of included ingredients (column 4) would, of course, change the optimal mixture of feedstuffs in the least-cost diet formulation. Most linear programming packages (including Singlemix) will give estimates of the prices that would need to prevail before a rejected ingredient is likely to be included.

Changes of ingredients

The least-cost diet formulation process infers that pigs will respond to analysed nutrients such as given in column 1 of Table 14.10. This is to suppose that pigs do not notice, nor react to, the tastes or textures of different feed ingredients, such as given in column 1 of Table 14.11. It is further supposed that pigs do not notice differences in diets compounded from either few or many ingredients. Although in general the desire for nutrients by pigs does seem to override matters relating to ingredient taste, this is not likely to apply so strongly to younger pigs and lactating sows. These classes of pigs may have their appetites noticeably disrupted by the appearance in their diets of new ingredients, or ingredients included at unfamiliarly high levels. This being the case it is undesirable to allow the dramatic fluctuations in ingredient levels that may follow from changes in comparative prices of the major ingredients. Restraint should therefore be imposed on the formulation to ensure that dramatic and abrupt changes in diet ingredients do not occur. It has to be accepted that such restraints will move the formulation away from optimization achieved solely on the basis of consideration of price per unit of nutrient.

The computer output of a least-cost diet formulation is best used as a guide for the nutritionist toward appropriate diet construction. Whilst the exact suggestions of a linear program may indeed be implemented, it is not appropriate to allow the computer to override all human judgement. Least-cost diet formulation is particularly helpful in showing up the real costs of nutritional decisions that may be made regarding both the degree of appropriateness of various feedstuffs, and the assumptions that are made regarding nutrient requirements and target diet nutrient concentrations.

CHAPTER 15

OPTIMIZATION OF FEED SUPPLY TO GROWING PIGS AND BREEDING SOWS

Introduction

Control of the feed allowance remains the most biologically, financially and managementally effective way of optimizing pig performance. Feed control is simple in the mechanical sense through the medium of feed-supply engineering systems. These may range from computer-controlled pipeline feeding, through auger-supplied individual pen (or pig) volume-controlled metering devices, to self-feed continual access *ad libitum* hoppers. The influence of feed supply upon the rate and composition of growth, and upon reproduction, is both substantial and direct (see Figure 15.1). There is a close relationship between feed allowance management and financial management which comes from (a) the influence of feed supply upon the level and quality of pig production output and (b) the cost of feed being by far the greatest single element of total production costs (usually about 75% of all costs, including fixed costs).

Growing pigs

The pig production manager has many points at which his operation may be controlled, but six are especially important: health control, environment control, diet formulation control, feed intake control, genetic control and product quality control (Figure 15.2). As the figure shows, feed intake

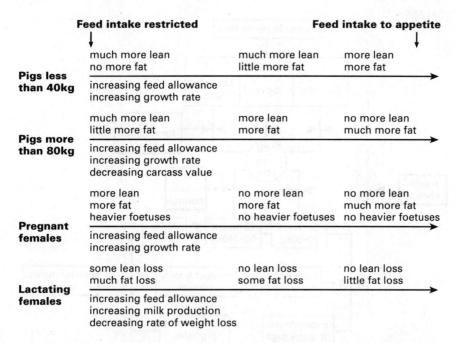

	Feed intake restricted		Feed intake to appetite
Pigs less than 40kg	much more lean / no more fat	much more lean / little more fat	more lean / more fat
	increasing feed allowance / increasing growth rate		
Pigs more than 80kg	much more lean / little more fat	more lean / more fat	no more lean / much more fat
	increasing feed allowance / increasing growth rate / decreasing carcass value		
Pregnant females	more lean / more fat / heavier foetuses	no more lean / more fat / no heavier foetuses	no more lean / much more fat / no heavier foetuses
	increasing feed allowance / increasing growth rate		
Lactating females	some lean loss / much fat loss	no lean loss / some fat loss	no lean loss / little fat loss
	increasing feed allowance / increasing milk production / decreasing rate of weight loss		

Figure 15.1 Influence of increase in feed allowance upon pig productivity.

control is pivotal to the effective usage of both feedstuffs and pig inputs. Through interaction with both genotype and diet nutrient specification, feed intake control determines the quantity and quality of the end-products. This is well illustrated by the case history shown in Figure 15.3. The grow-out unit concerned was showing a loss. Upon investigation, the following aspects of management, amongst others, were noted:

● The pigs were fed twice daily to appetite with a computer-controlled pipeline wet-feeding system.
● The carcasses were unacceptably fat.
● Growth rate was adequate, but not particularly impressive.
● Health and environment were satisfactory.
● The diet used was of relatively low nutrient concentration.

The obvious remedial action might have been to suggest that a diet of higher nutrient specification should be used, but this would have increased input costs at a time when such would have been financially hard to bear, and when other circumstances – such as feeding level – would have conspired to ensure that the positive effects of improving the diet would have been minimal. Rather, the recommendation was that the feed allowance be reduced. This had the apparently negative effect of reducing growth rate but

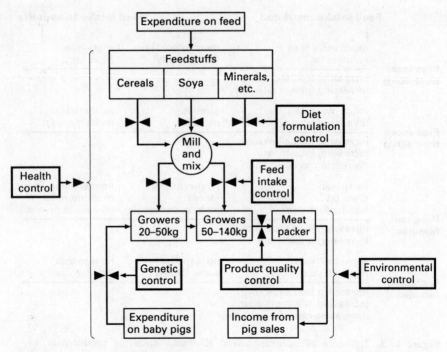

Figure 15.2 Flow diagram of a grow-out unit for pigs, showing inputs, outputs and points of decision-making and operations control.

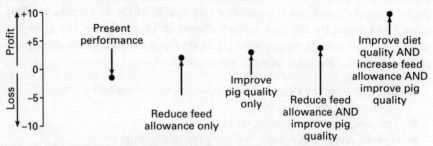

Figure 15.3 Production strategy decisions for profit optimization.

input costs were reduced, and carcass quality improved; thereby raising output value and margin. This helpful response was indicative of (a) the stock being of relatively low genotype in terms of potential daily lean growth rate and (b) the particular market for which these pigs were destined discriminating strongly against fat and meriting an improved genotype. A pig of improved quality *alone*, however, would not have maximized the response. Further actions need to be taken with regard to both diet quality (concentration of nutrients) and feeding level. Given an improved genotype with a high lean-tissue growth rate potential, the longer-term strategy recommended was not only to move to the use of a higher quality diet but also to reintroduce the original high-level (appetite) feeding system (Figure 15.3).

Interactions with level of feed supply

Slaughter weight

Product quality control (Figure 15.2) has many elements but a major aspect is the interaction between feed supply and slaughter weight. Pigs may usually be dispatched from the production unit at a weight which will target a particular market; for example, 120–160kg live weight for heavy pigs, 85–120kg for conventional processing and manufacture or 60–80kg for light, fresh pork. Even within these bands, more rigorous dead-weight limits might be set (for example, 65–72kg dead weight which would require a live-weight dispatch range of only 85–95kg). The manager must therefore decide which pigs should be dispatched at what weight, and to which market outlet, in order to maximize profitability. But the quality of the carcass and the quantity of carcass can be counteracting forces. The heavier the weight of the individual pig at sale, the greater will be its individual value and the lower will be the fixed costs as a proportion of total costs; but the heavier pig at sale is likely to be fatter, and this may reduce the value per unit weight of pig sold. This means that where carcasses are likely to be at risk of being judged unacceptably fat, there is impulsion not only to reduce daily feed input by rationing but also to reduce the weight at slaughter. If a range of 62–75kg dead weight is allowed, it is probable, all other things being equal, that pigs of 73–75kg will have a lower value per kilogram when a premium is paid for leaner pigs. This may be countered by giving a lower level of feed to those pigs which are more likely to fatten, or by selecting them for earlier dispatch. Figure 15.4(a) shows the relationship between weight at slaughter and subcutaneous P2 backfat depth, and the interaction with pig genotype, while Figure 15.4(b) shows the equivalent relationship between feed allowance and backfat depth. It may be inferred that lower feed intakes – by encouraging a lower P2 backfat depth – may allow a higher live weight at slaughter. Disbenefits in terms of carcass value due to pigs being unacceptably fat will be more evident when less-improved genotypes are allowed to eat higher quantities of feed (Figure 15.4(b)). In the absence of any price penalty for fat pigs (all pigs are equally acceptable regardless of fatness), the highest levels of feeding and the heavier slaughter weights will tend towards being more economic.

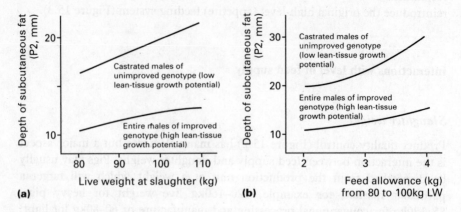

Figure 15.4 Influence upon carcass fatness of: (**a**) live weight at slaughter, and (**b**) feed allowance from 80 to 100kg live weight (the diet was 14MJ DE/kg).

Feed conversion efficiency

The contrary pressure to the conclusion that heavier (and fatter) pigs are to be preferred in the absence of any quality discrimination against fat is primarily the efficiency with which the most expensive input commodity (feed) is used. Feed conversion efficiency will inevitably decrease at higher slaughter weights due to the progressively greater costs of maintenance. Maintenance costs result in no product yield, and are related to the body mass of the pig. Thus at 60kg live weight there is a fixed maintenance cost of about 0.7kg of feed, at 90kg about 1.1kg of feed and, at 120kg, 1.3kg of feed. Feed conversion efficiency will also inevitably decrease at the highest levels of feeding when this brings about excessive fat deposition. It is salutary to note that reducing the feed input may not always improve feed conversion efficiency; this will usually happen only when extremes of fatness are restrained. More usually reducing feed intake will – by reducing growth rate – worsen the efficiency of utilization of feed for growth. As the amount of feed eaten increases, there is a beneficial reduction due to progressive savings in maintenance costs (Figure 15.5). Ultimately, however, efficiency deteriorates due to the deposition of excess fat at the higher feed intakes, and the fact that a unit of fatty-tissue growth requires more than three times as much feed as a unit of lean-tissue growth. Importantly, the figure also shows that there is relatively little change in the amount of feed required for each kilogram of live-weight gain over quite a wide range of amounts of feed eaten; changes in feed conversion ratio tend to be more evident at the extremes than in the middle of the range.

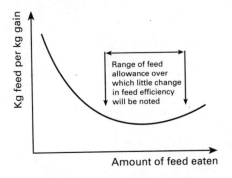

Figure 15.5 Relationship between feed conversion ratio (kg feed/kg gain) and feed intake.

Response to an increase in feed intake

Optimization necessitates quantification of response. Empirical measurements would suggest that a 100g increase in daily feed intake (approximately 1.5MJ DE daily) from 20 to 100kg will bring about an increase of 30–40g daily live-weight gain, an increase of 1mm or so in P2 backfat depth, an increase of 10–15g in the daily lean-tissue growth rate, an increase of 20–25g in the daily fatty tissue growth rate and a marginal deterioration in feed conversion ratio (Figure 15.6). Such statements of empirical response, although often much beloved, are dangerous. The nature of the response to an additional increment of feed allowance is dependent – *inter alia* – upon the absolute feed level, the lean-tissue growth potential of the pig, the weight range over which the response was measured and the environmental conditions. In the extreme, it is possible for a 100g feed increment to be used infinitely inefficiently to achieve no growth, or to be used for the single purpose of the growth of 100g of straight lean tissue (1kg lean tissue uses 15MJ of balanced DE), or to be used for the single purpose of the growth of 30g of straight fatty tissue (1kg fatty tissue uses 50MJ of DE) and, of course, any combination amongst these three extremes is entirely feasible.

Interaction with genotype and sex (potential lean-tissue growth rate and predisposition to fatness)

Increasing feed allowance will in general increase growth rate and fatness (Figures 15.4 and 15.6). The extent of these increases is highly dependent upon genotype and sex. Thus the same level of feeding which would induce inefficient fattening in a castrated male could give beneficial lean growth in a female. Castrated males and unimproved genotypes of lower lean-tissue growth potential will justify feed allowance restriction to a greater extent, and imposed at a lighter weight, than will females (or entire males) and improved genotypes. The later in the growing period that restriction is imposed, the faster will be the average rate of growth, with those animals of higher lean-

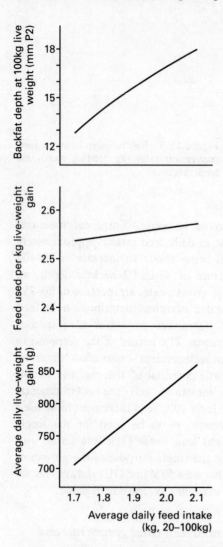

Figure 15.6 Responses to increasing average daily feed intake. The diet was of 14MJ DE, 200g CP and 11g lysine/kg. Pigs were fed from 20 to 100kg.

tissue growth potential benefiting more from delaying the live weight at which the feed allowance restriction occurs (Figure 15.7(a)). Similar growth response may be expected consequent upon the severity of restriction imposed (Figure 15.7(b)).

Figure 15.8 remains the most useful paradigm for considering the appropriate level of feed to be supplied to any pig in any given day. The first principle is that the overall growth response of fat plus lean is always positive to increasing increments of feed. The more feed that is given, the faster the pigs will grow. Secondly, at lower feeding levels this response is particularly steep and efficient because the next increment of feed is going towards

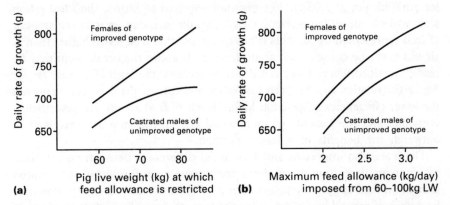

Figure 15.7 Influence upon average growth rate (g/day) of: (**a**) pig live weight at which the daily feed allowance is restricted to $0.10LW^{0.75}$/day (diet of 14MJ DE/kg), and (**b**) the level of maximum daily feed allowance imposed from 60 to 100kg.

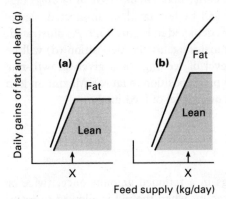

Figure 15.8 Daily gains of fat and lean in response to increase in feed supply: (**a**) for pigs of unimproved type; (**b**) for pigs of improved type.

growth of equal composition (i.e. equal fat:lean ratio) as the last increment; this composition is primarily of lean tissue. Thirdly, the response becomes broken and the slope less steep with higher levels of feeding when fattening is induced. The fourth principle, and of paramount significance, is that the point at which the reduction in efficiency occurs, and at which fattening begins, is dependent upon the lean-tissue growth rate potential of the pig (as may be seen by comparison of pig types (a) and (b)). Feed supplied to the amount X (Figure 15.8) will maximize lean-tissue growth rate. Above X growth rate will continue to increase but the pig will fatten. Below X lean-tissue growth rate is not maximized. The absolute amount of feed represented by X is highly dependent upon the type of pig (sex, genotype and lean-tissue growth potential). The absolute amount of feed represented by X will also increase as the pig grows. For a 40kg live-weight pig, X may be at a feed supply level of 1.75kg, for a 60kg live-weight pig at 2.25kg and for an 80kg pig at 2.75kg. For pigs not required to fatten, the ideal ration scale will be one which allows for each pig, on each day, exactly the amount of feed represented by X. This is not to suggest that feed levels higher than X should never be chosen. Not at all; feed levels above the break point may be readily justified where there is a desire to increase the level of carcass fat, or where the relationship between feed price and carcass price is such that even the lower efficiencies achieved at higher levels of feed supply are nevertheless economic and the benefits of more rapid pig growth rate are perceived to outweigh the benefits of a leaner product.

There are also important and substantial differences between pigs in their inherent predisposition to lay down fatty tissue *even* during nutrient-limited growth (at feed-supply levels less than X). While the normal expectation may be a ratio of around one of fat to four of lean, this may be both better (one of fat to five of lean) or considerably worse (one of fat to two of lean); and this characteristic is open to the influence of both sex and genotype. Thus castrates will be inherently fatter than entire males *at any level of feeding*. Pigs selected for their capacity to be lean will be less fat than unselected pigs *at any level of feeding*. This phenomenon is depicted in Figure 15.9. As illustrated, the pig type which inherently discriminates against fat deposition (d) will be more efficient and require a lower level of feeding for a given growth-rate response than the pig type which has a predisposition to fat (c). Maximum lean tissue growth rate is also reached at a lower level of feed intake.

Ration scales for growing pigs

Feed can be supplied to growing pigs in a rationed amount once, twice or many times daily; in an unrationed amount when the pig is allowed to eat to appetite as much as it wants but does not have access to feed 24 hours per

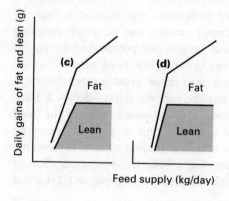

Figure 15.9 Daily gains of fat and lean in response to increase in feed supply: (c) for pigs with a predisposition to fatten; (d) for pigs which discriminate against fat, even when feed is in limited supply.

day; or in an unrationed amount where feed is continuously available. It should not be assumed that these systems are mentioned in ascending order of feed intake achieved. Maximum feed intake will probably occur when pigs are fed high levels of rationed amounts three or four times daily.

Most feeding scales which ration the feed allowance do so in the knowledge that:

● young pigs, because they convert efficiently and grow mostly lean tissue, are unlikely to be overfed;
● as pigs grow bigger they can use (and need) more feed;
● as pigs grow bigger, their appetites catch up with their lean-growth potential and they will tend to fatten.

The pig live weight at which feed restriction is imposed, and the extent of the restriction, will depend upon the strength of the desire to achieve a lean carcass. It can be assumed that there are few – if any – circumstances, other than control of gut diseases, which merit feed restriction in pigs of below 40kg live weight. Some fattening scales based on pig live weight are exemplified in Figure 15.10(a) and (b). Those in section (a) are smoothly curved, whilst those in section (b) bring about abrupt restriction. In practice, pigs are rarely weighed as they grow; but because the relationship between feed intake and growth rate is both close and direct, weight can be estimated readily from knowledge of feed intake and time. There is some danger in this procedure if it is assumed that pigs are lighter than they really are, for the pigs will be restricted more than is the intention and growth will be slowed below potential. It is good policy always to challenge the grow-out unit every 6 months or so with an increased feeding scale to determine the response of the pigs. On the other hand, assuming the pigs to be heavier than they really are will cause them to be fed more than originally intended. This may be no bad thing if the reason for being underweight has been some degree of illness, because some catching up may be called for; but if the potential of the pig has been overestimated the scale will be excessive and the carcasses will become fatter than intended.

Realistically, the weight-based scales in Figure 15.10(a) and (b) will actually be operated on the basis of feeding either *ad libitum* (self-feeding systems) or to appetite (controlled-feeding systems) up to the point at which the restriction is imposed, which will occur consequent upon a decision relating to the *time* passed since the pigs entered into the pen at a known starting weight. The restrictions themselves may be imposed either gradually (a) or abruptly (b). The figure shows three levels of restriction (i), (ii) and (iii) indicative of increasing severity which would relate to the relative quality of the genotype (propensity to become fat), and the need to achieve leanness (severity of price penalty for unacceptably fat carcasses).

441

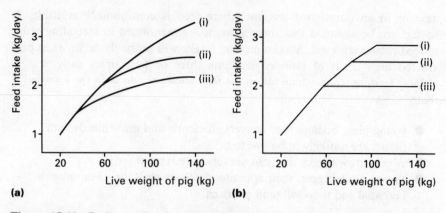

Figure 15.10 Ration scales for pigs based upon their live weight: (a) smooth transitions from appetite to restricted regime, (b) abrupt transitions. Scales (i), (ii) and (iii) represent progressive degrees of feed restriction.

Time-based scales for rationing pigs are shown in Figure 15.11. Section (a) shows weekly steps of different size from 20kg live weight onwards while (b) shows a system of appetite feeding up to a given time after entry into the grow-out pen, after which the time-based restriction is imposed. Scales (i) and (ii) in Figure 15.11 represent progressive degrees of feed restriction.

Ration scales may be usefully employed in automatic feeding systems, but not necessarily with the objective of restricting feed supply, reducing growth and controlling fattening. Rather feed intake may be increased above *ad libitum* levels if feed is provided fresh in discrete and limited meals two or three times daily, and offered in amounts that challenge the pig's natural appetite. It should not therefore be assumed that pigs with limited appetites, or known to yield lean carcasses, are always best fed through an *ad libitum* self-feed hopper system.

The choice of scale, whether it be purposely limiting or purposely generous, is highly dependent upon:

1. The nature of the product required (carcass fatness).
2. The importance (in financial terms) of growth rate.
3. The nutrient concentration of the feed used (a given response following from a lower quantity of a more concentrated feed).
4. The inherent potential of the pig for lean tissue growth (pig sex and genotype).

As has been explored previously, it may be presumed that:

1. young pigs should be encouraged to eat to the maximum of their appetite;
2. improved genotypes may have lower appetites;

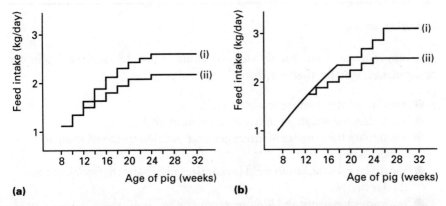

Figure 15.11 Ration scales for pigs based upon their age, or upon days from the point of entry into the grow-out unit (the pigs are presumed to enter at 8 weeks of age and 20kg live weight, or 10 weeks of age and 30kg live weight): (**a**) weekly increments from time of entry, (**b**) weekly increments following a period of *ad libitum* feeding.

3. improved genotypes may eat more feed without the likelihood of becoming fat;
4. at any given feed allowance, castrated males will fatten more readily than females, and females more readily than entire males.

Simplistically, the point at which lean-tissue growth rate is maximized but fattening avoided (X, Figure 15.8) may be empirically determined by trial and error. If the scale exceeds X, the pigs will fatten and this will be evident by fatter carcasses at the point of slaughter. On the other hand, if the ration scale provides feed at a level of less than X, then the pigs will not be maximizing their growth. This latter position can be made evident by an increase in the level of feeding being shown to bring with it improved daily growth rate without an increase in carcass fatness. The appropriate method for selecting a correct ration scale is therefore to increase feed allowance until an unacceptable proportion of the pigs are judged to have carcasses which are over-fat, and then to reduce the ration scale to the point at which acceptable carcass fatness is achieved.

Many growing pigs are fed *ad libitum* world-wide right through to the point of slaughter and no ration scale is imposed. This is because either genotypes of high quality are being used which do not require to have their feed restricted in order to be acceptably lean, or because high fat levels in the carcass are not considered to be unacceptable. The imposition of rationing scales becomes especially relevant for pigs slaughtered at higher weights and/ or where there is a desire for a lean carcass.

Breeding sows

The appropriate feed supply for breeding sows to achieve optimum performance will be that which:

- maximizes the number of piglets per litter;
- optimizes the weight of individual piglets at birth;
- maximizes the number of litters per year (minimizes the weaning to conception interval);
- maximizes sow lactation yield (maximizes piglet growth from birth to 28 days);
- optimizes longevity and lifetime productivity.

General pattern of live weight and fatness changes in breeding sows

Sows gain body weight in pregnancy and lose it during lactation. In addition, at parturition, they produce (and lose) the combined weight of the foetal load plus a further 50% of placenta and fluid (Figure 15.12). As is the case for body weight overall, sows also grow fatty tissue in pregnancy and lose it in lactation, but for fat there is not the same overall positive gain as the sow increases in age and body size. Rather, under well-managed sow-feeding programmes, average sow body fatness will remain between 15 and 20% or so (being lowest at weaning and highest at parturition) throughout breeding life (Figure 15.13). As a result, sows that are properly fed will become bigger as they age, but not necessarily fatter.

The greater the feed allowance in pregnancy, the greater will be pregnancy gains. The less the feed consumed in lactation, the greater the lactation losses. Figure 15.12 shows 25kg of gain, parity upon parity, up to the fourth pregnancy. During the early parities the sow is making maternal gains to achieve her target mature lean mass. Over this period of breeding life pregnancy gains may be assumed to be of both lean and fat. Later, however, pregnancy gains are likely to be almost entirely of fatty tissues. In all parities (regardless of sow age) lactation losses are most likely to be fat alone, as this is the prime substrate for which there is likely to be a shortfall in relation to the requirements for milk synthesis. There may also be some lactation losses of protein, but this will only occur in a protein-deficient diet, or to yield energy if the sow has used up most or all of her energy reserves. Table 15.1 shows *maternal* body gains achieved by adequately fed sows mated for the first time at 125kg live weight, and maintaining a pattern of body fat change of the sort indicated in Figure 15.13. The compositions of the pregnancy

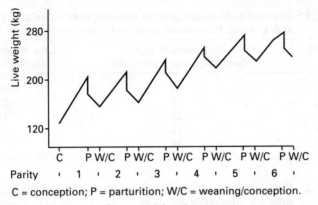

Figure 15.12 Expected pattern of sow weight change.

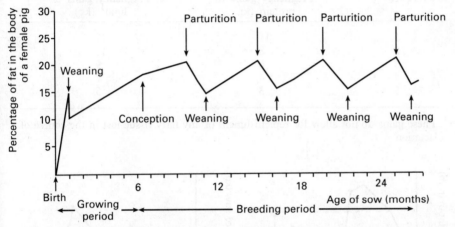

Figure 15.13 Changes in lipid content of the body of pigs consequent upon fatty tissue losses after weaning and during lactation.

gains in terms of chemical protein and lipid are given in Table 15.2. Figure 15.14 shows how weight gains (in this case in the first two parities) can take place at the same time as significant fat losses. In this case the feeding regime used was both inadequate for lactation feeding (being only 4.8kg per day) and also inadequate for the necessary fat recovery in pregnancy (the sows being fed only 1.8kg per day in the first pregnancy and 2.3kg per day in the second). This figure shows how in pregnancy both weight and fat can be gained, but more weight than fat; whilst in lactation both weight and fat are lost, but more fat than weight. In this case gains of 11kg of live weight in the course of a single parity appear to have taken place despite absolute losses of more than 3.5mm of P2 backfat depth, which is equivalent to losses of 4kg of fat. The responses shown in Figure 15.14 should not be interpreted to mean that sow weight gains and fat losses of this order are natural expectations; in

Table 15.1   Maternal body-weight gains of breeding sows mated at 125kg live weight and growing to maturity under adequate nutritional conditions

Parity	Live-weight gain conception–conception (kg)	Live weight at next conception (kg)
1	35	160
2	28	188
3	23	211
4	18	229
5	14	243

Table 15.2   Chemical composition of maternal body pregnancy gains of protein and lipid for adequately fed sows

Parity	Pregnancy gains of protein (kg)	Pregnancy gains of lipid[1] (kg)
1	11	15
2	8	11
3	6	9
4	4	7
5	3	5

[1] These gains do not allow for replenishment of any fatty tissue lost in the course of lactation.

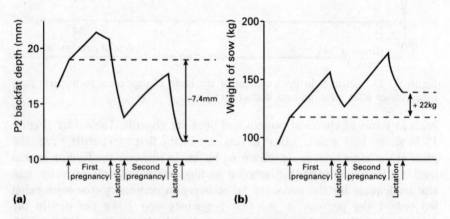

Figure 15.14   Changes in fat depths and live weights of sows over two parities (from Whittemore, C T, Franklin, M F and Pearce, B S (1980) *Animal Production* **31**:183).

fact, it has been clearly demonstrated that P2 backfat depths of less than 12mm are likely to be associated with reproductive inefficiencies.

Figure 15.15 shows the distribution of P2 backfat depths throughout the population of pigs on the same trial as that referred to above. There are three

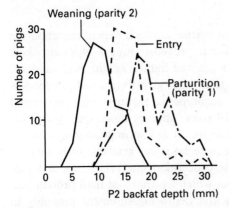

Figure 15.15 Distribution of fatness amongst breeding pigs.

distribution curves depicted. The middle distribution is that which prevailed upon the entry of the pigs into the breeding unit when they were around 90kg live weight. The pigs had an average P2 backfat depth of 15mm, and this was mostly distributed between 12 and 18mm. At parturition (the distribution curve to the right of the figure) the average fatness had increased to about 20mm P2. By the time of weaning at the end of the second parity (as shown by the left-hand distribution curve) the backfat depths had not only decreased to a dangerous average of below 10mm P2, but the variation in P2 backfat depth had spread to a range of between 6 and 18mm. Because sows are likely to show rebreeding difficulties at P2 backfat depths of below 12mm, a large proportion of these sows were clearly at risk. The degree of variation that has been created in the herd is a further and independent cause for serious concern. The standardized feeding regime (4.8kg per day in lactation and 1.8kg [parity 1] or 2.3kg [parity 2] per day in pregnancy) over all the sows has created *not* a standardized population of pigs but instead a highly variable and diverse one. It is evident that to achieve a standardized level of sow fatness at any point in the breeding cycle sows must be fed individually a non-standardized ration allowance calculated according to their needs. The optimum feed supply will therefore be sow-specific, individual and variable within the breeding herd.

If young breeding females start with about 20% of fat at mating, and subsequently increase this to 25% of fat at parturition parity 1, then designing fat losses into the feeding regime may be tolerable. However, the newer – leaner – hybrid strains may have fat levels nearer 15% at parity 1 mating and 18% at parturition so pursuit of a fat-depleting regime would be intolerable. For young breeding sows coming to first mating essentially lean, sow feeding strategies must maintain fatness throughout breeding life. This means that lactation fat losses must be minimized and pregnancy fat recovery encouraged.

Feed supply to first mating

The appropriate weight and age at first mating is much dependent upon the genotype of the pig. Asian breeds and unimproved European types are likely to be ready for breeding at a younger age and lighter weight than modern, improved hybrid genotypes with high lean-tissue growth rate and greater sow productivity potential. For improved genotypes, a mating weight of around 125kg live weight, and an age of 210 days, appears most appropriate for optimum subsequent fertility and longevity. It is also important that at the start of the breeding life there are adequate fat stores available to facilitate a good lactation and a short weaning-to-conception interval. It appears that adequate body composition may be that which has more fat than protein (i.e. greater than 17% lipid) and a P2 backfat depth measurement probably in excess of 18mm. Figure 15.16 shows that modern hybrid sows tend to be much less fat than their predecessors, and the required combination of fatness and age at sexual maturity is unlikely to be readily achieved by the same feeding techniques which would optimize growth rate and carcass quality of pigs grown for meat production. The slaughter generation of hybrid pigs, if fed for maximum lean-tissue growth, will achieve 125kg live weight in less than 180 days, and the P2 measurement at that time may be little more than 16mm! High levels of high-protein diet are not best suited to the rearing of breeding stock which should more properly be given a more conservative diet of adequate protein and relatively low energy. The formulation used for the lactating sows might be ideal. The level of feeding would normally be somewhat lower than that encouraged in finishing pens of slaughter pigs. Some rationing is likely to be needed if the growth rate is to be restrained to the required average of around 700g daily between 30kg and mating. It is difficult to generalize about what level of feeding will achieve

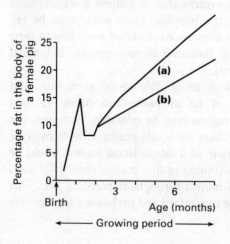

Figure 15.16 Modern hybrid strains of pigs (**b**) are much less fat than their predecessors (**a**).

optimum growth for breeding pigs up to the point of first mating (nor even to generalize about optimum age and weight at first mating), as so many other factors interact. The ultimate objective, however, is to mate pubertal gilts at their third heat, at which time they should have an adequate but not excessive cover of fat. Age and live weight are not themselves targets with respect to the right time for first mating. They arise as a consequence of the age pattern of puberty, the weight pattern of fatty tissue growth, and the relationship between ultimate mature size and the appropriate proportion of mature size that requires to have been reached before the phase of reproduction is initiated.

There is no evidence to support the idea of short-term 'flushing' before first mating, but a level of feeding adequate to support positive fatty-tissue gains (as well as positive lean-tissue gains) throughout the period prior to mating is essential. Particularly important is the status of the body fatty tissues between the first pubertal heat and the first conception. This is interpreted by some to mean an elevated feeding level at this time and this is not itself a bad thing.

Feed supply in pregnancy

The influence of feed allowance in pregnancy was the subject of a series of classical experiments by Professor F W H Elsley. At feeding levels below 2kg it was found that the maternal body would need to be catabolized for purposes of supporting foetal gain; litter size was reduced and individual piglets were of lower birth weight than those from dams which had received more appropriate levels of pregnancy feeding (Figure 15.17). The figure shows that the effect of feed allowance (ration) upon the live weight of the pigs at birth is significant but small. Individual piglet birth weight will increase by around 0.2kg in response to an extra 1kg of feed given daily throughout the 115 days of pregnancy, which is a feed:gain ratio of about 50:1. None the less pig birth weight does have significance when pigs of particularly low birth weight die simply because they are small; this often happens to piglets born at less than 1kg live weight. Percentage survival can be approximated as $55W_b^{1.3}$ where W_b is the weight at birth.

It has been suggested that birth weight can influence subsequent growth performance (Figure 15.18). This, of course, will not be a direct effect but mediated indirectly through factors such as post-natal health and position in the dominance order.

Because the growth of piglets *in utero* is exponential, dramatic increases in foetal weight can take place in the last 3 weeks of pregnancy. Increasing the feed allowance at this time is therefore likely to be more efficient in terms of

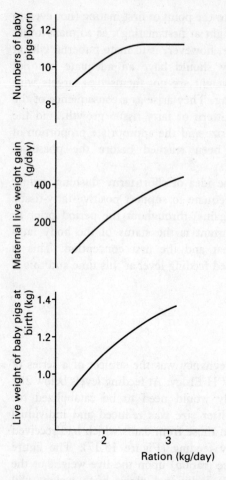

Figure 15.17 Responses of pregnant females to a change in feed allowance.

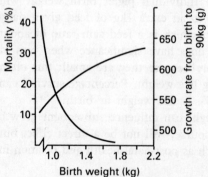

Figure 15.18 Relationships found between mortality, growth rate and weight at birth.

feed use for foetal piglet growth than increasing the allowance over the whole of the pregnancy. Whereas increased feed level over the whole of the pregnancy will tend to support mostly maternal gains, feeding subsequent to day 90 of the pregnancy will be more likely to support mostly foetal gains. This is especially the case for sows carrying inadequate body fat reserves at this time (as suggested by the propositions in Figure 15.19).

Figure 15.17 shows that the relationship between feed allowance and maternal live-weight gain is almost linear, and may be broadly represented by an efficiency of between 0.2 and 0.25. Positive responses to pregnancy feed intake in terms of maternal and piglet live weight do not, of course, continue to be linear at extreme levels; over-fat females waste feed and are also likely to have fewer young and problems at parturition.

The experiments of Elsley confirmed pregnancy feeding as needing to:

● establish the foetal number (ultimately litter size);
● grow the foetal load (ultimately piglet birth weight);
● grow maternal lean tissue mass (ultimately mature size).

Additionally, pregnancy feed is required to:

● grow mammary tissue (ultimately a relatively minor factor in milk yield);
● replenish maternal fatty tissue mass (ultimately a major factor in milk yield and in the weaning to conception interval).

Fatty-tissue deposition appears to be most active during the first three-quarters of pregnancy, whilst in the last quarter some fat may be catabolized if the sow goes into negative energy balance on account of the rapidly developing foetal load. Whether this loss of fat, even during pregnancy, is an unavoidable biological phenomenon or the result of underfeeding in late pregnancy is not clear. The picture of fat loss from the 85th day of pregnancy is particularly evident in young gilts, as may be seen from Figure 15.14.

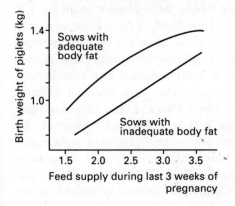

Figure 15.19 Influence of late pregnancy feeding upon individual piglet birth weight.

Weight gains in pregnancy are closely related to pregnancy feed intake (Figure 15.20). Due to individual sow variability there is a broad band of response. *Maternal* body gains of 40kg or so may be associated with a daily feed intake of 3kg, while 1.5kg will achieve approximate *maternal* weight stasis. The *total* weight gain (maternal body-weight gain plus the foetal load and the other products of conception) will be about 20kg greater than the maternal gains alone. Maternal gains of 25kg will therefore relate to overall body weight gains of 45kg. A guideline regression relating maternal live weight gains in pregnancy (δW) to daily feed intake (F) might be:

$$\delta W(kg) \simeq 25F(kg) - 27$$

(Equation 15.1)

This equation points to a daily feed allowance of 2.2kg in pregnancy giving maternal weight gains in pregnancy of 28kg, and confirms that the rate of live-weight gain in sows may be increased with an efficiency of feed use of 0.22 or a rate of feed usage of 4.6kg feed for every kilogram of live-weight gain.

The relationship between maternal pregnancy fatty tissue gain (δT) and daily pregnancy feed intake (F) is shown in Figure 15.21. The corresponding guideline regression for this illustration is:

$$\delta T(kg) \simeq 15F(kg) - 30$$

(Equation 15.2)

The relationship between fat change and feed intake may also be given in terms of backfat depth (P2) change in pregnancy:

$$\delta P2(mm) \simeq 4.60F(kg) - 6.90$$

(Equation 15.3)

by which equation a daily feed allowance of 2.2kg in pregnancy would give maternal gains of 3.2mm P2 backfat.

The broad approximations contained within the regression coefficients in equations 15.1, 15.2 and 15.3 imply that (a) depth of fatty tissue is a large component of weight gain in pregnant sows and (b) weight gains can proceed at lower feed intakes than those that would support fatty-tissue gains. A daily intake of 2kg of an average diet (which is often quoted as a feeding level appropriate for pregnant sows) can be associated with maternal gains of around 23kg, total body-weight gains of about 43kg and little fatty-tissue gain or replenishment. This response might well be considered appropriate for adequately fat sows with no requirement for maternal growth or replacement of lipid catabolized during lactation, but could be inadequate for sows in their first three parities (and which are therefore still growing) and for sows with fatty-tissue levels which are sufficiently low, unless replenished, to put

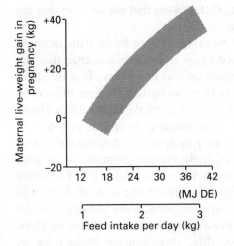

Figure 15.20 Response of maternal live-weight gain to feed intake in pregnancy. The diet is assumed to contain about 13MJ digestible energy (DE) per kilogram.

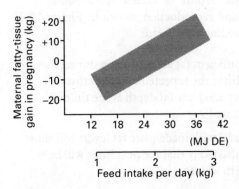

Figure 15.21 Responses of maternal fatty tissue gain to feed intake in pregnancy.

at risk subsequent reproductive success. Where inadequate pregnancy feeding results in sows carrying insufficient fatty tissue at parturition, the subsequent milk yield is likely to be adjusted downwards by the mother and the growth of the sucking pigs prejudiced. A recent experiment showed the necessity of allowing for losses of body fatty tissue and body weight in the subsequent lactation:

Litter weight (kg) at 28 days of age = 57 + 1.5 sow fat loss (mm P2) + 0.46 sow weight loss (kg)

(Equation 15.4)

Sows that are thin will also react to their unacceptable state by delayed rebreeding. This will show as an increase in the number of days that lapse between weaning and next conception. Days between weaning and conception have been found to be consistently negatively related to both sow

453

fatness and sow live weight at weaning; that is, sows that are too lean and too light rebreed less soon after weaning.

Although the fatness of the sow at weaning is more to do with lactation feeding level than with pregnancy feeding, it is nevertheless true that the chance for reinstating lost fat stores comes only in pregnancy. It is therefore relevant to pregnancy feeding, as well as lactation feeding, to note the crucial effect of sow body condition upon readiness to rebreed. Figure 15.22 shows the relationship between the interval from weaning to oestrus and the P2 backfat depth of the sow at weaning. Usually there are 3–5 unproductive days between weaning and oestrus. However, these escalate dramatically as sows become thinner and have backfat depths below about 12mm P2. The effect seems to be more serious for primiparous gilts at the end of their first lactation than for more experienced multiparous sows. Sows becoming too fat may exhibit similar problems. Whilst more attention may justifiably be given to the problem of sows which are too thin, there can on occasion be an equally important problem with sows that become too fat. There is no benefit in the breeding pig attaining backfat depths in excess of 25–30mm P2.

The relationship between fatness and reproduction shown in Figure 15.22 is pivotal to the success of the breeding sow herd:

- pregnancy feeding must allow sufficient fat accumulation during the pregnancy minimally to (a) facilitate the expected fat losses that will occur in lactation *and* (b) support adequate fat depth at the time of subsequent weaning (at least 12mm P2);

- sows weaned in poor condition and with inadequate fat levels will show reluctance to rebreed, and it is also likely that the next litter will be reduced in both number and birth weight.

Feed supply between weaning and conception

At one time or another, there has been reason to suppose that, after weaning, both a generous ration and a restricted ration might enhance the number of ova released and the number of embryos safely implanted into the uterine wall. It would now appear that heavy feeding before conception and in the first 3 weeks of the pregnancy is counter-productive but very high feeding levels after weaning are, in any event, rather difficult to achieve. It would also appear that problems of embryo implantation will only arise with feeding levels in early pregnancy that are at least as high as 3kg daily, and probably greater than that. A sow already thin at the point of weaning is unlikely to benefit from being further embarrassed by a restricted feeding regime. It may be concluded therefore that between weaning and conception sows may be

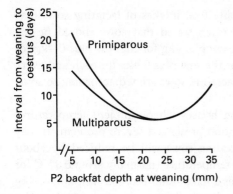

Figure 15.22 The relationship between the interval from weaning to oestrus and the depth of backfat in breeding sows.

fed to appetite for the duration of this 3–5 days period; and they will eat 3–4kg daily. If conception is delayed, some restriction will become necessary and the level of feed supply should relate to the condition of the sow but would normally be between 2 and 3kg daily. The first 3 weeks of pregnancy are sensitive because this includes the period of embryo implantation which has a large effect on ultimate litter size. It would appear that both particularly high and particularly low feeding levels at this time may compromise the number of foetuses that will settle into the uterine wall. Feeding levels in early pregnancy therefore usually range between 2 and 3kg daily – the higher levels for sows in poorer body condition.

Feeding in lactation

Lactation feed allowance requires to be provided at adequate level to:

● maximize milk yield (ultimately to maximize piglet weaning weight);
● minimize maternal fat and protein losses (ultimately to minimize the time between weaning and conception, and maximize subsequent litter size).

It is well recognized that the breeding sow will gain both fatty tissue and lean tissue in pregnancy and lose them (but mostly fatty tissue) in lactation. There will be some relationship between these two phases of the reproductive cycle. Sows of greater weight and fatness at parturition have more tissue available to be volunteered toward lactation and rearing of the young. Postpartum appetite is negatively related to pregnancy feed intake. A 0.5kg increase in daily allowance during pregnancy may result in as much as 0.5kg decrease in daily appetite during lactation, but usually the relationship is much weaker. Lactation feed intake is linked to the status of the sow's fat stores. The fatter the sow at parturition, the greater the appetite depression.

Under *ad libitum* conditions, the daily feed intakes of lactating sows may vary between 3 and 9kg daily. It is often stated that sows should be fed according to the size of the litter; for example, 4kg for the sow plus 0.4kg for each piglet over six, or perhaps 2kg for the sow plus 0.4kg for each and every piglet in the litter. But more usually lactating sows are fed according to some flat rate, regardless of litter size.

In warm climates, and in farrowing houses where ambient temperatures are high, adequate feed intake in lactation presents a severe problem. It may be estimated that the appetite of lactating sows will be reduced by about 0.2kg/day for each 1°C above comfort temperature (normally around 15°C for a lactating sow). In these circumstances a maximum feed consumption of 4kg per day or less is quite common. Reducing the heat output from the body by providing energy in the form of fat rather than carbohydrate can be helpful in maintaining appetite in the face of high ambient temperatures. Other opportunities to reduce the likelihood of appetite loss by heat stress are to increase ventilation rate, provide sprays and wallows, and enhance conduction through the floor by use of solid rather than perforated flooring. All these devices will increase the effective comfort temperature by facilitating a more rapid removal of heat from the body surface. Removal of heat from the body at a faster rate has the same effect as reducing the level of heat production in the first place, and thereby forestalls feed intake depression. The total effect of all these factors combined can offset appetite depression by as much as 1.5kg daily at an ambient temperature of 25°C, and therefore represents most important considerations for the breeding-herd manager.

Some large sows kept under cool conditions may eat 7–8kg daily, but most sow appetites rarely exceed 6.5kg during lactation. On the first day after parturition relatively little can be consumed (1–2kg) but this will rise by about 1kg daily to reach a maximum of 6–7kg or so at the end of the first week of lactation. Some reduction may occur after 21 days of lactation as yield falls. Primiparous sows often only eat 4kg or less in lactation, some larger and more highly productive multiparous sows may eat 9kg or more daily. Sows fed the diet wet will eat 10–20% more feed than when the diet is given dry.

Lactation weight and fat losses are directly related to lactation feed allowance (Figure 15.23). At the higher feed intakes maternal body weight gains of more than 10kg may be achieved in the course of a 28-day lactation, but at the lower intakes 20kg of sow maternal body weight may be lost. Fat losses as measured by the change in ultrasonic P2 measurement show similar trends. The less the feed, the greater the rate of fat loss, with lower feed allowances causing P2 losses in excess of 5mm over a 28-day lactation, whilst the higher feed allowances may almost achieve fatty tissue balance.

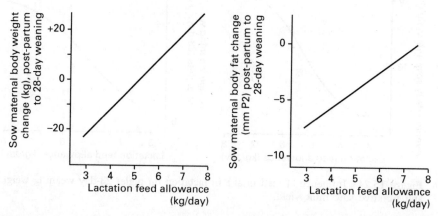

Figure 15.23 Influence of feed intake in lactation on sow body weight and body fat change.

The linear regressions for Figure 15.23 are:

Weight loss during 28-day lactation (kg) = 48.8 − 9.60 lactation feed intake
(kg/day)

(Equation 15.5)

Fat loss during 28-day lactation (mm P2) = 11.1 − 1.37 lactation feed intake
(kg/day)

(Equation 15.6)

An average feed intake in lactation of 6kg daily will therefore result, in the course of a 28-day lactation, in a weight loss of 9kg and a reduction in P2 backfat depth of 3mm.

These equations suggest:

● lactation weight loss may be largely prevented at feed intakes of above 5kg per day, which may be quite readily achieved;
● lactation fat loss may be largely prevented at feed intakes of above 8kg per day, which are not at all readily achieved;
● at weight stasis significant quantities of body lipid are lost;
● an extra 1kg of feed per day over a 28-day lactation will save about 10kg of maternal body weight loss and about 1.4mm of P2 maternal subcutaneous fatty-tissue loss.

Feed intake in lactation has a positive effect upon milk yield and the weight of piglets at weaning. Putative responses are shown in Figure 15.24. The

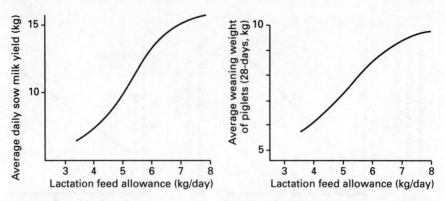

Figure 15.24 Influence of feed intake in lactation on piglet 28-day weaning weight and putative sow milk yield.

sigmoid curve suggested is a consequence of early increments of feed supply going primarily to satisfy the priority shortfall in maternal requirement. Sow milk yield, of course, is also highly dependent upon litter size. Further, large sows will have less feed available for lactation than smaller sows eating the same amount of feed, on account of differences in maintenance requirement. Last, as is the case with all responses expressed in terms of feed weight (rather than energy or amino acid intake), lactation responses are highly dependent upon the concentration of nutrients in the ration consumed.

Sows which are either unacceptably thin or unacceptably fat have less desirable reproductive performance. As has been presented in Figure 15.22, there is now clear evidence for negative relationships between sow body condition at the end of lactation and readiness to rebreed (days between weaning and oestrus). There are also negative relationships with conception rate (percentage of mated oestrous females conceiving), and ovulation rate (potential litter size) (Figure 15.25). It has been proposed that for each 1% of fat lost from the body of the sow in the course of lactation there will be 0.1 piglets less born in the next litter. This is consistent with the view that, for thin sows, delaying conception by 21 days can give an extra piglet per litter. Australian workers are unequivocal in support of evidence from a variety of sources that the greater the extent of fatty and lean tissue losses during lactation, the longer will be the weaning-to-conception interval. Figure 15.22 has represented these propositions.

Figure 15.26 may be used to depict the relationship between body-weight losses during lactation and the weaning-to-conception interval. Average intervals are generally about 10–12 days because although sows would usually be expected to show oestrus 3–5 days after weaning, some of these will return in the normal course of events. From the figure it is apparent that rebreeding problems are likely to become evident with maternal body-weight losses in

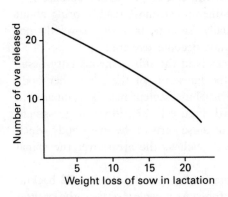

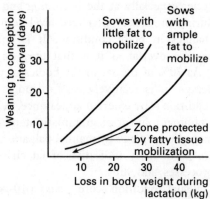

Figure 15.25 Influence of sow weight loss in previous lactation upon the number of ova released.

Figure 15.26 Influence of body weight loss in lactation upon rebreeding proficiency in sows.

lactation that are greater than 10kg per sow for sows with little fat, or 20kg for sows with ample fat. The implication here is that the adverse effect of weight loss upon rebreeding proficiency would be mitigated by the presence of adequate fatty tissue stores. Figure 15.22, alluded to earlier, is seminal to the understanding of the importance of fatty-tissue losses in lactation to subsequent reproductive efficiency. Failure to achieve at least 12mm P2 of backfat depth at the time of weaning will result in a significant increase in the weaning-to-oestrus interval. Similarly, failure to restrain backfat depth to levels of below 25mm P2 will have a like negative effect.

Condition scoring

The measurement of P2 backfat depth in pregnant sows is entirely feasible and relatively simple, but may not always be convenient. In this case sow condition may be used as a good indicator of sow body fatness. It has been well established that there is an optimum breeding fatness which is the most vital component of the feeding regime. This is more important than live weight or the slavish adherence to some generalized rationing scheme which stipulates given feeding levels at given times regardless of individual sow requirement. Feeding to sow body condition is thus an entirely reasonable concept. Feeding to condition demands that when a sow is seen to be in lower body condition than she should be, her feed allowance needs to be increased; but if the sow is in higher body condition than she need be, the feed allowance should be reduced. It therefore remains to define only what optimum body condition might be at the various stages of the reproductive cycle; especially at the beginning and at the end of pregnancy because it is during pregnancy that feed-level adjustments can most readily bring about changes in body condition. It is equally as important that sows do not become overfat as it is that they do not become overthin.

As 70% of the fat in the bodies of pigs is in the subcutaneous fatty tissue depots, it is relatively easy to judge the fat status of the sow – her body condition – by external appearance. A visual assessment may be compared to a photographic or diagrammatic standard (Figure 15.27). Physical assessment may also be made by manual palpation of those parts of the sow's body which are particularly indicative of fat change, such as the areas over the spinal processes and hips.

Visual condition score (using a 10-point scale) is well related to P2 backfat depth as measured with an ultrasonic probe. As a generalization, the backfat depth at P2 (mm) is about three times the numerical value of the condition score. The use of sow fatness, through the medium of condition scoring, as a monitor of nutritional adequacy for breeding proficiency, has the great advantage of being independent of feed level, health, environment and management. Proposals for feeding strategies made in the form of target-condition scores surmount the problems of sows in different environments reacting in a variable way to the same standard feed level. The condition score target demands that the animal be fed what it takes, not fed any previously stipulated amount. In consequence, a sow judged too thin but already on a high feed level will be identified as requiring to be given more; whilst a sow judged to be too fat, but already on a low feeding level, will be identified as requiring to be given even less.

Usually a condition score of above average (6) is an appropriate target for

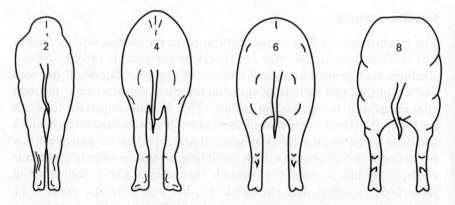

Figure 15.27 Visual condition scoring scheme using a diagrammatic standard. The appearance of the sow is compared with the picture and the animal given an appropriate score.

the end of pregnancy, and one below average (4) is acceptable at the end of lactation. Under normal circumstances these targets can be achieved by relatively minor adjustments to feed levels. Sows that are thinner than a condition score of 4, or fatter than a condition score of 6, merit special attention. It is evident that many modern hybrid sows are being weaned at condition scores of <4 consequent upon an inadequate level of feed supply during lactation, with inevitable negative results. Condition score at weaning is, of course, highly dependent upon both condition score at farrowing (influenced by the pregnancy feeding rate) as well as the rate of condition loss during lactation as a result of lactation feed intake. Problems with sow fertility should not, however, invariably be seen as soluble by nutritional means. Other vital factors are health, the hormone status of the sow, environmental conditions and, of course, the fertility of the boars.

An example of a feeding procedure for use on breeding units employing condition scoring techniques is given in Table 15.3. This procedure is now being taken up with enthusiasm by practical pig producers.

The problems with condition scoring systems are:

1. Condition scoring measures fatness only indirectly, and can be prone to error consequent upon variation in sow size and shape.

2. Condition score is dependent upon the subjective opinion of the scorer and standards can drift. Individuals often have different views about adequate condition states through the reproductive cycle.

3. The response of condition score to change in feed intake is not well documented and is variable. Values given in Table 15.3, for example, should be treated as guidelines and not rules.

Rationing scales for breeding sows

Sow feeding is mostly about long-term strategy and long-term trends in body weight, fatness and condition. It is the general status of the sow that requires to be maintained to a satisfactory level, and sow fatness changes are probably more important than sow weight. It has been established that young, modern hybrid gilts should be mated at live weights in excess of 125kg and ages in excess of 210 days. They should have been fed sufficient to make gains of around 700g daily from 30kg live weight onwards, and to gradually increase their levels of body fat over this time. Feeding in the 3-week period immediately prior to mating may well be accommodated by a generous *ad libitum* rationing regime.

It has further been established that excessively high feed levels in

Table 15.3 Feeding pregnant sows according to body condition

1. Using all the sows in the breeding herd that are pregnant or empty (that is, all sows not lactating), determine the average daily feed allowance per sow. Record this number.

2. Determine the condition score of all the sows in the herd that are pregnant or empty. Use whole numbers on a 10-point scale, with 1 as emaciated and 10 as grossly obese. The target average condition score for the whole herd should be about 5. Usually young sows should have a condition score of rather more than 5, while old sows should usually be rather less than 5. Newly weaned sows should score about 4, while sows due to farrow should score around 6.

3. If the average condition score for all pregnant and empty sows in the herd is above 5, then the average daily feed allowance (as per paragraph 1) should be reduced. If the average condition score is less than 5, then the average daily feed allowance for the sows should be increased. In the first event, increases and decreases in the average daily feed allowance should not be greater than ±0.2kg per sow per day.

4. Although the average condition might be satisfactory, this average can be made up of sows that are thin and sows that are fat. To deal with this, individual animals must be fed according to their individual condition scores.

5. For each individual sow, determine the condition score immediately after weaning the piglets. Feed according to the Table below until a condition score of 6 is reached for the particular individual in question. Upon reaching condition score 6, the sow should be returned to the average daily feeding level (as per paragraphs 1 and 3).

6. At weaning the sow should be changed immediately from her lactation feed allowance to a suitable regime to prepare her for mating. Feed allowances between weaning and conception are likely to involve feeding at least 3.0kg of lactating sow diet, but it is usually best to feed to appetite. It is unnecessary to present more than 3.5kg of food daily. Sows failing to be mated within 10 days of weaning should be fed until they are pregnant according to the Table below.

Increases or decreases to the average daily feed allowance for pregnant sows consequent upon their individual body condition score

Sow body condition score	Likely feed requirement in addition to the average daily feed allowance for sows in pregnancy (kg/day)
2	+0.6
3	+0.4
4	+0.3
5	+0.2
6	Average daily feed allowance[1]
7	−0.2
8	−0.3

[1] On exceptional units, or where a highly concentrated diet is used, average allowance may be around 2kg per sow daily. More usually, however, the average feed allowance is around or in excess of 2.2kg per sow daily.

pregnancy may reduce embryo survival, and that pregnancy feeding should be adequate to allow gains of sow maternal body weight and fatness to achieve a condition score of around 6 by the time of farrowing. This will be approximately equal to a P2 fat depth of 16–20mm and a total body gain from conception to farrowing of about 50kg. Under normal circumstances, feeding levels in pregnancy will vary, according to sow condition, around a mean of 2.3kg daily.

In lactation, sows will invariably lose fatty tissue and may also lose some lean tissue. If losses are severe not only will the lactation be compromised but also the success of the next parity. Lactation feed intakes may often average 6kg of feed daily, which will require some degree of fatty tissue replenishment in the course of the subsequent pregnancy. The lactation feeding regime should normally be one of intake encouragement and not intake limitation.

Overfatness will also result in both diseconomies and reproductive inefficiencies. Sows of condition score greater than 6 should be restricted. Generous levels of feeding in pregnancy should be avoided by ensuring the correct general strategy, which ought to obviate the need for short-term, crash-course tactics to fatten sows up or slim them down. Because sows will unavoidably lose fat during lactation, they should gain fat in pregnancy to compensate. Fat-level minima are probably in the region of 16mm P2 at the end of pregnancy and about 12mm P2 at the end of lactation. Sows with over 25mm may be considered too fat.

Lastly, it has been established that attemps to define general nutritional requirements and to prescribe general blueprints for feeding all the breeding sows in a herd will be disappointing. Sow variability in response to nutrients means that each sow must be rationed according to her own individual condition. Appropriate guidelines to ration scales for sows are to be found in Figure 15.28. The figure allows for a pattern of pregnancy feeding which is lower than average in the first 3 weeks, and higher than average in the last 3 weeks. But unless there is reason to believe that there is a problem of embryo mortality, or of low piglet viability due to inadequate birth weight, then a single feeding level based on condition score and perceived needs for fatty-tissue replenishment can be fed throughout pregnancy.

In practice:

- Sows should be condition scored frequently during pregnancy and appropriate adjustments be made to the ration.

- Feeding allowance in pregnancy will normally range between 2 and 2.8kg.

- Early pregnancy feeding is often set to a standard rate regardless of sow condition, say 2.5kg daily.

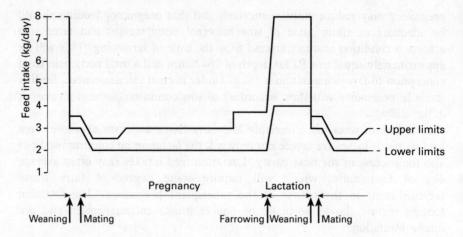

Figure 15.28 Guide to ration scales for breeding sows. The upper and lower limits to the allowances are those commonly expected; even higher or lower levels may on occasion occur in practice.

- The appropriate feed allowance during lactation is *ad libitum*. Should the litter size be particularly small (less than eight piglets), the feed allowance may be limited to 6kg daily. Nevertheless, the near-impossibility of overfeeding the lactating sow means that control of feed allowance is centred during lactation not upon feed restriction but upon appetite enhancement, and the maximization of feed intake by all possible devices.

- At weaning, feed supply may be set to a standard rate that is just below appetite, say 3.5kg daily.

- As a general guide, a good sow producing 2.3 litters per year will use annually at least 1.2t of feed of conventional nutrient concentration $[(115 \times 2.5 \times 2.3) + (13 \times 3.5 \times 2.3) + (28 \times 6.5 \times 2.3)]$.

- It will readily be appreciated that breeding sows, as also growing pigs, respond to quantity of nutrient, not weight of feed; the feed being potentially quite widely ranging in its nutrient concentration. The above estimates for optimum levels of feed supply refer in general to diets with around, or above, 13MJ DE and 150g CP (pregnant sows) or 13.5MJ DE and 165g CP (lactating sows) per kg fresh weight.

- By use of equations 15.3 and 15.6, it may be calculated that a sow farrowing at condition score 4, and with a P2 backfat depth of 14mm, and predicted to eat 5kg of feed daily over the period of a 28-day lactation, will come to weaning with an inadequate P2 measurement of

9.75mm and a condition score of less than 3. Feeding 2.8kg daily in pregnancy would raise P2 fat levels to 16mm and condition score to around 5 by the time of the next farrowing.

Farrowing at condition score 6, with 17mm of P2 backfat and eating 6kg daily, will result in the loss of 3mm of P2 and 9kg live weight, and a condition score of 4 by the time of weaning. A feed allowance of 2.2kg in pregnancy will replace the 3mm of fat lost and allow a pregnancy gain of maternal body weight of 28kg; or a net gain (conception to conception) of 19kg. The latter regime would appear just adequate for sows later in breeding life, but clearly marginal for those in their first three parities who would benefit from a higher level of feeding.

Feeding the sucking pig

Solid feed is usually offered to baby pigs from 14 days of age as a supplement to mother's milk. Substantial quantities are rarely eaten before 25 days of age. If weaning is at 28 days of age, it is beneficial to piglet growth for the same diet to be used before weaning as afterwards, and to continue until the weaned pig attains 10kg live weight.

In addition to being dependent upon the lactational ability of the sow, the actual milk received by an individual piglet is a function of the number of piglets sucking, the size and vigour of those piglets and the stage of lactation. Vigorous sucking pigs in smaller litters may readily attain live weights of more than 10kg at 28 days of age, showing the potential for growth when milk is in ample supply. However, usually most sucking pigs from litters of 10 or more weigh only 7–8kg at 4 weeks, indicating that there is a shortfall in milk supply at the udder. The potential for growth and the availability of sow milk begin to diverge at about 20 days of age, and it may be assumed that from this point on supplementary feeding is beneficial. Should the litter be greater than eight piglets, or the sow be yielding less than expected, then earlier feeding of sucking pigs is warranted – usually from 14 days of age (earlier, of course, in cases of milk shortage).

In many studies, piglets sucking sows of good milk yield have been shown to ingest little or no supplementary feed before 18–21 days of age. Generally, sucking pigs may begin to eat trivial amounts of supplementary feed at around 10 days of age, and 10–30g daily between 14 and 21 days. By the fourth week of life intake may have risen to 60g or so. The increasing relationship between the intake of supplementary feed and sucking piglet age could be expressed as:

$$Y = 0.0044X^{2.80}$$

(Equation 15.7)

where Y is feed per day per piglet (g) and X the age of the piglet (days).

Sucking pigs have digestive systems different from pigs eating solid feed, and the change-over at weaning represents a considerable digestive trauma. This period is counter-productive, and it is helpful if young pigs are persuaded to eat solids before they actually need them in order that they should be acclimatized to the post-weaning diet. This requires frequent provision of a highly palatable diet offered to the sucking pig in appealing form. Unless the highest levels of management skills are employed, the consequences can be lost growth, poor feed-conversion efficiency and diminished pig throughput in the post-weaning phase.

There are therefore two good reasons for offering solid feed to sucking piglets. First, for litters of reasonable size, supplementary feed will help to attain the greatest possible growth rate during the sucking phase. Secondly, all piglets (regardless of the litter size) are helped if they are accustomed to the nature of solid feed before they are abruptly weaned at 3, 4 or even 5 weeks of age.

The overriding requirement of a supplementary solid feed for sucking pigs is that it be readily eaten. The diet must therefore be interesting to the baby pig in non-nutritional (taste, texture and smell) as well as nutritional terms. For baby pigs to avoid a post-weaning growth check, they should consume around 300g of solid feed on the day following their removal from the sow. Such intakes are rarely, if ever, achieved by sucking pigs less than 30 days of age, even under the best conditions. *Ad libitum* feeding, indeed appetite encouragement to maximize voluntary feed intake, should continue from weaning through to the need for feed intake limitation and rationing, which is unlikely to occur until around 40kg live weight at the earliest.

Ration control by diet dilution

Nutrient intake is the product of feed nutrient concentration and level of feed supply. Where appetite is limiting nutrient intake can be enhanced by increasing nutrient concentration. Equally, a decreasing diet nutrient concentration can be countered with an increasing level of feed supply. However, when the feed allowance meets the bound of appetite, further decrease in nutrient concentration will result in a decrease in nutrient supply;

de facto an effective restraining ration scale is imposed. The possibility of allowing pigs to eat to appetite, but controlling nutrient supply by means of diet dilution, has some attractions – especially with respect to the simplicity of management. All pigs would be fed *ad libitum*. Where control of the allowance is needed, the diet is diluted pro rata. This apparently enviable scheme has been tried many times but invariably foundered because:

- the diluent itself is not free of charge, although nutritionally inert;
- a diluted diet is more expensive to mix, to transport, to store and to convey;
- the amount of diluent needed can be extraordinarily large before appetite restraint is achieved, as with a pregnant sow, for example, whose requirement can be met by 2kg of a concentrated diet, but whose appetite limit is probably three times that.

Relatively few classes of pig fall within the range appropriate to being nutrient-restricted by means of diet dilution (Table 15.4).

Conclusions

Optimization of feed supply for growing pigs and breeding sows is a hands-on activity requiring constant reappraisal and adjustment, managerial

Table 15.4 Possibilities for imposition of effective rationing through means of diet dilution with an inert (or nutritionally sparse) filler

Pigs appropriate for feeding to appetite (diet dilution contra-indicated)	Pigs appropriate for feed intake limitation but level of diluent required likely to be too high to be economic (diet dilution contra-indicated)	Pigs appropriate for feed intake limitation and for which the level of diluent required may be economic (diet dilution feasible)
Lactating sows Weaned sows Sucking piglets Weaned piglets Growing pigs Finishing pigs of high genetic merit	Pregnant sows Finishing pigs of high appetite and low genetic merit	Finishing pigs and castrated male pigs with conventional appetite but a predisposition to being unacceptably fat

dexterity, and a clear view of relative benefits. Rationing scales and recommended feed allowances are not open to generalized rules or blueprints, but are highly specific to individual production circumstances.

While broad guidelines as to expected feed regimes can be helpful, it is through knowledge of the likely responses to changes in nutrient supply that the optimum level of feed allowance will best be set.

Maximum appetite is usually sought in baby pigs, young growing pigs and lactating sows.

Effective control of the production process through rationing the feed supply is achieved primarily with only two classes of pig: (i) pregnant sows and (ii) finishing meat pigs. But effective control of feed supply at these points in the production process is crucial to efficient production in both grow-out and breeding herds.

CHAPTER 16

PRODUCT MARKETING

Introduction

The first phase of marketing begins with the creation of the product, follows with its effective description to a potential customer base and concludes with the product's sale. The second phase of marketing begins with the delivery of the product, is followed by its effective use and concludes with a repeat order. There are some ancillary requirements, not least that the product is demanded by a sufficient number of customers to justify its creation, and that the price of the product is acceptable to the customer.

The production of pigs involves three seminal marketing activities: (a) the movement into the pig unit of feed from the manufacturing and supply trades, (b) the similar movement into the pig unit of breeding stock from pig breeding companies and (c) the movement of pig meat product from the pig unit into the human food industry.

There are other marketing activities, such as obtaining buildings and equipment, but these tend to be less frequent in their occurrence and have less impact on the day-to-day decision-making processes than the buying of feed and breeding stock and the selling of finished meat pigs.

Marketing pig feeds

The proportion of total pig feed which is sold fully compounded to pig producers differs widely amongst nations and depends, *inter alia*, upon the extent to which a country may require to import feed ingredients. Where a large number of ingredients are imported, pig feeds will tend to be fully compounded – usually at the port of entry – by large-scale, high-technology feed mills. The final mixed diet is transported from the milling site to the (rural) location of the pig feeder. Particularly good examples of this behaviour are to be found in The Netherlands and Japan.

Because the main production inputs (feeds) are brought into the pig unit in their final form as balanced diets appropriate to each class of pig, the pig unit may function as an enterprise independent from other farming operations. This will lead to the development of large, self-contained, stand-alone, intensive pig units which, while integrated into the animal feed supply industry on the one hand and into the human food supply industry on the other, may not be integrated into the fabric of the farming community.

If the tradition for pig keeping has developed from the base of wider farming operations, such as grain production, there is an increasing likelihood that feeds will be compounded on site at the production unit rather than in a specialist feed-compounding mill. Pig-production units compounding their own diets on site will require to bring together the necessary feedstuff ingredients and to mix them in the right proportions. This may present problems where the number of ingredients are large, the amounts needed are small, and where the accuracy of the weighing of an ingredient is important. Many on-site pig-feed operations rely on weighing machinery with limited precision, and they have a limited capacity for receiving and holding different feedstuff ingredients. Home mill-and-mixer units will major on the utilization of energy-rich, home-grown cereals (maize, wheat, barley, oats, rye, triticale), and may possibly also use home-grown protein supplements (field beans, peas, sunflower, rape-seed, soya bean); although many protein-rich pig-feed ingredients require factory processing before use – either to extract valuable oils for human food or to detoxify antinutritional factors, or both. Because of this, the on-site compounding of pig feeds may often be limited to the milling of home-produced cereals, to which may be added purchased feedstuffs such as extracted soya bean meal, fish meals, meat-and-bone meals, processed pea, bean, rape and lupin meals and so on. Protein supplements may be purchased from a remote feed compounder as ready-mixed and balanced protein concentrates appropriate to each of the various classes of pigs on the unit. This will dramatically reduce the number of ingredients required on the home production unit and facilitate a simple on-site compounding operation requiring only the adding together of milled home-grown cereals and the purchased protein concentrate.

Home milling-and-mixing operations commonly try to *avoid* diets which:

● may be overly complex because of the need to include many ingredients;

- include ingredients needed in small amounts;
- require special drugs and additives;
- use ingredients which require special storage facilities, or which have a short shelf-life;
- need special ingredients which are difficult to handle or to incorporate into diets using only relatively unsophisticated engineering.

Pig production units mixing their own diets will therefore usually be content to buy in fully formulated:

- pre-starter (creep-feed) diets;
- starter diets;
- vitamin supplements;
- trace-element supplements.

The pig production industry can therefore be sold their pig feeds broadly in one of three ways:

1. A range of fully compounded, nutritionally balanced diets targeted to each particular class of stock on the unit.
2. A range of balanced protein concentrates together with vitamin and mineral supplements included to the appropriate level.
3. Vitamin and trace-element supplements and premixes.

Marketing fully compounded diets

The presumption that the fully compounded diet purchased from a national or multinational feed-compounding company is always the more expensive option may be mistaken. Speciality feed compounders of substantial size are able to:

1. compound more sophisticated diets using a broader range of ingredients;
2. use high-technology machinery which allows raw material treatment and improved presentation of the final product;
3. employ scientists with special knowledge of the nutrient requirements of pigs, the appropriate diet specifications, likely feed intakes and who have solutions to particular nutritional problems;
4. purchase a wide variety of feedstuffs and thereby facilitate optimum least-cost diet formulation;
5. purchase feedstuff ingredients in bulk and thereby negotiate lowest possible raw material prices;

6.   have world-wide communications to optimize commodity trading.

National feed compounders will find themselves in competition with each other and this, in itself, will help to maintain a low compound feed price and will help those who purchase feed to bargain with regard to the final compound price, especially if that purchaser is a single customer of substantial size.

Pig producers purchasing fully compounded diets for the complete range of pig classes on their unit will vary greatly in their attitudes toward feed price. Some will believe that any pig feed targeted to a named class of pig ('pig grower feed', 'sow lactation feed') will have equal merit to any other pig feed named as targeting the same pig class. In this case purchasing behaviour will be highly price-orientated. Other pig producers will believe that it is the nutrient specification and the feedstuff ingredient composition which best indicate the value of a compounded diet. These customers will be concerned with levels of DE, lysine, available amino acids, vitamins etc., and will be likely to compare unit values for nutrients across different feeds within-company and, indeed, between different feed companies. There is also likely to be concern about the relative inclusion levels of ingredients perceived to be of positive value (cooked cereals, oil, lysine) or negative value (manioc, cotton-seed meal, blood meal). So deep may be the cause for concern about ingredients that, if purchasing power is adequate, the customer may wish to specify to the compounder both the nutrient concentrations of the diet required and their ingredient restraints. This customer will require to be nutritionally knowledgeable.

The value of a compounded diet is ultimately not best measured in terms of its chemical – or even ingredient – composition but rather in terms of the response that comes from its use. Such responses may be seen in the form of parameters such as growth rate, carcass quality and piglets per sow per year. The fully integrated business will be sensitive to this fact and will be indifferent to the nutrient concentration or the price of the diet, these qualities being merely intermediaries in the creation of the overall financial benefit. Profitability for an integrated compounding and feeding operation is dependent upon the overall difference between input costs, in terms of the original purchases of feed ingredients, and output worth, in terms of numbers and values of pigs sold. This (realistic) philosophy may drive the purchaser of a compounded diet to consideration of the benefits of home-mixing. There is no reason, however, why the astute national compounder may not come to an appropriate arrangement with the pig producer to deal with this issue. Integrating arrangements are helped if: (a) the physical performance of the production unit is open to the feed compounder and (b) the feed compounder has some control over other input variables which interact with pig response to feed supply (such as health control, genotype and the environment). For example, if it be determined that, given the value of pig carcasses, profit can be assured when the feed costs are no greater than 30p per kg live-weight gain, then benefit will follow either from selling a feed costing (delivered) 14p per kg, used at an efficiency of 2kg feed per kg live-weight gain or from selling a feed costing 13p per kg, used with an efficiency of 2.2kg feed per kg live-weight gain, but not from selling a feed costing 14p per kg used at an efficiency of 2.2kg feed per kg live-weight gain. Provided that aspects of throughput and product quality were not compromised, the producer would be indifferent as to which of the first two diet alternatives were provided, having already determined that a satisfactory production margin is achievable at 30p per kg live weight gain. The compounder, however, would benefit from using the first-mentioned diet, but raising the price to 14.3p per kg, at which price parity with the second diet is attained.

The need to accept an increasingly realistic – if also complex – trading relationship between feed compounder and pig producer is evident from the following case history. Three diets, A, B and C, were available for purchase and use from 20 to 90kg live weight. Diet A was of low nutrient concentration and price, Diet C was of high nutrient concentration and price, while Diet B was intermediate. The pig producer also had the opportunity to market the pigs through two different outlets, one of which (Scheme 1) required the pigs to be leaner than the other (Scheme 2). In this case the primary interest of the feed purchaser was, quite properly, in his final financial margin rather than intermediate physical performance. Table 16.1 shows the results of feeding the three alternative diets at three different feeding levels when the pigs were placed to one or other of the two alternative grading contracts. With pigs produced for grade Scheme 1, low feeding levels of any of the three diets would have resulted in similar margins. In the case of pigs destined for grade Scheme 2, Diet A, although the cheapest, was overpriced, whilst low-level feeding of Diet C and medium-level feeding of Diet B yielded better margins. Buyers of Diet A would have needed to feed to the medium or high feed allowances to make optimum use of it. Table 16.1 also provides an example of the difficulty in predicting the outcome of one set of circumstances by knowledge of the results from another. A judgement as to which diet was best to buy from the compounder under conditions of high-level feeding and grade contract Scheme 1 would be useless for a production situation involving low-level feeding and grade contract Scheme 2.

It is essential, therefore, that feed-sales personnel take care to match the appropriate diet (or diet series) to a particular customer. Customers with different systems of feed provision, production objectives and genotypes of pigs will certainly need different diet specifications to be provided to their

Table 16.1 Margins over feed and weaner costs (£ per pig place per year) for three diets (A, B and C) of differing quality and price offered at three feeding levels to pigs destined for either of two bacon grade schemes

Feed allowance	Grade Scheme 1			Grade Scheme 2		
	Diet A	Diet B	Diet C	Diet A	Diet B	Diet C
Low	40	41	40	41	45	48
Medium	37	36	34	46	48	47
High	33	31	31	46	45	45

various classes of pigs. A producer feeding improved entire male pigs will optimize with a very different series of diet specifications for the grower and finisher feeds than would be optimum for a producer feeding utility castrates! Because of the importance of 'customizing' any recommended profile of feedstuff sales to a particular producer, the sales force should maintain a dossier of information on potential and actual customers (Table 16.2).

The proper price for any product is (naturally) that which the market will bear. Many customers for compound feeds, however, will be likely to have good intelligence about costs of bulk ingredient purchases, the costs of the milling and mixing process and of transportation. The purchaser is therefore in a position to challenge excessive prices. An unfortunate aspect of world commodity trading is that fluctuations in feedstuff ingredient prices may not bear any relationship to the prevailing value of pig carcasses. Acute financial embarrassment can result from elevated feed ingredient costs and depressed pig carcass prices. Nevertheless, when pig meat is highly valued, the market will bear a higher price for feed. As a general rule, feed manufacturers will have a precise view of the added costs of production above the basic costs of raw materials. To this must be added an element for fixed costs and a margin. Prices may be above or below that total, but if they are persistently below, then fixed costs must be cut and profits diminished. Most pig producers will not usually have an exact idea of the feed cost that they can afford per kilogram of live-weight gain. Retrospectively they will be aware of their profit levels, and at any moment in time they will also have a view not so much of a feedstuff price being excessive but of it being greater than that from an alternative source. It is competition amongst feed compounders, and the absence of strong customer loyalties, which as much as anything keep feed prices down.

Technical support

The type and extent of the technical support to be given by national feed compounders to pig producers purchasing a suite of fully compounded diets from them is a source of contention and has a chequered history. Technical

Table 16.2 Client profile: prospective client for sales opportunity

1. Description of client and of the business.
2. The primary purpose of the business in terms of its product.
3. Description of farm.
4. Statement of client's business and personal objectives.
5. Description of:
   (a) stock
   (b) feeding regime
   (c) diet type
   (d) health
   (e) management
   (f) final destination of the stock sold from the farm.
6. Description of performance of stock on the unit by presentation of key performance points for:
   (a) breeding unit
   (b) grow-out unit
   (c) end-product quality.
7. Important elements of the production system which would change this level of performance:
   (a) breeding
   (b) feeding
   (c) management.
8. How the introduction of the new product to the client would specifically help to change performance levels.
9. Assessment of the financial benefit to client which would follow from his purchasing the product.
10. Description of why the client is interested in buying.
11. Comment/quote from client.
12. Forward orders and scheduling of future deliveries.

support involves costs in terms of manpower (specialist technical consultants) and services (vehicles, travel, computers) which add to the price of the feed. Some customers may resent paying for advice that they do not need. Alternatively, the feed compounder may lose a customer because of that customer's misuse of a product through failure to appreciate how its particular benefits can be optimized. An expensive high-nutrient concentration diet, for example, may continue to be fed at the same high level as previously used for a diet of lower nutrient concentration, resulting not only in higher costs but also in poorer performance in terms of lost carcass quality.

The major element of support needed is clear user-guidance with regard to the:

(a) particular classes and types of pig for which a particular compounded pig-feed diet product should be used and, in consequence, the number of diets to be provided in a suite appropriate to a given customer;

(b)   particular levels of feeding which should be allowed for a particular pig feed;

(c)   means of reducing wastage;

(d)   optimization of pig marketing and maximization of the customer's financial margins;

(e)   appropriate storage and frequency of delivery of the product.

By encouraging simple on-farm comparative trials, technical support services may also encourage new customers to try a particular proprietary brand. Such farm trials usually involve the straightforward comparison of the new product versus the one currently used. These are particularly readily set up in weaner and grower pens where the amount of feed used and the growth achieved can be easily measured. The benefits of comparative tests include the realization by the producer that an alternative might be better and also provide interest and stimulation to the production staff who will be conducting the trial. The cynic would also point out that, for the company offering the new product, the trial presents a 'no-loss' situation if it is conducted with a new client. If the product is better than that presently used, the contract is won; if the product is worse, no business is lost; if the product is the same, there is still some likelihood that it will be shown to be better by chance alone.

The 'example farm' or 'example client' can be a potent medium for advertising compound feeds (Table 16.3). The performance achieved by a successful user, especially one who is named and perceived as a peer of a prospective customer, is often even more credible as a measure of product value than performance achieved on research stations or the in-house development farms of the compounding company. It is essential, of course, that the product is demonstrated to deliver the performance rates that it promises, and that it does represent value for money and will win profit for the client. In the initial stages of marketing a new product, reliance must of necessity be placed on trust (the worthiness of previous products from the same compounder) and on in-house performance figures. Only after the market-place has used a product for some time can the clients themselves be used to demonstrate product value.

Marketing proprietary protein concentrates

The pig producer who is formulating from a combination of home-grown (or purchased) cereals and a proprietary protein concentrate has firmly in his own hands the response of the pigs to nutrient provision and the financial consequences of that response. It is to be presumed that, being ready to mix

Table 16.3 Client profile: successful client for sales example

1. Description of client and of the business.
2. The primary purpose of the business in terms of its product.
3. Description of farm.
4. Statement of client's business and personal objectives.
5. Description of:
   (a) stock
   (b) feeding regime
   (c) diet type
   (d) health
   (e) management
   (f) final destination of the stock sold from the farm.
6. Description of performance of stock on the unit by presentation of key performance points for:
   (a) breeding unit
   (b) grow-out unit
   (c) end-product quality.
7. Important elements of the production system which have allowed this level of performance:
   (a) breeding
   (b) feeding
   (c) management.
8. How the introduction of the new product to the client specifically helped to change performance levels.
9. An assessment of the financial benefit to client which has followed from his purchasing the product.
10. Comment/quote from client.

his own diets, the producer will be nutritionally knowledgeable and interested in the cost per unit of protein concentrate supplement, and in the nature of the ingredients used in it. But this may not be the case at all, however, and the protein supplement may be used without detailed technical query, and even be wrongly used.

Usually the purchase of a protein concentrate by a customer would be on the basis of both its composition and its price. The competition will come from three sources:

1. Other protein concentrates of suggested better value.
2. Fully compounded feeds of suggested lower price (value of home-grown cereals sold rather than used for pigs; benefits of bulk ingredients purchased; improved growth or reproductive response).
3. Individual feedstuff 'straights' such as straight soya bean meal, together with mineral and vitamin supplements (readily available for purchase at a cheaper price; improved knowledge of ingredient quality).

It would appear that customer loyalty (consistency of pig growth, reproduction and product-quality response) is an essential element of this trade, together with the convenience that the purchased 'everything included' protein concentrate can bring to an on-farm mill-and-mix unit.

Technical support

Technical services are important to this sector. They must first provide those same elements of support mentioned as being required for fully compounded diets (a)–(e). But additional support is needed to the on-farm compounding operation itself:

(f)   choice of milling, mixing and other ancillary engineering equipment;
(g)   assistance to ensure correct means are provided for weighing the different feedstuff components so that they are mixed in the ratio required;
(h)   advice about cereal ingredient purchase if the home supply falls short;
(j)   nutritional knowledge to enable (i) correct usage of cereals of different type, (ii) correct choice of appropriate protein supplement from the proprietary range, (iii) appropriate supply of macro and micro minerals, vitamins, medicaments and additives.

The extent of support input to this market, the wide variety of customer types each with different requirements and the intense competition from fully compounded feeds on the one side and straight feedstuff ingredients on the other, tend to make this sector of the pig-feed market difficult.

Marketing supplements and pre-mixes

The on-farm or small-scale feed compounder will usually find it convenient to purchase vitamins and micro minerals in pre-mixed form. Sometimes the macro minerals, calcium and phosphorus, may also be included; but because their bulk is so much greater, calcium carbonate and/dicalcium phosphate are usually included in diets as separate ingredients.

The mineral and vitamin supplement is a relatively small part of the total diet cost, but is exceedingly complex in its composition. The commercial credibility of the supplement lies, therefore, not primarily in its unit price, nor in its component specification (which is often taken by the purchaser as being too complex to query and therefore assumed correct), but rather in those attributes which are associated with the product. These would include:

● security of knowledge that all the required minerals and vitamins will be included in the pig's diet in the correct proportions and at the correct level;

- the ready provision of a comprehensive technical support service which is available with the pre-mix;

- the availability of special additives which might not be so readily obtained elsewhere: for example, artificial amino acids, medicaments, cocktails of enzymes, growth-promoting additives, anti-microbials, micro-organism cultures, flavours, aromas, digestion enhancers, colours and the like.

Technical support

The technical support services are therefore the vital component of this sector of the industry. It will provide, first, those elements of support mentioned for the sale of fully compounded diets (a)–(e) and proprietary protein concentrates (f)–(j). Additionally the service is likely to provide:

(k)   advice about the beneficial purchase of all feedstuff ingredients;
(l)   nutritional knowledge to cover aspects of:
    i.   the nutritional specification of diets;
    ii.   ingredient composition of diets;
    iii.   nutritional restraints;
    iv.   appropriate ingredient restraints;
    v.   potential usage of new and unusual ingredients;
    vi.   least-cost diet formulation;
    vii.   optimization of feed allowance to various classes of pigs.

In short, the technical service provided by the manufacturers of supplements and pre-mixes must be fully comprehensive and provide the widest range of support in as many aspects as is possible of the practical – and, indeed, scientific – nutrition of the pig.

New products

New products are required to exploit new ideas which will increase profit margins, to keep up with scientific knowledge and technical advance, to attract the attention of prospective customers and to maintain the attention of

479

existing customers. New products may be manufacturer-driven or customer-driven (the latter usually by demand for the incorporation of new ideas and scientific knowledge).

A feed product may be perceived as new on account of:

● a new name and image;
● a new nutrient specification (higher nutrient concentration, increased amino acid content, elevated vitamin level);
● new ingredient inclusions, novel ingredient combinations, ingredient exclusions;
● improved performance;
● a new physical form (high-fat meal, crumb, pellet, nut, cob);
● innovative inclusions of special additives to aid health, growth, digestion, appetite and so on;
● more convenient packaging or delivery systems;
● increased value in terms of pig output per unit of feed-cost input.

The last of these is both the most important and the most difficult to promulgate. The buyer is likely to act against an increase in feed price, even if benefit is promised in the longer term. The buyer may equally react negatively against a decrease in feed price, finding it difficult to accept the notion that performance will *not* be more than proportionately diminished. Vital parts of a new product launch are: that the novel element is attractive to the likely purchaser and is timely; that the new product has a specific target (weaner pigs, lactating sows) and addresses a specific problem (poor feed intake, poor carcass quality); that there is adequate follow-up to ensure it is properly and effectively used so that the extra benefits are noted by the customer.

Sometimes a worthy new product may bring with it a need for improved management. This may introduce a need for sales to be restricted in the first event to especially competent customers in order to avoid misuse and the discredit of the product. The introduction of a high-protein specification diet into a market normally using a low-specification diet brings with it a need to ensure proper feed-intake control, improved feeding management and enhanced awareness in pig-product marketing if the customer is to see maximum benefit.

Customer complaints

Complaints about compounded feed at the time of delivery are rare; visual and olfactory loss of quality may be observed after manufacture and before despatch, so quality control is readily exercised. But occasionally, if a product which has been stored is subsequently transferred, there may be loss of quality to the extreme extent of visible fungal contamination, deterioration due to moisture or infestations such as with mites.

More often feed complaints relate to unexpected subsequent pig response:

1. Variability of response (interrupted growth performance, interrupted reproductive performance).
2. Lack of response (growth at lower than the expected rate, carcass quality of less than the expected level, reduction in reproductive performance).
3. Negative response (ill-health, failure to grow, failure to consume adequate amounts of feed, failure to breed, abortions, unthrifty litters).

Quality checks on the feed itself after a problem has arisen may illuminate: a mix which has been improperly labelled; a diet of incorrect nutrient specification; a specification which fails to meet the nutrient content indicated at the point of sale; the faulty inclusion of a particular ingredient; failure to include an essential ingredient. With regard to the major ingredients, microscopic examination of the diet can be helpful as a means of identifying the presence and approximate levels of individual feedstuffs, each of which has its own particular characteristics.

Whilst quality checks on the feed may give guidance to the justification (or otherwise) of a complaint, it is often the case that the particular batch of the diet in question is no longer available for quality assessments to be made. The implication of the feed in the causation of a problem must often be achieved, therefore, through circumstantial propositions that a particular feed problem would be the most likely cause of a particular response problem. This can be highly contentious, as most pig responses are extremely sensitive to health, environment and management, as well as to the nature of the feed supply. It is also often the case that intractable problems on a unit may be blamed on some shortcoming in the food, examination of all other causes failing to produce a resolution. It is quite unacceptable to find the cause of a problem by the default of other causes; if the feed is to be implicated justly as the cause of complaint, it must be implicated directly. Even when the feed has been – quite properly – found as the cause of the problem, it is often difficult to determine at what point the feed itself became faulty. Failure to include an essential ingredient, or the inclusion of a wrong ingredient, is clearly the fault of the manufacturer. But contaminations (which can be many and various) may occur after the feed has been delivered to the point of use. A further difficulty arises when a feed fault is (justly) implicated in the cause of a problem, but only to some limited degree – many problems on pig units result from interactions amongst a variety of causal forces, only one of which may have been the feed.

Feed manufacturers are encouraged by their customers to keep at the forefront of nutritional science, to maintain feed price as low as possible and to be fashionable. This can lead to diets being compounded with the best possible will but which nevertheless may lead to disappointing response on the production unit. The difficulty may be further exacerbated by such disappointments being liable to be farm-specific. Interactions between elements of the feed and elements of the genotype, environment and health mean that some diets will give excellent performances on a majority of units, but will fail to be successful on a small number of particular units whose special circumstances interact negatively with some component of the diet. Often both elements of this interaction can remain mysterious. It may be considered the manufacturer's responsibility to resolve such problems, but it would be difficult to apportion blame.

Breeding stock supply

The value of purchased breeding stock is in their genetic potential: the extent to which their use will improve productive performance and end-product quality. The commodity transacted is a combination of genes, and the value of those genes lies in the science of the genetic improvement programme that has created their particular pattern. To achieve real production benefit, it is necessary that there has been real genetic change toward a given goal and that this change is reflected in the ultimate performance of the stock sold.

The importance of the customer in product design

The customer for breeding stock is at the end of a long chain of commercial and scientific activity. Creation of improved breeding stock occurs at nucleus level. Numbers are bulked up to commercial quantities at multiplier level where there is also rejection of inferior types, in both male and female lines, before the product is offered to the customer.

The transfer of information up and down the pyramid between the nucleus pinnacle and the customer base is an essential part of breeding stock marketing. Because the genotype is 'assembled' at nucleus level, and because this is a continuous process, the breeding company geneticist must have complete information about real customer needs. Equally the customer must be aware, in realistic terms, of what the pigs he purchases can and cannot do.

Changing the genotype of pigs is readily achievable for many (most) of the important characters, but such changes take time, and once made would take more time to reverse. The breeding targets for the geneticist are heavily influenced by customer requirements. It is essential, therefore, that customer needs are clearly expressed, credible, based upon hard facts and long-term in nature and are not based on prejudice, fad, fashion or fantasy.

Marketing breeding stock is therefore about:

- identifying the need for particular characteristics to be improved in a particular way;
- creating a new product through the application of the science of genetics;
- demonstrating improved performance through the results achieved by the purchaser.

Improved genotypes are natural populations of variable individuals. This variation can be a source of frustration to both buyer and seller because achieved performance will show a distribution. Whilst the mean performance levels may show improvements that may readily justify the purchase of improved breeding stock, extremes of performance will range from the outstanding to the abysmal. Some of the performances shown by genuinely improved stock will actually be no better than (and sometimes worse than) those shown by the stock which have been replaced. Also, until such time as genetic material is transferred wholly through AI and embryo transfer (which is not imminent), the commodity itself is a delivery of live pigs. These pigs themselves are prone to biological variation and imperfections, such as in their health, locomotor abilities and shape. These variations and imperfections may or may not be related to their subsequent ability to perform.

Types of breeding stock

Traditional breeding-pig sales were of pure-bred males, the breeding females being retained from home-bred stocks. But the ability of the breeding companies to harness the forces of the science of genetics and achieve advances in performance so much quicker than the individual producer has created a substantial market for the sale of replacement gilts. These are invariably hybrid and carry genes for high-level reproductive, growth and carcass performance. This moves the breeder into the realms of the production, marketing and movement of large livestock masses. Breeding companies are now likely to offer:

- 'pure-bred' males;
- cross-bred (hybrid) males;
- cross-bred (hybrid) females.

Females will inevitably major on qualities connected with reproductive potential such as litter size, lactation yield and readiness to rebreed. The males will inevitably be judged at first sight on their enthusiasm to mate and their fertility. Both male and female lines will be expected to demonstrate superior qualities for growth rate, efficiency of growth, carcass quality and carcass yield.

Breed improvement programmes will target particular market demands. Meat packer requirements for high lean-meat yields and large muscle areas will encourage the use of specialized sires with large hams and loins and, especially, good conformation. A movement from indoor to outdoor production of weaners encourages the use of specialized hybrid females with the particular attribute of hardiness.

The various parent hybrids offered for breeding stock will come from different grandparent lines, according to market requirements. If a particular line of a particular breed is perceived to offer meat-quality characteristics, it may be included in the parent male; whilst if another is perceived to offer hardiness, it might be included in the female. It is not enough for breeding stock to be 'improved'. Customers require particular *types* of improvement, and some traits are more important to improve than others. Lean pigs may be considered to be 'improved' in some markets but to be 'unimproved' in others. It is necessary for the type of breeding pig to be identified as being improved in a specific way that is attractive to the specific customer at which it is aimed. The improvement requires to be readily visible to the customer and preferably not able to be achieved by the customer through means other than purchase from the breeding company. These principles are basic to effective marketing:

- The hybrid should be the result of crossing lines which have been successfully selected for improvement in the quality for which the stock are identified.
- Maximum benefit can only be achieved if the replacement stocks are always purchased from the source and not achievable by subsequent purchase avoidance (such as through the retention of slaughter

generation animals from the grow-out unit for use as replacement breeding stock).

Failure to meet these two criteria will in the first case present for sale a product which is no more than a combination of unimproved breeds (and which therefore will fail to show improvement in performance over competitive stock) and, in the second case, will create grave difficulty in achieving repeat sales.

The sale of performance expectations has special problems when the achievement of improved performance is dependent upon interactions of the genotype with the feeding, management and environmental elements of the production process. Not only will an improved genotype fail to solve the non-genetic problems of faulty feeding, housing, health and management, but some improved genotypes may actually require improvements in these aspects of production before the genetic potential can be fully demonstrated.

It is essential that the seller and the buyer are both quite clear as to the ways in which the stock are improved, and the particular aspects of beneficial performance which will come from the use of the stock. It is also essential that the seller and the purchaser have complete confidence in the stock and in their ability to perform up to the stated level. Where this does not happen, both parties should be willing to address those other aspects of production which may be standing in the way of the achievement of genetic potential. To win this confidence, the seller of the breeding stock must be knowledgeable about that stock, its various characteristics and the scientific methods used to achieve the genetic improvements. The sales force is the better for being well-informed about the production process and about the likely causes of limits to performance on production units. Specialized technical back-up is therefore an essential part of the successful selling of breeding stock.

Selection of breeding stock for sale

Some of the qualities of improved breeding stock may be immediately apparent, while others develop over a longer period. Relatively few of the genetic qualities are visible at the point of sale, these emerging later (Table 16.4).

The transaction which occurs at the point of sale of breeding stock is one that deals with future expectations rather than immediate perceptions. The relationship between the physical appearance of the pigs at the point of sale and their ultimate performance (for which they were purchased) may not always be close. While the major sales point must ultimately be improvements in performance, the immediate achievement of a satisfactory

Table 16.4 Qualities of breeding stock

Visible when purchased	Perceived through performance in the shorter term	Perceived through performance in the longer term
General appearance	Readiness to show oestrus	Meat quality
Straightness of back and legs	Readiness to conceive	Carcass yield
	Willingness to mate	Disease resistance
Body form	Growth rate	Efficiency of feed use
Shape of ham	Fatness	Lean-tissue growth rate
Shape of loin	Litter size	Weaner weight
Colour	Locomotor ability	Litters per sow per year
Ear position	Chronic ill health	Sow longevity
Body condition	Handleability	Aggression
Shape and position of legs		Fertility
Shape and angle of feet		
Slope of pastern		
Shape and positioning of toes		
Testicles		
Nipples		
External reproductive organs		
Skin blemishes		
Swellings on legs		
Abrasions on legs and toes		
Acute ill health		
Lameness and stiffness of joints		
Length of nose and tail		

transaction requires the assent of the purchaser (and this will relate to the immediately visible qualities of the animals). Selection of parent breeding stock for sale to customers (which usually occurs at 80–110kg live weight) therefore relates primarily to the visible qualities shown in the left-hand column of Table 16.4. It is to be expected that some 40% of potential breeding females, and 65% of potential breeding males, will be rejected before sale on the grounds of these visible criteria whilst, after sale, a further 5% complaint rate can be expected, to which may be added those females which fail to show oestrus or to conceive, and those males which fail to mate.

Description of breeding stock

Because of the relatively weak link between many aspects of performance and what a pig may look like, only some qualities are apparent at the point of sale. Ham shape, for example, is readily visible, whilst reproductive performance is invisible. The use of demonstration farms and performance figures from other customers are therefore valuable sales assets (see Table 16.3). Similarly, results from meat packers will confirm claims for high lean-meat yield and quality.

Health status

Because the trade in breeding stock – especially female gilt replacements – is large in volume, the health status of the stock is an important element of marketing strategy. Customers must be sure of the absence from the stock of disruptive disease. It is helpful if an individual customer is provided with stock always from the same breeding farm to ensure continuity of immunity. Equally, ill health may arise when stocks from different sources are mixed on a single customer farm. Farms free of diseases like enzootic (mycoplasma) pneumonia can expect to be kept free (as far as possible) by the supply of minimal-disease pigs. On the other hand, it is not helpful to supply minimal-disease pigs to farms with chronic respiratory problems for this would only ensure the infection of pigs with no immunity. It is helpful to match the disease status of the farm supplying the breeding stock to the disease status of the farm receiving the stock. In this respect health should be considered in comparative and not absolute terms.

Receipt of breeding stock on to the farm

Even although every effort may have been made to match the disease status of the farm from which the animals originated, and the farm to which they are destined, the period post-delivery is often a difficult one. It cannot be avoided that the health, environment, nutritional and management arrangements will differ from farm to farm and, following the trauma of transportation, the young gilts and boars newly delivered to the customer's farm will be in a relatively inadequate state to deal with the difficulties of change. It is unfortunately the case that excellent stock can be severely compromised due to unsympathetic management following delivery. Many aspects of post-delivery acclimatization and care have been dealt with in Chapter 7, but it may be iterated here that:

● Young gilts and boars should be fed ample amounts of high-quality feed after delivery to their unit of destination. A diet suitable for lactating sows would also be suitable for these pigs. A daily live-weight gain of around 750g should be achieved and fatty tissue accumulation should be positive. Feed should be supplied *ad libitum* in the 3 weeks before first mating to all but the fattest gilts.

- The period of some 40 days which should elapse between 100kg live weight and ultimate mating, at weights in excess of 130kg and ages in excess of 225 days, should be used for quarantine in semi-isolation and adjustment to the disease environment prevailing on the unit of receipt.

- Newly introduced pigs should not be exposed to new infections when placed upon their farm of destination. This means that either:

    i.   they should have been exposed to the relevant infections on their unit of origin some time (preferably a long time) before delivery; or

    ii.   they should be exposed slowly and by small degrees to the new infections; or

    iii. they should be vaccinated as appropriate and given suitable preventative medication routines; or

    iv.   they should not be exposed at all to infections which would in any event be better eradicated from the receiving pig herd.

- Trauma should be minimized by offering to newly entered gilts and boars the highest quality of management, including low stocking density, high comfort and an especially good environment.

New products

Product development since the 1950s has followed from both customer demand and leadership given by innovative breeders. There has been a succession of products which catalogues the success first of the pedigree breeders and subsequently of the hybrid-breeding companies.

1.   Initial interest was in perfecting breed type within established national races of pigs. This was mostly handled by small private breeders paying particular attention to body form.

2.   Possibilities for the reduction in backfat depth demonstrated by the improvement of the Scandinavian Landrace type generated interest in importations and possibilities for introducing new breeds with special attributes such as prolificacy and leanness. Genetic selection for feed efficiency and against fatness created new pig products and many European breeding stocks were sold on the basis of their ability to

improve carcass quality. Genetic selection requires the application of the science of genetics and is greatly assisted by a larger scale of operation; it was at this point that the initiative for the transfer of breeding stock tended to pass away from a multitude of small private breeders towards a lesser number of breeding companies.

3. Cross-bred breeding to enhance prolificacy led to the development of the hybrid breeding gilt as a product which combined improved piglet production with carcass quality in the slaughter generation.

4. At around this point in product development, high health status came to be identified as an essential element of marketing policy.

5. Increasing pressure arose from meat packers to concentrate customer interest on further reduction in subcutaneous backfat depth, and the sale of stock with particular ability to improve grading became a winning formula amongst breeding companies competing in the market-place. It should be noted, however, that not all pig-producing nations were following this trend toward extremes of carcass leanness. In some cases breeding stock was still being sold on the basis of perfection of breed type (as demonstrated by appearance and form), and any special attributes demonstrated were rather in characters such as hardiness and not in elements of carcass quality.

6. Given the achievement of improved carcass quality, new products next advertised improved growth rates, particularly for the rapid daily deposition of lean meat. Together with enhanced daily lean-tissue growth rate came the further possibility for developing pig-breeding stock products with special attributes such as larger eye muscle area, bigger hams and better carcass yield.

7. Continuing pressure on increasing litter size remained, and continues to remain, an important part of product description for hybrid female lines, together with qualities of fast growth rate and high lean-meat percentage in the slaughter generation.

8. Further product development at this point has been assisted by the separate identification of sire and dam lines. The sire line (with special meat, growth and carcass qualities and including further possibilities for muscle quality) has encouraged special marketing opportunities through the development of named male lines targeted to specific customer requirements. Proliferation of male terminal sire lines allows customers to choose amongst those which are especially lean, those which are of especially blocky shape, those which grow especially fast, those which have a particularly high libido and fertility or whatever.

9. Grandparent lines have often been viewed in the past as 'pure-bred', being the result of progressive selection for a given set of traits within a single population. More recently, however, grandparent lines may be found to have been made from mixtures of other lines, allowing combinations of desirable traits and rapid insertion into GP stock of characters only found readily in other (remote) populations. Many grandparent lines are now synthetic amalgams, and this offers a quicker response to customer demands within the context of relatively rapidly changing markets.

10. With the wider variety of lines of male and female stock becoming available, speciality products can be developed for animals especially suited to particular systems, such as outdoor pig-keeping or low intensity units, or to special market outlets, such as high-class supermarket chains requiring a particularly tender and flavoursome product.

11. In addition to continuing improvement in female fertility, future product development may well continue to introduce new breeds and synthetic lines with attributes of special interest and to stress quality aspects in the final product. The breeding-stock supply industry, like the feed industry, is under continual pressure to develop new products. This is not only in order to keep to the forefront of new technologies (perhaps the separation of the negative PSS and PSE aspects of the halothane gene from the positive attributes of lean-meat percentage), but also to maintain customer interest and confidence and to respond to customer demand (perhaps increased boar libido). Regular launches of new products will remain crucial to breeding-stock marketing policy. It would be unfortunate if product development were to deteriorate into nothing more than the remixing and rematching of different breed types, and the creation of new brand names and images for old products.

Pricing

Pig producers, having been in the habit of raising their own replacement breeding gilts together with the slaughter pigs, will expect the price of breeding stock to be little higher than the expected return from meat pigs of the same weight. The breeder, however, will identify many extra costs which have been incurred consequent upon the breeding company operation; not least the huge fixed costs of the genetic improvement programme itself, the technical staff, the sales staff, the stringent health requirements and the loss of stock rejected at pre-sale selection. The customer, on the other hand, will have accepted that the gilts carry presumed benefits of increased piglet production, enhanced efficiency of growth and improved carcass quality. An attempt can be made to value these improvements and that evaluation would set the upper limit on the price that could be paid.

The difference between slaughter price and market worth is very much greater for the purchased sire than for the parent gilt. A terminal boar will sire some 500 meat pigs (or 8,000 by AI) in a year, and each of these pigs may be expected to show an increase in profit consequent upon the qualities of its father. These can be measured in terms of growth rate, efficiency of feed use and carcass quality.

A grandparent sire must be valued in the wider contexts of (a) the grandparent boar's progeny being sold themselves for use as stock boars and (b) the very high cost of maintaining nucleus breeding herds, performance testing and selecting the grandparent boars, and the very high rate of rejection of candidate animals placed on to test but identified as being of inadequately marked superiority. Because GP boars have been subjected to individual test, information on performance is available, usually in the form of an index covering growth rate and fatness. Individuals can be priced according to their value as determined on the test, higher index boars commanding a higher price.

The breeding of improved stock by genetic selection requires high-level investment which can only come from the reinvestment of past profits or borrowing in the expectation of profits to come. If generation upon generation of genetic progress is to be made, it is necessary that the profits on current sales are adequate to fund an acceptably ambitious and continuous selection programme. For this a 'costs plus acceptable profit' pricing policy is not adequate but rather 'costs plus reinvestment in new products plus acceptable profit'. If this is not achieved, the competitive edge will be lost and sales fall.

Technical support

While it is true to say that improved stock will usually perform better than unimproved regardless of the circumstances of management, housing and feeding, it is also true that the full extent of the benefits from the improvement will not become evident unless standards of management, environmental control and nutrition are raised to match the levels required by the stock. Exploitation of the product by the customer, and proper return for the investment made, is therefore dependent, to a large extent, upon the ability of the customer and the extent of his technical knowledge. It behoves the seller to ensure that the buyer is satisfied with his purchase, and technical support is therefore a particularly important part of the marketing exercise with respect to breeding stock. The general principles of technical support provision after the sale of breeding stock are not materially different, however, from those appropriate for the marketing of pig feeds. In particular, the customer will need assistance in the correct choice of parent female and terminal sire from the product range. Intimate knowledge is required of the potential performances that can be expected from the various genetic lines, and confidence that performance levels predicted can be achieved. Genetic potential cannot be fully expressed when limited by inadequacies in aspects of health, environment, management and nutrition. It is prerequisite to the effective demonstration of improved performance that the non-genetic boundaries are identified and, where possible, removed. It is not unusual, therefore, for breeding companies to field technical support teams offering specialist consultancy in pig-disease problems, nutrition, feeding, management, building construction and environmental control. The more innovative companies may attempt blueprints and instruction manuals to help customers maximize the benefits of the product they have just purchased. It is necessary not just that the product be fully described in terms of the performances that can be achieved with it, but also that the means to achieve those performances are clearly elaborated. Amongst those means will be nutrition, feeding, environmental and health control and production (and reproduction) management. New customers may need special assistance in acclimatizing the new pigs to the demands of their unit and in acclimatizing themselves to the demands of the new pigs. Few attributes of improved pigs (with the possible exception of breeding for disease resistance) will be likely to suppress non-genetic production problems.

Supply schedules for replacement stock

Routine supply schedules are often compromised by transport costs, quarantine restrictions and weight limitations. In general, clients are likely to be asked to budget for a 40% annual replacement rate for females and a 50% annual replacement rate for males (or 100%, if used for AI).

The ideal replacement schedule for a 500-sow commercial herd producing slaughter pigs might be:

● 200 parent females per year, delivered at the rate of 16–17 monthly and despatched at 90–110kg.

- 12 terminal sire males per year for natural service (8–9 terminal sire males per year if using AI).

For a breeding-company multiplication unit supplying parent stock, replacement rates for females would be the same as for the commercial unit but they would need double the rate of turnover of males, that is, for a 500-sow herd, 25 males annually (two per month).

For restocking a whole new unit, or one which has been cleaned out and is receiving all new stock, deliveries can be phased (for a 500-sow unit) as follows:

- Week 1   140 gilts and 12 boars
- Week 5   120 gilts and 4 boars
- Week 10   120 gilts and 5 boars
- Week 15   120 gilts and 4 boars.

All gilts may be despatched at 70–110kg, and all boars should be greater than 110kg live weight. A planned 5–10% oversupply of stock may be a worthwhile insurance.

Marketing slaughter pigs

In many ways, optimization of the pig industry overall may be better served when the production and meat-packing sectors are integrated. This allows the competitive (exploitative) marketing interface between pig producer and meat-packing plant to be removed. Nevertheless, most producers do face difficult marketing decisions at the point of sale of their product, and the outcome of those decisions has a dramatic effect upon the financial viability of the operation.

Pig production is an entrepreneurial activity not dependent only on good husbandry and high output. Production, marketing and trading take place in a financial environment; managers need to know the consequences of an infinite variety of possible actions and interactions on the financial effectiveness of the business. The choice of pig-meat production strategy depends closely upon the prevailing circumstances which are variable between production units, geographical areas, and years.

Pig producers may adjust their marketing strategies for slaughter pigs through the media of:

- weight at sale (influenced by throughput targets, quality of pig, local market requirements);
- fatness at sale (controlled through feed level, feed nutrient specification, genetics);
- choice of meat packer (influenced by geographical location, price offered, method of payment, differentiation for quality);
- meat quality at sale (controlled through genetics, feed nutrient specification, management, handling).

The criterion for profitability by which various marketing decisions can be judged is always a moot point. Absolute profitability depends upon knowing the fixed and variable costs, while the gross margin depends upon knowing only the variable costs. Both of these characteristics are highly individual to a particular farm. The greatest singular variable cost elements are, however, those for feed and for weaners. Perhaps the most useful measure of profit is the margin between the price received for the slaughter pig and the costs incurred to produce (or buy) the weaner, and to feed it through to the point of slaughter. The financial margin over feed and weaner costs can be expressed on the basis of an individual pig (per pig) or on the basis of the pen space available in a year (per pig place per year). The latter value has the advantage of allowing for the real rewards of increased pig throughput which can come from increasing the rate of growth or selling the pigs at a lighter weight. The example in Figure 16.1 shows, for this particular case, that although the best margin per pig is achieved when selling at a higher slaughter weight, the best margin per pig place per year occurs at a lower slaughter weight. Whilst a pig producer may take pride in a particularly generous margin calculated on an individual pig basis, his effective income is

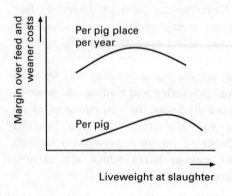

Figure 16.1 Comparison of profit margins expressed per pig, or per pig place per year.

far better calculated on the basis of the annual yield from the whole unit. Analysis of marketing strategies for slaughter pigs is therefore probably best undertaken by examination of margin over feed and weaner costs per pig place per year.

Weight at sale for carcass meat

Optimum weight at sale interacts particularly with the severity of any penalty there may be against fat. The less severe the penalty, and the less likely the pigs are to become fat, the higher will be the optimum slaughter weight (Figure 16.2). Within any given scheme for quality payment there is a further interaction with feeding level. Figure 16.3 shows the position for pigs dispatched to a meat packer which discriminates against fat carcasses. The figure lends support to the benefits of controlled versus appetite feeding, particularly at the higher slaughter weights. However, controlled feeding may not always be appropriate just because the grading scheme is severe. The imposition of too strict a ration scale can militate against margin optimization through excessive reduction in growth rate and throughput. This negative effect of ration control becomes particularly marked at the higher live

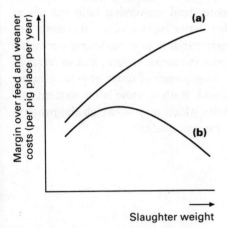

Figure 16.2 Influence of weight at slaughter upon financial margin, as affected by the fatness of the pigs: (a) pigs with low predisposition to become fat offered to a market with low discrimination against fat; (b) pigs with high predisposition to become fat offered to a market with a high discrimination against fat.

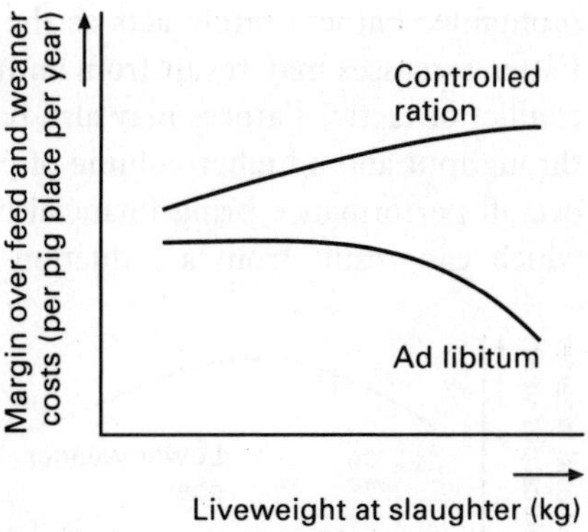

Figure 16.3 Profitability of pigs slaughtered at various weights having received different feed allowances.

weights. The pattern of response shown in the figures would also be different following an increase in the price of animal feed ingredients – particularly cereals – which would tend to favour a reduction in weight at slaughter and a decrease in feed allowance. Alternatively, in some countries it is common practice for flat-rate payments to be made for all pigs regardless of their level of fatness. This will stimulate higher rates of feeding, heavier slaughter weights and fatter pigs.

The cost of producing or buying weaners has the same predictable effect of any unavoidable cost upon a production system; as the cost of weaners goes up, it becomes progressively more beneficial to increase the slaughter weight of individual pigs – thereby spreading the fixed weaner cost over a greater amount of product sold (Figure 16.4).

Fatness at point of slaughter

It is evident that for the same or lesser input costs, a leaner pig would be more profitable than a fatter one if a buyer pays a premium for leanness. It is equally evident that if the buyer is indifferent to fatness, the relative benefits of a fat or lean pig at any given weight will depend upon the feed costs per kilogram live-weight gain – regardless of the composition of that gain. A cheap feed and a fat pig may give a poorer feed conversion ratio but not a poorer monetary conversion ratio; this latter option may therefore be profitable. Fatness rarely acts as the single variable in a marketing decision. Fatter carcasses may result from an increase in carcass weight, and so create a conflict of tactic. Fatness may also be a consequence of faster growth, greater throughput and a higher volume of pigs sold. Both of these latter elements of overall performance bring financial benefits which may outweigh the penalty which can result from a reduction in carcass quality.

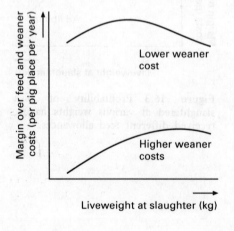

Figure 16.4 Influence of 20kg weaner costs upon optimum slaughter weight.

Given the background of a strong and practically linear response of fatness to feed intake itself, the interrelationships between pig type (growth rate, carcass quality) and feed intake may be seen in Figure 16.5. While pigs not predisposed to being fat will always be more profitable than those that are, there is an interaction to the extent that an increase in feeding level and fatness improves the position in the former case (pigs not likely to become fat) but worsens it in the latter (pigs likely to become fat).

Both backfat depth and percentage carcass lean will show a distribution (skewed for the former, and normal for the latter) about the mean value. Increase in the proportion of animals falling into the top quality category will often require sacrifices in terms of a lower weight at the point of sale and/or lower growth rates and/or higher specification (higher cost) diets. In order to encourage the pigs at the poorer end of the distribution curve to fall into the top quality category for carcass leanness, the benefits of greater growth rates, weights at sale etc., will require to be forgone in the case of all those pigs already positioned at the better end of the distribution curve and which did not require extremes of treatment to achieve success. For the benefit of the few, therefore, the majority are unlikely to have been performing optimally. There may therefore be financial disadvantage in a production strategy which places a very high proportion of all the pigs into the best grading classification – even when premiums are high (Figure 16.6). To achieve 95% top-grade pigs may bring maximum producer satisfaction but not maximum producer profit.

Choice of market outlet

Pig slaughterers and meat processors differ amongst themselves as to their preferred weight of pig and preferred level of fatness. There will be further differences in the average prices offered and the extent of the financial benefit

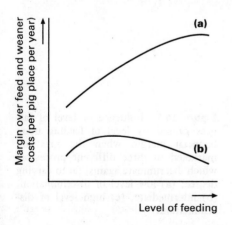

Figure 16.5 Influence of pig type and choice of feed level upon profitability when there is a penalty against carcass fatness: (**a**) pigs not predisposed to become fat, (**b**) pigs predisposed to become fat.

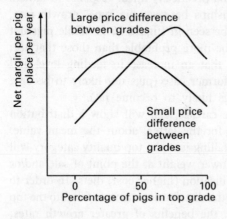

Figure 16.6 Where the percentage of pigs in top grade depends upon fatness, and where fatness is controlled by reducing feed intake and growth rate, the net margin per pig place per year is sensitive to the size of the price difference between grades.

that might be available for meeting quality criteria. The importance of choosing the right processor for pigs of different levels of fatness is exemplified by Figure 16.7. The figure relates to an increasing level of fatness at sale which has been achieved by increasing the level of feeding of a balanced diet. It shows the influence of pig fatness upon financial margin under three different payment schedules: (a) with a low penalty for carcass fatness, (b) intermediate and (c) with a high penalty for carcass fatness. For pigs destined for meat packer (a), the benefits of faster growth rate clearly outweigh problems of increased level of fatness; whereas the exact contrary is the case for pigs destined for packer (c). It should not be assumed, however, that the position depicted in Figure 16.7 is in any way static; quite the contrary, it is highly dynamic and dependent, amongst other things, on the relative price of feedstuffs as compared to pig-carcass meat. If pig prices stay high, the quality differentials remain the same, but feed prices fall, then the response to schedule (b) will take on the shape of that now shown for

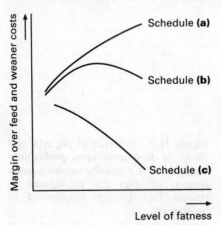

Figure 16.7 Influence of level of fatness caused by level of feeding upon financial margin when the pigs are marketed to three different processors which discriminate against fat to varying degree: (a) low level of discrimination, (b) intermediate, (c) high level of discrimination.

schedule (a); it will become profitable to produce fatter pigs. The system is therefore particularly sensitive to changes in pig and feed price.

Marketing opportunities: case calculations

Case 1: An alternative contract

An attractive flat-rate contract for 70kg pigs is offered to a producer already doing well with a quality premium scheme for 90kg pigs. A 17.5% crude protein diet is fed to a restricted scale and the farm margin was £7,000 per year. The lighter-weight flat-rate contract may be determined to be unattractive (Table 16.5) unless and until the pigs are fed generously, while the best solution for slaughter at either weight includes the use of two (rather than one) diets. The best solution seems to be for all pigs to be fed generously, but those with a predisposition to be fat (castrated males) being drawn out for slaughter at 70kg while the others remain to grow on to 90kg.

Case 2: Pigs which are too fat

A producer is alerted by the meat processor that the pigs are no longer acceptable unless something is done to make them less fat. Upon investigation of the accounts, it becomes apparent also that the unit is running at a loss. Mixed female and castrated commercial-type pigs are fed a limited feed allowance using two diets: one of a higher quality from 20 to 50kg, and one of a lower quality from 50 to 90kg live weight. Table 16.6 shows at (o) the current performance. Grading could be improved in five obvious ways: (a) reduce slaughter weight, (b) further increase protein

Table 16.5 Calculation of farm margin (£ per year) when pigs are slaughtered at 70kg or 90kg live weight under different production circumstances

	70kg slaughter	90kg slaughter
Restricted feeding of a single medium-quality feed	5,500	7,000
Generous feeding of a single medium-quality diet	9,000	4,000
Generous feeding of 2 diets (grower + finisher)	12,000	9,000

Table 16.6 Calculations to examine the comparative advantages of various tactics and strategies for a 600-place fattening house to resolve a problem of poor carcass quality: (o) presents the current performance; (a) reduces slaughter weight; (b) increases diet quality; (c) reduces feed; (d) ceases castration; (e) improves the quality of the pig

	(o)	(a)	(b)	(c)	(d)	(e)	(cde)	(bde)
Daily live weight gain (kg)	0.69	0.68	0.68	0.59	0.71	0.70	0.60	0.73
Feed conversion ratio	2.79	2.75	2.82	2.84	2.69	2.74	2.78	2.57
Pigs with P2 <14mm (top grade) (%)	26	34	33	59	62	43	86	91
Annual throughput of pigs through unit ('000)	2.11	2.23	2.09	1.81	2.15	2.13	1.84	2.23
Margin over feed and weaner costs (£/pig)	7.50	7.00	6.10	10.32	12.03	9.84	12.75	13.24
Annual margin for unit (£'000)	-3.43	-3.83	-6.44	0.05	6.54	1.66	4.78	10.12

content of the diets and incur extra costs, (c) reduce daily feed allowance, (d) find a market for entire male pigs and stop castrating or (e) use an improved strain of pig.

Reducing slaughter weight (a) is no help because it is not sufficient to reduce backfat depths by an adequate amount; throughput is increased, however, resulting in more pigs sold (at a loss!). Nor is increasing the cost and quality of the diets (b) any help. But reducing the feed allowance (c) seems to be an appropriate short-term tactic that can be put into effect immediately and will prevent the present rate of financial loss, whilst also substantially improving the grading. Longer-term strategies would need to include finding a market for entire males (d) and purchasing improved hybrid stock (e). Combining actions (c), (d) and (e) appears attractive but if this strategy were to be implemented it would not be nearly as attractive as the combination of actions (b), (d) and (e). Of all the possibilities open, increasing the diet quality even further, stopping castration and improving the quality of the pig (b, d, e) whilst maintaining slaughter weight and feed levels achieves maximum annual margin for the unit.

Case 3: Improving upon excellence

A breeding-company multiplier supplies stock to a second commercial unit also under his ownership. The meat processor consistently praises the high quality of the pigs received. Good quality diets are fed at restricted levels to mixed entire and female stock. The commercial unit is in profit (Table 16.7, (o)), but only just. The simple expedients of increasing the feed scale and slaughtering some 3kg heavier (a) are able to double the profits without materially harming the good grading performance, the latter being important to sales of hybrid breeding females from the parent multiplier herd.

Table 16.7 Calculations to examine the possibility of improving the profit of a high-quality herd requiring to maintain an excellent carcass quality record: (o) the current position; (a) increases the feed allowance and slaughters at a slightly higher weight. The unit of calculation is a 600-place fattening house

	(o)	(a)
Daily live weight gain (kg)	0.65	0.72
Feed conversion ratio	2.52	2.49
Pigs with P2 <14mm (%)	98	91
Annual throughput of pigs through unit ('000)	1.99	2.11
Margin over feed and weaner costs (£/pig)	14.03	15.58
Annual margin for unit (£'000)	3.95	8.60

Case 4: Difference between diets

A producer feeding the pigs *ad libitum* from self-feed hoppers is offered four alternative diets to be given from 20 to 90kg (Table 16.8). The feeds are of different qualities and different prices. Feed 3 is calculated to be the best option but the carcass quality and grading profile are not really good enough. On the unit, it is ascertained that feed intake can be restricted if the slides on the hoppers in the finishing house are closed down to reduce the rate of feed flow. When the feed supply is restricted, diets 3 and 4 both calculate to perform equally well. Growth rates fall to around 0.75kg per day but up to 80% of pigs can now fall into top grade, and the margin rises to around £10,000 for the year. The choice between diets 3 and 4 has thus become relatively insensitive.

Sensitivities

Comparing sensitivities is a vital part of assessing marketing opportunities, and these will change with fluctuations in the price of pigs, the price of pig feeds and the premium paid for quality with respect to both diet inputs and carcass outputs. The above cases relate to carcass grade premiums of around 10% or more of the average price per kilogram dead weight. Greater or lesser grade differentials would result in quite different conclusions to the various calculations.

The essential qualities of marketing opportunity calculations are their dynamism and sensitivity to prevailing circumstance. Such calculations are complex but nevertheless need to be repeated often if they are to be useful. As with all business forecasting of this nature, management decision-taking is

Table 16.8 Calculations to examine the comparative advantages of four diets of differing quality and price

Diets	1	2	3	4
Crude protein (g/kg)	150	175	200	225
Crude fibre (g/kg)	60	50	40	30
Oil (g/kg)	30	45	55	60
Price (£/t)	140	155	165	175
Responses				
Daily live-weight gain (kg)	0.68	0.78	0.84	0.86
Feed conversion ratio	2.97	2.62	2.49	2.42
Pigs in top grade (%)	26	50	55	57
Annual throughput ('000)	2.11	2.38	2.52	2.61
Margin over feed and weaner costs (£/pig)	6.21	9.39	10.14	9.00
Annual margin for the unit (£'000)	−6.13	2.60	5.49	3.24

made considerably simpler if the calculations are completed through the medium of computer simulation models, as indeed has been the case here.

Conclusion

The position of pig producers marketing their product to the meat processor is different from the position of the feed manufacturer or the pig breeder selling the product to the producer. But in each case it is the producer who must take an active – and not a passive – role in marketing if the value of the purchases and the sales is to be optimized. While having little direct influence upon the meat slaughterer and processor (unless integrated) the producer has many options through which optimum production and marketing can be achieved, including the genetic quality and cost of the pigs, the nutritional quality and price of the feeds, the ways in which the pigs are grown, choice of slaughter weight and the choice of fatness at the point of sale. It is perhaps surprising how, within the pig industry, there is such variation between producers in their willingness to exploit the marketing flexibility that is open to them.

CHAPTER 17

THE ENVIRONMENTAL REQUIREMENTS OF PIGS

Introduction

Feral pigs frequent neither thick forest nor open range but rather the intermediate ground of open woodland, scrub and the forest margin. They build nests of sticks, grasses and rushes which are used for shelter. It may be assumed that pigs will benefit from being protected from the elements of wind, sun and rain and the provision of a warm, dry bed. In providing for these requirements, and in providing also for an individual supply of feed, housing systems are many and various. Housing for pigs also fulfils functions other than the provision of shelter and feed. Included amongst these are the prevention of bullying, the protection of sucking piglets, the facilitation of mating, the prevention of physical trauma whilst pregnant, the handling and the removal of excreta, and the allocation of quarters in which pigs may rest and grow, free of predators, parasites and disease. These provisions, being costly, tend to encourage intensive livestock systems in which the pigs must be, to a greater or lesser extent, densely stocked into relatively small spaces and restrained in enclosures and buildings. This consequence may be seen as having some disadvantages in terms of welfare, population-density, disease and capital cost.

Pig buildings may accumulate heat by radiation through the roof, the drawing in of warm outside air, internal supplementary heating and from the pig occupants themselves. Pigs lose heat to their environment through radiation, conduction (mostly through contact with the floor), convection (from the body surface) and latent heat (through the medium of evaporative water loss from the lungs). The removal of heat – and water vapour – from the internal environment of the building is aided by ventilation (convection) but there can also be some control of conductive and radiant heat losses through insulation.

Temperature

Ambient temperatures at which pigs are comfortable are related strongly to body weight (Table 17.1). These values are much affected by the heat which has been generated in the course of energy metabolism, and which must be dissipated from the animal body. This heat helps to combat the exigencies of

Table 17.1 Comfort temperatures (Tc) for pigs of different body weights. The comfort zone will usually range 1°C above and below the values given. The pigs are assumed to be fed to appetite and able to lie on a floor whose thermal resistance is not different from that of air

Live weight of pig	Comfort temperature (Tc) (°C)
Sucking pigs	
<1kg	32
<5kg	28
Weaner pigs	
<8kg	28
<10kg	26
10–15kg	22
Growers	
15–30kg	20
Finishers	
30–60kg	18
60–120kg	16
Pregnant sows feed restricted	18
in groups on straw	15
Lactating sows	16
Boars	18

a cold environment. Indeed, there is a necessity for the heat to be carried away from the body if heat stress is to be avoided. The dependence of comfort temperature (Tc) upon heat output (H) from the body can be broadly expressed for pigs of 10kg or more as:

$$Tc = 27 - 0.6H$$

(Equation 17.1)

where H is positively related to the live weight, the feed intake and the growth rate of the pig. The value of H is generated from within the pig as the sum of the work energy used for maintenance, the growth of protein and fatty tissues and the production of milk. H will therefore be lowest for smaller pigs growing slowly, intermediate for smaller pigs growing faster and for larger pigs and highest for heavy pigs growing fast and lactating sows. The relationship between feed intake and comfort temperature is shown in Figure 17.1.

Once estimated, Tc must be modified by those factors which may materially influence the rate of heat loss from the body of the pig, especially conduction and convection, and pig behaviour.

Floor surface

A major influence upon effective comfort temperature is mediated through the nature of the floor surface upon which the pig may lie and heat be conducted away. All substances have some degree of resistance to thermal transfer (including air). Table 17.2 shows that losses of heat through deep bedding will occur at a much slower rate than through a concrete floor. Ambient comfort temperature on bedding is therefore much lower than on solid concrete.

Behaviour

Pig behaviour is not without its further moderating influences upon Tc. Pigs can lie recumbent, and by maximizing the body surface in contact with the floor maximize the rate of heat loss through it. Equally, losses can be minimized by pigs lying on the sternum. Pigs in groups have lower comfort temperatures than individuals; this is because they can lie together to conserve warmth. The lying position of pigs in pens can give a good indication of the acceptability of the ambient temperature. Pigs should normally not be huddled together (too cold), nor all laid apart and recumbent (too hot).

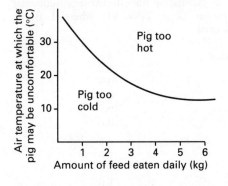

Figure 17.1 The ambient temperature at which a pig is comfortable is dependent upon the amount of feed it is eating (the illustration assumes a diet of about 13MJ DE/kg).

Table 17.2 Thermal resistance of flooring materials in comparison to air = 100

Air	100
Wire-mesh floor	100
Deep straw bed	500
Wooden slats	200
Concrete slats	50
Solid, wet concrete floor	25

Convection

The rate of air movement across the body controls the movement of heat from the body surface to the air. At low air speeds this effect is not great, but at higher air speeds (draughts) of above about 0.5m per second the consequences for comfort temperature can be significant, and in this range every 0.2m/s increase in air speed adds about 2°C to Tc. It is not surprising that, when not in heat stress, pigs will volunteer to lie away from draughts.

Taking full quantitative account of the influence of the quality of the environment upon comfort temperatures has proved difficult, and the solution appears not to lie (yet, at any rate) in ever-increasing levels of mathematical sophistication and implementation of the detail of the laws of physical science. Meanwhile a pragmatic approach may be to offer a score for environmental quality along the lines of Table 17.3. The factors Ve and Vl may be used to modify the ambient temperature (T) and allow calculation of the effective temperature (Te) which may be estimated as:

$$Te = T(Ve)(Vl)$$

(Equation 17.2)

The effective temperature Te can now be compared with Tc to assess any disparity there may be between what the pig requires and what the house provides.

Table 17.3 Scores for Ve and Vl for use in calculating the effective environmental temperature (Te) in the equation Te = T(Ve)(Vl), where T is the ambient temperature as measured

Rate of air movement and degree of insulation	Ve
Insulated, not draughty	1.0
Not insulated, not draughty	0.9
Insulated, slightly draughty	0.8
Insulated, draughty	0.7
Not insulated, draughty	0.6
Floor type in lying area	*Vl*
Deep straw bed	1.4
No bedding on solid insulated floor	1.0
Slatted floor with no draughts	1.0
No bedding on solid uninsulated floor	0.9
Slatted floor with draughts under	0.8
No bedding on wet, solid, uninsulated floor	0.7

Cooling and heating (Table 17.4)

It may be surmised that pigs can be cooled by:

- decreasing the number of pigs per pen (not only to allow recumbent lying but also to reduce the total number of pigs in the total air space of the house and thus reduce the total heat output (H); rule of thumb suggests that half a dozen average-sized growing pigs will throw off energy at the rate of about 1kWh);
- providing flooring with low thermal resistance (high ability to draw away heat);
- increasing the rate of air movement.

Also helpful would be the protection from radiant heat (by the provision of an insulated roof), humidification of the air, presentation of the feed in wet form, provision of water sprays and drips and wetting floors and providing wallows. Pigs are relatively indifferent to changes in relative humidity between 60% and 90%, although the structure of the building will not be indifferent to condensation which may take place at higher relative humidities and colder temperatures. Pigs do not sweat, but evaporative heat losses are possible from the lung surface and pigs in heat stress may pant; at this point cooling strategies have most certainly failed.

Warming pigs up may be achieved by tactics contrary to the above, plus the actual input of energy into the house by means of supplementary heating. Pigs are themselves a ready source of heat as witness the equation:

$$H = ME - (24Pr + 39Lr)$$

(Equation 17.3)

Table 17.4 Cooling and heating

Too hot	Too cold
Pigs will lie apart and maximize floor contact	Pigs will lie together and minimize floor contact
Increase ventilation rate (volume and speed)	Decrease ventilation rate
Decrease stocking density	Increase stocking density
Increase air space	Decrease air space
Provide solid floors	Provide suspended floors
Provide shade	Maintain floors dry and with bedding
Provide water sprays	Provide supplementary heat
Insulate roof and walls	Insulate roof and walls
Feed high-nutrient-dense diets	Give more feed
	Feed low-nutrient-dense diets

which is simply to say that feed metabolizable energy (ME) will end up as heat (H) to be lost from the body, with the exception of that which is retained as protein or lipid accretion (Pr and Lr). The amount of heat generated from body metabolism (that is, the effective ME yield from energy digested) is dependent to a small extent upon the form in which dietary energy is provided. Dietary fibre has an especially high heat of fermentation, while dietary fat can be used particularly efficiently. High-fibre diets will therefore be heat creating, whilst high-fat diets will aid cooling.

Whatever, the ambient temperature of a pig building can be most effectively increased by increasing the number of pigs in it (increasing the stocking density) and/or by increasing the amount of feed given to the pigs (raising the allowance; encouraging appetite).

Supplementary heat from oil, electricity or gas may be provided by conventional means, including convectors, floor-heating systems, overhead radiant heaters and the like. Hot-water radiators may be found in sophisticated pig buildings in cold climates, while overhead gas or electric heaters are normal for sucking pigs, weaners and some young growers. The great disparity between the comfort temperature for a newborn sucking pig (about 28°C) and its suckling mother (about 15°C) exemplifies the impossibility of providing an ambient room temperature ideal to widely differing classes of pig. In the case of the sow and her litter, a local source of heat and the creation of a warm microclimate for the sucking piglets is prerequisite. The consequence for growing pigs is that different house types maintained at different ambient temperatures, and requiring different degrees of sophistication in their control systems, are required for weaners, young growers and finishing pigs.

An intermediate step in heat input to a building between the utilization of pig heat and the introduction of supplementary energy is to do all that is possible to conserve the heat already present. Insulation, recirculation and heat exchange technology may be readily adapted to pig buildings and can make a material contribution to energy saving in cold environments.

Metabolic response to cold and heat

Pigs are cold when the effective ambient temperature (Te) is less than the comfort temperature (Tc). The pigs' response to cold is to direct ingested energy from productive functions into the creation of heat (that is, cold thermogenesis). The energy cost of cold thermogenesis (E_{h1}) for each degree of cold may be estimated as:

$$E_{h1}(MJ) = 0.012W^{0.75} (Tc - Te)$$

(Equation 17.4)

where W is the live weight (kg), Tc is the comfort temperature and Te is the effective environmental temperature (°C).

Thus, although body temperature is maintained, the pig will grow more slowly and the feed-conversion efficiency will fall. As an approximation, 50g of feed can be used up in this way per 100kg of pig live weight (W) per 1°C of cold (Tc − Te). There may, however, be more cost-effective means of inputting energy into the environment than the conversion of pig feed into heat: heat conservation certainly, but in many circumstances also the direct input of energy from (non-renewable) fossil fuels. The balance of benefit is dependent upon the comparative costs of pig feed (plant carbohydrate) energy and fossil fuel energy. It will be recalled that a higher feed intake decreases Tc (Figure 17.1) and fortunately there is also an appetite response to an environmental temperature which is less than the comfort temperature. This latter response may go some way to offsetting the extent of the reductions in productivity due to diversion of energy away from the accretion of protein and lipid and towards the generation of heat.

Pigs are hot when the effective ambient temperature (Te) is greater than the comfort temperature (Tc). The pig is embarrassed by its inability to dissipate heat into the environment and in consequence appetite falls by about 1g per kg W per 1°C of heat above comfort (Te − Tc). Growth rate and feed-conversion efficiency fall.

The influence of environmental temperature upon growth rate mediated through the metabolic responses of cold thermogenesis and appetite depression is depicted in Figure 17.2.

Plate 17.1 Outdoor pig production: pregnant sow paddock. Note the lack of vegetation. *Courtesy*: Cotswold Pig Development Company

Plate 17.2 Outdoor pig production: hut for lactating sow and litter. *Courtesy*: Cotswold Pig Development Company

Plate 17.3 Final stages of slurry drying procedure by use of banks of large fans. Japan

Plate 17.4 Naturally ventilated house for tethered pregnant sows. Sows are fed manually from a feed trolley passing down the passage. Spain

Plate 17.5 Natural side ventilation for pig grower house. The flexible wall is raised half-way up; it may be wound down to ground level. Within the flexible wall there are plywood boards which may also be removed at particularly high temperatures as there is a metal bar barrier within. Japan

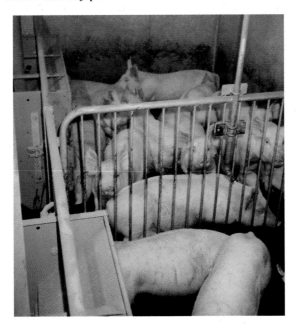

Plate 17.6 Young growing pigs on a suspended metal mesh floor with *ad libitum* self-feed hoppers. Europe.

Plate 17.7 Conventional fully-slatted floor house for growing pigs with self-feed *ad libitum* hoppers (manually filled). Southern Europe

Plate 17.8 High welfare straw-yard and kennel housing for finishing pigs or young gilts

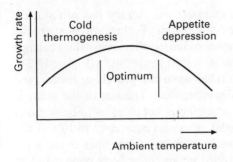

Cold
thermogenesis

Figure 17.2 The influence upon growth rate of metabolic response to increasing ambient temperature.

Influence of insulation

The degree of insulation influences both the rate of heat loss and the rate of heat gain by a building. Given conventional densities of stocking with pigs and an outside temperature of 0°C, an uninsulated building would be unlikely to be able to maintain an internal air temperature much above 12°C, and, in addition, condensation would occur. However, effective insulation (especially of the roof) would both allow a satisfactory internal air temperature of 20°C and prevent condensation. Equally an outside air temperature of 20°C would be associated, in an uninsulated building, with an internal temperature for the living quarters of the pigs of above 25°C. But when insulated, excellent ventilation could restrain the internal temperature to no greater than 2°C above that outside.

Ventilation

Pigs need a minimum air space of around $0.1m^3$ per kg metabolic weight ($W^{0.75}$). This air must move if:

1. evaporative moisture losses from the lungs are not to cause a build-up of relative humidity and condensation;
2. pigs are to lose heat from their bodies as is required;
3. the air is to be kept fresh and free of noxious gases. Acceptable levels of gases are 5 parts per million (ppm) carbon monoxide, 2,000ppm carbon dioxide, 0.1ppm hydrogen sulphide and 10ppm ammonia. Levels above these are indicative of inadequate ventilation, although levels may reach 3,000ppm CO_2, 5ppm H_2S and 20ppm NH_3 without undue alarm; but levels higher than these may give cause for concern.

Air movement by ventilation that is excessive will run the risk of reducing the effective temperature (Te) below the comfort zone (Tc) in climates where the outside temperature may be low. In temperate climates there is usually a balance to be struck between the freshness of the air and its temperature (Figure 17.3).

Natural ventilation (Figure 17.4(a)) relies upon the prevailing breeze; the inlets may be manually or automatically controlled. The size of the inlets is dependent upon outside temperature, requiring in hot climates to be as much as 20% of the floor area, with a ridge vent of an area equivalent to 10% of the floor area. This is normally achieved with a wide gap at the apex of the roof, and sides to the house which may be rolled up and effectively removed. The ridge vent may not be required with houses whose side ventilation is generous. In any case, the roof should be well insulated.

In 'controlled ventilation' houses, air movement is usually achieved by electric fan, activated according to the temperature of the house. The fans operate when the upper end of the comfort temperature zone is reached and stop when ambient temperature reaches the lower end of the comfort zone. Thermostat sensitivities therefore require to be plus or minus around 1°C.

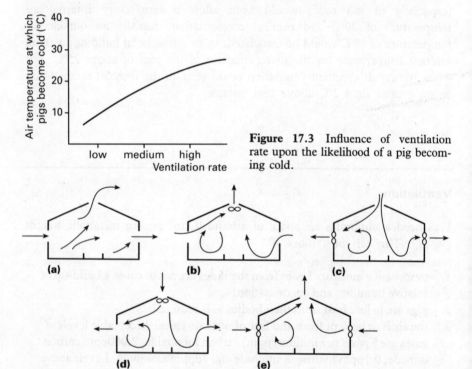

Figure 17.3 Influence of ventilation rate upon the likelihood of a pig becoming cold.

Figure 17.4 Patterns of ventilation: (a) natural ventilation, (b)–(e) different configurations for fan ventilation systems (preferred circulation systems are depicted to the left of the buildings).

Temperature fluctuations can be as counter-productive to pig health and well-being as can low or high temperatures themselves – especially in the case of weaner pigs. Variable speed fans, or a mix of smaller and larger fans, can be helpful in achieving smooth temperature transitions.

General examples of fan ventilation systems are shown in Figure 17.4(b)–(e). Air-flow patterns depicted on the left-hand side of the drawings are to be preferred over those to the right. The latter allows cold air to fall directly upon the pigs, while the former encourages faster air movements along the roof line and slower, more general, air mixing in the body of the building and around the pigs.

Usually a minimum ventilation rate (air movement) of around $0.2m^3$ per hour per kilogram pig live weight is needed in pig houses, even when the exterior temperature is low; although in cold weather this minimum may have to be further reduced in the interests of pig warmth. Maximum ventilation rates for warmer weather in temperate climates require to be around $2.0m^3$ per hour per kilogram live weight of pig. For hotter climates even more than this rate of air movement may be needed at peak.

Table 17.5 shows examples of the capacity of fans to deliver air. Knowing maximum and minimum needs of air movements, and maximum and minimum stocking densities of the room to be ventilated, a profile of the fan requirements of individual pig rooms can be determined.

Air speeds across the bodies of pigs are normally maintained between 0.1 and 0.5m/s. An air speed of 0.15m/s is considered suitable for smaller pigs maintained at higher room temperatures or for larger pigs maintained at lower temperatures; 0.3m/s is suitable for larger pigs if temperature rises above comfort levels whilst 0.5m/s may be considered appropriate for hot climates. At speeds above 0.5m/s the movement of air becomes quite apparent. Houses maintained at suitable temperature will normally have air moving at around 0.2m/s. Where the temperature is below the comfort zone, then air speeds will fall below 0.2m/s, whilst active cooling (when Te is above Tc) requires air movements above 0.2m/s. Air speeds along the roof line and near inlets may, of course, reach levels of 0.5m/s or above. Fans deliver air at speeds of up to 10m/s. Avoidance of draughts in conditions of high-density stocking and the need for positive ventilation may be assisted by diffusing the

Table 17.5 Capacity of fans to deliver air

Diameter of fan	Revolutions per minute	Air delivery (m^3/h)
500mm	1,400	8,500
	1,000	6,000
	700	5,400
1,000mm	800	35,000
	400	20,000

air before it enters the pig living space. This may be achieved by provision of a false roof or a ducting system, both of which give spaces into which the fresh air is drawn and from which the air is spread through very many smaller air holes.

Ventilation of buildings remains an imperfect science, and practical experience of effective systems is invaluable. It is difficult to maintain temperature and achieve adequate air movements in colder climates and equally difficult to achieve adequate cooling in hotter climates (even where water spray systems are fitted). Forced (fan) ventilation is expensive in terms of both equipment purchase and recurrent energy costs. More expensive still is the evident need to put energy into pig houses to keep them warm, whilst simultaneously moving the air out in the interests of proper ventilation. Minimum ventilation rates almost invariably require supplementary heating to operate in all houses where young pigs are to be kept.

Space

The linear dimensions of pigs are found to have the following relationships with their weight (W, kg).

- Pig length (m) approximately equal to $0.30W^{0.33}$;
- Pig height (m) approximately equal to $0.15W^{0.33}$;
- Pig width (m) approximately equal to $0.06W^{0.33}$.

The volume of a pig may thus be expressed with reasonable accuracy as a function of its weight. A pig lying recumbent with its legs outstretched (Figure 17.5(a)) occupies a space of $0.045W^{0.66}m^2$. If it is lying on its sternum (or standing) (Figure 17.5(b)) the space occupied is $0.018W^{0.66}m^2$. The actual occupancy of the $0.045W^{0.66}$ overall space by the pig's physical body form is between 0.025 and $0.030W^{0.66}$ (see Figure 17.5(a), the overall space frame not fully filled with pig; free space being found particularly under the jaw and between the legs).

From these simple elements it may be surmised that if pigs are each allowed a space in a pen of $0.05W^{0.66}$ then they may all lie recumbent simultaneously and about one-third of the floor will remain uncovered (vacant). This calculates to about $1m^2$ per 100kg of pig weight in the pen. If the pen floor is fully slatted or of suspended wire mesh, and the house is closely environmentally controlled, this may be considered adequate. The

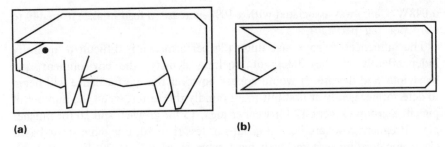

Figure 17.5 Physical space occupied by pigs (a) lying recumbent, (b) standing or lying on sternum.

industrial standard for intensive housing of pigs may even be as low as $0.025W^{0.66} + 25\%$, which is $0.032W^{0.66}$. In other circumstances, as for example for solid floored and straw bedded pens, an extra allowance of 25% may be added to the more generous standard of $0.05W^{0.66}$ to facilitate further movement and also to give specially designed areas to be used only for excretion.

Use of pig dimensions in the above way to determine pen-space allowances must be modified by the number of pigs in the pen. The fewer the pigs, the more space required per pig – and vice versa. Guide lying-area requirements are given in Table 17.6 which assumes group size of 8–15 pigs per pen. To this lying-area allowance will require to be added a further 20–60% of space for movement and excretion depending upon house type; the lower addition for fully slatted floors and the higher for pens with bedded dunging areas. In warmer climates, or for pigs with especially high levels of heat output, such as rapidly growing pigs with high appetites, the greater space allowances are beneficial. Adult sows kept in groups in straw yards may use 2–4m² of total space per sow.

There is a significant influence of space allowance on feed intake and growth rate – and a further interaction with disease level. In general, raising the total space allowance in a fully slatted pen from 0.5 to 1.0m²/100kg of pig live weight will increase growth rate by about 10% in a broadly linear manner (Figure 17.6). In a recent experiment with 10kg commercial pigs, an increase in space allowance from 0.14 to 0.22m² per pig ($0.031W^{0.66}m^2$ to

Table 17.6 Space requirements for lying areas

	Lying area (m²) per pig
Sows	1.5 –3.0
Piglets 5–20kg	0.10–0.20
Growing pigs 20–60kg	0.20–0.40
Finishing pigs 60–120kg	0.40–0.80

$0.048W^{0.66}m^2$) was associated with a 10% increase in feed intake (from 440 to 481g per pig per day).

The influence of group size upon pig performance is difficult to quantify independently of the effects of stocking density, the environment, pig behaviour and disease. It would appear however that performance will begin to deteriorate, gently at first but more rapidly thereafter, when the number of pigs in a group exceeds 12 for weaner pigs, 15 for growers and 20 for finisher pigs. Pregnant sows are best in groups of less than 10, but some straw-based electronic feeding systems may have sows in groups of 20–40 animals.

Space required for feeding differs with the feeding system. Solid floors may accept feed dropped from above and thus no special space allocation for feeding is made. Floor feeding is however both dusty and wasteful, nor does it help to maximize appetite, nor decrease pig variability. Self-feed hopper systems for the delivery of dry meal or pellets will normally be 300–400mm in width and linear floor space should be allowed to the extent of about 75mm of hopper length per weaner piglet, 60mm of linear hopper length for each growing pig and 70mm for each finisher pig. Pens with 10–20 growing pigs in them will normally have 4–6 feeder spaces provided in the hopper. Single-space feeders (usually one for eight pigs) naturally require a lesser area of floor space to be provided. When pigs are fed from a long trough, and all pigs in the pen require to eat simultaneously, it is evident that the absolute minimum allowance would be in excess of pig shoulder width ($0.06W^{0.33}$), whilst the optimum would probably be twice that amount. Usual recommendations fall between these limits at 150mm per pig for young growers, 250mm for growers between 25 and 75kg live weight and 300mm for finisher pigs. Sows each need a trough of around 500mm in length.

Summary of housing requirements

A summary of temperature, ventilation and space requirements is to be found in Table 17.7.

Waste

Excreta in the form of mixed faeces and urine is of about 10% dry matter (DM) and the average production per pig on a unit which both breeds and

Table 17.7 Summary of housing requirements

Quantitative

Space · About 0.8–1.6m²/100kg of pig; total area for pigs in groups.
Fully slatted or mesh floors, 0.5–1.0m²/100kg of pig.
Adult pigs loose-housed, 1.0–3.0m²/100kg of pig.
Young weaned pigs loose-housed, 1.0–2.0m²/100kg of pig.
Suckling adult with piglets, 4–7m².
Breeding males 9–12m².
Crates for adult females, 2,200 × 600mm (approx.).

Air speed · About 0.1m/s at pig height. 0.3–0.5m/s constitutes a draught.

Ventilation · A minimum of around 0.2m³/h per kg of pig weight. Maximum about 2.0m³/h per kg of pig.

Humidity · 60–80% relative humidity.

Temperature · In the region of:
26–30°C for pigs of <10kg
22–26°C for pigs of 10–15kg
18–22°C for pigs of 15–30kg
16–20°C for pigs of 30–60kg
14–20°C for finishing pigs and adult females.

Feeder space · 0.15m/pig when <25kg.
0.25m/pig when 25–75kg.
0.30m/pig when >75kg.

Excreta · Average of about 5 litres/pig/day for a mixed breeding/growing unit.

Housing insulation · To prevent more than 1 watt passing through the walls, ceiling or floor per m² per °C difference in temperature between one side and another.

Qualitative

Space · Enough lying area for all pigs to lie stretched out side-by-side, plus 20–60% for excretory and activity functions.
If pigs are likely to fight, enough space for the vanquished to escape.
When pigs standing, half the floor area to be visible.
When pigs lying, one-third of the floor area to be visible.
Animals restrained with tethers or in crates to have as much space as possible consistent with prevention of escape and attacks on neighbours.
Excretion in the feeding area to be prevented. Body sores to be absent and the animal clean and dry (able to rest with body away from faeces and urine).

Ventilation · Air movement over the surface of body not readily detected. Air not fresh, but not stale nor unduly odorous.

Humidity · Not so dry and dusty as to cause respiratory distress, not so damp as to cause condensation.

(contd.)

Table 17.7 (continued)

Temperature	Pigs <10kg: man comfortable with no clothes on.
	10–20kg: man comfortable in vest and pants.
	20–50kg: man comfortable in shirt and trousers.
	50–100kg: man comfortable in overalls but without coat.
	Adult females: man comfortable in overalls but without coat.
	Pigs lying huddled together are cold.
	Pigs lying apart are too hot.
	As pigs become colder, the lying area becomes dirtier.
	As temperature increases, likelihood of tail biting increases.
Feeder space	To allow all pigs in pen to eat simultaneously and eat their fair share.
Excreta	25% of feed eaten and all water drunk.
Floor insulation	Slatted or mesh floors to be free of draughts. Solid floors to be dry, insulated or bedded; wood shavings sufficient to form solid mat of 10–22mm between pig and floor; straw to allow pigs to lie both on and in it.
General	Pigs clean and dry, disease at low level, behavioural problems (cannibalism etc.) absent. House readily entered and worked in by staff and management.

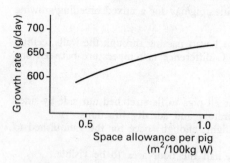

Figure 17.6 Influence of space allowance on the growth rate of young pigs.

rears the pigs up to slaughter weight is about 5kg per pig per day (1.8t per year). The rate of excretion is less for younger pigs and a great deal more for lactating sows; 2kg, 6kg and up to 25kg for pigs of 30kg, 100kg and lactating sows respectively. At 4–8% DM the excreta is a readily pumpable slurry (a 250-sow breeding and grow-out unit will use 20,000 litres of water daily). At 20% DM or above it can be handled as a solid. Slurry can be stored under the slats within the house, or extracted for storage in tanks and lagoons. It can be aerobically and anaerobically digested before disposal to the land. Anaerobic digestion will yield methane as a useful energy source. The solids separate naturally from the liquids in a lagoon. The latter is pumpable directly to land, whilst the solids can be taken for mechanical spreading. Solids and liquids may also be separated mechanically by centrifugation or sieving, whilst the solids may then be even further dried to a friable material. Pig excreta contains high levels of nitrogen (N), phosphorus (P) and potassium (K) and this can be both a useful fertilizer (when used judiciously) and a serious pollutant (when used to excess). In the normal course of events some 30% of the N is lost to the atmosphere but only 10% of P and K.

At 10% DM pig slurry contains about 4.2gN/kg, 1.7gP/kg and 2.5gK/kg. One hundred growing pig places on the farm (the output from ten sows) will produce in a year some 180,000kg of slurry: 750kgN, 300kgP, and 450kgK. On the basis of the average rates of application acceptable to a mixture of arable and grassland, without causing run-off or excessive losses through the soil, 1ha of farm land can handle the waste output of 50 growing pigs per year; or two sows and their progeny to slaughter weight. The amount of land required per pig does however depend greatly upon what is considered as an acceptable application rate. Only if 150kg N/ha is considered acceptable, can 1ha receive annually the waste output from two sows and their progeny taken to slaughter. Equally evidently, the calculation is sensitive to the amount of N and P excreted by pigs which is highly dependent upon the correctness of the amino acid balance and amino acid supply (N) and the availability of the minerals (P) supplied.

In terms of volumes to be spread, the output of slurry is calculated at 1.8t ($1.8m^3$) per year per pig place and it is estimated that arable land can accept a slurry volume dose of about $25m^3$ per ha, while grassland can accept repeated doses up to a maximum total delivery of $100m^3$ per ha. Physical limits for annual slurry spreading therefore vary between up to 15 pig places per hectare (arable) and up to 60 pig places per hectare (grassland).

Possible legislation in some countries relating to the maximum applications of nitrates and phosphates to the land could bring embarrassment to large, intensive pig farms on small land holdings; their waste would require expensive treatment and/or transportation. Were this to be the case then it may bring a move towards smaller-sized units, and these located on arable and grassland farms as part of a mixed farming enterprise.

Calculation of housing requirements

Pen requirements depend upon house type selected, the weight at which the pigs move from one type to the next and the performance of the pigs. Nevertheless the various housing requirements are readily calculated for each individual set of circumstances. Example calculations are laid out in Table 17.8 which deals with a unit of 100 sows. Allowing 115 days for pregnancy, 12 days for weaning to conception, 28 days for lactation, 10 pigs of 8kg weaned per litter and 9.6 pigs per litter surviving through to the point of slaughter, some 2,200 pigs will be sold per year, or something over 40 per week. The sows would produce 2.3 litters per year (365/[115 + 12 + 28] = 2.3), and from these will come the 2,200 grown pigs (2.3 × 9.6 × 100). Where the pregnant sows are housed in stalls then 75 of these should be provided ([115 × 2.3]/365 = 0.72), together with another four for replacement gilts. Mating pens may be in the form of either loose housing or stalls. Some 15 sows will need to be accommodated in the mating pens for about 22 days (12 + 10).

If the replacement rate for the herd is 40%, then some 40 gilts will come in annually. They will probably arrive in batches of eight every 2 months (to allow a safety margin). These will need quarantine quarters and then additional loose-housing pens or stalls whilst they are awaiting mating. Boars are best used around three times weekly. On the basis of five sows being weaned per week, and each sow perhaps needing to be mated as many as three times, one may calculate a minimum requirement of five boars per 100 sows. However, the farrowing rate will not be 100% (more likely 80%), and 20% of sows will return and need to be remated. The minimum requirement is therefore for six boars for every 100 sows.

There are a plethora of mix-and-match systems to take growers on from weaning through to slaughter. In the example shown in Table 17.8 the flat-deck weaner pens are followed by a short-stay grower pen (20–40kg). Often this is omitted. Pigs may be kept from weaning through to 30kg in weaner pens, and then moved straight into finishing accommodation. In either case the method of calculation is the same, and the number of pens required is dependent upon the daily weight gain that is assumed for the pigs and the number of pigs to be placed into each pen.

The major confusion factors in allocating housing are:

● bunching;
● variation in growth rates and reproductive performance between batches;
● variation in growth rates and reproductive performance within batches.

Bunches of breeding females reaching the end of pregnancy simultaneously cause accommodation shortages in the maternity ward. Avoidance of peaks and troughs in pen usages is an essential part of the efficient utilization of expensive capital buildings. Serious problems can result when, for one reason or another – usually disease – young growers which have been allowed a 30-day stay in a rearing house take 50 days before they are ready to move on.

Table 17.8 Example calculations for housing requirements (per 100 breeding females), allowing 115 days for pregnancy, 12 days for weaning to conception (5–20 days, as appropriate to circumstance), 28 days for lactation (21–56 days, as appropriate to circumstance), and 10 pigs of 8kg weaned per litter with a further loss of 0.4 pigs/litter during growth. Some 2,200 pigs will be sold per year, or just over 40 per week

Number of litters/year/sow	$365/(115 + 12 + 28) = 2.3$
Number of pigs/year/100 sows	$2.3 \times 9.6 = 22 = 2,200$
Stalls for pregnant females/100 sows	$(115 \times 2.3)/365 = 0.72 = 72$ (allow 75)
Farrowing pens (4.5 farrowings/week)/100 sows	$(28 + 10$ (safety margin)$) = (38 \times 2.3)/365 = 0.24 = 24$ (allow 25)
Mating pens/100 sows	$(12 + 10$ (safety margin)$) = (22 \times 2.3)/365 = 0.14 = 14$ (allow 15)
Boar pens and hospital pens/100 sows	6 boar pens and 4 hospital pens = 10
Replacement females/100 sows	10 pig places or 3 group pens and 4 sow stalls
Quarantine quarters/100 sows	3 group pens
Weaning to 20kg; 0.4kg gain/day	$20 - 8 = 12/0.4 = 30$ days
	30 days for 2,200 pigs = 66,000/365 = 180 pig places at 10 pigs/pen = 18 pens (allow 20)
	or, 30 days = 365/30 = 12.1 batches per year at 10 pigs/batch, 2,200/(12.1 × 10) = 18 pens
20–40kg; 0.6kg gain/day	$40 - 20 = 20/0.6 = 33$ days
	33 days for 2,200 pigs = 72,600/365 = 200 pig places at 16 pigs/pen = 13 pens (allow 15)
	or, 33 days = 365/33 = 11 batches per year at 16 pigs/batch, 2,200/(11 × 16) = 13 pens
40–100kg; 0.8kg gain/day	$100 - 40 = 60/0.8 = 75$ days
	75 days for 2,200 pigs = 165,000/365 = 452 pig places at 16 pigs/pen = 28 pens (allow 30)

521

This will mean that pigs ready to be weaned will have nowhere to go, while places in the finishing house remain unfilled. Pen planning must allow for some peaks and troughs in production.

Because of individual differences in performance, pigs in a batch will reach slaughter weight at different times. The manager must choose between emptying the pen all at once (accepting that the biggest pig might be much heavier than the smallest) or waiting for each pig to come up to correct weight (which might take a spread of some time). Sale contracts which allow pens to be emptied all at once, and accept variable pig weight, have much to commend them.

It is self-evident that pigs grow but their pens do not. This incompatibility means that pens are normally either too empty or too full. In order to reduce this problem to a minimum pigs are either moved from one pen size to another or the number of pigs in the pen is adjusted. In most systems all three of these things happen (the pens are not optimally filled, the pigs are moved, the number of pigs in the pen is adjusted). Pens which are insufficiently densely stocked predispose to the pigs being cold, whilst pens which are stocked too densely will bring about reductions in growth rate and efficiency. Each time a pig is moved, however, there is a growth check consequent upon that move. The greater the change in circumstances the greater the check. Thus the solution to one inefficiency can readily lead to the creation of another. Growth checks upon moving and remixing pigs will usually be equivalent to losses of 1–5 days in time through to slaughter; but sometimes lost days can add up to weeks. Pig mixing is encouraged by the need for matched batches of growing pigs and efforts made to try to reduce pen emptying spread. In many systems it is particularly during early growth that pigs tend to be mixed with strangers. This causes an especially high level of tension, confusion and quarrelling, with consequent reduction in performance.

Mixing and moving pigs is a managemental convenience, but one which reduces efficiency and is best kept to a minimum.

Outdoor pig production

Pigs kept in outdoor systems do not range freely. They are batched into groups, given shelters for minimal protection from the elements and confined to fenced paddocks. The system depends upon:

- low capital costs (about 0.3 of the equivalent of an intensive breeder herd);
- the relatively low environmental temperature requirements of the lactating sow;
- the ability of little pigs (and pregnant sows) to keep warm by huddling together in groups in deep straw bedding.

The paddocks may not be assumed to yield any significant nutrient supply. Quite the reverse, feed usage by sows will be significantly greater (usually 1.4–1.5t/sow/year, in contrast to 1.2–1.3t/sow/year for conventional protected environment), and this is on account of:

- feed wastage;
- use of feed for cold thermogenesis.

The pigs will root and tread the ground area, destroying all vegetation in a relatively short period of time. The paddocks will therefore tend toward becoming either dry earth (in hot dry weather), or mud (in wet weather).

The system is not usually considered as suitable for weaned piglets which are taken into relatively conventional housing for rearing from 8kg at weaning to point of sale at 30kg. Occasionally pigs can be fattened 'outdoors' through the provision of a generous paddock area ($5m^2$ per pig) outside of the grower house. Even with a relatively high specification for this (necessarily portable) building, and ample straw, the efficiency of feed usage can be poor for pigs' growth to slaughter in outdoor systems.

Whilst there is unquestionably more freedom of movement for sows kept outdoors as compared to those kept in farrowing crates and pregnant sow stalls, it would be wrong to assume that there is always an elevation in sow welfare. Outdoor pigs require the highest levels of management and avoidance of:

- cold and wet;
- heat stress and sunburn;
- bullying;
- unequal food provision;
- variability in sow condition;
- physical injury;
- piglet mortality;
- loss of fertility;

- excessive feed usage;
- inadequate nutrient supply;
- inadequate water supply.

Outdoor systems, in addition to using around 16% more feed (an extra 0.5kg per day), also tend to need more expensive feed compounded into large (16–24mm) cobs or rolls which are required to reduce wastage when scattered on the ground or in heaps. Sow productivity will be around 1.5–2 piglets less weaned per sow per year, and sow replacements will be about 10% higher than in a conventional system. The sows chosen for outdoor pig breeding require to exhibit special characteristics relating to their need to rebreed readily, and to withstand the rigours of outdoor life. Robust pig strains tend to be rather slower growing and more predisposed to be fat than other strains suitable for more highly protected environments. This can result in less efficient performance in the grow-out unit and lower carcass quality.

Fifteen to 20 sows rearing their young to 30kg on an outdoor system will require 1ha of dry free-draining ground (for example sand and gravel or chalk), in a temperate climate not prone to high rainfall, snow, severe frost, high humidity, high temperature, high radiation levels or high winds. The near presence of trees acting as shelter belts is advantageous.

The whole set-up requires to be moved annually to new ground which is not used more than once every 4 years. The system is therefore ideal as part of a 5-year arable rotation, moving on to grass established the year previously and followed by a cereal crop. The presence of arable crops will also facilitate the ample amounts of straw that will be used (around 60t per 100 sows).

For every 100 sows an outdoor system may be expected to comprise:

- One hectare of service area with six boars working around four paddocks. Each hut (6–6.5m^2) will hold five sows (four or five sows will be weaned weekly), making total provision to allow for 20 sows in the mating area. There may also be a separate fifth paddock (and huts) to receive (and mate) replacement gilts.
- Pregnant sows will occupy an area of 4ha with 15 huts found in 3–5 paddocks, making a total provision which will allow for 75 sows.
- Farrowing and suckling sows will be accommodated in 24 farrowing huts (one sow per hut of 3–4m^2) found in 4–6 paddocks in a total area of 1ha.
- Out of the total 6ha will be found the required space for equipment, services, feed hoppers, paths and so on; these may add up to as much as 1ha in themselves, the pigs occupying 5ha.

Young pigs are weaned at 21–28 days from the outdoor system usually into weaner huts. These may have well protected and highly densely stocked indoor kennel areas allowing 0.2m^2 of space or less per pig. There may be an outdoor yard allowing 0.4m^2 per pig, which may or may not be strawed. With an 8-week stay to grow to 30kg at 12 weeks of age, weaning 50 piglets (five litters per week) would require eight 50-pig huts or, more realistically, twenty 20-pig huts.

The principal redeeming features of outdoor pig systems are their low capital setting-up costs and a public perception of greater sow freedom. The latter is likely to be rivalled by improved pregnant sow housing now available to replace stalls and tethers. But, by definition, the building costs of the highly protected environment of the intensive unit will always greatly exceed those of simple sheet-metal and wood huts placed upon the bare ground. The outdoor system also has the advantage of being free of all the waste management costs and problems associated with intensive production units.

Feed delivery systems

Sucking pigs

Feed freshness and ingredient palatability are paramount to encourage intake, while the utilization of investigative behaviour demands that the feed be placed where the piglets would normally expect to find it – on the floor. A clean dry area is required in the creep upon which feed can be scattered frequently throughout the day. Meal, pellets or crumbs may be used. Various shallow dishes might help to avoid excessive wastage, but troughs can be counter-productive if they encourage feed to be held for – at most – 9 hours, during which time it will become stale and unpalatable. Equally important is to disallow the possibility of feed to be contaminated with excreta. Sucking piglets usually eat supplementary creep feed immediately after they have been suckled; access is therefore required for all the pigs in the litter to eat simultaneously. *Ad libitum* self-feed hoppers, even of reduced size, are not appropriate for offering creep feed to sucking pigs as they fail on most, or all, of the above criteria. Best are shallow troughs, mats or simply a clean area of floor from which uneaten food can readily be brushed away. Water is prerequisite to the consumption of dry feed and may be provided by nipple drinkers or in special water troughs.

Weaned pigs

The weaner is delivered its feed invariably from a self-feed hopper in the form of a dry pellet or a crumb, or sometimes a 'high-fat' meal. Because of the need to encourage intake to the maximum possible extent, the early feeds after the moment of weaning should be delivered by hand into the bottom of the trough, and preferably also on to a solid base-plate covering the floor to the fore of the hopper trough itself. Three or four times a day meal feeding in this manner may continue for 1, 2 or 3 days until the piglets are all eating enthusiastically. Even then the hopper should not be filled with more than a 24-hour supply of feed. Any contamination by excreta must be completely removed. The need for constant attention tempts the placement of the bulk feed receptacle (trolley or bag) into the weaner room itself ready for immediate use. This will, however, be self-defeating as the feed – being in the pig's environment – will become stale. Creep and weaner diets in particular should be stored cool and dry, and away from pig odours.

After 10kg live weight weaners will likely change on to a less expensive diet, and this can be provided in the hopper in a more conventional way. Automatic delivery to hoppers by auger and downpipe is helpful in ensuring feed freshness as delivery can be twice daily and the pipe placed low into the trough to avoid excess supply.

The self-feed hopper principle assumes feed to be available *ad libitum*, that is 24 hours daily. In the early stages this is to be avoided, and the trough should be cleaned and empty for at least 1 hour every day. Neither must it be assumed that because feed is available it is always accessible to the pigs. Young pigs would prefer to eat and lie together; there is a tendency therefore for the more timid pigs in the group to not get their fair share. At a provision of 75mm of trough length per piglet and a shoulder width for an 8kg piglet of 120mm, it is evident that at best only half of the pigs could access the trough at any one time.

Water is usually supplied through nipple drinkers provided at the rate of one for every eight piglets.

Weaners will happily eat their food wet, and intake may even be encouraged in this way. The problem, however, is one of feed freshness and contamination. It is unavoidable that wet feed supply will exceed that which the piglets can eat up immediately. Between meals wet feed tends to become unpalatable. If, however, very frequent feeding is possible, and the piglets are eating vigorously (normally when they are 10kg or more in live weight), then a wet feeding system might be usefully considered.

Growers and finishers

Dry pellet or meal feeding through *ad libitum* self-feed hoppers is the most widespread feed delivery system for growing pigs. It is usual to provide a trough length allowance of 60mm for younger pigs and 70mm for those which are heavier. The system lends itself to automation with auger and downpipe delivery from a bulk feed hopper situated outside the house. Some degree of feed intake restraint can be achieved by allowing only a small gap at the base of the hopper through which the feed can flow only at a limited rate. Feed intake can also be limited by increasing the number of pigs per hopper and raising the level of between-pig competition. *Ad libitum* self-feed hoppers have their trough length divided into individual pig feeder places by bars or solid divisions. These help the pigs to stand close together, but feed individually. Usually, one place is allowed for every three pigs in the pen (or more for larger pens). Feed intake encouragement is more problematic and predisposes to loss of feed from the trough at the base of the hopper and consequent wastage. The trough may also be readily contaminated with excreta if not regularly inspected and cleaned.

Floor feeding systems dispense with any trough, and deliver dry feed (pellets or meal) directly on to the floor of the pen. This can be done either by hand or by an overhead auger and hopper system. In both cases the amount delivered can be controlled and the pigs' daily feed allowance allocated as required. The floor must be clean and dry, and the feed presented frequently (4–8 times) throughout the day to ensure its rapid clearance. The advantages of floor feeding systems are their cheapness and the saving of both equipment (trough expenses) and pen space (allowing another pig or two per pen). The disadvantages are that the wastage is high, the house atmosphere tends to become dusty and it is difficult to arrange for all pigs to eat their fair share so that between pig variability in growth rate tends to be increased.

Feed given damp will be more readily eaten than that given dry, and wastage will also be reduced. *Ad libitum* feeders (both single space – one space per every 8–12 pigs, and multispace – one space per every three pigs) can be provided with nipple drinkers over the trough. This mechanism will allow pigs to drink and eat at the same time, and to mix their feed with water.

Full-blown wet-feeding systems deliver the feed into the trough at dry matter contents ranging from 15 to 25%. It is self-evident that for meal feeding all the pigs in the pen must eat simultaneously. This is usually arranged by providing for each pig in the pen a trough length of 150mm (young growers) to 300mm (pigs up to 100kg). Feed may be placed into the trough in dry form, to have water added on top, or it may be mixed at some more central point and then pumped. Central wet mixing and pumping is a highly flexible system which lends itself to the utilization of wet feed sources such as milk products, brewery waste, vegetables and by-products from the human food industry. Feed ingredients are compounded according to required nutrient specification and a pumpable wet mix is created. Here the mix may be held until pumped out to the chosen pig-house site.

The feed allowance to each pen within the house can be controlled manually or automatically. For manual systems there may be a tap controlling the flow of wet feed from the main pipe into the trough. This tap may be turned open for a given time consistent with the delivery of a known quantity of dry feed, or consistent with what the pigs are known to be able to eat up within 30 minutes or so of having been fed. Feeding is usually twice daily. Computer-controlled delivery allows each tap to be set at the central control point, and to deliver a given amount of feed to any particular trough (pig pen). Scale feeding by weekly increment is thus simple and automatic, and adjustable down to the level of an individual pen of pigs. Automatic systems can of course deliver the feed as frequently as is wished. Pipeline wet feeding systems have many advantages in addition to those of freedom of ingredient choice and infinite diet formulation alternatives. The system can be used either to restrict feed intake (where this is deemed necessary to avoid overfatness) or to encourage feed intake where growth rate is to be maximized. Given full access to the trough, a palatable feed, a dry matter above 18%, and at least twice-daily feeding, delivery of wet feed up to the level of appetite (that which the pigs will clear up) will achieve daily nutrient intakes considerably in excess (up to 10–20%) of dry *ad libitum* self-feed hopper systems.

Lactating sows

Individual feeding is essential to maximize intake of lactating sows and prevent extremes of fatness and emaciation as may rapidly occur when lactating sows are housed and fed as a group. Dry feeding, twice daily, of meal, pellets or nuts is common – delivered by hand or auger and pipe to a trough situated at the front of the individual feeder or sow farrowing crate. Feed allowance control is an essential part of encouraging intake without oversupply of feed and creating – by provision of excess – inappetence. Because the voluntary feed intake of lactating sows is so highly variable it is difficult to achieve automatic control and manual delivery remains the optimum system. Wet feeding will increase feed intake (as will three times daily feeding) and this may be achieved by delivery from pipeline, or by the simple expedient of adding water on to the top of the dry feed once placed into the trough. Sow feed troughs need to be capacious to hold 3–4kg of feed plus water added at a ratio of about 2:1.

Pregnant sows

Stalled sows may be provided with an individual allowance in a trough about 500mm wide placed to the fore of the pen. As there is no impulsion to maximize intake, dry-feeding systems are common. Feed may be delivered by hand or automatically by auger into temporary containers above each sow trough. These may be adjusted to allow delivery of variable quantities and the provision of a different allowance to each sow – which facilitates allocation of individual feed supplies. As pregnant sows are often fed only once daily, and they may be especially hungry before the daily feed becomes due, noise can be reduced if all the pigs are fed instantaneously and simultaneously. Accumulation of the feed allowance in a small (1.5–3.5kg) container above the trough allows the activation of a release mechanism to feed at the same time all the sows in any given row of stalls. Alternatively, all the sows can be given a single flat-rate level of feeding, those needing extra being given it by a manual top-up. Stall-housed pregnant sows can also be fed through wet-feeding systems; but where the trough is not divided between each stall, care must be taken to ensure equality of supply.

Alternatives to tethers and stalls for pregnant sow housing have led to a variety of novel feed delivery alternatives. But it was, at least in part, because previous feeding systems were found not to satisfactorily supply the individual nutrient needs of pregnant sows, that first the individual feeder, and next the sow stall, came into being.

Pregnant sows can be fed by provision of cobs or nuts directly on to the ground or into the straw bedding. Groups of sows fed in this way tend to become variable in body condition due to between-sow aggression. It is difficult to make individual adjustments to feeding levels other than through the creation of smaller groups of sows drawn out for special treatment. The same aggression in loose-housed sow groups may also bring about a further reduction in reproduction efficiency due to physical damage.

Groups of sows in yards may each be provided with an individual feeder. Sows entering these may be fastened in, and there can receive (and consume) their due feed provision. This system remains the most effective way of ensuring the individual feeding of sows kept in groups. Automation may be achieved with an auger and a holding container above each feeder. But as the occupant of each stall is not known before entry, a flat rate must be allowed, and individual rationing achieved by a subsequent manual top-up. Volunteer feeding stalls in which the sows are not fastened suffer from problems of fast eaters exiting and attacking the rear of the slower eaters. This problem may be alleviated by 'trickle feed' delivery. In this latter case all the sows are fed at once when feed is trickled via an auger and downpipe system to a trough at the front of each voluntary stall. Pellets are delivered at the rate of 100g per minute over a period of 10–15 minutes, twice daily. The slowness of the delivery ensures that all the pigs remain at a single feeding place for the period of the feeding session – thus ensuring equality of provision and prevention of between-sow aggression. All sows receive the same flat-rate ration. Some individual feeding stalls into which sows needing special provision can be fastened may be a useful adjunct to a trickle-feeding system.

Individual identification of sows by means of an electronic tag, and the provision of a central electronically controlled sow feeder, appear to offer the best of all worlds. Sows may be kept in groups of 20–40 in yards (preferably strawed). The sows can be allocated, through the computer control system, any individual level of daily feed allowance as may be required. Thin sows can therefore be readily given a more generous allowance and fatter sows restricted. The level of daily allowance can be adjusted according to the stage of pregnancy. Changes in diet nutrient density (and thereby the required level of feed provision for all of the sows) can be automatically accommodated through the computer control panel. The computer printout will give information on how much feed each sow has eaten in the course of every day. Upon visiting the feeding station the sow is identified and the allocation, specific to herself, is dropped into the trough for her to eat. The drop may be only one each day, or it may be divided over a number of possible visits to the feeding station. Appropriate entry and exit mechanisms to the station can prevent other pigs attacking whilst one is feeding.

Electronic sow feeders, though an excellent feed delivery system, are unlikely to be the universal method of feeding pregnant sows for they are not without disadvantages. Training is needed before a sow can join a group. A constantly altering group composition predisposes to agonistic behaviours, and this is particularly stressful for sows newly entering. There can be high levels of between-sow encounters. Pigs may spend a significant amount of time queuing for access to the feeder. During this queuing time sows are hungry, impatient and aggressive – vulva biting being an outward sign. Sows may queue and visit the feeder having already eaten their full allocation for the day. Sows may wait for long periods in the feeder preventing access by others. Boss sows may occlude the entrance to the feeder and discourage sows lower in the social order from visiting. The feeder station and associated equipment can be expensive to buy and requires sophisticated mechanical and computer maintenance. The system benefits greatly if group size is relatively low, ample space is provided per sow and there is a copious allowance of forage (or straw) in addition to the concentrate feed, so that hunger can be alleviated and excessive aggressive encounters avoided.

Housing layouts

Accommodation for pigs must be laid out in such a way as to facilitate control of temperature and ventilation, to control disease, to allow adequate space, to deal with waste, and to deliver feed. Means toward these ends are legion. It is remarkable, given the few definitive requirements of pig accommodation, how often the building actually constructed fails to meet one or other of the basic tenets (although perhaps meeting less important – if fashionable – criteria).

Effective pig housing is most likely to result from local experience and expertise, together with rigorous application of basic principles and knowledge of successful designs operating in other regions and countries. Examples of some house layouts which have been found to operate effectively are shown in Figures 17.7–17.37, measurements are given in metres.

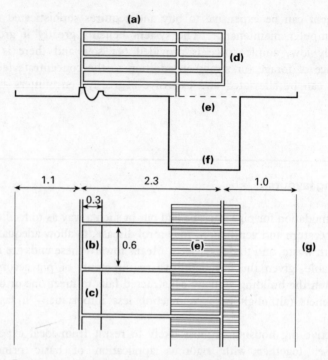

Figure 17.7 Pregnant sow stalls: (**a**) metal framework to confine newly weaned and pregnant sows (access to water is usually through a nipple (**b**) located over the feed trough – if there is spillage the pipeline may be opened for limited periods only), (**c**) feed trough, (**d**) gate to rear of stall to confine the animal or a tether may be attached to the neck or girth of the sow and fastened to the floor or the metalwork, (**e**) slats through which excreta falls into a slurry chamber (**f**). Boar pens are also located in the house (**g**), to which sows may be walked to be mated, or from which boars may be walked to the sows. Straw is not provided for bedding but may be given as a forage into the feed trough.

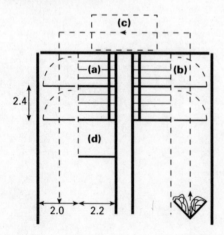

Figure 17.8 Loose housing for pregnant sows: (**a**) covered kennels providing both individual lying and feeding places for pregnant sows held in batches of four (access to feeders is through a hinged lid in the kennel roof – the kennel may or may not be strawed), (**b**) yard from which excreta may be scraped automatically into a slurry tank (**c**). Sections may be arranged in every third position (**d**) for the boar or for sows in need of special care. Water is best provided by nipple drinkers in the yard to avoid spillage over the lying area.

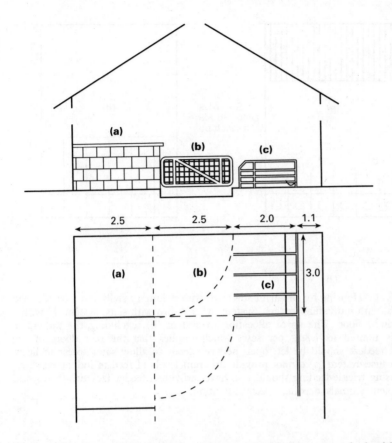

2.5   2.5   2.0   1.1

Figure 17.9 Loose housing for pregnant sows: (a) deep straw bed with covered kennel over, (b) yard, which may be strawed and the excreta removed as a solid, or left unbedded and scraped off, (c) individual feeding stalls. Each section is designed to hold five animals. Water bowls or nipple drinkers may be placed in the yard. If sows are moved here immediately after weaning an adjacent pen may be given over to house the boar. The same design may be used for a multisuckling system where sows are moved at 7–10 days post-farrowing to suckle their litter in communal groups of five sows through to weaning at 28 or 35 days.

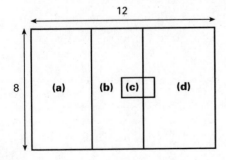

Figure 17.10 Strawed yards with electronic sow feeding station, suitable for 40 sows: (a) covered kennel area, (b) strawed lying area, (c) electronic transponder sow feeder with rear entry and forward exit, (d) dunging and queuing area.

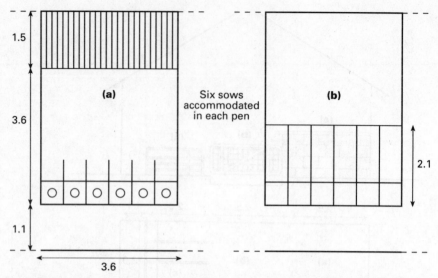

Figure 17.11 Housing for pregnant sows: (a) with voluntary stalls and a trickle-feed system, (b) with individual feeders. System (a) is shown with slats, system (b) with a solid, strawed floor. The space allocation in system (b), excluding the individual feeders, is limited to 1.8m² per sow, which implies that the rear doors of the individual feeders should be left open between feeds to allow sows access to lie in them. Because system (a) cannot provide different levels of feed to individual sows, systems using trickle-feeding should also have individual feeder facilities (b) to deal with problem animals needing special attention.

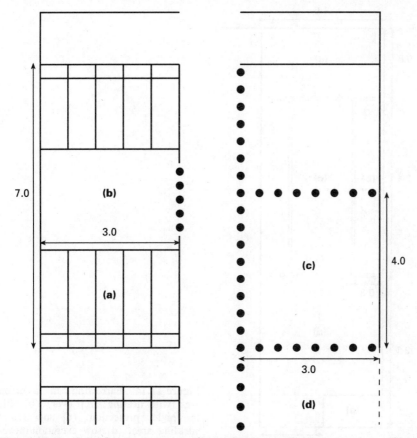

Figure 17.12 Solid floor arrangement for mating house allowing easy access of boars to sows, adequate mating areas and the possibility of stalled or group penning: (a) stalls, (b) boar working area or group pen, (c) boar/mating pen, (d) group pen for sows or boar pen.

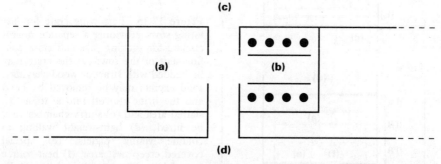

Figure 17.13 Facilities for washing, weighing and measuring fat depth of sows: (a) working office area, (b) crate and wash area, (c) farrowing house, (d) gestation/mating house.

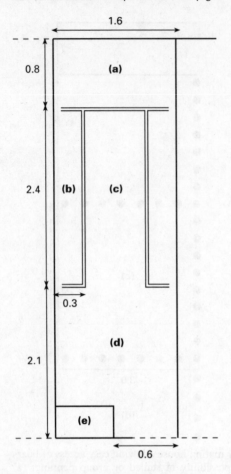

Figure 17.14 Farrowing pen: (**a**) creep area (with heating lamp over), (**b**) rails for piglet protection, (**c**) sow lying/suckling area, (**d**) sow excretion/movement area, (**e**) trough.

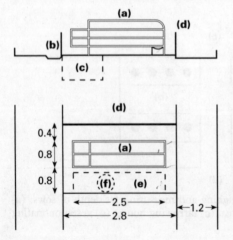

Figure 17.15 Farrowing crate for lactating sows providing a separate microclimate for sucking pigs and close confinement of the sow: (**a**) the crate may be bedded with straw or wood shavings, solid excreta may be removed by hand and the urine run off into a drain (**b**), slatted area and (**c**) slurry chamber may be fitted, (**d**) light-weight walling to confine young piglets, (**e**) special covered creep nest area, (**f**) heat source for piglets. Water and feed would be provided in the creep area for young pigs from 14 days of age.

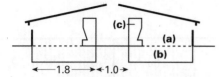

Figure 17.16 Intensive housing for newly weaned pigs: (**a**) the floor is entirely of mesh, slotted metal or slats, with a slurry pit below (**b**). No bedding can be provided. *Ad libitum* feed hoppers (**c**) form the front of the pen and water troughs or nipple drinkers are provided at the rear. Pigs are usually housed in batches of 8–16. Pens are 1m wide.

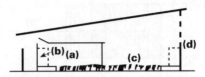

Figure 17.17 Traditional straw yards suitable for young weaned pigs or growers (particularly to 50–70kg): (**a**) deep-strawed kennel which may also contain the *ad libitum* hopper, set on a small plinth (**b**). Alternatively, or in addition, the feed hopper could be situated in the yard (**c**) with access from the front (**d**). Water is provided next to the feed troughs. Excreta are removed from the yard with the straw. Pigs are housed in groups of 15–30.

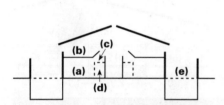

Figure 17.18 House for growing pigs: (**a**) solid floor lying area which could be bedded with straw or shavings, (**b**) kennel roof provided over lying area with hinged panel allowing access (**c**) to *ad libitum* hoppers (**d**). Pigs excrete onto a slatted (or mesh) floor area (**e**) which may be outside as shown, or covered. Watering points are usually situated over the slats.

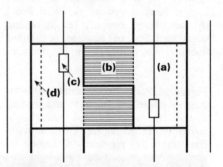

Figure 17.19 Pens for growing pigs: (**a**) with a solid-floor lying area and (**b**) with a slatted area for excretion. Animals may be floor-fed automatically from hoppers (**c**) held above, or troughs (**d**) could be provided for pipeline feeding.

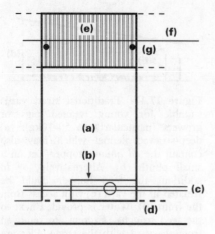

Figure 17.20 Part-slatted grower/finisher pen with automatic *ad libitum* hopper feeding system: (a) solid floor, (b) *ad libitum* hopper, (c) automatic auger feed line, (d) passage, (e) slatted floor, (f) water line, (g) nipple drinker.

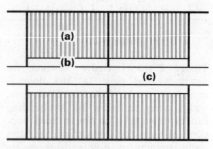

Figure 17.21 Pens for growing pigs: (a) fully slatted floor, (b) feed trough suitable for pipeline wet feeding, (c) central feeding and access passage. Each pen may contain 8–20 pigs. Slurry is pumped out from below the slats.

Figure 17.22 Straw-based grower/finisher house with pipeline wet feeding system: (a) straw-bedded kennel, (b) movement/excretion area (with or without straw), (c) pipeline, (d) trough.

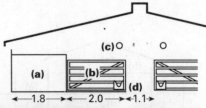

Figure 17.23 Pregnant sows group-housed in a deep straw-bedded yard.

Figure 17.24 Electronic sow feeder available to group-housed sows lying in straw.

Figure 17.25 Group housing for sows. The sows are protected from the heat within but may come out into the yard to excrete, drink and eat from individual feeders. Each pen is for six sows, and the space allowance is generous.

Figure 17.26 Fully slatted group pens for pregnant sows.

Figure 17.27 Farrowing crate for lactating sow.

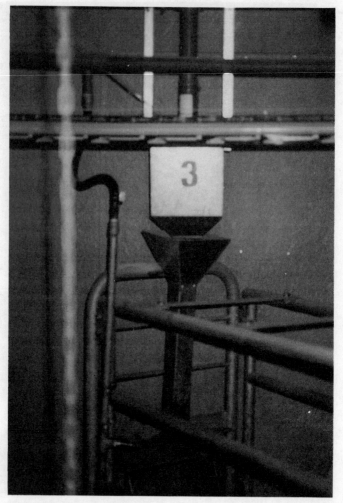

Figure 17.28 Metering device for measuring ration allowance for lactating sow held in a farrowing crate. The container is replenished by an auger system twice daily.

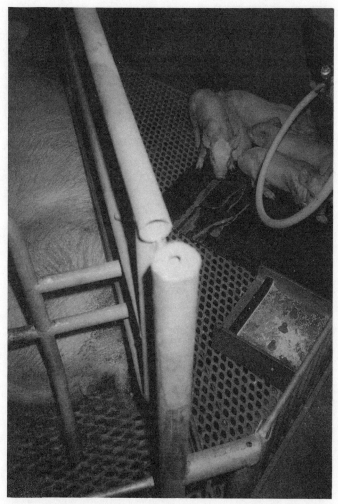

Figure 17.29 Fully slatted, plastic-coated, metal suspended flooring for lactating sow and litter. Note the solid pad for piglets to lie on (with heater over) and the flat dish for early provision of supplementary creep feed.

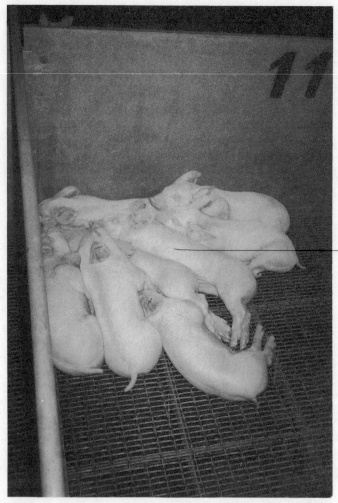

Figure 17.30 Newly weaned piglets on suspended wire mesh floor demonstrating that they are cold.

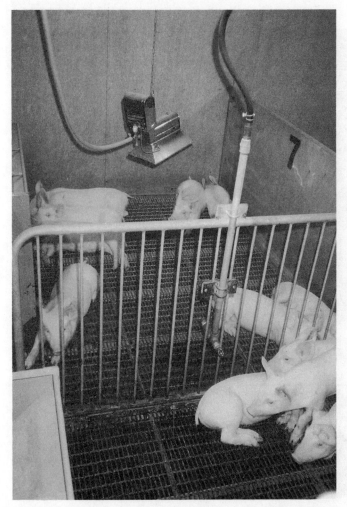

Figure 17.31 Newly weaned piglets on suspended wire mesh floor and provided with a heater over.

Figure 17.32 Growing pigs on solid floor with slatted area for excretion to the rear. The trough is positioned down the side of the pen for automatic liquid feeding.

Figure 17.33 Growing pigs on solid flooring provided with some straw bedding, scraped/solid floor passage to the rear for excretion, four-space self-feed *ad libitum* hopper replenished by auger and downpipe.

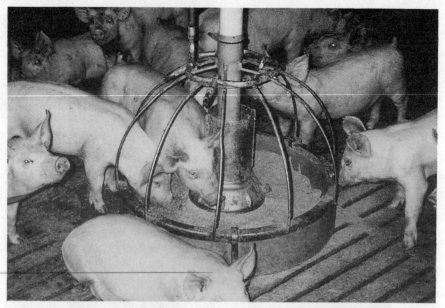

Figure 17.34 Young growing pigs on fully slatted floor with self-feeder trough provided with water nipples above. The feed is delivered into the trough through the central downpipe. Water is allowed to flow on to the feed, upon activation by the pigs. There are eight spaces provided around the circular feed bowl. The pen was designed to hold 16–18 pigs.

Figure 17.35 Growing pigs on solid floor pens with access to outside runs. Feed available to appetite in simple troughs replenished twice daily through an auger and downpipe system.

Figure 17.36 Electronic feeder for growing pigs. This system allows the recording of individual feed intake of pigs housed in groups, and it is especially appropriate for nucleus breeding herds pursuing genetic improvement objectives where it is helpful to know individual candidate boar feed intake, in addition to growth rate and fatness.

Figure 17.37 Young gilts newly introduced to an outdoor system. The simple arcs are placed in pairs in each paddock. Note also bulk feed hoppers and office area to the rear.

CHAPTER 18

PRODUCTION PERFORMANCE MONITORING

Introduction

The objectives of records to monitor production performance are:

1. to inform as to the present status of the production unit and the individual pigs upon it;
2. to inform as to the past performance of the production unit and the individual pigs upon it;
3. to provide a basis for production management, for diagnoses and for problem-solving;
4. to provide factual data to allow considered forward policy planning.

Amongst the particular problems of any recording scheme are those associated with individual animal identification, with the physical exertion of taking the record, and with the difficulty in retrieving the information when it is needed from the place where it was put (be it on paper or in a computer).

Records – on the pig unit

Management information should be held, for as long as it may be needed, near its point of use. The expected date of parturition is essential knowledge required about a pregnant sow so that she can be moved timeously to farrowing quarters. This may be written on a card which is associated with the pen in which the sow resides. Similarly, the prospective date of weaning is essential knowledge required about a lactating sow. After weaning, the placement of the sow in the mating pen adjacent to the boars is evidence of her empty and unproductive state. Upon mating, the next piece of relevant information is the prospective return date should she not have conceived and, for those sows which do return to oestrus, the number of times this has happened.

Weight at weaning is useful information but, for management purposes, not essential for all litters. The expected and achieved dwell-time of pigs in the weaner, grower and finisher pens is essential information for smooth management and throughput control but, equally important, this statistic indicates the growth rate which is being achieved. Dwell-time in pens may be readily calculated from dates and weights, in and out.

Signs of ill-health, inappetence and, of course, veterinary treatments must be noted at first sight and recorded at the pen side. It is especially important to register all routine veterinary procedures and inoculations, such as anthelmintic administration and parvovirus, *E. coli* or erysipelas vaccination.

All such records, and indeed many more, can also be entered into paper recording systems and carried to the office. Office-based recording schemes can be used to deliver, on a daily basis, management information which can be taken each morning on to the production unit and acted upon. Paper records in the office, however, are always best used as an adjunct to, and never in place of, the record that can be read face-to-face with the pig it concerns.

Naive records

Much information can be derived from simple and readily available records:

- A description of the stock type can inform as to potential growth rate, quality of carcass and reproductive performance.
- The returns from the slaughterhouse/meat packer will inform about actual weight at finish, predisposition to fat, variability of end-weight, and variability of carcass quality.

- The date of weaning together with the date of dispatch will inform about growth rate. Failing even this, the number of pigs sold per year, together with the number of pig places available on the grow-out unit, can do the same.
- The number of pigs marketed per year, together with the number of breeding sows on the unit, will inform about the productivity of the breeding herd as well as mortality in the grow-out unit.
- The total annual tonnage of sow feed purchased, together with breeding-herd size, informs about sow feed usage.
- The total annual tonnage of grower/finisher feed purchased, together with the number of pigs sold to the slaughterhouse/meat packer, informs about feed-conversion efficiency.
- A visual appraisal of the sucking pigs at weaning informs about lactation performance.
- A visual appraisal of the stocking density in pens informs about the likely performance of these pigs and about management efficiency.
- A visual appraisal of the growing pig, at the time of movement from the weaner house to the grower pens, informs about disease incidence and post-weaning growth.
- A visual appraisal of the condition (and the variation in condition) of the growing pig at the point of dispatch, informs about growth, disease and likely carcass quality.
- A visual appraisal of the condition (and the variation in condition) of the breeding sows at the points of weaning and farrowing informs about nutrition and feeding management.

A great deal of intelligence about a pig unit can be gained from (a) simply looking and counting and (b) referring to gross records such as total feed deliveries and total feed dispatches. The following illuminates the possibilities:

A pig farm, with both a breeding and a grow-out unit, records only the feed purchased for the pigs. The unit brings in extra pigs transported from a neighbouring producer at 28 days of age, and these enter the weaner accommodation on the grow-out unit. From the meat packer's grade returns it is apparent that (a) the average sale weight is 100kg, (b) 80% of pigs receive top grade price and (c) 10,500 pigs are sold for slaughter annually. Feed deliveries weekly are: sow feed deliveries 12t, starter feed deliveries 2.5t, grower feed deliveries 12.5t, finisher feed deliveries 50t. Piglets are weaned at 4 weeks of age (26–30 days) and pigs are moved from the weaner house to the grower house at 20kg and from the grower to the finisher house at 60kg. Sows are moved from the mating pens to the pregnant sow house immediately after they are mated for the last time and are no longer in standing heat. Sows go from pregnant sow house to farrowing house 7 days before the date they are due to farrow.

The pigs found present on the farm are as follows:

Total number of breeding sows either pregnant, lactating or awaiting mating	500
Sows awaiting dispatch as culls	20
Maiden gilts	60
Boars	35
Sows in pregnant sow house	350
Sows in mating pens	80
Sows in farrowing pens	70
Piglets in farrowing pens	690
Pigs in weaner houses	1,323
Pigs in grower houses	1,614
Pigs in finisher houses	1,909

From this basic evidence, the performance of the breeder and the feeder herd can be calculated and compared with expected performance, as may be perceived from Tables 18.1 and 18.2. On the right-hand side of Table 18.1 are presented the actualities pertaining to the unit in terms of those parameters which are immediately determinable on site, together with those that can be calculated. To the left of the table are given the expected statistics which might reasonably have been presumed. Expectations and actualities for the grow-out unit are similarly presented in Table 18.2.

Records – in the office

Paper records form a vital part of management control: they are the permanent record of the day-to-day activities and performances of particular pigs or groups of pigs. It is easy to gather an excess of paper records of the wrong sort and an insufficiency of the right sort. A wrong record is one that

Table 18.1 The sow breeding herd. Calculations of expectations and determination of actualities

Expectation	Actuality
Sows: 500	Sows: 500
Sow to be culled: 17 per month	Culls: 20
Maiden gilts: 17 per month × 2 = 35	Maiden gilts: 60

Sows in pregnant sow, mating and farrowing pens, if 2.25 litters/sow/year are expected: 365/2.25 = 162

115 − 7	= 108	66.7%
28 + 7	= 35	21.6%
19	= 19	11.7%
	162	100

per 100 sows expect:
  67 pregnant sow places
  22 farrowing places
  12 mating pen places

per 500 sows expect:

335 in pregnant sow places	Sows in pregnant sow house 350
110 in farrowing pens	Sows in farrowing pens 70
60 in mating pens	Sows in mating pens 80

Expectation		Actuality
Percent of time in pregnancy pens	67%	Percent of time in pregnancy pens 350/500 = 70%
Percent of time in farrowing pens	22%	Percent of time in farrowing pens 70/500 = 14%
Percent of time in mating pens	12%	Percent of time in mating pens 80/500 = 16%

Expectation	Actuality

Days in pregnancy pens
  0.67 × 365 = 245 days
	Days in pregnancy pens
	  0.70 × 365 = 256 days

Days in farrowing pens
  0.22 × 365 = 80 days
	Days in farrowing pens
	  0.14 × 365 = 51 days

Days in mating pens
  0.12 × 365 = 44 days
	Days in mating pens
	  0.16 × 365 = 58 days

If sows spend 28 + 7 = 35 days per litter in farrowing pens, then 80/35 = 2.25 litters/sow/year

If sows spend 28 + 7 = 35 days per in farrowing pens, then 51/35 = 1.5 litters/sow/year

If pregnant days are 256, then expected number of pregnancies is 256/(115 − 7) = 2.4 per year. As only 1.5 are achieved, then non-pregnant sows must be in the pregnant sow places

Sows should be in mating pens for 13 days or about 13 × 2.3 = 30 days per year, not 58. So either sows are not being mated or they are not coming on heat

(*contd.*)

Table 18.1 (continued)

Expectation	Actuality
Piglets per sucking litter 10	Piglets in farrowing pens/sows in farrowing pens = 690/70 = 9.9
Sow feed usage ≃ 1.3t/adult/year	Total adult pigs on unit = 500 + 20 + 60 + 35 = 615
	12t sow feed per week × 52/615 ≃ 1t/adult/year

is never used (for example, individual sow records of litter size when only 10% of all culls occur because of low litter size). A correct record is one which addresses a particular problem pertinent to the unit concerned (for example, days between weaning and conception for a herd selling less than 22 piglets per sow per year).

Records for selection

Breeders selecting for improved performance need information on which to base that selection. As breeding is a long-term and disciplined effort, the records need to be comprehensive. Breeding for faster growth and leaner carcasses can only be achieved if days to slaughter, weight and fat depth are carefully recorded for all candidate animals for selection within the various breeding lines. This sort of recording does not usually present any question of principle – whatever criteria are appropriate to measure, are measured. Problems that arise are more often of logistics and cost. In performance tests for males an estimate of feed-conversion efficiency might be thought necessary. This means recording the feed intake of each pig on each day. Recording of sow breeding performance, if comprehensively done to include birth weight, litter size, litter weight, weaning weight, lifetime performance, breeding regularity and so on, also consumes time and effort.

Commercial records

It is not sensible to mount complex recording schemes for commercial herds. Apart from the costs involved, there is a major problem of individual identification of large numbers of animals. For most producers, recording should be kept to a minimum and probably utilize a computer-based system.

There are two outstandingly vital statistics for the commercial herd:

Table 18.2 The grow-out unit. Calculations of expectations and determination of actualities

Expectations	Actualities
	Total slaughter pigs sold annually = 10,500
	With assumed 5% loss in grow-out unit, total pigs entering unit annually = 10,710
	Pigs in weaner houses 1,323 27% Pigs in grower houses 1,614 33% Pigs in finishing houses 1,909 39% Total growing pigs 4,846 100%
7.5–100kg; 2.6 batches/year	10,710/4,846 = 2.21 batches/year
28-day weaning weight = 7.5kg	If weaner weight is taken as 7.5kg, then 100 − 7.5 = 92.5kg of growth is achieved on the grow-out unit
365/2.6 = 140 days per batch	365/2.21 = 165 days per batch 165 days for 92.5kg = 0.561kg daily live weight gain
Given 140 days per batch, then:	Given 165 days per batch, then:
7.5–20kg at 0.43kg/day = 30 days in weaner house	27% of 165 = 45 days for 12.5kg gain in the weaner house (20 − 7.5) = 0.28kg daily gain
20–60kg at 0.67kg/day = 60 days in grower house	33% of 165 = 55 days for 40kg gain in the grower house (60 − 20) = 0.73kg daily gain
60–100kg at 0.80kg/day = 50 days in finisher house	39% of 165 = 65 days for 40kg gain in the finisher house (100 − 60) = 0.62kg daily gain
	Showing greatest problems in the weaner house, and next in the finisher house
Age at 100kg = 140 + 28 = 168 days	Age at 100kg = 165 + 28 = 193 days
	Feed usage 50 + 12.5 + 2.5 = 65t/week = 3,380t/year
	Total pig gain 92.5kg × 10,500 pigs = 971.3t
Feed conversion ratio dependent upon diet density, but less than 3.0	Feed conversion ratio = 3,380/971.3 = 3.48

- Pigs sold annually per breeding female. This encapsulates numbers born, rebreeding regularity and mortality. It is calculated from the knowledge of number of pigs leaving the farm divided by the total number of breeding females in the herd.

- Feed used per pig sold. This is an overall feed conversion factor, and measures the efficiency of utilization of the major raw material in the production process. It is calculated from the total tonnage of feed transferred into the unit, per month or per year, divided by the total number of pigs leaving the unit over the same period.

Compendium efficiency measures will indicate current performance levels and show changes in productivity, but cannot identify the causes of either of these. For diagnosis, closer examination needs to be made of specific aspects of performance. Pigs sold per breeding female may decrease owing to reduction in the numbers born, a lengthening of the breeding-to-weaning interval, an increase in mortality, a high proportion of unmated herd replacements or a slowing down in growth rate. Inefficiencies of feed use might be in the growing or breeding herd, and might be caused by wastage, cold, pig quality, too high a level of feeding, too low a level of feeding or feed quality.

Targets

Pig producers are helped by knowing how they are performing in comparison with their peers. In this regard, centralized computer-based recording schemes can provide seminal information of the type shown in Table 18.3, which refers to one particular system used in the UK. Other similar conflated statistics may relate to reasons for culling sows (Table 18.4), to feed ingredient usages (Table 18.5) (particularly useful to those compounding feeds) or to the breakdown of the financial structure of the production unit (Table 18.6).

More specific data may come from particular example herds and serve to give encouragement through the medium of demonstrating that excellence is achievable (Table 18.7).

From such databases as these more generalized targets can be set. Some

Table 18.3 National UK average pig statistics[1] for herds despatching at 90kg live weight (per 100 sows in herd)

Stocks		*Monthly entries*	
Sows[2]	100	Gilts into the breeding herd	4
Maiden gilts	7.1[3]	Boars into the breeding herd	0.25
Sucklers	153		
Boars	5.3	*Monthly sales*	
		Average live weight at sale (kg)	89
Sow performance		Number of pigs sold	177
Farrowing %[4]	83		
Born alive per litter	10.9	*Monthly exits*	
Total born per litter	11.7	Culled boars	0.25
% mortality birth to weaning	10.7	Culled sows	3.5
Reared per litter	9.7	Deaths	0.5
Weaned per sow per year	22.3		
Sold per sow per year	21.2	*Feeder herd performance*	
No. of sows served monthly[6]	23	Stocks (7–89kg)	707
Successful services (%)	85	Batches per year (7–89kg)[5]	3.0
Litters per sow per year	2.3	Feed/gain (feeder herd)	2.4
Empty days per year[8]	44	Feed/gain (total breeder and feeder	
Weaning weight (kg)	6.5	herd)	3.0
Age at weaning (days)	25	Weaned per sow per year	22.3
		Mortality (%)	5.2
Annual financial performance per 100 sows		Daily gain (kg) from birth	0.60
Pigs sold per productive sow	21.2	Total feed used (kg monthly)[7]	48,660
Sales of sows and boars (£)	5,970		
Sales of finished pigs[9] (£)	142,000		
Purchases of boars and gilts (£)	9,900		
Increase in valuation	800		
Feed costs (£):			
sow and boar	18,500		
piglet	440		
growing/finishing	68,700		
Variable costs (£):			
vet./med.	2,600		
transport	1,180		
power and water	3,750		
bedding	490		
others	770		
Fixed costs (£):			
labour	13,500		
contract	570		
repairs and maintenance	4,250		
rent and leasing	1,020		
depreciation and interest	4,030		
others	4,320		

[1] Drawn from Easicare Pig Management Year Book.
[2] Average herd size in 100 herd sample = 350.

Table 18.3 (continued)

[3] Maiden gilts reside in the breeding herd for an average of 45 days between entry and service.

[4] Number farrowed/number mated in equivalent period 4 months earlier.

[5] 148 days to despatch; less 25 days to 7kg; ([365/123] = 3.0).

[6] Average number of ejaculations per sow served is 2.2.

[7] 5.840t per sow per year (of which 1.2t is used for the sow) for the production of 21.2 pigs sold at an average live weight of 89kg.

[8] Empty days = days between weaning and oestrus plus days lost through mated sows returning to oestrus.

[9] 2,120 pigs of 67kg DW and 100p/kg.

Table 18.4 Reasons for culling sows

Reason for culling	Age at culling (years)	Percentage of total culls
Old age	4	10
Litter size	2.5	10
Infertility: including anoestrus and failure to breed after two returns to oestrus	2	35
Legs, feet and associated locomotor problems	2	25
Others	2	20

Table 18.5 UK pig feed ingredient usage (1991) (%)

Wheat	33
Barley	12
Maize	1
Wheat feed	14
Grain by-products (barley and maize)	3
Bakery by-products (wheat and maize)	1
Soya bean meal	17
Rape-seed meal	2
Beans and peas	4
Meat-and-bone meal	2
Fish meal	3
Molasses	4
Oils and fats	2
Others	2

Table 18.6 Example financial analyses (as percentage of total expenditures)

Breeding-herd production of 30kg weaner pig		Grow-out unit 30–100kg live weight	
Total expenditures	100		100
Total income	110		110
Expenditures as percentage of total:			
Feed for sows and boars	30	Purchase of 30kg weaners	50
Feed for piglets to 30kg	30	Feed	35
Veterinary	4		1
Power	4		2
Others	2		2
(Total variable costs	70)	(Total variable costs	90)
Labour	18		4
Buildings and equipment	10		5
Others	2		1
(Total fixed costs	30)	(Total fixed costs	10)

such average targets are listed in Table 18.8. But even targets of this sort may be inappropriate when:

- the unit is highly sophisticated and already has performances as good as the average target, even although above-average performance is essential for the unit to be profitable;

- the target is so far in advance of the current performance of the unit that its attainment is not considered credible by the unit staff;

- the individual unit does not fit in comfortably with the average production systems or objectives, perhaps having widely different ages at weaning, feed quality, house types, breeding stocks etc.

The best targets may often be derived internally to each particular unit. Current performances can be identified and an achievable percentage added on. This will make for a gradual step-wise progression upwards. Individual aspects of production efficiency can be singled out one-by-one for special attention. Current performance levels can be compared to targets by means of histograms, graphs or charts. These aids to management – serving to encourage the staff of the unit on to greater things – must, of course, reside in the clear view of that staff. Progression towards the target should be monitored and, when reached, rewarded.

Table 18.7 Six-parity performances of a well-managed breeding herd – 1990

	Parity						
	1	2	3	4	5	6	All
Number born alive	10.1	10.3	11.5	12.2	11.3	11.4	11.0
Born dead	0.14	0.28	0.23	0.79	1.37	1.90	0.51
Piglet birth weight (kg)	1.45	1.47	1.45	1.40	1.37	1.24	1.43
Number weaned	9.2	9.8	10.1	10.6	10.0	9.9	9.8
Average weaning weight (kg)	6.56	7.17	6.98	6.83	6.66	6.19	6.82
Days to weaning	22	21	21	22	21	21	21
Mortality (%) (to weaning)	9.2	5.5	12.1	13.5	12.3	12.7	10.7
Lactation feed (kg)	110	119	122	137	137	127	122
Condition score (farrowing)*	3.67	3.64	3.56	3.96	3.56	4.31	3.70
Condition score (weaning)*	2.95	2.98	2.90	3.13	3.14	3.53	3.02
Sow weight (kg) (farrowing)	196	230	251	262	267	285	236
Sow weight (kg) (weaning)	174	207	226	238	245	263	212
Backfat (mm) (farrowing)	23.8	21.2	20.8	20.5	18.3	18.3	20.9
Backfat (mm) (weaning)	17.9	16.6	15.7	15.2	14.9	15.1	16.0

*5-point scale

Table 18.8 Example performance targets. Values relate to 100 breeding females (sows + mated gilts)

Maiden gilts	5–7
Sows served weekly	4–6
Boar ejaculations weekly	3–5
Matings per sow per oestrus	>2
Number of boars, including young boars	>5
Sows farrowing/sows mated	>0.8
Sows anoestrous for up to 14 days post-weaning	<5
Farrowings per week	4–5
Litters per breeding female per year	>2.2
Pigs born per litter	>10
Pigs weaned per litter	>9
Pigs weaned per month	>180
Birth weight of pigs (kg)	>1.2
21-day weight (kg)	>6
28-day weight (kg)	>8.5
Interval from weaning to mating (days)	3–5
Interval from weaning to conception (days)	<12
Number of farrowings per month	18–21
Pig sales yearly	>1,900
Pig sales monthly	>170
Breeding herd replacements per quarter year	<12
Feed used per breeding female per year (t)	1.1–1.4
Total feed usage by the breeding herd per month (t)	10–11
Feed used by growing pigs yearly (t)	<550
Feed used by growing pigs monthly (t)	<45
Feed used by each growing pig in one month (kg)	<50
Age at 100kg (days)	<160
Feed conversion efficiency to 100kg using medium-nutrient-density diet	<2.8
Feed conversion efficiency to 100kg using high-nutrient-density diet	<2.4
Mortality post-weaning to slaughter (%)	1–2
Growth rate from 5 to 15kg live weight (g/day)	>350
Growth rate from 15 to 50kg live weight (g/day)	>600
Growth rate from 50 to 100kg live weight (g/day)	>800

Routine control

Management checks are best designed around the individual characteristics of the staff, buildings, problems to be solved and production processes. On large units even the vital statistics are hard to keep up with: whether or not a female is pregnant, when a litter is due for weaning, how often a male has been used, how old is a pen of growing pigs.

Maintaining control is simplified if information is incorporated into the physical structure of the unit; for example, a special area of the house set aside for females not yet pregnant, weaning always on the same day of the week, marking pens of weaners with expected moving-on date, pens of growers with expected slaughter date and so on.

Day-by-day tactical management requires both a computerized (or paper) recording system and close day-by-day observation of the current physical status of all the animals on the unit. The tables which follow are presented as examples of some of the control points which management may feel essential to the day-to-day routine of the production unit (tables 18.9, 18.10 and 18.11).

Table 18.9 Example management points for farrowing and lactating sows and sucking pigs

Observe	Consider
Sows	
Days to parturition	Move to maternity ward after 108th day of pregnancy.
Cleanliness of quarters	Disinfect and rest for 7 days between pigs.
Cleanliness of pig	Wash and scrub, treat with dressing against internal and external parasites and mange.
Temperature of house	Different requirements of lactating sows and baby piglets.
Welfare of pig	Correct pen arrangements.
Onset of parturition	Cost-benefit of attending the birth.
Post-parturient problems	Lack of milk for baby pigs, high temperature (above 40°C) fever in mother.
Availability of water	Drinkers, regularly checked.
Days to weaning, reproductive performance of female	Record data in preparation for decisions to move, cull or remate.
Disease	Routine injections.
Sucking Pigs	
Completion of routine tasks within 3 days of birth	Administration of iron compound. Clipping teeth. Ear mark (to record at least the week of birth).
Completion of routine tasks within 15 days of birth	Castrate if necessary. Provide highly palatable creep feed diet for baby pigs; keep fresh by twice-daily provision and attention to cleanliness of trough. Water supply to baby pigs.
Signs of diarrhoea	Treatment. Improve standards of cleanliness.
Environment	Bedding, temperature, draughts.
Growth performance	Reductions in suckled milk supply can be offset by increased intake of supplementary creep feed.

Table 18.10 Example management points for boars, sows awaiting mating and pregnant sows

Observe	Consider
Boars	
Size	Feed allowance. Age.
Willingness to mate	Age. Health. Feed allowance. Frequency of use.
Numbers born	Frequency of use. Use of 2 males on same female.
Disposition	If aggressive, check staff, check breed. Check water supply.
Sows	
Days between weaning and conception	Feed allowance, feed specification, disease, management. Culling policy. Housing type.
Body condition	Feed allowance in lactation.
Body condition	Pregnancy feed allowance.
Confirmation of pregnancy	Check with males 18–25 days after previous matings. Use pregnancy diagnosis equipment.
House temperature	Use max./min. thermometer. Draughts. Behaviour of animals.
Welfare and restlessness	Environment; space allowance. Concentration of diet, provision of roughage, feed allowance.
Cleanliness	Penning arrangements.
Physical abrasions	Penning arrangements. Group antagonisms.
Time since mating	Movement to maternity quarters.
Feed and water supply	Automation.
Disease	Routine vaccinations.

Table 18.11 Example management points for weaned and growing pigs

Observe	Consider
Adequacy of environment	Provision of plenty of space but not so the animals are cold. Possibilities of deep, clean straw. Housing quality.
Diarrhoea	Medication of feed.
Feed provision	Supply of more feeding space and fresher, more palatable feed.
Growth rates	Temperature, respiratory disease levels, intestinal disease, welfare, feeding arrangements, size of group, stocking density, water provision.
External and internal parasites	Dress and dose.
Cannibalism	Environment, bedding material, diet.
Feeding arrangements	Adequate space.
Slow growth, variable growth rates within pens	Disease. Too frequent mixing and moving. Ration or diet. Stocking density. Environment. Disease. Water.
Weight for slaughter	Check weight.

CHAPTER 19

SIMULATION MODELLING

Introduction

Modelling fulfils a primary scientific need: that of placing within the grasp of the human mind a description of the world – biological and economic – that would otherwise be too complex to be comprehensible. The multitude of interactions which make up the holistic pig production process can only be understood through analysis of their reduced component parts; but this is inadequate from both the scientific and the industrial standpoint, neither of which is interested only in parts. Modelling allows the components to be reassembled, and the whole to be studied. The creation of metaphors, paradigms, algorithms and models is, of course, not new as scientific devices to help explain complex phenomena; they are basic to scientific method. What is novel is the role of the computer in the process. This allows many layers of quantitative and qualitative descriptors of actions, reactions and interactions to be synthesized. The model thereby creates a system that is similar in scale to the industrial process itself (that is, pig production), thereby facilitating effective simulation. Simulation offers prediction; prediction offers planning and control.

Simulation modelling is a satisfactory scientific endeavour in its own right, involving the creation and validation of understanding through the development and testing of hypotheses. But, equally as important, once an effective simulation model has been created it becomes a highly useful part of production management.

Development of quantitative response prediction simulation models

The ultimate drive for response prediction modelling is – through the application of science – to increase the profitability of production. Production managers must decide upon production strategies that will give the greatest profit within the wider context of the business environment. Only if the biological and monetary outcomes of a change in production practice are known can a sensible decision be made as to what optimum production practice might be. .

Broadly, responses to changes in management practice can be predicted by either one of two methods. The first is on the basis of historical precedence – previous experience leading to the conclusion that a certain action will result in a certain response. Unfortunately, in pig production this is not particularly useful, as both production and financial circumstances change rapidly. The second approach requires an understanding of the causal forces of responses. If the nature of the driving mechanisms for a response is understood, then responses may be predicted. It is this approach that leads to the building of simulation models. Models can represent the individual production unit, and allow themselves to be manipulated in a similar way as would occur to the unit in real life. If the model is good, then the simulated responses and outcomes will mirror the happenings that are likely to occur in reality. By this means managers may reject courses of action that are considered likely to be counter-productive and may select appropriate courses of action from amongst various suggested feasible and winning solutions.

Response prediction is most effective when the causes of the responses are relatively few and the system abides by laws which are relatively simple. Causal forces are fewest, and operational activities closest to simple laws, at the lower, more fundamental levels of biology. Consequently applied experimentation, being at the higher biological levels, is not the best route to the creation of algorithms for applied response prediction. It is better to use data relating to the lower layers of the active elements of the system, and to operate underneath the levels of the interactions.

Production managers previously making decisions on the basis of a blueprint or historical approach to husbandry practice are likely to experience an inversion of their preconceptions when they move to decision-making on the basis of response prediction. What were previously assumed to be given rules (nutrient requirements, carcass quality targets and the like) are now variables open to manipulation. Success is no longer measured in terms of closeness of fit between biological production targets and biological production achievements, but now in terms of profitability. Methods of production previously seen as ends are now placed where they should always have been – into the category of means. For example, the blueprint approach to grow-out production begins with a definition of recommended nutrient requirement. Diets are mixed to a given specification and presented to the pigs in a given amount. Such rules have been delineated to provide for pigs of a predetermined quality. The production manager works on the assumption that the given feed composition is optimal and that if the target of pig quality is successfully met, then the production system has been well-controlled and profit will automatically follow. This is not true. On the contrary, the pig-production manager requires, first and last, to target profit as the criterion of success. The nutritional specification of the feed, and the feed allowance, now become flexible variants open to manipulation, and their optimal definition becomes dependent not on any given recommendation but on the specific circumstances of the production unit and its entrepreneur. The concept of nutrient requirement moves from the domain of the nutritional biochemist to the domain of the nutritional economist. The task of the nutritionist is to predict the outcome of changes in nutrient specification of the diet. The actual specification used should be variable, not given.

Pig producers are well aware of the complexities caused by interrelationships between the factors impinging upon pig growth and sow reproduction. For example, increased feeding levels accelerate growth rate and increase throughput, but they also reduce carcass quality. Maternal fatty tissue is a fine source of nutrient for the support of lactation, but subsequent fertility may be impaired if it is used in profligate amounts. The cost-benefit of increasing the feed allowance to growing pigs or breeding sows is a profound calculation which cannot be completed until the façets of the production process are linked interactively. Doing only one part of the calculation (such as the relationship between feed intake and fatness) can lead to erroneous conclusions. Because of interactions, the manager must be aware simultaneously and quantitatively of all the factors impinging upon the production process. The provision of production recommendations on the basis of intermediate goals, such as growth rate, feed-conversion efficiency, carcass quality and sow weight change, is inadequate from the manager's point of view. By the by, it is also an unsatisfactory way of conveying the sum of scientific knowledge that is available.

To achieve the goal of margin maximization, the best possible route must be found through the whole production system, not through merely one of its component parts. That optimum route will change with changing production circumstances. Appropriate feed allowances and diet specifications, for example, are likely to vary year by year.

By simulating the production process, models can allow the manager to make both short-term (tactical) and long-term (strategic) decisions on the basis of improved information – and with proportionately reduced risk. The financial and production consequences of alternations in management practices are deserving of equal consideration, but the bottom line of a management model will always be financial rather than biological. A model does not need to come to a single optimum solution; rather, it is better if it lays out the options that are open to a decision-maker and gives guidance as to the likely outcome of taking those options. The main purpose of a model is to show the direction and magnitude of the response and the sensitivity of the production system to a tactical or strategic change. For example, altering from *ad libitum* to restricted feeding might be shown by a growth model to reduce growth rate from 900g to 750g/day, and to decrease backfat thickness from 15mm to 14mm. What is important is not so much that the simulated growth rate is shown to be identical to that in real life, but rather that (a) both growth rate and fatness were reduced, (b) growth rate was reduced to a substantially greater degree than fatness and (c) these effects are open to being assessed in terms of their financial consequences.

Nevertheless it is helpful if a model can be tuned in as closely as possible to the current production situation. The simulation should give results similar to, but not identical with, those found in real life. This is because responses do change according to their position on the scale (Figure 19.1). For example, growth rate is more sensitive to change in feeding regimes at lower growth rates than at higher growth rates; 1mm of backfat depth would be more important for pigs near the upper limit of fatness.

In general, therefore, the results from a simulation model should reflect reasonably closely the real-life situation being simulated. However, exact replication should not be looked for, nor is exactitude a necessary prerequisite for the model to fulfil its purpose.

A model might be referred to by the decision-maker when:

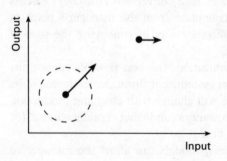

Figure 19.1 The model should initiate from a position within range of the existing real-life circumstances; i.e. within the broken-line boundary. An exact overlay is rarely achieved and not required; it is only necessary for the model to be started from a similar point to that perceived in real life. However, if the starting point for the model differs widely from that in real life, the magnitude and direction of the response can be wrongly predicted.

- production is to be optimized, but it is not presently known if that is so;
- performance being achieved is different from that required;
- a change in the production system is contemplated but the consequences of the change are not clear;
- the circumstances of production change;
- the market to which the product is directed changes;
- the markets supplying input to the pig unit change.

A model might be referred to by the pig technician or scientist when:

- experiments are to be planned;
- short-falls in knowledge require to be identified;
- there is a need for a repository of information to collate knowledge from disparate sources;
- biological phenomena and/or their economic consequences are to be investigated;
- relationships between inputs and outputs require to be demonstrated.

Diagnosis and target setting

A simulation model predicts a future outcome. Given an adequate definition of inputs (especially feeding, pig type and environment) and an understanding of the relationships between inputs and outputs, a picture of the resultant response is derived. Model responses may thus serve as targets for the production process: a description of what level of performance is possible within any given set of defined circumstances. A mismatch between predicted outcome and achieved outcome may be due to: inadequacy of definition of inputs; the incursion of perturbing forces not accounted for in the model; or, indeed, errors in the model itself. But, most frequently, a difference between the predicted response and what is achieved in practice simply identifies, and measures quantitatively, a shortcoming in some aspect of management. In addition to indicating the costs of failing to optimize the production system, a model may thus also be used as a diagnostic tool to identify the most likely area of sensitivity which requires priority attention and resolution.

Information transfer

As well as to acting as a decision-making aid, simulation models may also be used to transfer information and identify gaps in knowledge. Computer simulation models can contain within them a full range of current scientific awareness, and bring together separated disciplines and experimental results. A model may serve as a library that does not require the books to be either read or understood – but merely used. A sophisticated methodology can by this means be made available to end-users who need not be concerned with the detail.

Short-circuiting the need to understand deep science may also open such models to misuse. The model builder will have had to make choices and to have included information of varying degrees of validity. Because models require information in a chain of events, the modeller must sometimes interpolate in order to fill gaps and the user may be (dangerously) unaware of such weaknesses. Nevertheless, whilst it is important that model builders are honest with model users about the weaknesses (and strengths) of their products, a model need not necessarily be exactly right in order to be helpful. It is sufficient merely that a new methodology is some improvement upon that which it supersedes.

There is little doubt that from now on computers will ensure that the vast proportion of information transfer between science and practitioners will be through the medium of simulation models and other similar computer packages. If the need to complete information strings is a shortcoming in a simulation model, this is an advantage inasmuch as the gaps in knowledge are identified and not ignored. One of the beauties of a model is that it requires quantification of production parameters which may only be known qualitatively. Researchers may be pointed towards areas of ignorance where further experimentation is needed. Sensitivity analysis of model components will throw light upon which aspects of a system require to be known accurately and which only approximately. It is sometimes the case that the pursuit of science may encourage ever more precise knowledge in areas where an approximation would do. On the other hand, matters of primary importance for investigation may (wrongly) be left on one side because they are perceived as being too difficult; or worse, scientifically unfashionable. Computer models can assist in organizing research resources to avoid the study of non-problems, and to prevent the refinement of measurements that do not need to be further refined.

Model-building approaches

Simulation models usually exercise themselves through the medium of computing, and are made up of algorithms comprising mixtures of text statements and mathematical equations. Relationships within models can range from empirical regression derived from feeding trial data (for example, live-weight gain on feed intake, litter size on parity number) to deductive sequences of mathematical descriptions relating to the fundamentals of biochemistry and physiology. Empirical regression tends to be static and inflexible, and strictly best used to find intermediate points within a given data set. There is no reason to believe that a regression relationship will necessarily throw light on the nature of the causes of the responses achieved.

Relationships linking inputs to outputs can be of three general types:

- the empirical linking of parameters of production with high statistical accuracy, but using constants and coefficients that are biologically nonsensical;
- the empirical linking of parameters of production through the use of constants and coefficients which are reconcilable with biological expectations;
- the linking of inputs and outputs through the medium of their causal forces; that is, by the employment of deduction.

For deductive methodology to be effective, a relatively complete knowledge of the system is required. Therefore, although more logical, the deductive approach is likely to contain a mixture of hard fact, soft fact, hypothesis and even inspired guesswork. Empiricism, on the other hand, is often an undeniable and exact expression of a past observation; but that expression may or may not be useful in terms of understanding or of predicting a future response.

Working models tend to be made up of mixtures of empirical and deductive relationships; but the greater the extent to which the model relies upon the deductive approach, the more likely it is to be truly informative. Deduction – building up input:output relationships from a knowledge of the causal forces – increases the flexibility of a model and allows prediction outside the circumstances within which the information was collected. Such a model can therefore account for a wide range of interactions. The same deductive approach also helps in the understanding of the system: it highlights ignorance and allows sensitivity analysis and definition of crucial components.

Deductive modellers should be judged on their ability to form hypotheses about the nature of life, because these will control the algorithms. Modelling is best pursued at the level of basic principles. Preconceived notions on the use of empirical analyses of feeding-trial data are likely to lead to less-effective simulations. Deductive models employ what is known of the science of pig production, and the necessary hypotheses reflect the nature of the science that goes into them. The deductive model thus goes further than merely redescribing phenomena previously observed. It avoids simplistic recounting of history and attempts to foresee the outcome of future activities. There remain nevertheless many phenomena that need to be described within a model but for which causal forces are not understood. In this event, relationships can only be approached in an empirical way. It is evident from preceding chapters that it is easier to be deductive about the growth responses to nutrition than about the equivalent reproductive responses. Of necessity, therefore, sow reproduction models are much more empirical in their structure than·growth models.

The empirical and deductive approach may be contrasted with the following example. It is accepted that as pigs grow bigger they become fatter. Regression relationships can be drawn up with fatness as a function of live weight. Using this relationship, increasing fatness may be predicted as a consequence of increasing weight. This empirical relationship leads to the conclusion that slaughter at a heavier weight will invariably result in a fatter carcass. On the other hand, to obtain a deductive function between fatness and live weight, the causative forces require to be identified. One reason why animals may get fatter as they get bigger is because their appetite increases, and the extra feed consumption has gone into the production of fat. This will only happen if the potential rate for lean growth has been reached. Therefore there are interactions between pig type, feeding level and the likelihood of fattening. By considering the causal forces, fatness is seen not to be a direct function of live weight itself but manipulable according to feed intake and pig type. At the extreme, there need be no relationship between fatness and live weight, animals of high lean-growth potential being able to eat to appetite and remain equally as thin at 100kg as they were at 20kg. The deductive and the empirical approaches present the production manager with quite different sets of propositions as to how pig carcasses may be made more lean and as to what the consequences of an increase in slaughter weight might be.

It is evident that the empirical statement of the growth response to feed intake shown in Figure 19.2, whilst being a wholly correct report of an experimental result, is quite inadequate for effective simulation of the type required by production managers. The deductive approach is shown in Figure 19.3 in the form of a flow diagram relating feed to growth. This second approach is more complicated, requires a deeper understanding of the relationships and is therefore more likely to simulate real-life responses under a wide range of circumstances, including those different from the original conditions under which data were collected. The flow diagram shows how information is required about the relationships in the system. So, in wishing to define growth response to feed intake, relationships between pairs of

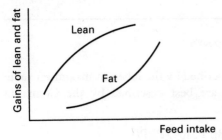

Figure 19.2 The relationship between feed intake and growth: an empirical approach.

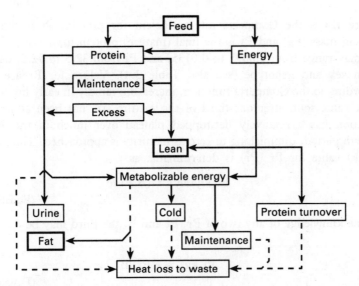

Figure 19.3 The relationship between feed intake and growth: a deductive approach.

factors such as feed and maintenance, feed and cold thermogenesis, feed and lean tissue growth potential, feed and fatty tissue production, must all be understood at the level of their causal forces.

A model forces its builder toward better understanding of the system being modelled; otherwise it will not work effectively. The scientist must describe a living system systematically, and ensure all the parts are interlinked in a way that properly reflects nature. Once built, however, good models may stand in the place of real-life experiments. Traditional nutritional response trials are no longer required now that the causal forces of these responses may be modelled effectively. Where possible, modelling approaches discussed here are deductive (deal stepwise with causal relationships in a biologically coherent manner), dynamic (respond sensitively to changes in inputs in a way reflective of real-life pig production) and deterministic (simulate response by arrival at a single quantitative prediction of an outcome).

575

Potential growth: limitations to growth

The limits to the rate of growth, as described by the inherent maximum value for daily protein retention (Pr̂, kg) are best described by the Gompertz function:

$$\text{Pr̂(kg/day)} = \text{Bp.Pt.log}_e(\text{Pt̂/Pt})$$

(Equation 19.1)

where Bp is the Gompertz rate-coefficient for protein, Pt is the present protein mass (kg) and Pt̂ is the final (mature) protein mass (kg). Values for Bp may range from 0.0095 to 0.0135, and Pt̂ from 32.5 to 52.5, dependent upon sex and genotype (see also Table 3.4). Values for Pr̂ are variable according to the Gompertz function, increasing rapidly in early life to reach a peak value soon after one-third of the mature size has been attained. The function has a relatively flat-topped plateau over much of the 20–120kg growth period, diminishing to zero as maturity is approached. The maximum (peak) value for Pr̂ (Pr̂̂) is determinable as:

$$\text{Pr̂̂(kg/day)} = (\text{B.Pt̂})/e$$

(Equation 19.2)

In the knowledge of any two of Pr̂̂, Pt̂ and B, the third may be determined, thus:

$$\text{Pt̂(kg)} = e(\text{Pr̂̂})/\text{B}$$

(Equation 19.3)

$$\text{B} = e(\text{Pr̂̂})/\text{Pt̂}$$

(Equation 19.4)

Fat growth may be perceived as having its own appropriate estimates for Lt̂ and B (mature lipid mass and the Gompertz coefficient) from which a value for Lr̂ (the maximum daily limit to lipid retention pertaining at any given point in growth) can be derived. However, such an approach is not appropriate for total body fat, which may often fulfil the role of body energy storage and the accommodation of excess dietary energy provision over daily need. It is probably realistic to suggest that no upper limit to lipid growth need be set independent of that imposed by appetite.

It is more reasonable to require in a model some view of a minimum level of lipid growth. Here it can be proposed that under normal conditions of positive growth, and when nutrient supply is insufficient to maximize protein growth (Pr<Pr̂), the pig will nevertheless wish to deposit some fatty tissue. This will set the minimum fatness for the animal. The proposition can be

framed in terms of a minimum ratio of lipid to protein in the daily live weight gain $(Lr:Pr)_{min}$ and may range from 0.4 to 1.2, depending on sex and genotype (Table 3.5). This ratio may depend to a small extent upon the size of the pig and its degree of maturity, thus:

$$(Lr:Pr)_{min} = (Lr:Pr)_{min} + b(Pt)$$

(Equation 19.5)

where b is presently an unknown small number.

Unless $(Lr:Pr)_{min}$ is particularly high and pigs are reaching slaughter weight with excess levels of fat and failing to grade satisfactorily, it is to be expected that under normal circumstances Lr will exceed $(Lr:Pr)_{min}$. Where $(Lr:Pr)_{min}$ is less than 0.9, and body composition at slaughter is such that Lt<Pt, it is unlikely that pigs would be considered overfat. This lower boundary to fat growth is significant to carcass quality, therefore, with pigs of fatty type (see Table 3.5) destined for meat markets which discriminate against fat.

In times of energy shortage, not only can pigs grow lean tissue without accumulating fatty tissue but lipid may actually be catabolized (Lr is negative) to support protein growth (Pr is positive). In these circumstances (of fatty tissue catabolism), it may be important to present in a model some view not of Lr:Pr (the ratio of lipid to protein in the growth) but of Lt:Pt (the ratio of lipid to protein in the total body mass). In the absence of better information, values for $(Lr:Pr)_{min}$ in Table 3.5 may also be used to approximate settings for $(Lt:Pt)_{min}$. It is notable in this respect that Lt:Pt>1 is considered necessary for effective reproduction in breeding sows.

Suffice it to say that an effective simulation is crucially dependent upon full and exact characterization of the pig which is to be simulated. The correct estimation of Pr and $(Lr:Pr)_{min}$ is essential for the modelling of growth, as these parameters describe the bounds within which the nutritional response will function.

Nutrient (energy and protein) yield from the diet

Energy

Where the digestible energy content of a diet (DE) is not known from direct measurement of the diet itself or of the ingredients, it can be postulated (Tables 8.4 and 8.5) that:

$$DE(MJ/kg\ DM) = 0.016\text{'Starch'}\ (g/kg) + 0.035Oil\ (g/kg) + 0.019Protein$$
$$(g/kg) + 0.001Fibre\ (g/kg)$$

(Equation 19.6)

which is probably a little generous in its assumption about the digestibility of the diet components. Fibre in this regard is not, of course, crude fibre (CF) but more properly the larger non-starch polysaccharide fraction, as may be represented by neutral detergent fibre (NDF). Empirical regressions offered in Chapter 8 have included the useful:

$$DE(MJ/kg\ DM) = 17.5 - 0.015NDF + 0.016Oil + 0.008CP - 0.033Ash$$

(Equation 19.7)

and

$$DE(MJ/kg\ DM) = 17.0 - 0.018NDF + 0.016Oil$$

(Equation 19.8)

where CP is crude protein. If CF has been determined rather than NDF, it is possible to use the rather less efficient:

$$DE(MJ/kg\ DM) = 17.0 + 0.011Oil - 0.041CF$$

(Equation 19.9)

Daily energy intake is:

$$DE(MJ/day) = n(DE[MJ/kg\ DM])(F[kg/day])$$

(Equation 19.10)

where n is proportion of dry matter in the air-dry feed and F is the air-dry feed intake. Digestible energy is an adequate descriptor of the energy value of a feedstuff but not of the ultimate yield of dietary energy to the metabolic processes of the pig. First, DE embraces the 23.6MJ/kg of energy contained within ileal-digested dietary protein moieties ($D_{il}CP$, kg) which are not immediately available for energy transfer, and indeed much of which will, hopefully, be used for protein growth (Pr) or milk synthesis, and therefore not available for energy production at all. Secondly, there is a small element of energy 'digested' but not 'captured' (such as in gases produced from hind-gut fermentation). An estimate of protein-free energy (Epf) can be determined as:

$$Epf = DE - (23.6D_{il}CP) - (0.05DE)$$

(Equation 19.11)

As only a part of the digested protein appears in growth as Pr, or in the milk, there will be some protein energy available from deamination. Where Qd represents the energy yield from protein deamination, then:

$$Qd = 23.6Pm - (7Pm + 5Pm) = 11.6Pm$$

(Equation 19.12)

where Pm is the protein deaminated and which assumes urine to carry 7MJ/ kg of protein deaminated, and the work energy needed for urea synthesis to be 5MJ/kg protein deaminated.

The total metabolizable energy (ME_t) available for use by the body of the pig may now be summed:

$$ME_t(MJ/day) = Qd + 23.6Pr + Epf$$

(Equation 19.13)

the term 23.6Pr representing that energy which, although metabolizable, is nevertheless deposited as retained body protein. The rate of protein deamination (Pm) is dependent first upon the level of supply of protein (possible oversupply) and next upon the quality of supply of protein (efficiency of transfer of digested protein to body protein). This parameter is best measured by difference:

$$Pm = DCP - Pr$$

(Equation 19.14)

that is, what has been digested but not retained must be excreted through the processes of deamination.

The available metabolizable energy (ME_t) may now be offered to the animal for its use to support the needs of maintenance, protein retention, lipid retention, the products of conception, lactation and cold thermogenesis.

Protein

The dietary content of protein may be determined from chemical analysis for nitrogen (CP = 6.25N). The digestibility of the CP is variable amongst feedstuffs and it is further apparent that determination of the ileal digestibility will avoid errors of hind-gut disappearance. Ileal digestibility of crude protein ranges from 0.3 to 0.9, as discussed in Chapter 9 and shown in Tables 9.1 and 9.5.

In addition to knowing the yield of ileal digested crude protein ($D_{il}CP$):

$$D_{il}CP(kg/day) = (D_{il})(CP[kg/kg \text{ diet}])(F[kg/day])$$

(Equation 19.15)

A model to simulate protein use requires knowledge of the ileal-digested amino acids and their balance, one to the other, in order to estimate V, the protein value. The total ideal protein available for metabolism (IP_t), and the support of maintenance, protein growth and reproduction, may be determined as:

$$IP_t(kg/day) = (D_{il})(CP)(F)(V)(v)$$

(Equation 19.16)

where v is the efficiency of use of ileal-digested ideal protein for body processes. While D_{il} may usually average around 0.75 for mixed pig diets, V may range from 0.80 for young pig diets to 0.65 for sow diets (see Tables 11.3 and 11.4), and v is usually around 0.85 (see also Chapter 11).

Nutritional requirement for energy

Figure 19.3 has described a flow-diagram guide to the general principles for the predictive simulation of growth response to nutrient input by means of deductive modelling.

Maintenance (E_m, MJ ME) can be quantified as:

$$E_m(MJ\ ME) = 1.75Pt^{0.75}$$

(Equation 19.17)

where Pt is the total protein mass at the given moment in time when the maintenance requirement is to be estimated.

Maintenance needs may be inflated by a cold thermogenesis response to an insufficient environmental temperature. The energy cost of cold thermogenesis (E_{h^1}) may range from 0.012 to 0.018MJ ME for each kilogram of metabolic body weight ($W^{0.75}$) for each degree of cold; the higher value for pigs housed singly, including sows:

$$E_{h^1} = 0.012W^{0.75}(Tc - Te)\ (pigs\ in\ groups)$$

(Equation 19.18a)

$$E_{h^1} = 0.018W^{0.75}(Tc - Te)\ (for\ individual\ pigs)$$

(Equation 19.18b)

thus estimates energy use in cold environments when Tc is the comfort temperature and Te is the effective temperature (°C). Comfort temperature is heavily dependent upon heat output (feed intake and rate of production) and has been expressed as:

$$Tc = 27 - 0.6H$$

(Equation 19.19)

for pigs of greater than 10kgW, where H is the total body heat output resulting from work associated with maintenance, growth, pregnancy and lactation. H is least for smaller pigs growing slowly, and greatest for larger pigs growing rapidly. H may be determined by difference as all digested energy not lost with urine, or as gas, must either be deposited in the body in the form of protein or lipid (or in the milk in the form of protein, lipid and milk lactose), or leave the body as heat:

$$H = DE - (23.6Pr + 39.3Lr + 5.4My + 7Pm)$$

(Equation 19.20)

where My is the milk yield (kg milk/day) in lactating sows (see later); or:

$$H = E_m + 31Pr + 14Lr + 2.3My + 5Pm$$

(Equation 19.21)

for the DE must be used up for body processes or lost in urine:

$$DE = E_m + 54.6Pr + 53.3Lr + 7.7My + 12Pm$$

(Equation 19.22)

These equations have (purposely) excluded the small variable losses of DE as gas.

Meanwhile the estimate of H, having allowed the calculation of Tc, now leaves only the determination of Te for the full costs of maintenance (E_m) and cold thermogenesis (E_{h^1}) to be accumulated. Te, the effective temperature, is calculated from the ambient temperature (T, °C) adjusted for the rate of air movement and degree of house insulation (Ve), and for the floor type upon which the pig must lie (Vl):

$$Te = T(Ve)(Vl)$$

(Equation 19.23)

Values for Ve and Vl are found in Table 11.1 and range between 0.6 and 1.4.

The likely rate of energy use for protein retention depends on the energy cost of protein retention (E_{Pr}) and the daily rate of protein retention (Pr). Pr is, of course, a function of the supply of nutrients (metabolizable energy and ideal protein) available to support it. The outer boundary for protein retention (Pr̂) has been described earlier. Taking the efficiency of conversion of energy for protein retention (k_{Pr}) as 0.44, and fixed rather than variable, then the simple equation:

$$E_{Pr} = 23.6Pr + 31Pr = 54.6Pr$$

(Equation 19.24)

expresses the energy cost of protein retention (E_{Pr}), 23.6MJ being deposited and 31 leaving as heat following the work of protein synthesis.

Taking the efficiency of conversion of diet ME to body lipid (k_{Lr}) as 0.75, again a simplification, then the equation:

$$E_{Lr} = 39.3Lr + 14Lr = 53.3Lr$$

(Equation 19.25)

expresses the energy cost of lipid retention (E_{Lr}), 39.3MJ being deposited in the fatty tissue and 14 leaving as heat.

The rate of energy use for lipid retention depends on the energy cost (E_{Lr}) and the daily rate (Lr). Lr is primarily a function of energy availability. The order of priority for calls upon available metabolizable energy (ME_t) is likely to be E_m, E_{h^1}, E_p, E_{My}, E_{Pr} and finally E_{Lr}. Once maintenance, cold thermogenesis and the target rate for protein retention (P$\hat{r}$, provided ideal protein is also in adequate supply) are satisfied, the remaining energy may be used for fat deposition. Thus in the growing animal:

$$Lr = (ME_t - [E_m + E_{Pr} + E_{h^1}]/53.3)$$

(Equation 19.26)

If P$\hat{r}$<Pr on account of a limitation of dietary ideal protein, then fatty deposition will take place even although the priority demand for Pr is not yet satisfied. If, however, Pr<P$\hat{r}$ on account of a limitation of dietary metabolizable energy, then fatty deposition will be curtailed to the lower limit set by (Lr:Pr)$_{min}$, as already discussed.

In summary, the available metabolizable energy (ME_t) which has been corrected for losses of energy as gases and in the urine, and for the heat of deamination, may be apportioned in the growing pig:

$$ME_t = E_m + E_{Pr} + E_{Lr} + E_{h^1}$$

(Equation 19.27)

The energy requirements for reproduction in the sow may be divided between maintenance ($E_m = 1.75Pt^{0.75}$), retention of energy in the foetal load, growth of mammary tissue and lactation.

$$E_u(MJ/day) = 0.107e^{0.027t}$$

(Equation 19.28)

describes the rate of energy deposition in the uterus (E_u), where t is the day of pregnancy. In like manner:

$$E_{mamm}(MJ/day) = 0.115e^{0.016t}$$

(Equation 19.29)

describes the rate of energy deposition in the mammary gland (E_{mamm}). As both k_u and k_{mamm} are around 0.5, then the total cost of pregnancy (E_p) is:

$$E_p = (E_u + E_{mamm})/0.5$$

(Equation 19.30)

Values for E_p are trivial in early pregnancy but at day 110 E_p totals 5.5MJ ME/day.

Milk yield (My) in the lactating sow will be a function of the needs of the litter according to the equation:

$$My(kg/day) = (L_{tg})4.1$$

(Equation 19.31)

where L_{tg} is the total litter daily live-weight gain (kg) (or daily gain per piglet × number of piglets).

A limit to yield requires to be set according to the potential of the sow. The boundary to the lactation curve of a sow may be described as:

$$M\hat{y}(kg/day) = a.e^{-0.025t}.e^{-e^{[0.5-0.1(t)]}}$$

(Equation 19.32)

where t is the day of lactation and a varies according to sow type, and may range between 18 and 30.

The energy value of milk is 5.4MJ/kg and $k_l = 0.7$, thus the energy cost of milk production (E_{My}) is:

$$E_{My}(MJ\ ME/day) = (My)(7.7)$$

(Equation 19.33)

In the absence of adequate energy supply from the diet, maternal body lipid may be made available with an efficiency of 0.85, which is to say that 1kg of maternal body lipid may be used to create $(39.3 × 0.85)/5.4 = 6.2$kg milk.

In extremis maternal body protein may be used to provide energy for milk synthesis; the efficiency in this case is probably no greater than 0.5.

Nutritional requirement for protein

The total nutrient yield of protein from the diet, expressed in terms of the total available ideal protein (IP_t):

$$IP_t(kg/day) = (D_{il}CP)(F)(V)(v)$$

(Equation 19.34)

is used for purposes of maintenance (IP_m), protein retention in growth (IP_{Pr}), protein retention in the uterus (IP_u) and protein deposition in milk (IP_l):

$$IP_t = IP_m + IP_{Pr} + IP_u + IP_l$$

(Equation 19.35)

Because IP_t is constructed from knowledge of the ileal digestibility and the value v for crude protein as a whole, there is an (unnecessary) approximation when the ileal digestibility (D_{il}), and perhaps the value v, is known for individual amino acids (aa) in individual feedstuffs. IP_t would be better constructed on the basis of the nine essential amino acids appearing at tissue level in utilizable form:

$$Utilizable\ aa_{(1-9)} = (D_{il}aa_{(1-9)})(vaa_{(1-9)})$$

(Equation 19.36)

These may be summed, and the total aa spectrum assessed for the value V by comparison with the ideal amino acid spectrum for the activities of protein maintenance and protein production.

Maintenance protein requirement (IP_m) may be given as:

$$IP_m(kg/day) = 0.0040Pt$$

(Equation 19.37)

Ideal protein required for production is that actually used, as the factors V and v have accounted already for all inefficiencies:

$$IP_{Pr} = Pr$$

(Equation 19.38)

and in the growing animal all available ideal protein will go to protein retention:

$$Pr = IP_t - IP_m$$

(Equation 19.39)

until $Pr = P\hat{r}$.

Protein deposition in the products of conception (Pr_u) proceeds at an exponential rate consequent upon the day of pregnancy:

$$Pr_u(kg/day) = 0.0036e^{0.026t}$$

(Equation 19.40)

and

$$IP_u = Pr_u$$

(Equation 19.41)

Likewise for deposition in the mammary tissue (Pr_{mamm}):

$$Pr_{mamm}(kg/day) = 0.000038e^{0.059t}$$

(Equation 19.42)

and

$$IP_{mamm} = Pr_{mamm}$$

(Equation 19.43)

Both Pr_u and Pr_{mamm} are ultimately lost from the body, Pr_u at parturition and Pr_{mamm} (in the urine) after weaning.

Protein secreted in milk may be determined as:

$$IP_l = My \times 0.055 = Pr_l$$

(Equation 19.44)

there being 55g protein per kilogram of milk.

In the absence of adequate protein supply from the diet, maternal body protein may be made available with an efficiency which may be expected to be around 0.85.

Feed intake

Where pigs are rationed to a given allowance of feed, the (known) value for feed intake (F, kg/day) can be simply programmed into a model. This position is often normal for pregnant and lactating sows, and is becoming more likely for growing pigs with the increasing popularity of fully automated, computer-controlled systems which deliver carefully measured and predetermined amounts of feed, either to individual pigs or to pens of pigs. Feed levels may be incremented weekly with varying degrees of restriction, as befits optimum tactics. Automatic feeding systems may be used to feed pigs to appetite; that is, offer as much as the pigs will eat, and while so doing also measure feed intake. Pigs fed through automatic wet-feeding systems delivering predetermined meal sizes twice or three times daily will usually consume more feed than pigs given continual access to dry feed in an *ad libitum* hopper.

None the less, many pigs are fed *ad libitum* and no measurement of feed intake is made. As feed intake is a crucial force in the growth response, where feed intake is not known its accurate prediction is essential.

Usually young pigs feed *ad libitum* between 5 and 50kg, and often up to 100kg live weight or more. After weaning, 3-week old piglets may eat no more than 50g daily, increasing their intake by about 30g daily until 20kg live weight. If weaning is delayed until 4 weeks of age, initial consumption will be nearer 100g daily, and the rate of increase will be about 40g daily. By 8 weeks and 20kg live weight, expected feed intake will be in the region of 1.2kg per pig daily. Subsequently the amount of feed consumed may be calculated in various possible ways.

On-farm determination

Because of the extremes of variation in feed intake between pig types, farms and environments, and because of the great importance of the correct estimation of the value F, there may often be merit in undertaking specific investigations to measure the parameter on-farm. This can be done by spot checks on the daily feed intake of sample pens of pigs of different weights, and a feed intake curve derived. Or, more crudely, the total feed supply to the various categories of pig (weaners, growers, finishers) over a given period may be used to construct a simple curve for daily feed use which is consistent with the overall usage rate.

From calculation of perceived nutrient (energy) needs

The notion that a pig will eat what it needs to fuel the nutrient requirements of maintenance, cold thermogenesis, growth, pregnancy and lactation leads to the proposition that:

$$F(kg/day) = (E_m + E_{h^1} + E_{Pr} + E_{Lr} + E_p + E_{My})/ME$$

(Equation 19.45)

where ME is the energy concentration of the feed (MJ ME/kg, or 0.96DE, as fed). This proposition clearly signals enhanced appetite to result from larger body size (maintenance), a cool environment (cold thermogenesis), a high rate of growth (protein and lipid retention), the state of pregnancy (particularly in the latter stages, and as opposed to barrenness) and (markedly) lactation.

Whilst target values for Pr are accessible through determination of $P\hat{r}$ (Equation 19.1), and minimum values for Lr may be estimated from $(Lr:Pr)_{min}$, a target value for the likely maximum value of Lr is more problematic. Whilst the view of Lr as a storage medium, absorbing excess of energy consumed over the direct needs of maintenance and growth, is satisfactory for dealing with a given energy intake, it is not satisfactory to predict energy intake. Some feeling is therefore needed of the Lr:Pr ratio desired by pigs who may eat as much as they wish, including for the satisfaction of the desire to fill body storage tissues with fat. For present purposes it might be assumed that $(Lr:Pr)_{max}$ is dependent on sex (being greater for the castrate than for the entire male, and intermediate for the female) and genotype (being greater for the unimproved), and range between the extremes of two (improved entire) and four (unimproved castrate).

From knowledge of gut capacity limits

Calculation from perceived nutrient needs represents the upper limit to feed intake as controlled by the metabolism of the animal. This boundary may need to be curtailed if the bulk of the daily feed intake is beyond the physical capacity of the gut, which may be determined as:

$$F(kg/day) = 0.013W/(1 - \text{Digestibility coefficient})$$

(Equation 19.46)

Diet digestibility may vary from 0.90 for milk and 0.80 for baby pig diets to 0.70 for finisher diets, and even down to 0.65 for diets given to pregnant sows. Where diet digestibility is not known, it might be estimated. The gross energy of a diet is likely to be related to its composition:

$$GE(MJ/kg) = 39.3\text{lipid}[kg/kg] + 23.6\text{protein}[kg/kg] + 17.5(\text{starch+fibre}[kg/kg])$$

(Equation 19.47)

while the DE(MJ) is similarly calculable (Equations 19.7, 19.8 and 19.9). The ratio between DE and GE is, of course, the digestibility coefficient.

Taking the purist (deductive) view of feed intake, it may be proposed that within the limits of gut capacity the pig will eat enough feed to satisfy its perceived nutrient needs; that is, Equation 19.45 pertains provided its solution for F(kg/day) is less than that derived from Equation 19.46. In the event of that not being the case, Equation 19.46 applies. This position requires modification in order to account for the many factors other than needs and capacity which impinge upon appetite; not least environmental temperature, stocking density, management and disease. The latter two cannot be accommodated by deduction, although some empirical 'gearing' could be attempted, particularly in the light of data from on-farm determination of feed intakes actually achieved. It is evident – and a major problem in feed intake prediction – that inadequate feeder space provision, ill-considered self-feeder flow control, the number of pigs in the pen, the wholesomeness of the diet and a whole host of other factors under direct management control can make a nonsense of attempts to model feed intake on the basis of pig biology. Expression of the former two (temperature and space) can be attempted, however.

Influence of environmental temperature

Because the intake of feed results in heat losses to the environment, a hot environment which reduces the rate of heat losses from the body will induce a negative feedback which will reduce appetite:

$$\text{Feed reduction (g/day)} = W(Te - Tc)$$

(Equation 19.48)

or, more logically, in terms of digestible energy (DE) at about 14MJ/kg feed:

$$\text{Reduction in DE (MJ/day)} = W(Te - Tc).0.014$$

(Equation 19.49)

where W is the live weight, Te is the effective environmental temperature (Equation 19.23) and Tc is the comfort temperature (Equation 19.19).

Influence of stocking density

The area (m^2) occupied by a pig may be expressed in terms of the value k in the equation:

$$\text{Area occupied (m}^2) = kW^{0.67}$$

(Equation 19.50)

Usually k will be around 0.05, which gives 1m^2 per 100kgW. A reduction in k of 0.01 below 0.05 (i.e. k = 0.04) may be associated with a fall in feed intake of the order of 8%.
  Where:

$$k^1 = 0.05 - k$$

(Equation 19.51)

then:

$$F(kg/day) = F(kg/day) - 8(F)k^1$$

(Equation 19.52)

the reduction in k being represented by k^1 and Equation 19.52 giving the final (adjusted) food intake.

Empirical estimates

Measurement of feed intakes of pigs of various weights, and in various experimental and practical conditions, has yielded a plethora of empirical estimates for F. As has been indicated earlier, estimation of the feed intake of young weaned pigs is problematic due to the great influence of management factors at this time. Most estimates therefore relate to pigs growing positively and of 10kgW or greater.

The equations:

$$DE \text{ intake (MJ/day)} = 3.0W^{0.63}$$

(Equation 19.53)

and the rather better:

$$DE \text{ intake (MJ/day)} = 55(1 - e^{-0.0204W})$$

(Equation 19.54)

serve as empirical measures of intake according to nutrient need. Where DE is 14MJ/kg diet:

$$F(kg/day) \simeq 0.13W^{0.75}$$

(Equation 19.55)

The use of metabolic body weight may be naive but it is practical, $0.14W^{0.75}$ being a realistic view of appetite limits.

Under commercial conditions, achieved feed intakes are usually lower than the limits described above, often:

$$F(kg/day) = aW^{0.75}$$

(Equation 19.56)

where a ranges between 0.09 and 0.11, or:

$$DE \text{ intake (MJ/day)} = 2.4W^{0.63}$$

(Equation 19.57)

It may be helpful to adjust these estimates for sex, as castrated pigs may eat 5–15% more than entire males and females.

Breeding sows

Pregnant sows are usually fed a restricted ration, and this quantity would normally be known, obviating the need for prediction. Nevertheless calculation of feed intake according to nutrient need (Equation 19.45) limited by gut capacity (Equation 19.46) is probably also quite effective for breeding sows. The feed intake of pregnant sows is likely to be influenced by their body fatness. The drive towards fat accumulation may be accommodated adequately through the equation calculating F according to nutrient need (Equation 19.45), within which the value for E_{Lr} may be quite high. A pregnant sow may have an Lt:Pt ratio of only around 1:1 but be seeking an Lr:Pr ratio of 3:1 or more in order to achieve a satisfactory Lt:Pt in the final body mass for reproductive processes (probably Lt:Pt = 1.5–2; or Lt = $1.1Pt^{1.1}$).

It is also often the case that lactating sows are fed a restricted ration allowance, and their feed intake will be known and not require prediction. But lactating sows should be allowed to express their appetite if lactation yield is to be maximized. Failure of lactating sows to reach daily feed intakes suggested by Equations 19.45, 19.46 and 19.55 (that is, $0.13W^{0.75}$) may be due to their being in their first parity (all other things being equal, gilts may consume up to 1kg/day less than multiparous sows), and (more likely) due to the influence of heat.

In general:

$$F(kg/day) = (0.013W/(1 - Digestibility\ coefficient)) - (W[Te - Tc]/1000)$$

$$(Equation\ 19.58)$$

may prove a useful guide to the upper limit to lactating sow appetite.

A short route to nutrient need may be through knowledge of litter size:
$$F(kg/day) = 0.033W^{0.75} + 0.7(n)$$

$$(Equation\ 19.59)$$

where n is the number of piglets in the litter and $0.033W^{0.75}$ serves to estimate sow maintenance. In the event, sows may often achieve no more than 0.4kg increase in feed intake for each piglet in the litter, suggesting a more pragmatic and practical:

$$F(kg/day) = 0.033W^{0.75} + 0.4(n)$$

$$(Equation\ 19.60)$$

Equation 19.45 (nutrient need) may therefore often be irrelevant in the face of tissue catabolism to support lactation demand. The influence of management factors on lactation feed intake is well known and substantial (Chapter 13).

Feed intake in lactation may also require adjustment for pregnancy feed intake or sow body fatness at parturition. Sows better fed in pregnancy and fatter at term have lower lactation feed intakes:

$$F(kg/day) = F - 0.13(P2 - 15)$$

$$(Equation\ 19.61)$$

where Equation 19.61 gives the adjusted daily feed intake in lactation, and P2 is the depth of subcutaneous fat (mm) measured at the P2 site at the time of parturition.

Composition and growth of the live body of growing pigs

Simulation of growth and its component parts requires characterization of the live weight (W) and composition at the start, which may be immediately after weaning or at some subsequent point. Young pigs may contain:

$$\text{Ash (At)} = 0.03W \text{ (or, } 0.20Pt)$$

(Equation 19.62)

$$\text{Protein (Pt)} = 0.17W \text{ to } 0.15W$$

(Equation 19.63)

$$\text{Fat (Lt)} = 0.06W \text{ to } 0.18W$$

(Equation 19.64)

$$\text{Water (Yt)} = 0.74W \text{ to } 0.64W$$

(Equation 19.65)

where W is the live weight and >5kg. Unless negative growth is to be modelled, it is assumed that the growth to be predicted is positive in general nature. Pigs which have lost fat (especially after weaning) will have higher values for Yt and lower values for Lt; an Lt value of <0.08W is indicative of a thin pig, an Lt value of >0.20W is indicative of a fat pig. At the point of weaning, pigs will usually contain 0.16W of protein and 0.16W of lipid (Lt = Pt).

Given a satisfactory prediction of the daily rate of protein retention (Pr) and lipid retention (Lr), the body of the pig may be accumulated daily:

$$Pt = Pt + Pr$$

(Equation 19.66)

$$Lt = Lt + Lr$$

(Equation 19.67)

$$At = 0.20Pt$$

(Equation 19.68)

$$Yt = 4.9(Pt^{0.855})$$

(Equation 19.69)

The water:protein ratios are about 5:1 at birth, 4.5:1 at 10kg live weight, and 3.5:1 at 60kg. The exponent in the estimate for Yt reflects the decreasing water:protein ratio as the pig ages. This ratio most probably relates to the degree of maturity of the muscle mass, and at higher values for mature size ($\hat{Pt}$) it is possible that a slight increase in the coefficient may become apparent for any given value for Pt.

The whole empty body of the pig (without the content of the gut) (We) is the sum of its parts:

$$We(kg) = Pt + Lt + At + Yt$$

(Equation 19.70)

As the gut contents usually comprise about 5% of the body weight, then W, the live weight, is:

$$W(kg) = 1.05We$$

(Equation 19.71)

Gutfill may, however, be better estimated with a knowledge of the fibre content of the diet concerned. Fibre both attracts water into the gut and increases the volume of dry matter in the intestine as a result of a reduction in the rate of digestion and of passage. There can be 1–2% difference in carcass yield due to the effects of variation in gutfill alone. So where:

$$W(kg) = We + Gutfill$$

(Equation 19.72)

gutfill may be estimated as:

$$Gutfill(kg) = 0.05We + (0.05We)(0.008[CF\ (g/kg) - 40])$$

(Equation 19.73)

Carcass quality

Quality may be determined (and rewarded) in terms of fat thickness (subcutaneous fat depth – usually at the P2 site or equivalents) and/or percentage of lean. The amount of subcutaneous fat is related to the total body lipid (Lt):

$$P2(mm) = 0.81Lt + 0.5$$

(Equation 19.74)

It would appear illogical for the relationship between lipid mass (Lt, kg) and backfat depth (P2, mm) to be independent of body size (expressible as W, the live weight). Thus a relationship may be offered:

$$P2(mm) = (73Lt)/W + 0.5$$

(Equation 19.75)

which for W = 90kg is the same as Equation 19.74, but which spreads the fat more thinly over weights greater than 90kg and more thickly over weights less than 90kg. However, some data sets have shown that a simple relationship, such as in Equation 19.74, may hold over a wide weight range (W = 70–110kg). If this were so, then it would be because lighter pigs deposit more lipid internally whereas heavier pigs deposit more lipid subcutaneously, thus making the final relationship between P2 subcutaneous fat depth and the body mass of lipid appear to be independent of the body weight.

For any given level of body lipid, it is likely that the 'blocky' pig types will have slightly greater subcutaneous fat depths – and, of course, greater eye muscle areas, killing-out percentages, and percentages of lean meat in the carcass side. Pigs may be scored for 'blockiness', or better, 'meatiness', according to a 5-point scale; with a score of 1 being appropriate for conventional Large White/Landrace pig types, 5 for Pietrain/Belgian Landrace types, 2.5 for the first cross between these two types, and intermediate values as appropriate. Some recently bred strains of Large White/Landrace origins, especially selected for meatiness, should also be given a good score on a 5-point scale. Where Sm is the meatiness score ranging from 1 to 5:

$$P2(mm) = (P2)(0.98 + [Sm/50])$$

(Equation 19.76)

The percentage of lean meat in the carcass (Lean meat[%]) may be predicted from P2 together with carcass weight or muscle depth (the latter accounting for the meaty pig types):

$$\text{Lean meat } (\%) = 65.5 - 1.15P2 + 0.076Wc$$

(Equation 19.77)

where Wc is the carcass weight (kg). The inclusion of muscle depth (M_d, mm) (at the P2 site) allows:

$$\text{Lean meat } (\%) = 59 - 0.90P2 + 0.20M_d$$

(Equation 19.78)

593

M_d may range around 45–55mm for a pig of 100kgW, depending much on type.

In the absence of muscle measurement, the equation:

$$\text{Lean meat (\%)} = bP2^{-0.21}$$

(Equation 19.79)

may be tried; b is a coefficient that can be created from the meatiness score as:

$$b = 90 + 2(Sm)$$

(Equation 19.80)

The simulation of the mean value for P2 or percentage lean for a population of pigs may allow prediction of the average grade achieved by the pigs in a quality assessment scheme, but not the all-important distribution of pigs through the various grade categories (Figure 2.20). This may be undertaken by spreading the predicted mean through a distribution (skewed) on the basis of a known value for the variation. The variation is likely to be specific to particular production units, and is best obtained by on-farm measurement.

The killing-out percentage (KO, %) – that is, the carcass yield after killing and evisceration – depends, amongst other things, on the live weight (W, kg), the fatness (P2, mm) and pig type (meatiness score, Sm).

The simple equation:

$$KO(\%) = 66 + 0.09W + 0.12P2 + (Sm/2)$$

(Equation 19.81)

may be sophisticated to account for effect of dietary fibre (CF, g/kg) upon digestibility and gut loading so that, in place of the term (0.09W), a corrected estimate for gutfill may now be inserted, namely (0.09[W − 0.05We{0.008(CF − 40)}]). In this We is the empty body weight taken at an initial value of 0.95W, which reflects an average gut content equivalent to 5% of the live mass. The carcass weight (Wc, kg) is, of course:

$$Wc = KO(W)/100$$

(Equation 19.82)

Should they be required, carcass dissected lean (Ld, kg), carcass dissected fat (Fd, kg), and carcass dissected bone (Bd, kg) might be approximated from the total body protein and lipid mass (Pt and Lt, kg):

$$Ld = 2.3Pt$$

(Equation 19.83)

$$Fd = 0.89Lt$$

(Equation 19.84)

$$Bd = 0.5Pt$$

(Equation 19.85)

Other useful relationships are:

$$Blood\ (kg) = 0.036We + 0.22$$

(Equation 19.86)

$$Head\ (kg) = 0.053We + 1.2$$

(Equation 19.87)

Flow diagram

A flow diagram for a putative pig growth model is shown in Figure 19.4.

Attempting a simulation model to predict the performance of breeding sows

Whereas much of a simulation model to predict the growth response to nutrient supply may be deductive (as well as dynamic and deterministic), the reproductive response of breeding sows must be predicted almost entirely empirically. This is often because the causal forces of the phenomena that are to be simulated are imperfectly known. Modelling theory and practice are also far more advanced with growing pigs than with breeding sows. As the simulation of sow reproduction has not benefited from validation, through both scientific appraisal and industrial use, in the same way as has been possible for meat pigs, it might be considered ambitious to describe the precepts of sow modelling here. Nevertheless enough is understood to put forward statements concerning relationships between one factor and another, and to draw attention to those elements of the system which can be understood as some function of other elements of the system. Qualitative descriptions alone are not most helpful, so here an attempt has been made to quantify the various relationships presented. However, such quantification is presented only as a guide to the sizes of the various forces impinging upon

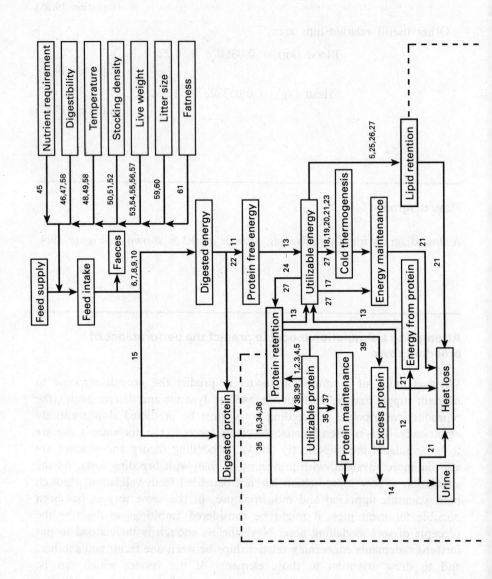

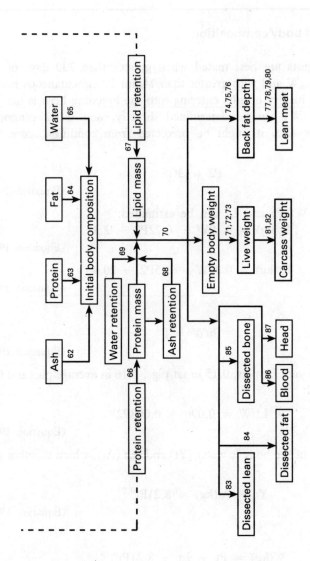

Figure 19.4 Flow diagram for a putative pig growth model. The numerals refer to equation numbers used in the text.

sow reproduction. The constants, coefficients and exponents used should be considered only as estimates of the magnitudes of the factors involved.

Sow weight and body composition

Modern hybrid gilts are best mated when greater than 220 days of age, greater than 120kgW and with greater than 14mm P2 subcutaneous backfat depth. Often the history of gilts entering into the breeding herd is not well known. Weight (W) can be determined directly, and if P2 cannot be measured directly also, it might be predicted from condition score (CS, 1–10):

$$P2 \simeq 3CS$$

(Equation 19.88)

From P2 and W, Pt and Lt can be estimated:
$$Pt(kg) = 0.19W - 0.22P2 - 2.3$$

(Equation 19.89)

$$Lt(kg) = 0.21W + 1.5P2 - 20.4$$

(Equation 19.90)

or:

$$Pt/W = f$$

(Equation 19.91)

where $f = 0.14$ in very fat pigs, 0.15 in fat pigs, 0.16 in average pigs and 0.17 in thin pigs; and:

$$Lt/W = 0.096 + 0.008P2$$

(Equation 19.92)

The remainder of the body is water (Yt) and ash (At), which together may be estimated:

$$Yt + At(kg) = 8.21Pt^{0.79}$$

(Equation 19.93)

thus:

$$W(kg) = Pt + Lt + 8.21Pt^{0.79}$$

(Equation 19.94)

From the point of conception, potential protein growth (Pf) is curtailed such that the Gompertz growth rate parameter B is about half that found for uninhibited growth. The expression:

$$P\hat{t}(kg/day) = B.Pt.log_e(P\hat{t}/Pt)$$

(Equation 19.95)

will serve, where $P\hat{t}$ may commonly range between 35 and 45kg (Table 3.4) and B is around 0.005 (see also Figure 3.13).

Sows may be considered to have become overfat if Lt>2Pt, and overthin if Lt<Pt. Recent measurements with hybrid sows have suggested that ideally the relationship between Lt and Pt should be:

$$Lt(kg) = 1.1Pt^{1.1}$$

(Equation 19.96)

The actual rate of accumulation of fatty tissue in sows is highly dependent upon nutrient supply, and the body fat status is a vital component of reproductive response prediction (of which, more later). Equation 19.96 is not, therefore, predictive, although it may be used for analysis and diagnosis.

Given the deposition of fat (Lr) according to the balance of nutrient demand and supply, P2 can be predicted from Lt or the Lt:Pt ratio (although direct measurement of P2 in breeding sows that are already restrained is a relatively trivial task):

$$P2(mm) = 125Lt/W - 12$$

(Equation 19.97)

$$P2(mm) = 15.3(Lt:Pt) - 6.3$$

(Equation 19.98)

Piglet birth weight

Piglet birth weight has a positive influence on individual piglet survivability and on individual piglet weaning weight; while one (substantial) data set points to a sustained positive influence upon growth throughout life to slaughter weight:

Daily live weight gain (birth to slaughter, kg/day) = $0.42 + 0.12W_b$

(Equation 19.99)

where W_b is piglet birth weight (kg).

Birth weights usually range between 0.8 and 1.6kg, and although conditions for the whole litter may appear identical *in utero*, differential foetal nutrition and uterine environmental effects (including disease) will create a normal distribution of the birth weight of pigs within a litter. Where this variation is required to be modelled (rather than simply the mean), the mean value can be spread through an appropriate distribution curve whose variation is likely to be farm-specific, and may best be judged by on-farm measurements. Meanwhile the average piglet birth weight for the litter (W_b) can be estimated from an equation which has been derived from data collected from hybrid sows of around 180kgW:

$$W_b(kg) = 0.89 + 0.013DE(MJ/day)$$

(Equation 19.100)

where DE is the daily intake of digestible energy by the sow and which indicates a positive relationship between birth weight and maternal pregnancy feed intake. Sows given 2.2kg of a feed with 13.0MJ DE/kg daily on average throughout pregnancy will be expected, all other things being equal, to provide piglets of average birth weight, $W_b = 1.26$.

It is known that late pregnancy feeding has a greater effect on piglet birth weight than early pregnancy feeding and, given proper feeding in the first two-thirds of pregnancy, a suitable equation to adjust W_b for the effects of feeding level in late pregnancy might be speculated to be:

$$W_b(kg) = W_b + 0.008(DE-28)$$

(Equation 19.101)

Average piglet birth weight is markedly influenced by litter size, larger litters tending to have in them piglets of smaller than average size (see also Figure 4.11). This dimension may be accommodated by use of correction factors (Table 19.1) or:

Table 19.1 Corrections for predicted average birth weight of piglets to account for the influence of litter size

Number of piglets born (live and dead)[1]	Correction (kg) to average piglet birth weight (W_b)
6	+0.10
8	+0.06
10	0
12	−0.10
14	−0.22
16	−0.40

[1] Only perinatal demise to be included.

$$W_b(kg) = W_b + 0.04\text{Pigs born} - 0.004(\text{Pigs born})^2$$

(Equation 19.102)

It would also appear sensible to allow expected piglet birth weight to relate in some way to maternal body size, as it is well appreciated that the birth weight of piglets from large breeds of pig is greater than that from small breeds. Quantitative information on this point is not, however, especially safe. It has been suggested that:

$$W_b(kg) = 0.43 + 0.0053W$$

(Equation 19.103)

where W is the maternal body weight at parturition, which referred to sows of around 180kgW and which is almost certainly optimistic when used for sows with live weights above 220kg.

Perhaps a 10% reduction in birth weight should be allowed for each 40kg of maternal body weight below 180kg at parturition:

$$W_b = W_b - (0.0025W_b)(180 - W)$$

(Equation 19.104)

The strength of the logic of this equation is not equalled, however, by that of its coefficients!

If a sow given 2.2kg of feed and calculated to produce piglets of W_b = 1.26kg is now known to have been of 160kg maternal body weight at parturition and to have produced a litter of 11 live-born and 1 dead-born pigs, then the final value for W_b would now be = 1.26 − 0.10 − 0.06 = 1.10kg.

Survivability

Not all live-born piglets will survive through to weaning. Minimum achievable losses appear to be around 5%, with 8–12% common in well-managed herds, and 15–25% in poorly managed herds where facilities for protection from crushing are not provided, where the standard of care is low and where disease levels are high. It is largely futile to determine with accuracy piglet survivability in particular circumstances, but the major issue should at least be raised. Lighter piglets at birth are much more likely to fail to survive than those which are heavier:

$$\text{Survivability}(\%) = 55W_b^{1.3}$$

(Equation 19.105)

If piglets are susceptible, the likelihood of their being crushed is dependent upon the extent to which they are protected (Table 19.2). Similarly, the likelihood of piglets succumbing to a disease challenge is dependent upon the extent to which they are challenged (Table 19.3). It is evident that these latter estimates are nothing more than 'gearing' factors which can only be satisfactorily elucidated by on-farm observation and collection of relevant data. It is also evident that Equation 19.105 and Tables 19.2 and 19.3 cannot be used additively; for example, piglets of small birth weight are those most likely to succumb to disease. Equation 19.105 is helpful in showing the likelihood of survival of smaller rather than larger piglets in a litter; but mean piglet birth weight as predicted earlier is not the required information for this equation. (It is the piglets of below average weight in the litter which are prone to die, not all the piglets in litters of below average weight.) It is vital to the judgement of survivability that the variation in piglet birth weight

Table 19.2 Estimate of numbers of newborn piglets per litter which may be crushed and fail to survive

Protection score	Number of piglets crushed per litter
1	2.5
2	2.0
3	1.5
4	1.0
5	0.5

Table 19.3 Estimate of numbers of newborn piglets per litter which may be lost due to disease challenge

Disease challenge score	Number of piglets lost per litter
1	0.5
2	1.0
3	1.5
4	2.0
5	2.5

within the litter is known for the production unit in question, and that this is used to spread the mean predicted value for birth weight through a suitable normal distribution curve. In this way likely individual birth weights can be determined and equations such as 19.105 utilized.

Litter size (numbers of piglets) at birth

The number of piglets born per litter is highly dependent upon breed type and the extent of heterotic effects. Base-level expectations of the type given in Table 19.4 must therefore be obtained from objectively measured performance data. A well-known phenomenon receiving the attention of reproductive physiologists has been the influence of the pubertal oestrus at which conception occurred on the size of the primiparous litter. Where conception occurs at the first pubertal oestrus, litter size is invariably substantially smaller than when conception occurs at the third (or subsequent) pubertal oestrus. Where X is the number of the pubertal oestrus at which conception occurred, and LS is the litter size, then:

$$LS(\text{number of live-born piglets}) = 7.6X^{0.23}$$

(Equation 19.106)

As parity number increases, so do the number of ova released:

$$\text{Ova released} = 14.5 + 1.5 \text{ Parity number}$$

(Equation 19.107)

but the incidence of foetal and perinatal deaths also increases. The overall influence of parity number on the litter size of live-born piglets is

Table 19.4 Estimated litter size depending on breed type

Breed type classification	Number of piglets in litter born alive
Low prolificacy	8
Medium prolificacy	10
Good prolificacy	11.5
Excellent prolificacy	12.5

approximately according to Table 19.5. The expected percentage of ova released which reach term, to be delivered as live-born piglets, is about 55.

Litter size is also diminished by early weaning and enhanced by later weaning. It may be approximated that:

$$LS^1 = 0.09Days - 3.0$$

(Equation 19.108)

where LS^1 is the correction (number of piglets) to live-born litter size which may be made for the days between parturition and conception (Days; a number between 15 and 45).

There is also evidence of effects of nutrition on litter size, although this is much weaker than nutritional effects on, for example, birth weight or weaning-to-oestrus interval. Nutrition during lactation, and between weaning and conception, can influence both ovulation rate and embryo survival. Thus where ΔW is the total lactation weight change of the sow, it may be suggested:

$$Number\ of\ ova\ released = 23.5 + 0.7\Delta W$$

(Equation 19.109)

Table 19.6 uses sow fatness as a common measure of the variable consequences of absolute feeding levels.

Lactation

The equation

$$M\hat{y}(kg/day) = a.e^{-0.025t}.e^{-e^{(0.5-0.1[t])}}$$

(Equation 19.110)

Table 19.5 Corrections for the influence of parity number on live-born piglets per litter

Parity number	Correction to number of live-born piglets per litter (no. of piglets)
1	−1.0
2	−0.5
3	0
4	+0.5
5	+1.0
6	+0.5
7	0
8	−1.0

Table 19.6 Corrections for the influence of sow fatness at conception on the number of live-born piglets per litter

Sow fatness at conception (mm P2)	Corrections to litter size (no. of piglets)
<10	−0.5
10–14	0
14–18	+0.5
18–22	0
>26	−0.5

sets the outer boundary to lactation yield where Mŷ is the limit to potential milk yield, t is the day of lactation, and a is a characteristic of litter size and of breed type (ranging between 18 and 30). But within that boundary My is a function of the piglets' rate of milk withdrawal, which in turn depends upon their growth rate:

$$My(kg/day) = (4.1)(LS)(Piglet\ daily\ gain\ [kg])$$
(Equation 19.111)

The argument is circular as piglet growth is a function of milk yield!

Limits to the growth of sucking piglets may be best considered apart from post-weaning growth because it appears that if a Gompertz function is to be used, B values are somewhat higher:

$$Piglet\ daily\ gain\ (kg) = B.A.e^{-e^{-B(t-t^*)}}.\log_e(A/A.e^{-e^{-B(t-t^*)}})$$
(Equation 19.112)

where t* can be calculated from:

$$t^*(days) = \log_e(-\log_e[W_b/A])/B$$
(Equation 19.113)

In the above equations A is the mature live body weight (Table 3.3 and Equation 3.1), t is the day of piglet age, W_b is the weight at birth and B is the growth coefficient which is likely to range from 0.014 to 0.019, according to the inherent potential of the piglet to grow. Potential piglet gains derived from these equations are shown in Table 19.7. Estimation of piglet growth rate, approached empirically from the aspect of likely milk supply, leads to an estimation, determined with white cross-bred pigs, such as:

$$Piglet\ daily\ gain(kg) = 0.191 - 0.18\Delta P2 - 0.05\Delta W$$
(Equation 19.114)

Table 19.7 Potential piglet daily live-weight gain for various values of B_g, and potential daily milk supply from the sow for various values of a. Derived from equations in the text (From C.T. Whittemore and C.A. Morgan (1990) *Livestock production science* 26, 1)

	Days post-partum						
	0	7	14	21	28	35	42
Potential piglet live weight (kg)							
$B_g = 0.0143$	1.2	2.0	3.3	5.0	7.4	10.6	14.5
$B_g = 0.0165$	1.2	2.2	3.7	6.0	9.3	13.5	19.0
Potential piglet weight gain (kg/day)							
$B_g = 0.0143$	0.10	0.15	0.21	0.29	0.39	0.51	0.63
$B_g = 0.0165$	0.11	0.18	0.27	0.39	0.53	0.60	0.86
Potential milk supply (kg/day)							
a = 18	3.5	6.7	8.4	8.7	8.1	7.1	6.1
a = 24	4.6	8.9	11.3	11.6	10.8	9.5	8.2

where $\Delta P2$ is the change in P2 (mm/day) which takes place during lactation, and ΔW is the change in maternal sow body weight (kg/day) that takes place during lactation. The constant is crucial to effective prediction, and is likely to vary amongst breed types and, indeed, farms. Additionally, it is not helpful always to imply that sows losing no weight or fat will cause their piglets to grow slowly. This will be the case where the sow is eating less than is needed to satisfy the nutrient demands of My, but it would not be the case where ingested nutrients supplied all requirements for My. Equation 19.114 is often a useful guide, however, as most sows do not eat enough feed to supply the needs of lactation, and so milk yield (and piglet growth) is often dependent upon the rate of catabolism of body mass in general, and body lipid in particular.

Achieved daily milk yield will be a result of either piglet demand (Equation 19.111) or the boundary limit (Equation 19.110), whichever is the lesser; provided that there is no nutritional limitation upon lactation yield. When there is a nutritional limitation upon lactation yield, it is reasonable to start from the proposition that:

$$My = (Me_t - [E_m + E_{Pr} + E_{Lr} + E_{h^1}])/7.7$$

(Equation 19.115)

The lactation feed intake of sows is often (usually) inadequate to supply ME_t in sufficient quantity to supply the needs of My as calculated either from sucking piglet need or sow yield potential. This does not mean, however,.that F is the invariable controller of My, because the sow will readily use body stores of lipid (and protein) to supply the shortfall between the needs of milk biosynthesis and the provision of energy and protein from the diet.

$$\Delta My(kg) \equiv 6.2\Delta Lr(kg)$$

(Equation 19.116)

which is to say that the use of 1kg of retained body lipid will result in the creation of 6.2kg of sow's milk.

The conversion of body protein to milk protein takes place with an efficiency of 0.85, and sow's milk contains 55g protein/kg milk. The protein retained in 1kg of milk may therefore be resourced from 65g of body protein tissue.

Body lipid and body protein may thus be used in support of ME_t and IP_t from dietary sources in order to meet My as calculated according to piglet need within sow potential (Equations 19.111 and 19.110). Such support can only take place, however, if body lipid and protein stores are available to so provide, and, furthermore, the *level* of support from the sow by using her own body tissue to grow that of her piglets will be in proportion to the extent of the body stores that she has present. Sows in good body condition may lose up to 1kg of lipid and 0.25kg protein daily. The ratio of losses appears to range from 2.5 lipid to 1 protein to 10 lipid to 1 protein, depending on the dietary nutrient balance and the status of sow body-nutrient stores.

Broadly, when Lt<Pt, sows may be judged as thin, whilst when Lt>(2Pt), sows may be judged as fat. It may be proposed, if Lt/Pt<0.7 then no fat will be made available to resource My, but for sows with fatness levels above this the rate of daily lipid loss (ΔLt) available to support My, should the need arise, might be:

$$\Delta Lt(kg/day) \text{ to support milk yield} = 0.50 - 0.7(Lt/Pt)$$

(Equation 19.117)

this latter equation having an upper limit on loss of 1.0kg daily.

Losses of Pt may be put at a maximum of 0.25kg daily and probably at a minimum value of 0.1 of the lipid losses, because it appears normal for there to be some loss of protein when fatty tissue is being catabolized, even when the shortfall is only in energy.

Weaning weight

The weight of the litter at the point of weaning is a function of the litter size, the birth weight of the piglets and their subsequent growth. The relationship to birth weight may be of the form:

$$\text{Piglet growth rate (kg/day, birth to weaning)} = 0.150 + 0.08W_b$$

(Equation 19.118)

but the constant and coefficient would be highly farm-specific and dependent upon many other impinging factors. Early growth, however, is primarily a function of (a) potential growth rate (Equation 19.112) and (b) milk yield (Equation 19.110), these being related.

Expected growth rates are deduced in Table 19.7, while Equation 19.114 is an empirical description relating to sow body and fat losses. Equally as sound would be an empirical relationship with maternal lactation feed intake:

$$\text{Piglet growth rate(kg/day)} = 0.150 + 0.012F$$

(Equation 19.119)

where F is the daily feed intake during lactation. The shortfalls of such equations have been discussed earlier in this section.

Weaning-to-conception interval

The length of pregnancy is a fairly invariable biological constant of 114–116 days. The length of lactation is at the command of man. Prediction of the number of litters that a breeding sow may have in a year is therefore dependent upon simulating the weaning-to-conception interval. Leaving aside the possibility of lactational oestrus and conception, the weaning to oestrus interval has a minimum value of 4 days, and assuming a 100% conception rate the same value applies to the weaning-to-conception interval.

The major factors extending the weaning-to-conception interval above the minimum value are maternal body condition, as indicated by the extent of maternal body fat and protein depletion. The relationship between the weaning-to-oestrus interval and the depth of subcutaneous fat on the sow at weaning (P2, mm) may be expressed as:

Primiparous sows:

$$\text{Weaning-to-oestrus interval (days)} = 31.3 - 2.03P2 + 0.043(P2)^2$$

(Equation 19.120a)

Multiparous sows:

$$\text{Weaning-to-oestrus interval (days)} = 19.3 - 1.27P2 + 0.030(P2)^2$$

(Equation 19.120b)

The Australian work reviewed by R H King points to the importance of the amount of tissue lost during lactation, as well as the absolute levels of protein and lipid in the body of the sow at the time of weaning; thus for primiparous sows:

$$\text{Weaning-to-oestrus interval (days)} = 7.3 - 0.39\Delta W$$

(Equation 19.121)

$$\text{Weaning-to-oestrus interval (days)} = 9.4 - 0.59\Delta Lt$$

(Equation 19.122)

$$\text{Weaning-to-oestrus interval (days)} = 9.6 - 3.44\Delta Pt$$

(Equation 19.123)

where ΔW is the total live-weight change (kg) in lactation, ΔLt is the total fat change (kg) in lactation, and ΔPt is the total protein change (kg) in lactation. If Equation 19.123, which examines the specific influence of protein losses during lactation, is seen as an additional factor influencing the weaning-to-oestrus interval established in Equation 19.121 or 19.122, then the coefficient $0.344\Delta Pt$ could be used as a correction factor:

$$\text{Weaning-to-oestrus interval (days)} = \text{Weaning-to-oestrus interval (days)} - 3.44\Delta Pt$$

(Equation 19.124)

or, for every kilogram of protein lost from the body of the sow during the course of lactation, another 3.4 days require to be added to the weaning-to-oestrus interval.

Lactation length also modifies the weaning-to-oestrus interval (see Figure 4.20) but this is likely to be mediated through sow body-tissue losses in lactation and therefore should not be accounted for independently.

The loss of maternal body weight in lactation has a lesser effect on weaning-to-oestrus interval for sows which are fat than for sows which are thin. The strength of the above relationships will therefore be dependent upon sow body condition; the lower the level of internal body stores, the greater the negative effects of weight loss in lactation upon rebreeding (see also Figure 15.26).

Any effect of the feeding level post-weaning may be accounted for by a further correction:

$$\text{Weaning-to-oestrus interval (days)} = \text{Weaning-to-oestrus interval (days)} + 40F^{-3.0}$$

(Equation 19.125)

where F is the daily feed intake between weaning and oestrus.

Some workers have shown a negative relationship between litter size and weaning-to-oestrus interval, which is reasonable, but it is difficult to think that this effect could be independent of effects accounted for elsewhere, such as maternal tissue losses.

Whilst most information from trial data relates to the more easily monitored weaning-to-oestrus interval, the key parameter is in fact the weaning-to-conception interval. Conception rate is highly variable between farms and greatly influenced by mating management and the presence of chronic (and acute) reproductive disease. The effect of lactation length upon conception rate, although less than the effects of management and disease, is better documented:

$$\text{Conception rate (\%)} = 54 + 1.3 \text{ lactation length (days)}$$

(Equation 19.126)

which may well be independent of body condition loss and a function, at least in part, of the strength of circulating lactation hormones if the sow is weaned before the natural peak of lactation is reached at 3–4 weeks. Similarly the effects of number of matings (1–3) which occur at oestrus:

$$\text{Conception rate (\%)} = 63 + 7.5 \text{ number of matings during oestrus}$$

(Equation 19.127)

Normally the proportion of sows mated which farrow is around 80%. It is possible for farm data to yield a value for conception rate, or it could be approximated. The additional days that require to be added to the weaning-to-oestrus interval in order to estimate the weaning-to-conception interval can be calculated as:

$$\text{Weaning-to-conception interval (days)} = (\text{Weaning-to-oestrus interval}) + 37 - 0.37 \text{ Conception rate (\%)}$$

(Equation 19.128)

It may be possible to determine at least some of the contributions to the number of additional days between weaning-to-oestrus interval and the weaning-to-conception interval, from the two factors which appear to influence it most after the major – and impossible to model – effects of management and disease. These are the rate of body condition loss in lactation and the length of the lactation. For body condition loss (as fat) in lactation:

Primiparous sows:

$$\text{Additional days} = -2\Delta P2 - 4$$

(Equation 19.129a)

Multiparous sows:

$$\text{Additional days} = -2\Delta P2 - 6$$

(Equation 19.129b)

For each equation a minimum value of -2 requires to be set, and $\Delta P2$ is the total change in the maternal measurement (mm) which has occurred over the course of the whole lactation.

The modifying effect of lactation length can be accounted as:

$$\text{Additional days} = 191L^{-1.63}$$

(Equation 19.130)

where L is the lactation length in days. Thus:

$$\text{Weaning-to-conception interval (days)} = \text{Weaning-to-oestrus interval (days)}$$
$$+ (-2\Delta P2 - 4) + (191L^{-1.63})$$

(Equation 19.131)

These two factors cannot account for all conception failures of mated sows and until more and better information is forthcoming on this point a 'gearing factor' for conception rate appears the only likely means of reflecting on-farm performance. Such gearing factors are best measured and used on the farm units for which the simulation is attempted.

Maternal weight and fatness

The importance of absolute maternal body weight and fatness, and of the change in maternal body weight and fatness, to the reproductive performance of the sow has been well described in the foregoing sections. These may be deduced from knowledge of nutrient inputs and product outputs, as has been shown. However, the relative fragility of the deductive approach gives ample reason for the use of empirical equations where these may be helpful.

Let it be assumed that sow live weight at first conception is 125kg, and the backfat depth (P2) is 15mm and the DE content of the feed is 13.2MJ DE/kg; and let it be further assumed that the sow growth expectations (which may or may not be achieved) are as given in the earlier section on sow weight and body condition; then, the total change in P2 backfat in pregnancy (P2, mm) may be estimated as:

$$\Delta P2 = 4.60F(\text{kg}) - 6.90$$

(Equation 19.132a)

where F is the daily pregnancy feed intake, or:

$$\Delta P2 = 4.14F(kg) - 9.3$$

(Equation 19.132b)

which may be especially appropriate for sows kept in environments where the environmental temperature is less than the comfort temperature.

Change in total body lipid (ΔLt) has been estimated as:

$$\Delta Lt = 15F(kg) - 30$$

(Equation 19.133)

The total live weight change in pregnancy (ΔW, kg) may be estimated as:

$$\Delta W = 25F(kg) - 27$$

(Equation 19.134)

The daily change in P2 backfat depth ($\Delta P2$) in lactation may be estimated as:

$$\Delta P2 = 0.049F - 0.396$$

(Equation 19.135)

where F is the daily lactation feed intake. The daily change in maternal live weight (ΔW) during lactation may be estimated as:

$$\Delta W = 0.343F - 1.74$$

(Equation 19.136)

Better simulations may be forthcoming from more complex but more complete prediction equations:

$$\Delta P2(mm/day) = -0.0101$$
$$-0.0095 \ (P2, \ mm \ at \ parturition)$$
$$+0.037 \ (F, \ kg/day)$$
$$-0.0178 \ (number \ of \ piglets \ in \ litter)$$

(Equation 19.137)

and:

$$\Delta W(kg/day) = -0.1360$$
$$-0.00535 \ (maternal \ W \ at \ parturition)$$
$$+0.362 \ (F, \ kg/day)$$
$$-0.119 \ (number \ of \ piglets \ in \ litter)$$

(Equation 19.138)

A 125kg gilt with 15mm P2, fed 2.5kg daily in pregnancy, may gain 4.6mm P2 (Equation 19.132a) and 35.5kgW (Equation 19.134). If she eats 5kg feed daily and suckles eight piglets in lactation, 0.15mm P2 (Equation

19.135) and 0.025kgW (Equation 19.136) will be lost daily, or 4.2mm P2 and 0.7kgW over the whole 28-day lactation. Lactation losses over the 28-day lactation, calculated from Equations 19.137 and 19.138, are 4.3mm P2 and 3.7kgW, the gilt ending lactation with $125 + 35.5 - 3.7 = 157$kgW and $15 + 4.6 - 4.3 = 15.3$mm P2. But if she were next to receive 2.2kg over the ensuing pregnancy, and 5kg/day during the subsequent lactation to suckle 11 piglets, then the sow would wean her second litter with $157 + 28 - 0.6 = 184$kgW and $15.3 + 3.2 - 8.4 = 10.1$mm P2, which implies that the (unhappy) weaning-to-conception interval which is now about to follow will be 20.4 days (Equations 19.120b and 19.129b). These responses, although described in terms of 'a sow', refer not to an individual but to the mean response of a breeding herd being treated in the way described. Again, the variation in the response – the extent to which an individual sow's description should be spread around the mean – is crucial to the use of models in the practical environment; but information on this aspect of simulation is as yet both highly farm-specific and largely undocumented.

Flow diagram

A flow diagram for a putative breeding sow model is shown in Figure 19.5.

Physical and financial peripherals

Previous sections have dealt with the concepts, and the deductive and empirical relationships which can be used in the creation of biological cores for models which simulate the production response of growing/slaughter pigs and breeding sows. Before a model can predict the consequences of changing production circumstance within the context of a business or production unit, however, further information is needed about the system.

Thus far the 'biological core' has discussed the simulation of:

● protein and lean growth;
● fatty tissue growth;
● uterine deposition;

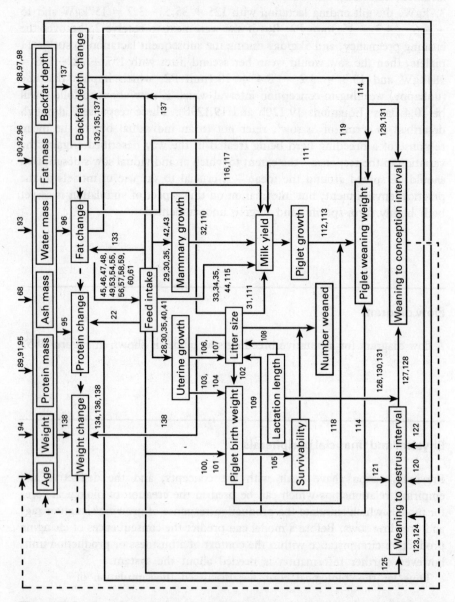

Figure 19.5 Flow diagram for a putative breeding sow model. The numerals refer to equation numbers used in the text. A complete model would require the use of additional equations given in Figure 19.4, and care to avoid double-counting the influence of various factors on a single character.

- lactation yield;
- energy yield from the diet;
- protein yield from the diet;
- energy use by growing pigs;
- energy use by breeding sows;
- protein (amino acid) use by growing pigs;
- protein (amino acid) use by breeding sows;
- feed intake (including the influence of environmental factors);
- the composition of growth;
- carcass quality;
- the body composition of breeding sows;
- piglet birth weight;
- piglet survivability;
- litter size;
- piglet growth and weaning weight;
- weaning-to-oestrus and weaning-to-conception interval;
- sow maternal body weight changes;
- sow maternal fat changes.

To these now need to be added the physical and financial peripherals, such as:

- gearing factors for the general production environment;
- gearing factors for the quality of management;
- gearing factors for the level of disease;
- fixed costs for the production unit (see Tables 18.3 and 18.6);
- variable costs for each pig produced (see Tables 18.3 and 18.6);
- numbers of pigs available;
- cost of pigs available;
- management factors, such as castration policy, split sex feeding, refinement of feed allowance control, feed wastage etc.;
- breed type and genetic quality of growing pigs and breeding sows;
- quality of housing;
- climate, weather and microclimate of buildings (temperature, relative humidity, air movement, air quality);
- numbers of pigs per pen, pen size, density of pigs in pens, space allocation, availability of pens for pigs of different classes, utilization rates for pens;
- minimum and maximum dead-weight limits for carcass pigs;
- method of carcass quality assessment;
- method of carcass quality payment;
- value of culled breeding sows.

Given this type of information with which to surround the biological core, a model may be constructed which can simulate growth, carcass quality and reproductive performance, and which can place the resulting predictions into the orbit of the management of the business. This is what is required for the science and the practice of pig production to be brought effectively together.

CHAPTER 20

CONCLUSION

The pursuit of knowledge about the biology and economics of meat production from pigs is worthy, not just because the profession of agriculture is helped and because more people can be fed, but also on account of the absolute benefit which comes from the winning of scientific advance. It is hoped that all three of these objectives may have touched the reader in the foregoing chapters.

If the science is found good, and the production practices found effective, it will not be through the implementation of rules but through the understanding of the forces – economic and biological – which impinge upon pig production; especially through the understanding of the causes of the animal's responses to various aspects of the nutritional, genetic, physiological and production environments.

Presentation of a blueprint instruction manual would, of course, have been easier for both the author and reader; but less useful. This is because pig production world-wide is rapidly changing, and not necessarily in the one direction. For example, while the move towards intensivism is strong in many countries where the pig industry is young and the demand for meat is high, elsewhere may be found a move toward extensivism, and a perception of quality in pig products which relates as much to systems of animal care as to the nutritional value of the resultant meat.

While the production characteristics of the industry may diverge with geographic location, many aspects of science, and of scientific goals, are common:

- the importance of creating pig-meat products which the market-place

demands, rather than merely producing that which is most convenient for the farmer;
- the possibility for improving the reproductive performance of sows and the growth performance of young pigs, not so much to raise their potential as to achieve the potential which already exists;
- the need to optimize growth by attending realistically to the ratios between costs of input and values of output, and not just the seeking of high biological performance;
- creating through genetic improvement a diversity of pig types which may be used best to exploit the prevailing wide range of production circumstances (there are substantial genetic goals yet to be reached in reproductive performance, growth rate and carcass quality, and there may be more novel goals in the form of disease resistance);
- understanding better the relationships between nutrient input and growth and reproductive response (a briefly stated goal but a discipline requiring immense scientific effort and investment);
- being able to determine effectively the real nutritional value of feedstuffs;
- attacking the problems of disease, both chronic and acute; not only through the development of protective or curative drugs but also by the determination of means of avoiding disease, and of means of giving to the pig ways of bettering its own defence mechanisms against disease;
- the need to enhance feed intake of pigs; especially highly productive young growers and lactating sows.

Interest in animal welfare is not only a matter of humankind's responsibility toward the rights of animals which are used for food, it is also a matter of good and practical husbandry. Regrettably, the issues of pig behaviour and welfare may sometimes be subjugated in the quest for ever more efficient production systems. If so, there is benefit for pig producers in heeding the voice of the caring customer.

It is evident that housing systems are not yet optimal. The microclimate is not understood, while the relations between stocking density, health and performance are fraught with complexity. Even if there were full understanding of the pig's environmental needs (which there is not), there remains a lack of proper understanding of the housing systems that would be needed to provide for them.

Above all, regardless of the nature of any individual system, the production of pigs must be cost-effective, or it cannot take place. Optimization requires knowledge and manipulation of the market and a readiness to be flexible because of the continually changing nature of that market. Here the simulation model represents a novel resource that science response of both the breeding herd and the growing pig to changes in production circumstances – especially nutritional and genetic. Models can incorporate the body of scientific knowledge and this strengthens further their potency in aiding good production practice. The science of computer modelling will come, covertly and overtly, to be the major force influencing the practice of feeding, breeding and management of pigs.

This book has been written mostly while in Edinburgh but also in the United States, Asia, Japan and many countries on the European Continent. It so happens that these final sentences are written in Northern Italy, in the home of dry-cured Parma ham made from pigs grown to 160kg live weight. The pig serves mankind best when we understand most clearly how to use its supreme adaptability and diversity, and how to exploit compassionately its innate facility for domestication.

NUTRITIONAL GUIDE VALUES OF SOME FEED INGREDIENTS FOR PIGS

Nutritional guide values of some feed ingredients for pigs: cereals[1]

	Barley	Wheat	Maize
Dry matter (g/kg)	860	860	880
Crude protein (g/kg)	105	110	90
Faecal digestibility of protein	0.76	0.82	0.78
Ileal digestibility of amino acids	0.72	0.75	0.73
Crude fibre (g/kg)	45	30	23
Neutral detergent fibre (g/kg)	175	120	90
Oil (g/kg)	15	16	36
Ash (g/kg)	24	18	15
Lysine (g/kg)[5]	3.5 (0.70)	3.0 (0.73)	2.6 (0.70)
Methionine + cystine (g/kg)[5]	3.8 (0.75)	4.0 (0.82)	3.7 (0.82)
Threonine (g/kg)[5]	3.4 (0.63)	3.0 (0.65)	3.3 (0.70)
Tryptophan (g/kg)[5]	1.5 (0.73)	1.3 (0.75)	0.8 (0.70)
Isoleucine (g/kg)[5]	5.1 (0.73)	5.0 (0.80)	3.4 (0.78)
Linoleic acid (g/kg)	9	8	21
Calcium (g/kg)	0.5	0.3	0.2
Phosphorus (g/kg)[2]	3.5	3.4	2.4
Digestible energy (MJ/kg)	12.9	14.0	14.5
Inclusion in starter diets[1]	<0.4	<0.5	<0.5
Inclusion in grower diets	<0.6	<0.6	<0.6
Inclusion in finisher diets	<0.7	<0.6	<0.6
Inclusion in sow diets	<0.7	<0.6	<0.6

[1] Individual cereals may be included at a minimum of 250g/kg to maintain continuity in diet mixtures. Cooked cereals are more palatable and should be used in starter diets; cooking avoids DE depression with young pigs, while for other pigs DE is increased by 5% and protein digestibility decreased by 5%.

[2] The phosphorus in cereal grains and by-products is usually only 50% (or less) available.

[3] May be contaminated with ergot.

Oats[15]		Rice		Rye[3]		Sorghum[4]		Cassava/manioc	
860		870		860		860		900	
103		66		110		95		28	
0.75		0.70		0.75		0.75		0.40	
0.70		0.65		0.65		0.70		0.25	
115		15		20		25		45	
270		170		140		100		80	
42		4		16		35		5	
30		8		20		20		70	
3.8	(0.67)	2.4	(0.70)	3.8	(0.67)	2.0	(0.75)	0.9	(–)
3.8	(0.78)	2.3	(0.72)	3.5	(0.78)	3.4	(0.85)	0.4	(–)
3.4	(0.58)	2.5	(0.70)	3.5	(0.60)	4.0	(0.75)	0.5	(–)
1.5	(0.65)	0.6	(0.70)	1.4	(0.66)	1.1	(0.75)	0.2	(–)
5.0	(0.72)	3.0	(0.72)	3.3	(0.70)	3.6	(0.85)	0.6	(–)
18		7		7		10		2	
1.0		0.7		0.5		0.4		1.0	
3.2		1.0		3.5		3.5		1.4	
11.4		15.0		13.3		13.0		13.2	
<0.4		<0.2		0		0		0	
<0.5		<0.3		<0.2		<0.1		<0.1	
<0.6		<0.4		<0.3		<0.4		<0.4	
<0.6		<0.4		<0.2		<0.3		<0.2	

[4] Low tannin varieties. High tannin varieties will have lower digestibility, especially for ileal digestibility of amino acids. DE may range from 12 to 14 (yellow varieties).

[5] Ileal digestibility is given in brackets (all measurements of ileal digestibility require verification).

[15] Dehulled oats have digestibility and inclusion characteristics similar to wheat.

Nutritional guide values of some feed ingredients for pigs: by-products

	Distillers' dark grains[7]		Brewers' grains		Wheat feed (middlings)	
Dry matter (g/kg)	900		900		870	
Crude protein (g/kg)	290		180		155	
Faecal digestibility of protein	0.60		0.70		0.70	
Ileal digestibility of amino acids	0.55		0.60		0.60	
Crude fibre (g/kg)	100		120		90	
Neutral detergent fibre (g/kg)	275		450		300	
Oil (g/kg)	105		80		40	
Ash (g/kg)	40		36		46	
Lysine (g/kg)[5]	7	(0.50)	5.2	(0.68)	5.8	(0.65)
Methionine + cystine (g/kg)[5]	8	(0.65)	5.6	(0.70)	4.5	(0.75)
Threonine (g/kg)[5]	10	(0.55)	5.5	(0.65)	4.8	(0.60)
Tryptophan (g/kg)[5]	1.5	(0.50)	2.1	(0.70)	2.0	(0.70)
Isoleucine (g/kg)[5]	12	(0.64)	6.0	(0.85)	5.3	(0.74)
Linoleic acid (g/kg)	45		30		20	
Calcium (g/kg)	0.5		2.9		1.4	
Phosphorus (g/kg)[2]	8.0		7.0		7.5	
Digestible energy (MJ/kg)	12.5		8.0		11.3	
Inclusion in starter diets	0		0		<0.1	
Inclusion in grower diets	<0.03		<0.05		<0.2	
Inclusion in finisher diets	<0.05		<0.1		<0.3	
Inclusion in sow diets	<0.05		<0.1		<0.3	

[2] The phosphorus in cereal grains and by-products is usually only 50% (or less) available.

[5] Ileal digestibility is given in brackets (all measurements of ileal digestibility require verification).

[6] Can cause soft and/or yellow carcass fat.

[7] Some samples may have a high copper content (100mg/kg).

Wheat bran meal		Maize germ meal		Maize gluten meal		Maize gluten feed		Cooked oat	
880		880		900		900		900	
150		100		600		210		125	
0.60		0.70		0.70		0.65		0.90	
0.50		0.63		0.65		0.55		0.80	
120		42		25		86		20	
450		160		25		350		100	
42		100		25		31		60	
60		36		20		70		30	
4.0	(0.65)	4.8	(–)	11	(0.72)	5.5	(0.50)	4.5	(0.80)
4.2	(0.70)	3.3	(–)	25	(0.80)	8.0	(0.50)	3.7	(0.90)
4.4	(0.60)	4.0	(–)	21	(0.80)	7.0	(0.50)	4.5	(0.75)
2.5	(0.70)	1.2	(–)	3.0	(0.70)	1.3	(0.33)	1.5	(0.80)
6.3	(0.70)	–	(–)	25	(0.80)	5.5	(0.60)	5.2	(0.84)
20		70		17		14		18	
1.3		0.2		0.5		2.8		0.6	
11.0		6.6		3.2		8.0		4.2	
9.0		14.1		13.7		11.8		15.5	
0		<0.1		0		<0.05		<0.5	
<0.1		<0.1		<0.1		<0.1		<0.4	
<0.2		<0.1		<0.1[6]		<0.1[6]		<0.4	
<0.2		<0.1		<0.2		<0.2		<0.4	

Nutritional guide values of some feed ingredients for pigs: by-products (*continued*)

	Oat hulls		Sugar beet pulp		Rice bran	
Dry matter (g/kg)	880		870		900	
Crude protein (g/kg)	35		94		145	
Faecal digestibility of protein	0.60		0.50		0.50	
Ileal digestibility of amino acids	0.50		0.45		0.40	
Crude fibre (g/kg)	280		150		150	
Neutral detergent fibre (g/kg)	650		265		320	
Oil (g/kg)	18		5		20	
Ash (g/kg)	52		62		135	
Lysine (g/kg)[5]	0.8	(–)	2.5	(–)	5.0	(–)
Methionine + cystine (g/kg)[5]	0.8	(–)	1.6	(–)	3.4	(–)
Threonine (g/kg)[5]	1.5	(–)	3.2	(–)	4.4	(–)
Tryptophan (g/kg)[5]	0.4	(–)	0.1	(–)	1.7	(–)
Isoleucine (g/kg)[5]	–	(–)	1.0	(–)	4.5	(–)
Linoleic acid (g/kg)	–		–		4	
Calcium (g/kg)	1.4		5.5		1.9	
Phosphorus (g/kg)[2]	2.0		0.5		11.0	
Digestible energy (MJ/kg)	3.5		10.0		8.0	
Inclusion in starter diets	0		0		<0.05	
Inclusion in grower diets	<0.05		<0.1		<0.1	
Inclusion in finisher diets	<0.1		<0.2		<0.1	
Inclusion in sow diets	<0.2		<0.2		<0.2	

Cooked potatoes		Grass meal pulp	
900		900	
91		140	
0.70		0.36	
0.75		0.30	
20		173	
70		475	
4		30	
40		90	
4.8	(−)	8.0	(0.50)
1.8	(−)	5.3	(0.40)
3.5	(−)	6.4	(0.50)
1.0	(−)	3.2	(−)
3.0	(−)	4.5	(0.50)
−		3	
0.8		9.5	
1.6		3.0	
14.4		6.5	
<0.2		<0.1	
<0.3		<0.1	
<0.3		<0.2	
<0.3		<0.2	

Nutritional guide values of some feed ingredients for pigs: fats, minerals, molasses and animal proteins[16]

	Soya oil	Tallow	Molasses	Fish meal[8]	
Dry matter (g/kg)	990	990	640	900	
Crude protein (g/kg)	0	0	30	650	
Faecal digestibility of protein	0	0	0.30	0.92	
Ileal digestibility of amino acids	0	0	0.25	0.80	
Crude fibre (g/kg)	0	0	0	1	
Neutral detergent fibre (g/kg)	0	0	0	0	
Oil (g/kg)	990	990	0	60	
Ash (g/kg)	0	0	80	180	
Lysine (g/kg)[5]	0	0	0.3	47	(0.82)
Methionine + cystine (g/kg)[5]	0	0	0.2	24	(0.85)
Threonine (g/kg)[5]	0	0	0.2	26	(0.82)
Tryptophan (g/kg)[5]	0	0	0	7	(0.75)
Isoleucine (g/kg)[5]	0	0	0.2	32	(0.85)
Linoleic acid (g/kg)	600	25	0	1.8	
Calcium (g/kg)	0	0	6.0	65	
Phosphorus (g/kg)	0	0	0.5	32	
Digestible energy (MJ/kg)	36	29	10.0	15.0	
Inclusion in starter diets	<0.1	0	<0.05	>0.08[17]	
Inclusion in grower diets	<0.1	<0.05	<0.05	>0.05[17]	
Inclusion in finisher diets	<0.03	<0.05	<0.10	<0.1	
Inclusion in sow diets	<0.05	<0.05	<0.08	<0.1	

[5] Ileal digestibility is given in brackets (all measurements of ileal digestibility require verification).

[8] Highly variable in both composition and the utilizability of the ileal digested amino acids (40–90%), according to source and treatment.

[14] Especially variable in protein content according to source, and when provided in the liquid form variable in dry matter content.

[16] Artificial amino acids may be assumed to have faecal and ileal digestibilities of 100%. L-lysine HCl contains 780g lysine/kg, L-threonine contains 980g threonine/kg, DL-methionine contains 980g methionine/kg and DL-tryptophan contains 800g tryptophan/kg.

[17] Minimum levels if highest quality only.

Herring meal		Meat-and-bone meal[8]		Separated milk (skim)		Whey[14]		Limestone	Dicalcium phosphate
900		900		950		930		980	975
680		480		340		125		0	0
0.92		0.65		0.95		0.90		0	0
0.80		0.58		0.90		0.85		0	0
1		20		0		0		0	0
0		0		0		0		0	0
85		110 ·		10		42		0	0
100		300		80		85		980	950
52	(0.88)	25	(0.65)	25	(0.95)	8.9	(0.93)	0	0
26	(0.85)	12	(0.73)	12	(0.95)	5.0	(0.93)	0	0
30	(0.80)	16	(0.58)	15	(0.90)	7.3	(0.90)	0	0
6	(0.80)	3	(0.50)	5	(0.85)	2.1	(0.85)	0	0
34	(0.85)	16	(0.67)	24	(0.85)	8.5	(0.85)	0	0
1.5		2.5		0.14		0.42		0	0
29		100		10		8.5		360	230
20		50		8		7.0		0	180
17.5		10.0		16.0		15.5		0	0
>0.08		0		>0.1		>0.05		<0.1	<0.1
>0.05		0		0		<0.2		<0.1	<0.1
<0.1		<0.1		0		<0.2		<0.1	<0.1
<0.1		<0.05		0		<0.2		<0.1	<0.1

Nutritional guide values of some feed ingredients for pigs: vegetable proteins

	Soya bean meal (extracted)		Full fat soya (extruded)		Single cell protein		Rape seed meal (extracted)	
Dry matter (g/kg)	890		900		900		880	
Crude protein (g/kg)	440		360		720		360	
Faecal digestibility of protein	0.87		0.85		0.90		0.70	
Ileal digestibility of amino acids	0.80		0.75		0.80		0.60	
Crude fibre (g/kg)	67		50		5		120	
Neutral detergent fibre (g/kg)	140		105		0		250	
Oil (g/kg)	15		175		80		20	
Ash (g/kg)	55		50		100		77	
Lysine (g/kg)[5]	29	(0.85)	24	(0.80)	45	(–)	20	(0.72)
Methionine + cystine (g/kg)[5]	13	(0.85)	11	(0.75)	21	(–)	12	(0.80)
Threonine (g/kg)[5]	17	(0.78)	15	(0.70)	32	(–)	16	(0.68)
Tryptophan (g/kg)[5]	6.4	(0.75)	4.9	(0.70)	9.9	(–)	4.5	(0.70)
Isoleucine (g/kg)[5]	22	(0.80)	19	(0.70)	30	(–)	13	(0.75)
Linoleic acid (g/kg)	5		80		–		3	
Calcium (g/kg)	2.7		2.5		10		6.6	
Phosphorus (g/kg)[2]	6.0		5.5		22		13	
Digestible energy (MJ/kg)	14.5		17.0		17.0		11.8	
Inclusion in starter diets	<0.1		<0.1		<0.08		0	
Inclusion in grower diets	>0.1		<0.2		<0.08		<0.05[9]	
Inclusion in finisher diets	>0.1		0[6]		<0.08		<0.1[9]	
Inclusion in sow diets	>0.1		<0.2		<0.08		<0.05[9]	

[2] The phosphorus in cereal grains and by-products is usually only 50% (or less) available.

[5] Ileal digestibility is given in brackets (all measurements of ileal digestibility require verification).

[6] Can cause soft and/or yellow carcass fat.

[9] Rape-seed meal contains various antinutritional factors including glucosinolates, tannins, complexed fibres, complexed carbohydrates and erucic acid. These have toxic and appetite depressant effects. Some new varieties are available with low, and double-low, toxin levels and it is these which should be used. Some 'low' glucosinolate rape seeds may contain up to 20 micromol, levels which have been found to cause performance reduction. Treatment processes show promise in further detoxifying the material and removing its appetite depressant effects; these sources can be included at higher levels (add 0.05 to limits in table).

[10] Higher levels of angustifolius.

[11] Of which only 60–70% actually available (total available about 40% for cotton-seed meals and 50% for lupin meals).

[12] Crude protein content varies with degree of dehulling: fully dehulled, 400; partially dehulled, 350; with hulls, 280.

[13] Of which only 80% actually available (total available about 60%).

Sunflower meal (extracted)[12]		Field beans		Field peas		Full fat sunflower		Lupin meal (extracted)		Cotton-seed meal	
900		860		860		900		900		920	
300		260		190		150		300		430	
0.75		0.75		0.75		0.75		0.70		0.70	
0.65		0.68		0.68		0.65		0.68		0.65	
200		80		54		250		150		120	
370		150		120		–		–		300	
20		13		14		300		60		60	
70		34		30		–		30		70	
12	(0.68)[13]	14	(0.78)	15	(0.80)	5.1	(–)	12	(0.70)[11]	18	(0.70)[11]
12	(0.85)	5.0	(0.75)	5.0	(0.70)	6.0	(–)	5	(0.65)	12	(0.70)
12	(0.70)	10	(0.70)	9	(0.75)	5.5	(–)	9	(0.75)	14	(0.75)[11]
4	(0.77)	2.3	(0.70)	2.0	(0.70)	2.0	(–)	2	(–)	5	(0.70)[11]
14	(0.77)	10	(–)	9.0	(0.74)	6.4	(–)	11	(0.75)	14	(0.68)
12		7		12		280		40		25	
3.8		1.4		0.7		2.0		2.0		2.1	
10.3		4.9		3.9		5.0		3.5		13.0	
9.0		13.0		13.6		17.0		13.0		13.2	
0		0		0		0		0		0	
<0.05		<0.05		<0.2		<0.1		<0.05		<0.05	
<0.1		<0.2		<0.4		$<0^6$		$<0.10^{10}$		<0.10	
<0.1		<0.1		<0.3		<0.1		<0.05		<0.05	

GUIDE NUTRIENT SPECIFICATIONS FOR PIG DIETS

Nutrient specification (chemical analysis) (g/kg)

	Up to 15kg[1]	Up to 30kg[1]	Up to 100kg[1] (A)	Up to 160kg[1] (B)	Breeding gilts/lactating sows[1]	Pregnant sows[1]
Crude protein	220–260	210–240	160–200	150–180	150–180	130–170
Crude fat	50–120	50–100	20–70	20–50	20–70	20–40
Crude fibre	10–30	10–40	20–80	20–80	30–80	30–80
DE (MJ/kg)	14–17	14–16	13–15	13–15	12–15	11–14
Lysine[2]	13–17	12–15	9–13	6–10	6–10	5–8
Ca	9–15	9–10	8–10	8–10	8–12	8–12
P	7–11	7–10	6–8	6–8	6–8	6–8
Na	1.4–2.5	1.3–2.5	1.2–2.5	1.2–2.5	1.2–2.5	1.2–2.5
Linoleic acid (approx.)	10–50	10–50	5–20	5–20	5–30	5–25
Methionine + cystine[2]	6–10	5–8	5–7	4–6	4–6	3–5
Lysine (g/MJ DE) (approx.)	0.95	0.85	0.75	0.60	0.60	0.55
Crude protein (g/MJ DE) (approx.)	16	15	14	12	12.5	12

Vitamins and trace elements (to add, per tonne of feed)[3] (approx.)
Vitamin A 15mill i.u. (4.5g retinol or 50g betacarotene), D_3 2mill i.u. (0.05g cholecalciferol), E 0.10mill i.u. (60g D-alpha-tocopherol)[4], K3 (Phytylmenaquinone) 1–3g, Thiamin (B_1) 2g, Riboflavin (B_2) 6g, Nicotinic acid 20g, Pantothenic acid 15g, Pyridoxine (B_6) 3.0g, B_{12} 20mg, Biotin (H) 100–500mg[5], Folic acid 1g[11], Choline 600g[6], Zinc 50–150g[8], Magnesium 30g[7], Manganese 10–50g, Iron 50–200g[8], Cobalt 1g, Iodine 0.5g[9], Selenium 0.25g, Copper 5–175g[10], Butylated Hydroxytoluene 125g. A need for vitamin C is under consideration.

1   Individual diets will vary according to the circumstances of the unit, the genetic type of the pig, the number of different diets acceptable, and the economic cost-benefit. Diets for pigs up to 30kg and for lactating sows should be fed to appetite. Diets for pigs up to 100kg may be fed to appetite or restricted during the later stages of growth; diet A is of higher specification than B and is appropriate for improved genotypes. To achieve the required intake of nutrients, a greater quantity of diet of lower nutrient density is needed.

2   The required provision of other essential amino acids may be achieved by attaining the following minima which are expressed in relative terms as proportions of the lysine requirement (1.00): histidine 0.36; isoleucine 0.57; leucine 1.14; methionine + cystine 0.57 (at least 50% as methionine); tyrosine + phenylalanine 1.00; threonine 0.64; tryptophan 0.18; valine 0.75.

3   Levels vary widely according to the risk of loss of potency and local knowledge. In-feed growth promoters, medicines, diet acidifiers and probiotics may be included according to local need.

4   Higher levels if significant quantities of fat in the diet (especially if unsaturated, a rule of 30g for every 1% linoleic acid has been suggested); range 0.02–0.25mill i.u. Breeding sow diets may merit the higher levels.

5   500mg to maintain reproductive performance, up to 1,000mg for breeding sows with hoof problems.

6   Widely ranging between 0 and 1,000g. A quarter to a half of requirement is added; total requirement is about 1,500g.

7   Widely ranging between 0 and 500g; adequate provision is likely from main ingredients.

8   The higher level for diets with added copper; usually 100–150g are included.

9   0.5–2.0g, the higher levels if goitrogens in the diet ingredients.

10   The higher level for growth promotion. Background copper levels are 10–20g, the limit for breeding sows is 30–40g.

11   Up to 5g for breeding sows.

APPENDIX 3

GUIDE FEEDSTUFF INGREDIENT COMPOSITIONS OF PIG DIETS

Example ingredient composition (kg/t) (ingredients may differ according to national availability and price)

	Up to 15kg[1]	Up to 30kg[1]	Up to 100kg[1] (A)	Up to 100kg[1] (B)	Breeding gilts/lactating sows[1]	Pregnant sows[1]
Barley[2] ⎫	0–200	20–300	0–500	0–500	0–500	0–500
Wheat[2] ⎬ or other cereal	50–300[3]	20–300[3]	0–500	0–500	0–500	0–500
Maize[2] ⎭	10–500	10–500	0–400	0–500	0–500	0–500
Full fat soya	0–50	50–150	0–100	0–100	0–100	0–100
Ext soya bean meal	0–50	10–100	100–250	50–200	50–200	50–200
Fish meal[4]	50–150	100–200	0–100	0–100	0–100	0–100
Dried whey powder[5]	100–200	50–150				
Dried skim milk[5]	100–200	0–150				
Feed oils and fats[5]	20–80	10–50	0–30	0–30	0–50	0–30
Millers' by-products	0–100	0–150	0–200	0–250	0–250	0–300
Grass/alfalfa products			0–100	0–100	0–100	0–100
Limestone	1–15	2–15	2–15	2–15	1–15	1–15
Dicalcium phosphate	1–10	2–15	2–15	2–15	1–15	1–15
Salt[6]	0–5	0–5	0–5	0–5	0–5	0–5
Lysine HCl	1–2	1–2	1–2	0–2	0–2	0–1
Vitamins/Minerals	+	+	+	+	+	+

[1] Individual diets will vary according to the circumstances of the unit, the genetic type of the pig, the number of different diets acceptable, and the economic cost-benefit. Diets for pigs up to 30kg and for lactating sows should be fed to appetite. Diets for pigs up to 100kg may be fed to appetite or restricted during the later stages of growth; diet A is of higher specification than B and is appropriate for improved genotypes. To achieve the required intake of nutrients, a greater quantity of diet of lower nutrient density is needed.

[2] A good proportion to be cooked up to 15kg, a small proportion to be cooked up to 30kg.

³ Dehulled oats are a useful cereal in diets up to 30kg. Whole oats may be used subsequently.

⁴ Or other sources of animal protein. In some cases a minimum of 25 may be advisable.

⁵ Or a fat-filled milk replacer. The higher levels are justified in starter diets for pigs from 14 days of age to 10kg live weight.

DESCRIPTION FILES FOR RAW MATERIALS

Description files for raw materials

Name: Barley (1)		**Name: Soya 44 (6)**	
Currency cost:	110	Currency cost:	148
Minimum weight (kg):	1	Minimum weight (kg):	1
Rounding (kg):	10	Rounding (kg):	10
Analysis		**Analysis**	
(Volume)	100.00	(Volume)	100.00
DE	12.90	DE	15.00
CP	107.00	CP	435.00
DCP	77.00	DCP	410.00
Lysine	3.50	Lysine	28.90
Methionine + cystine	3.70	Methionine + cystine	13.20
Methionine	1.7	Methionine	6.6
Threonine	3.30	Threonine	17.10
Tryptophan	1.50	Tryptophan	6.40
Linoleic acid	9.00	Linoleic acid	4.00
Starch + sugar	482.00	Starch + sugar	108.00
Copper	4.10	Copper	22.50
Phosphorus	3.40	Phosphorus	5.90
Av. Phosphorus	2.00	Av. Phosphorus	3.00
Sodium	0.20	Sodium	0.40
Oil	15.00	Oil	13.00
Crude fibre	45.00	Crude fibre	67.00
Ash	24.00	Ash	60.00
Neutral detergent fibre	202.00	Neutral detergent fibre	135.00
Selenium	0.03	Selenium	0.14
Dry matter	875.00	Dry matter	881.00
Pelleting quality factor	5.00	Pelleting quality factor	4.00

BIBLIOGRAPHY

AFRC 1990 Nutrient requirements of sows and boars. *Nutrn Abs & Revs, Series B: Livestock Feeds & Feeding* **60**:383–406

AFRC 1990 *Advisory booklet, nutrient requirements of sows and boars.* HGM Pubs, Bakewell 31 pp

AFRC 1991 Theory of response to nutrients by farm animals. *Nutrn Abs & Revs, Series B: Livestock Feeds & Feeding* **61**:683–722

ARC 1967 *The nutrient requirements of farm livestock no. 3: Pigs.* Technical reviews and summaries. ARC, HMSO, London 278 pp

ARC 1981 *The nutrient requirements of pigs*: Technical review by an Agricultural Research Council Working Party. CAB, Farnham Royal 307 pp

Baxter S 1984 *Intensive pig production: environmental management and design.* Granada, London 588 pp

Brent G 1982 *The pigman's handbook.* Farming Press, Ipswich 230 pp

Brent G 1986 *Housing the pig.* Farming Press, Ipswich 248 pp

Brody S 1945 *Bioenergetics and growth.* Reinhold, New York 1023 pp

Buttery P J, Haynes N B, Lindsay D B (eds) 1986 *Control and manipulation of animal growth.* Butterworth, London 347 pp

Campbell R G 1988 Nutritional constraints to lean tissue accretion in farm animals. *Nutrn Res Revs* **1**:233–253

Cole D J A, Haresign W (eds) 1985 *Recent developments in pig nutrition.* Butterworth, London 321 pp

Cole D J A, Foxcroft G R, Weir B J 1990 Control of pig reproduction III. *J Reprod & Fertil* Supplement **40**:1–396

English P R, Smith W J, MacLean A 1977 *The sow: improving her efficiency.* Farming Press, Ipswich 311 pp

English P R, Fowler V R, Baxter S, Smith W J 1988 *The growing and finishing pig: improving efficiency.* Farming Press, Ipswich 555 pp

Evans M 1985 *Nutrient composition of feedstuffs for pigs and poultry.* Queensland Dept of Primary Industries, Queensland 134 pp

Falconer D S 1989 *Introduction to quantitative genetics.* Third edition, Longman, London 438 pp

Fisher C, Boorman K N (eds) 1986 *Nutritional requirements of poultry and nutritional research*. Butterworth, London 224 pp

Forbes J M 1986 *The voluntary food intake of farm animals*. Butterworth, London 206 pp

Fraser A F, Broom D M 1990 *Farm animal behaviour and welfare*. Baillière Tindall, London 437 pp

Gardner J A A, Dunkin A C, Lloyd L C (eds) 1990 *Pig production in Australia*. Butterworth, Sydney 358 pp

Hughes P E, Varley M A 1980 *Reproduction in the pig*. Butterworth, London 241 pp

Hunter R H F 1982 *Reproduction of farm animals*. Longman, London 149 pp

Legates J E, Warwick E J 1990 *Breeding and improvement of farm animals*. McGraw-Hill, New York 342 pp

McDonald P, Edwards R A, Greenhalgh J F D 1988 *Animal nutrition*. Longman, London 543 pp

MAFF 1988 *Codes of recommendations for the welfare of livestock: pigs*. MAFF Publications, Alnwick 16 pp

NRC 1979 *Nutrient requirements of swine*. National Academy of Sciences, Washington 52 pp

NRC 1988 *Nutrient requirements of swine*. National Academy Press, Washington 93 pp

Pond W G, Maner J H 1974 *Swine production in temperate and tropical environments*. W H Freeman, San Francisco 646 pp

Sainsbury D 1972 *Pig housing*. Farming Press, Ipswich 212 pp

Sainsbury D, Sainsbury P 1988 *Livestock health and housing*. Baillière Tindall, London 319 pp

Stranks M H, Cooke B C, Fairbairn C B, Fowler N G, Kirby P S, McCracken K J, Morgan C A, Palmer F G, Peers D G 1988 Nutrient allowances for growing pigs. *Res & Dev in Agric* 5:71–88

Taylor D T 1989 *Pig diseases*. Burlington Press, Cambridge 310 pp

UFAW 1988 *Management and welfare of farm animals*. Baillière Tindall, London 260 pp

Verstegen M W A, Henken A M (eds) 1987 *Energy metabolism in farm animals: effects of housing, stress and disease*. Martinus Nijhoff, Dordrecht 500 pp

Walton J R 1987 *A handbook of pig diseases*. Liverpool University Press, Liverpool 166 pp

Whittemore C T, Elsley F W H 1976 *Practical pig nutrition*. Farming Press, Ipswich 190 pp

Whittemore C T 1980 *Lactation of the dairy cow*. Longman, London 94 pp

Whittemore C T 1980 *Pig production: The scientific and practical principles*. Longman, London 145 pp

Whittemore C T 1987 *Elements of pig science*. Longman, London 181 pp

Whittemore C T 1989 *The prediction and understanding of nutritional response in pigs*. DSc Thesis, The University of Newcastle upon Tyne

Wiseman J 1986 *A history of the British pig*. Duckworth, London 118 pp

Wiseman J, Cole C J A (eds) 1990 *Feedstuff evaluation*. Butterworth, London 456 pp

Wood J D, Fisher A V (eds) 1990 *Reducing fat in meat animals*. Elsevier, London 469 pp

INDEX

Abattoir, 41
Aberrant behaviour, 158, 162
Abortion, 257, 274
Abrasions, 154, 249
Abrasiveness, in feedstuffs, 309, 313
Accelerating phase of growth, 53, 55, 67
Accessory fluids, 100
Acclimatization
  importance of, 466
  of breeding stock, 487
Accommodation for boars, 166
Accredited breeder, 225
Accretion of lean and fat, 63
Acetate, 386
Acid detergent fibre and lignin (ADF),
  300
Acid oils, 324
Acidity
  of muscle, 16
  of gut, 389
Actinobacillus, 270
Activities of pigs, 149
Ad libitum
  feeding, 19, 79, 149, 221, 380, 392,
    443
  self-feed hoppers, 525, 585
Additive genetic variance, 179, 184, 191
Additives in diets, 416
Additivity, of DE values, 294
Adrenal glands, 85
Adrenalin, 95

Adrenocorticotrophic hormone
  (ACTH), 86, 106
Adzuki bean, 316
Aerobic digestion, 518
Aflatoxins, 251
African swine fever, 261, 277
Agalactia, 107, 255
Ageing, of ova, 97
Aggression, 145, 163. 257
Agility of pigs, 150
Agonistic behaviour, 149, 530
Agricultural
  Revolution, 167
  Societies, 168
Air movement, 147, 336, 507, 513
Air space, 269, 509, 511
Air speed, 507, 513, 517
Air, freshness of, 248
Alkaloids, 314, 315
Allometric relationship, 51
Alpha tocopheral (Vit E), 325
Alveolus, 109, 111
Ambient temperature, 248, 335, 361,
  505
American breeds, 169
Amines, 306, 316
Amino acid, 280, 304, 421
  anabolism, 82
  availability, 320
  balance, 306, 312, 405
  composition of meat and milk, 343
  content of pig feedstuffs, 322

637

Amino acid (*cont.*)
  rates of absorption, 317
  ratio of digestibilities, 319
  requirements for pigs, 631
  spectrum, 342
Ammonia, 256, 269, 306, 310, 316, 388, 511
Amperage, for stunning, 18
Amylases, 387
Anabolic hormones, 81
Anabolism of fatty tissues, 67
Anaemia, 256
Anaerobic digestion, 518
Androgens, 81, 87
Androstenone, 26
Animal,
  fat, 8
  identification, 550
  protein sources, 409
  proteins, 305
  rights, 167
Anoestrus, 257
Anorexia, 372
Anterior pituitary gland, 85, 113
Anthelminths, 260
Anthrax, 251
Anti-nutritional factors, 314, 385
Antibiotics, 390
Antigenic reaction, 266
Antioxidants, 391
Apathy, 158
Apparent digestibility, 289, 307
Appetite, 46, 74, 79, 160, 182, 370, 466, 586
  depression/enhancement, 372, 385, 416, 510
  limitations, 411
  potential, 378
Appleby, 160
Arable rotation, 524
Arachidonic acid, 323
ARC Working Party, 379
Area occupied by pigs, 588
Aromatic oil, 391
Arthritis, 250
Artificial
  amino acids, 307, 321, 344, 377
  insemination, 98, 99, 181, 194, 255, 483
  insemination centre, 234
  lysine, 398, 421
  selection, 169

sow, 100
Ascaris, 260
Ash, 28, 50
  in body of pig, 591, 598
  in pigs' milk, 112
Asian or Siamese pig, 168, 172
Aspergillus, 251
Atresia ani, 252
Atrophic rhinitis, 270
Attachment, of embryos, 102
Auger delivery, 526
Aujeszky's disease (pseudorabies), 259, 264, 273
Automatic delivery, 526
  feeding systems, 585
  wet-feeding systems, 363
Availability, of amino acids, 321
Available amino acids, 340, 347
Available essential amino acids in ideal protein, 345
Available ideal protein, 347, 583
Available lysine, 313
Avermectin, 260
Aversive response, 131, 134

B, the growth coefficient, 57
Babyrousa, 168
Back-crossing, 189
Backfat depth, 14, 30, 47, 196, 435, 593
  change in lactation, 612
  measurement of, 33
Backfat in pregnancy, 611
Bacon, 9, 11
Bacterial
  carbohydrases, 386
  degradation of amino acids, 317
  fermentation, 386
  population, 151
Bakewell, 167, 172
Balance
  of amino acids, 317
  of energy and protein, preferred, 134
Balancer meal, 292
Bar-biting and bar-sucking, 159
Barking, 150
Barking/thumping, 269
Barley, 290, 316
  meal, 307
Barley/fish diets, 424
Basal balancer, 293
Batterham, 321
Beef meat consumption, 8, 11

Behaviour
  description of, 129
  influence on comfort temperature, 506
Behavioural needs, 162, 164
Belgian Landrace, 175, 189, 190
Berkshire pig, 168
Berkshire, 177
Best linear unbiased prediction (BLUP),
    182, 210, 216, 243
Beta-adrenergic agonists, 82, 370
Beta-glucanases, 387
Beta-glucans, 385
Bichard, 212
Bile, 308
Bio-availability, of vitamins, 368
Biological variation, 483
Biosynthesis of milk, hormones in, 113
Biotechnology, 237
Birth weight, 104, 105, 128, 256, 599,
    602
Birth, 106
Blastocyst, 101
Blocky pig types, 22, 593
Blue ear disease, 275
Blueprint, 568, 617
Boar
  cross-bred, 187
  energy and protein requirements of,
    355
  hybrid, 187
  influence of age, 254
  pens, 532
  working rate, 254
Body composition, 44
  change, 62
  of weaners, 69
Body condition, 258
  of sows, 598, 607
Body lipid
  change in sows, 612
  content, 65, 71
  in sows, 65
Body mass, 48
Body protein of sows, 607
Body tissue loss in sows, 340
Bone, 49, 50
  dissected, 594
Bordetella, 270
Botulism, 251
Boundary fences, 262
Brascamp, 212

Breed
  creation, 191
  crossing, 191
  mixing, 191
  substitution, 61
  type, 615
Breeder herd performance, 561
Breeder-finisher operations, 6
Breeding
  animals, growth of, 60
  companies, 197, 204, 225–31, 484
  life, 84
  lines, 232
  nucleus, 197
  plans, 224
  pyramid, 216
  stock, marketing of, 482
  stock, types of, 483
  value, 197, 213
British breeds, 171
British Saddleback, 176
Brody, 53
Brucellosis, 261
BSE, 251
Buffers, for semen, 100
Bulk feed, 526
Bulking characteristics of feedstuffs, 377
Bullying, 523
Butting, 139
Butyrate, 386
By-products, 1
  in compounded diets, 393
  of human meat preparation, 251

Cabbage, 290
Caecum, 289, 310
Calcium soaps, 364
Calcium, 364
Calcium:phosphorus ratio, 365
Calorimeter, 281
Campylobacter, 267
Canadian rapes, 315
Candidates for selection, 196, 199, 214,
    227
Cannibalism, 138, 256
Capital costs, for low intensity systems,
    523
Carbohydrate fraction, of feedstuffs, 287
Carbohydrate, 280
Carbon dioxide stunning, 17
Carbon dioxide, 256, 283, 284, 511
Carbon monoxide, 511

Carcass
  composition of, 51
  dissection, 215
  fat, 62
  fatty acids, influence of diet, 329
  lean meat, percentage for grading, 32
  lean percentage, 21, 497
  lean, 34, 47
  quality, 4, 29, 592
  weight, 35, 41, 594
  yield, 35
Cardiovascular disease, 8
Cassava, 315
Castrates, 24
  growth potential of, 76
Castration, 25, 156, 166
Catabolism
  of fat, 68, 577
  of muscle protein, 73
Catalysts, 357
Catch-up growth, 73
Catecholamines, 81
Catheter, 100
Cell enlargement and multiplication, 48
Cell walls, 387
Cellulases, 386, 387
Cellulose, 299, 385
Central nervous system (CNS), in
      appetite control, 371
Central testing, 198, 211, 227
  stations, 212, 224
Cereal grains, in compounded diets, 393
Cerrano ham, 6
Cervix, 99, 106, 135
Chain
  chewing, 159
  length, of fatty acids, 324
Charts, 560
Chasing behaviour, 150
Chelates, of minerals, 315
Chemical components, influence on DE,
      298
Chemical composition
  of live weight change in sows, 446
  of pig diet, 630
  of the pig, 28
Chester White, 178
Chick peas, 316
Chilling media, for semen, 100
Chinese Meishan breed, 174
Chinese pig, 168, 172
Chlorine requirement, 366

Cholecystokinin, 371
Choline, 369
Chorionic gonadotrophin, 92
Chromosomes, 239
Chymotrypsin inhibitor, 315
Cimaterol, 82
Citric acid, 389
Classical swine fever, 261
Clausen, 212
Clenbuterol, 82
Client profile, 475
Climate, influence upon sows, 456
Clippers, for teeth, 155
Cloning, 241
CNS, 113
Cobalt requirement, 366
Coconut meal, 307, 319
Codes of recommendations for welfare
      of livestock, 163, 164
Coitus, 136
Cold
  pigs, 523, 544
  thermogenesis, 336, 510
Colibacillosis, 265
Colitis, 252
Collagen, structural, 108
Colostral milk, 107, 112, 265
Colour, of meat, 15
COMA, 8
Comet, 170
Comfort
  and shelter, 165
  temperature, 335, 456, 505, 580
Commercial
  breeder/grower, 234
  growers, 236
  product evaluation, 225
  records, 555
Comminution
  of proteins, 314
  of feedstuffs, 292
Commodity trading, 472
Comparative advantage, 500
Comparative slaughter technique, 283,
      285, 311
Compensatory growth, 73, 75
Competition, for feed, 145, 146, 152
Composition of body
  changes in, 62
  chemical, 51, 591, 592
  physical, 50
Composting systems, 388

Compound traits, 211
Compounded feeds, DE determination of, 295
Compounding of pig feeds, 470
Computer print-out, 530
Computer, for modelling, 567
Computer-based recording, 557
Computer-controlled feeding systems, 222, 528
Computerized diet formulation, 428
Concentration of nutrients in the diet, 354, 404
Conception, 257
    failure, 125
    influence on growth, 59
    rate, 91, 123, 610
Concrete floors, effect upon health, 249
Condensation, 508
Condition scoring, 459, 598
Condition, of the sow's body, 125, 454
Conduction, 504
Conformation, 213
Constipation, 253
Consumption, of pig meat, 8
Contaminants of feeds, 250, 481, 526
Continential breeds, 171
Contractions at parturition, 138
Controlled feeding, 495
Controlled ventilation, 512
Convection, 504
Cooking, 9, 316
    of protein, 313
Cooling, 508
Copper requirement, 366
Coppice, 137
Copulation, 90, 94, 135, 254
Coronavirus, 268
Corpora lutea, 87, 102
Correlated traits, 202
Corticosteroids, 106
Corynebacterium, 259
Cost
    per kilogram of carcass, 39
    per unit, of energy, protein or lysine, 414
    reduction in diet formulation, 430
Cotton seed meal, 307, 319
Cotton seed oil, 323
Coughing, 269
Courtship, 134, 135
Coventry, 168
Cow peas, 316

Crates, 249
Criss-crossing, 187
Critical temperature, 336
Cross-bred, 185
    females, 236
    lines, 228
Cross-breeding, 172, 186
Crude fibre, 299
Crude protein, 305
Crumbs, 392
Crushing, 166, 601
Culling, 6, 197, 559
Curve of growth, 53
Curvilinearity, in digestibility, 293
Customer
    complaints, 480
    for pig feeds, 472
    loyalty, 477
    needs, for improved pigs, 483
Customization, 474
Cutting techniques, 13
Cyanogenic glycosides, 315
Cystine, 304
Cystitis, 259

Daily
    fatty tissue growth, 371
    feed intake, 47, 438
    feed intake, of lactating sows, 456
    gains of sucking piglets, 605
    lean tissue growth, 371
    live weight gain, 47, 57, 371
    protein retention, 76
Dam lines, 229
Danish Landrace breed, 173, 175
Danish pig industry, 173
Danish slaughterhouses, 212
Darwin, 169
Days empty, 126
DE see Digestible energy
    value, from chemical analysis, 302
Dead weight, 29
Deamination, 334, 345, 579
Deceleratory phase of growth, 55
Deep-frozen semen, 101
Deep-litter systems, 388
Defecation, 153
Defects and deformaties of sperm, 96
Dehydration, 266
Delivery
    of breeding stock, 487
    systems, for feed, 527

Denmark, 211
Density
  of stocking, 511, 522
  restraint, in feed formulation, 411
Depopulation, 188
Deposition of fat by sows, 599
Depot fat, 63, 68
Depression, 162
Deprivation experiments, 360
Deprivation, 162
Deterministic models, 575
Detoxification, 315
Developmental changes in body, 48, 51
DFD, 17
Diagnosis, 550, 571
  of disease, in relation to welfare, 165
Diarrhoea, 256, 265, 361, 425
Dickerson, 212
Diet
  balancing by pigs, 130
  choice, 502
  concentration of digestible energy,
    399
  crude protein, growth responses to
    increasing supply, 403
  density, effect upon value, 411
  dilution, 466
  energy concentration, 394
  fortification, 415
  imbalance, influence on appetite, 374
  ingredients, effects on feed intake,
    385
  mineral & vitamin density, 358
  specifications, 396
Diet formulation
  and construction, 406
  changes in, 410
  control, 432
  package, 427
Diets for pigs, 397, 630
Difference method for DE
  determination, 292, 294
Digestibility
  of amino acids, 422
  of calcium and phosphorus, 364
  of fats and oils, 326
  of fatty acids, 326
  of minerals, 360
  of protein, 306, 307, 312
Digestibility coefficient, 292, 378
Digestible
  amino acids, 309

crude protein, 312, 395, 421
Digestible energy (DE)
  content of diets, 394
  determination, 293, 295
  fibre corrected, 290
  in feed, 281, 288, 578
  intake, 379
  of fibrous and fatty feeds, 420
  prediction from chemical
    components, 298
  values for fats and oils, 326
  value from chemical analysis, 302
  versus net energy, 419
Digestion, site of, 316
Digestive
  juices, 307
  system, enzyme activity in, 124
  tract, diseases of, 265
  trauma, 466
Diluent
  for diets, 467
  for semen, 100
Disease
  monitoring, 264
  prevention, 263, 277
  resistance, 224, 239
Disinfectant spray, 262
Disinfection, 269
Dissectible lean, 50
Distress, 164
DNA, 239
  recombinant technology, 237
Dominance, 145, 147, 154, 183, 257,
    392
Dose response experiments, 360
Draughts, 248, 507
Drip loss, 22
Droplet, spread of disease by, 248
Dry feeding systems, 417, 528
Dry matter content of pig diets, 394
Duct, 109, 119
Duroc, 20, 22, 176, 190
Dust, 269, 363, 392
Dutch Landrace, 175
Dynamic models, 575
Dystocia, 257

Eschericha coli, 259, 265, 273
Ear
  biting, 147
  notching, 156
Early growth, 58, 65

Early maturity, 171
Early-life markers, 242
Eating quality of pig meat, 19, 20
Eating to energy, 327, 377
Eating to protein, 377
Economic weightings, 204
Edible meat, 28
Edinburgh Family Pen System of Pig
    Production, 149
Edinburgh Pig Park, 149
Effective temperature, 580
Efficiency measures, 557
Efficiency
    of conversion, of energy for protein
        retention, 581
    of feed conversion, 49, 79, 371
    of pig production, 11
    of retention of protein, 311, 336
    of transfer of absorbed balanced
        amino acids, 321
    of use of diet protein, 343
    of use of ileal digested ideal protein,
        346
Eggs, 8
Ejaculate from boar, 96, 100, 135, 255
Ejection, of milk, 120, 140
Elasticity of demand, 7
Electrical stunning, 17
Electricity, 509
Electrolytes, 266
Electronic feeder, 549
Electronic feeding systems, 147, 148,
    222, 530, 549
Electronic tag, 530
Elimination behaviour, 153
Elsley, 291, 449
Embryo, 85
    growth, 54
    implantation, 454
    loss, 94, 146, 257
    survival, 100, 257
    transfer, 241
Empty days, 125, 559
Endocrine system, 80, 85
Endogenous
    enzyme activity, 386
    faecal losses, 291, 304
    hormone, 241
    loss of minerals, 360, 365
    protein, 307
    secretions, 308
Energy, 44, 280

density, 395, 400
intake, 284
limits to lean growth, 402
requirements of growing and breeding
    pigs, 333, 349–54, 418, 582
retention, 283, 284
supplementary for heating, 509
use in the body of the pig, 282
value of milk, 339, 583
value, of feedstuffs, 287
Energy balance, 284
Energy concentration, 416
Energy content
    of diet, prediction of, 578
    of feedstuffs, 287, 303, 418
Energy cost
    of cold thermogenesis, 580, 581
    of lipid retention, 338, 582
    of maintenance, 286
    of piglet live weight gain, 339
    of protein deposition, 336, 337, 581
Energy deposition
    in the mammary gland, 339, 583
    in the uterus, 339
Energy yield
    from diet, 577
    from protein deamination, 578
    from slurry, 518
Energy : protein ratio, 395
English, 147
Entire male pigs, 24, 26
Entire male, growth potential of, 76
Environment control, 432
Environment, 162, 167
    commercial, 201
    nutritional, 201
Environmental effects, 178, 193, 216
Environmental temperature, 336, 508,
    510
    influence on feed intake, 588
Environmental variation, 192, 199
Enzootic pneumonia, 269
Enzyme, 341, 386
    attack, by proteases, 306
    cocktails, 387
    in feeds, 385
    systems, 123
Epidemic diarrhoea, 268
Epilepsy, 274
Epinephrine, 81
Epistasis, 183
Epithelial cells, 341

Epithelium, 109
Eradication, of diseases, 246
Ergot, 315
Erucic acid, 315, 323
Erysipelas, 276
Essential amino acid requirement, 421
Essential amino acids, 304, 340
  in meat and milk, 343
Essential fat, 63
Essential fatty acids, 326
Ethics of pig production, 4
European
  Community, 8
  Community grade scheme, 32
  domestic breeds, 168
  hybrid, 30
Evaluation of diet cost, 411
Excreta, 517
Exogenous
  enzymes, 386
  hormone, 108, 241
Exponential growth, 106
Export markets, for improved pigs, 214
Extensivism, 617
Exudative epidermitis (greasy pig
  disease), 146, 276
Eye muscle, 23

F1 hybrid
  females, 186
  parents, 231
Factorial calculation of requirements,
  359
Faecal digestibility, 307, 316, 318
Faeces, as a source of energy loss, 283
Falconer, 178, 212
Family groups, 145
Family Pen, 149, 153
Fan ventilation, 512, 513
Farinaceous products, 1
Farm margins, under different
  circumstances, 499
Farm trials, 476
Farrowing crate, 138, 158, 257, 536, 541
Fat, 26, 48–50
  additions to pig diets, 326, 328, 407
  change in pregnancy, 451
  consumption of, 9
  depth comparison, during testing, 219
  depth, 14, 22, 29, 31, 41, 446
  dissected, 594
  essential, target and surplus, 63

firmness, 15
for cooking, 171
gains by pigs, 401
gains, influence of feed intake, 439
growth in sows, 451
growth response, for tested
  candidates, 220
growth, 62
in body of pig, 69, 445, 591
in meat, 9
in pigs' milk, 112
loss in lactating sows, 68, 72, 457,
  609, 612
loss, 67, 68, 69
mass of sows, 458, 598, 599
percentage, 19
quality, 29
quality, in feedstuffs, 331
reduction, 5
reduction, by selective breeding, 208
Fat-free carcass, 304
Fat-level minima, 463
Fat : lean ratio, 45, 439
Fatness control, 62
Fatness, 4, 41, 200
  in relation to reproduction, 454
  influence of protein:energy ratio, 402
  influence on profit, 498
  of carcasses, 496
Fats, 377
  and oils in diet compounding, 328,
    330
  in pig feeds, 323
Fattening, 44, 53
Fatty acid, 9, 323
  composition of carcass fat, 329
  composition of feedstuffs, 325
  transfer to body fat, 330
Fatty tissue
  composition of, 52
  growth, energy requirement for, 49
  loss in sows, 340
Feed
  allowance, 399, 432, 436, 450
  characteristics, influence on feed
    intake, 385
  composition tables, 303
  compounders, 393, 471, 473
  compounding mills, 407
  conversion efficiency, 79, 436
  conversion efficiency, rate of
    improvement in, 3

costs, 39, 558, 560
delivery systems, 525
description, 290
evaluation, 295
fats, 327
hopper, 527
level provided, 40, 47, 147
manufacturer, 503
reduction of in early pregnancy, 103
requirement for sows, 462
salesmen, 473
supply to pigs, 432
supply, between weaning and
    conception, 454
supply, influence on growth and its
    composition, 439
troughs, 529
Feed ingredients
    choice of, 408
    number and variety of, 406
    nutritional values for, 620
    usage, 559
Feed intake, 75, 80, 437, 585
    achieved, under practical conditions,
        589
    control, 40, 42, 46, 62, 372, 432
    curve, 586
    depression, 456
    during testing, 219
    enhancement, 374
    from gut capacity, 587
    in growing pigs, 378
    in practical circumstances, 381
    in relation to nutrient specifications,
        405
    of lactating sows, 590
    of sows, 382
    of weaned pigs, 379
    quantification of response to, 437
    reduction, 374
Feed-back, 81
Feeder-herd performance, 562
Feeder space, 517
Feeding
    time spent, 152
    to body condition, 459, 462
Feedstuff
    price, 424
    protein, 304
    raw material description file, 634
Feedstuffs, 620
    energy composition of, 303

protein composition of, 322
Female
    growth potential of, 76
    lines, 232, 484
    selection, 234
Feral pigs, 137, 504
Fermentation products, 316
Fertility, 97, 127, 258
    loss of, 523
Fertilization, 94, 258
Fibre content, influence on DE, 296
Fibre, 280, 299
    in feedstuffs, 377
Fibrous feeds, 300
    DE corrections for, 289
Field beans and peas, 316
Fighting, 146, 154, 276
Financial analysis of production unit,
    560
Financial management, 432
Financial performance of pig herds, 558
Fineness of grinding, 314
Fire, 165
Firmness, of meat, 15
First limiting amino acid, 343
First mating, feed supply at, 448
Fish meal, 26, 319
Fish oils, 323
Fish silage, 394
Fish, 8
Five freedoms, 164
Fixed costs, 39, 558, 560, 615
Fixed feed, fixed time-scales, 218
Fixed ingredient formulations, 424
Flavour technology, 372
Flavours, 391
    fishy and off, 324
Floor
    feeding, 516, 527
    space, 166
    space, effects on appetite, 382
    suspended, 509
    type, 336, 508, 581
    type, influence on welfare, 165
Flow of milk, 142
Flushing, 93, 449
Fodder beet, 290
Foetal death, 259, 274
Foetal growth, 54, 104, 451
Foetuses, 85, 103
Folic acid, 369
Follicle rupture, 96

Follicle stimulating hormone (FSH), 86, 89, 113
Follicles, 88, 90, 92
Foot and mouth disease, 261
Foot, ill health of, 249
Foraging, 150
Forced ventilation, 514
Format International, 428
Formic acid, 389
Formulation of diets, 412, 418
Fostering of piglets, 139
Fredeen, 212
Free fatty acids, 280, 324, 329
Freedom, 162
    of movement, 165
Freedoms, 130
French Landrace, 175
Freshness, of feed, 526
Frustration, 160
Fumaric acid, 389
Fumigation, 262
Fusarium, 251

Gametes, 90
Gamma-linolenic acid, 9, 326
Gas, 509
Gaseous losses, 311
Gases in the environment, 388
Gastric pH, 386
Gearing factors, 587, 602, 615
Gene
    addition, 191
    distribution, 181
    importation, 191, 208
    transfer, 239
General purpose blends of fats, 331
Generation interval, 180, 195, 197, 209
Genes, importation and selection of, 178
Genetic
    control, 432
    improvement, 62, 180
    lag, 216, 231, 242
    selection, 18, 489
    variance, non-additive, 183
Genome, 240
Genotype, 435
Genotype : environment interaction, 199, 216
German Landrace, 175
Gland cistern, 109, 117
Glandular tissue, 111
Glans, 135

Glassers disease, 271
Gloucester Old Spot, 177
Glucagon, 81, 372
Glucocorticoids, 81
Gluconeogenesis, 280
Glucose loading, 372
Glucose, 420
Glucosinolates, 315
Glucostatic theory, 372
Gluttony, 372
Glycerol, 280
Glycogen, 280
Goading, 157
Goat, 8
Goitrogens, 314
Gompertz function, 56, 59, 66, 76, 576
Gonadotrophic releasing hormone
    (GnRH), 86, 113
Gonadotrophins, 86, 89, 102, 113
Gossypols, 315
Grade, 30
Grading of pig carcasses, 29, 31, 32
Grading-up, 189
Grandparent lines, 227, 231, 236, 490
Grass, 290
Grazing, 149
Grinding, influence on appetite, 392
Gristle, 167
Gross energy, 280, 288
Group size, 515
Group-housed sows, 146, 148, 161, 539
Growth, 48, 61
    analysis, 42, 74
    constancy of, 55
    curve, 53, 61, 67
    hormone, 81, 82, 240
    impulsion, 66
    limits to, 576
    of piglets, 605
    of sucking pigs, 66
    percent of body weight, 55
    post-weaning, 67
    potential and response, 48, 55, 61, 66, 75, 378
    promoters, 390, 393
    rate, 497
    rate, influence of feed intake, 439
Grunt rate, 140, 144
Gut
    capacity, 370, 376, 587, 590
    content, 35, 592
    flora, 389

secretions, 341

Haemophilus pleuropneumonia, 270
Haemorrhage, 268
Halothane gene, 17, 21, 22, 23, 183, 189
Ham, 30
Hampshire, 22, 176, 190
Hardiness, 20
Hazel, 212
Health control, 432
Health status
  in breeding stock for sale, 487
  in relation to central testing, 214
  of replacement pigs, 188
Heart disease, 8
Heat
  damage, 313
  loss from the body, 281, 286, 508
  of digestion, 286
  of metabolism, 509
  output, 506, 581
  standing, 88
  stress, 100
  stress, influence on appetite, 374, 383
  supplementary, 509
Heaters, for pigs, 509
Heating, 508
Hectare requirement for outdoor pigs,
    524
Helminths, 260
Hemicellulose, 299
Herd size, influence on rate of
    improvement, 197
Heritability, 179, 192, 196
  estimates for pigs, 194
Hernia, 253
Herpes, 273
Heterosis, 183, 210
Heterozygote, 23, 183, 190
Hierarchical structure, 145
Hierarchy, 257
High density diets, 412
High-fibre diets, 35, 287
High-health status, 198, 233, 245, 262
High-welfare, 163
Hind gut digestion, 317
Histidine, 304
Histograms, 560
Hogs, 168
Holding pens, 17
Home mill and mix, 470
Homozygote, 183, 190

Hopper length, per pig, 516
Hormonal complex, 81, 85
Hormone dosing, 238
Hot pigs, 545
Housing
  allocation of, 520
  for pigs, 504
  lay-outs, 531
  requirements of pigs, 516
  requirements, calculation of, 519, 521
Human chorionic gonadotrophin
    (HCG), 92
Human diet, 8
Humidification of air, 508
Humidity, 256, 517
Hunger, 153, 160
Hybrid strains of pig, 447
Hybrid vigour, 20, 172, 178
Hydraulic fluid, 361
Hydrogen sulphide, 256, 388, 511
Hydrogenated fats, 330
Hydrogenated tallows, 329
Hydrolysis
  of fats, 324
  to peptides, 306
Hydroxyapatite, 357
Hypothalamus, 81, 85, 106

Ideal protein, 341, 583, 584
  for metabolism, 579
Identification of pigs, 156
Ileal digested amino acids, 345
Ileal digested crude protein, 579
Ileal digestibility, 307, 318, 584
  of protein, 316, 318
Ileal digestible amino acids, 317, 340
Ileum, 289, 310
Ill-treatment of animals, 130
Immune system, 123, 158
Immunization, 81
  therapy, 239
Immunosuppression, 373
Implanation, of embryos, 259
Importation of genes, 188
Importations, 173
Improved pigs, 53
Improvement by selection, 191, 202,
    484
In-breeding, 185
In-feed enzymes, 386
Inclusion rates, of feedstuff ingredients,
    409

Incremental levels, of test ingredients, 294
Independent culling level, 202
Index
  for boars, 205
  for gilt selection, 205
  selection, 182, 201, 202, 217
Indigenous breeds, 168
Indigestibility of feedstuffs, 377
Indigestible feed protein, 309
Indirect measurements, of genetic traits, 199
Individual
  feeders for sows in yards, 529
  feeding, 147, 148, 528
  selection, 180
  variation, 131
Industrial
  processing, 5
  Revolution, 167
  waste, 394
Inert filler, 467
Infertility, 257, 559
Inflection, 56
Information
  file, 427
  strings, 572
  transfer, 571
Ingredient
  compositions of diets, 410, 410
  inclusion, in least-cost diet formulation, 430
  limits, 430
  quality, 372, 425
  rejection, in least cost diet formulation, 430
  restraints, 410, 415
Inherited characters, 179
Inhibiting hormones, 86
Injury, 154, 249
Input : output, 570
Insects, 151
Inspection, 166
Inspectorates, 163
Insulation, 336, 337, 508–17, 581
Insulin and insulin-like growth factors, 81, 372
Intake, of feed & water, 585
Integrated arrangements for marketing, 472
Intensity of selection, 180

Intensive
  house for weaned pigs, 537
  livestock systems, 504
Intensivism, 617
Intermuscular fat, 49
Interval from weaning to conception, 128
Intestinal secretions, 308
Intramuscular fat, 19
Intromission, 134
Iodine requirement, 366
Ionic balance, 360
Iron 256
  requirement, 366
Isoleucine, 304
Ivermectin, 260

Johansson, 212
Joints, 154
Jonsson, 212
Jostling, 141

Kidneys, 361
Killing-out percentage, 21, 35, 36, 47, 593, 594
King, 185, 212
Kyriazakis, 133

Lacombe, 178
Lactate, 386
Lactation, 65, 84, 112, 604
  anoestrus, 116
  fat loss, 65, 68
  feed intake, 383, 444–55, 590, 606
  length, 121, 122, 127, 609, 610
  peak of, 114
  weight and fat changes, 457
  yield, 114, 606
Lactational
  anoestrus, 113, 257
  metabolism, 115
  oestrus, 608
Lactic acid, 16, 389
Lactobacillus, 389
Lactogenic hormonal complex, 112, 113, 121
Lactose, 280
  in pigs' milk, 112
Lacy fat, 14
Lagoon, 518
Lairage, 17, 157

Lameness, 249
Land, required per pig, 519
Landrace breeds, 22, 169, 175, 230
Lard, 4, 167, 323
Large Black, 177
Large White pig, 22, 169, 173, 175,
    190, 230
Large White, Dutch, 22
Lauric acid, 323
Lawrence, 160
Lean, 50, 78
    composition of, 52
    dissected, 594
    gains, by pigs, 401
    gains, influence of feed intake, 439
    growth , 47
    growth potential, 441
    growth response, for tested
        candidates, 220
    mass, 48
    mass at maturity, 63, 171
    meat percent, 593
    percentage, 19, 30, 402
    tissue growth, energy content of, 49
    tissue growth, potential for, 46, 77, 79
    tissue growth rate, 200, 223, 399
    tissue growth response, of tested
        candidates, 219
Leanness in relation to lean tissue
    growth rate, 78
Learning, 131, 132
Least-cost diet formulation, 426
Leather, 167
Lacombe, 190
Lectins, 315
Legs, 202
Leptospirosis, 251, 261, 275
Leucine, 304
Libido, 100, 187
Library, 572
Lice, 260
Licensing of premises, 163
Light, hours of, 165
Lighting intensity, 256
Lignin, 299
Limb, injuries to, 154
Limitations
    to feed intake, 74
    to growth, 576
Linamarin, 315
Lincolnshire pig, 169
Linear

growth, 56, 76
    programme, 426
Linear/plateau response, 76
Linoleic acid, 9, 14, 323, 325
Linolenic acid, 323, 332
Lipases, 387
Lipid, 28, 50, 280
    content of sows, 60
    in body, 64
    in piglets, 70
    loading, 372
    loss by sows, 607
Lipid : protein ratio, 52, 64, 577, 587
Lipolysis, 82
Lipostatic theory, 372
Litter size, 104, 559, 602, 604
    influence on feed intake, 384
Live weight
    and fatness changes in breeding sows,
        444, 446, 612
    at maturity, 57
    gain, 61
    gain, energy value of, 75
    of piglets, potential, 606
Lobes, 110
Locomotor activity, 150
Locomotor problems, 154, 559
Longevity, selection for, 224
Longissimus dorsi, 12, 14
Longman, 168
Loose housing, 161, 520, 532, 533
Lop, 177
Low, 168
    density diets, 412
    intensity pig-keeping units, 490
Lubrication, 361
Lupin meal, 316, 319
Lush, 212
Luteinizing hormone (LH), 86, 88, 89,
    92, 113
Luteolysis, 89
Luteotrophic hormone, 102
Luxury
    consumption, 360
    energy status, 74
Lying area, 153, 247, 269, 515
Lying position of pigs, thermal
    resistance, of flooring, 507
Lying, time spent, 150
Lysine, 304
    artificial, 344
    as first limiting amino acid, 398
    availability of, 312

Lysine : energy ratio, 416

Magnesium requirement, 364, 366
Maintenance, 38, 49, 335
  costs, 436
  energy requirement for, 281, 284
  nutrient needs for, 405
  of protein, 348, 584
  requirements, 66
Maize, 316, 319
  gluten feed, 319
  meal, 307
Maize/soya diets, 424
Male lines, 232, 484
Mamma, 109, 139
Mammary
  alveoli, 95
  gland cells, 110
  gland, 108
  stimulation, 140
  tissue, 105
  tissue, protein deposition in, 585
  tissue growth, 104, 110, 451
Management, 563
  factors, 615
  influence on feed intake, 378
  information, 550
  points, for boars, 564
  points, for growing pigs, 566
  points, for sows, 564
  strategies, 58
Manganese requirement, 366
Manipulation
  of forage, 153
  with the mouth, 138
Manual systems of feeding, 528
Marbling fat, 20
Margin
  maximization, 569
  over feed and weaner costs, 474, 494
Marker genes, 242
Markers, use in digestibility studies, 291
Market
  for pig meat, 4, 5
  outlet, for slaughter pigs, 497
  place, 173
Marketing, 5, 39, 171, 469
  of improved genotypes, 482
  of pigs, 473
  opportunities, case studies, 499
  strategies, for slaughter pigs, 493, 495

Masking agents, 391
Massage, 141
Mastication, 137
Mastitis, 273
Maternal body
  lipid, 60, 583
  protein, 60, 583
  size, 601
  weight change in sows, 446, 457, 609
Maternal
  fat and protein losses, 455
  fatness, 611
  live weight gains, 451
  weight, 611
Mathematical equations, 573
Mating, 84, 97, 258
  quarters, 136
Matings per oestrus, number of, 98, 610
Matrix, 427
Mature
  body weight, 61
  protein mass, 58
  size, 6, 38, 56
Maturity, 44, 48, 61
Maximisation of protein retention, 73
Maximum
  growth potential, 57, 76
  lean tissue growth rate, 46, 440
  restraints, in feed formulation, 427
ME value, from chemical analysis, 302
ME:DE ratio, 335
Meal
  feeding, 526
  size, 585
Meals, 392
Meat and Livestock Commission, 213
Meat and bone meal, 307, 319
Meat
  consumption, per person per year, 11
  fat content of, 4
  packer, requirements of for high lean
    meat yield, 484
  packers and processors, 7, 42, 503
  packing plants, 492
  production strategy, 493
  production, 4
  quality at sale, 494
  quality, 208
Meat-line, 21
  sires, 229
  type, 230
Meatiness, 593

Mechanical
  efficiency, of protein metabolism, 321
  spreading, 518
Medicines, in compounded diets, 393
Meishan, 177, 189
Melatonin, 87
Merit, 197
Metabolic
  body weight, 336, 589
  by-products, 362
  faecal losses, 304
  faecal protein, 309
Metabolisable energy, 289, 579
  in feed, 281
  yield from fats, 327
  yield, from fibre, 334
  yield, from protein, 334
Metabolism crate, for balance studies,
  291
Metering device, 432
  for sow feeding, 542
Methallibure, 91
Methane, 283, 289
Methionine, 304
Metritis, 107
Micellar formation, 329
Microclimate, 509, 618
Micro-injection, 239
Micro-organisms, 389
Microbial
  degradation, 306
  dermentation, 385
  synthesis, of vitamins, 368
Microflora of the hind gut, 318
Mid-parent value, 191
Middle White breed, 174, 177
Milk, 8
  by-products, 394
  composition of, 112
  ejection reflex, 95, 117 118, 119, 120,
    140
  from the lactating sow, 289
  production, 114
  protein deposition in, 585
  protein, 607
  secreting cells, 110
  solids yield, 114
  spot, 260
  supply for piglets, potential, 606
  yield, 403, 458, 581, 583, 605, 606
  yield, in relation to litter size and
    parity number, 115

Mill and mix, 470
Miller's Offal, 167
Milling, 470
Mineral
  and vitamin supplements, marketing
    of, 478
  composition of pigs, 358, 360
  deficiencies, 249
  requirements for pigs, 357, 630
Minimal disease, 264
Minimum
  levels of fat, 46, 78, 79
  restraints, in feed formulation, 427
Minnesota, 190
Mites, 260
Mixing and moving pigs, 522
MMA, 107, 273
Model Pig, 596, 597
Model
  for breeding sows, 595, 613
  for growing pigs, 595
Model-builders, 572
Modelling, 40, 567
Monitoring
  for disease, 261
  of pig performance, 550
Morphology of sperm, 100
Mortality, 22, 127, 274, 523
  in live-born piglets, 256
  in neonates, 255
  in relation to weight at birth, 450
Motility of semen, 98
Mould growth, and its inhibition, 391
Muirhead, 279
Mulberry heart disease, 277
Multi-national feed compounding
  companies, 471
Multi-suckling system, 533
Multiple
  characters, 203
  regression analysis, 301
Multiplier herd, 197, 234
Mummified feotuses, 259
Mung beans, 316
Muscle, 49
  area, 22
  damage, 249
  depth, 29, 31, 34
  fibres, white in proportion to red, 18
  quality, 29
  shape, 21
Mutilation, 154, 165, 166

Mycoplasma pneumonia, 261, 264, 269
Mycoplasmal arthritis, 250
Mycotoxins, 251, 259
Myoepithelial cells, 109, 112
Myristic acid, 323
Mystery reproductive syndrome (MRS), 275

N x 6.25, 306
NAS-NRC working party, 379
National
    feed compounders, 472
    pig improvement schemes, 198
    schemes, for breed improvement, 214, 224
    schemes, for performance testing, 215
Swine Improvement Federation, of US, 204
Natural
    immunity, 245, 261
    suckling, 117
    ventilation, 512
    weaning, 115
Neapolitan pig, 168
Negative
    fat growth, 67, 68, 70
    feedback, 374
    feedback, in relation to appetite, 372
    protein growth, 73
Nest
    area, 150
    building, 137
Net energy (NE), 287–9
    versus digestible energy, 419
Neuro-endocrine milk ejection reflex, 95, 117
Neuropeptides, 372
Neutral detergent fibre (NDF), 299
New
    genes, 210
    products in relation to breeding pigs, 488
    technology, 128
Nicotinic acid, 369
Nipple, 108, 202
    drinker, 362
    preference, 139
Nitrogen retention, 321
Nitrogen-free extractives (NFE), 297
Nitrogenous output, in the form of slurry, 306, 519
Non-additive genetic change, 184

Non-protein nitrogen, 340, 420
Non-starch polysaccharide, 289, 386
Norfolk and Suffolk pigs, 169
Normal growth, 78
Nosing, 139
Noxious gases, 511
Nucleus, 168, 228
    herd, 197, 230
Numbering systems, 156
Numbers born, 127, 196, 600
Nursing, 139
Nutrient
    balance, in diets for sows, influence on feed intake, 385
    concentration of diets, 397, 415
    cpntent of diet, 147
    content of feedstuffs, 404
    demand, influence upon appetite, 373, 379
    density of diets, 375
    density, 410
    intake in relation to nutrient need, 376
    need, 130, 586
    requirements for pigs, 333, 405
    specification, 405, 428, 630
    supply from outdoor pig paddocks, 523
    supply, influence on carcass quality, 42
    supply, influence on lean and fat, 52
Nutritional
    deficiencies, 256
    imbalances, 74
    requirement for energy, 580
    requirement for protein, 583
    requirements for pigs, 630
    specification, 410, 413
    values of feed ingredients, 620
Nutritionally limited growth, 79

Oat feed, 300
Oats, 316
Obesity, 372
Objectives for within-population selection, 223
Obligatory losses, of minerals, 360
Ochratoxins, 251
Odour reduction, in slurry, 388
Oestrogen, 81, 87, 89, 12
    from the embryo, 102
    secretion, 113

Oestrous cycle, 84, 88, 89, 116, 134
Oil
    blends, 332
    content, influence on DE, 297
Oils, 377
    added to diets, 301
Oleic acid, 323, 332
Omnivorous feeding, 151
On-farm testing, 225
Open-pen pig house designs, 138
Opiate-like activity, at farrowing, 107
Opioids, 90, 372
Optimization, 571
    of feed supply, 467
    of growth, 618
    of pig performance, 432
Optimum
    performance of sows, 444
    slaughter weight, 62
    water intake, in relation to feed, 363
Organic acids, 386, 389
Osteomalacia, 365
Osteoporosis, 365
Out-crossing, 172, 186
Outdoor pig-keeping systems, 161, 190,
        522, 524, 549
Outflow rate, 370
Ova, 84, 97
    number released, 459, 604
Ovarian follicles, 113
Ovaries, 85, 99
Over-fatness, 463
Oviduct, 99
Ovulation rate, 93, 128, 258
Ovulation, 92, 116, 126, 257
Oxidation, 280
    of amino acids, 305
    of fatty acids, 325
Oxygen, 284
    consumption, 283
Oxytocin, 89, 95, 96, 106, 107, 112,
        121, 139

P2, 15
Packing plants, 6
Paddocks, 523
Palatability, 327, 381
    negative & positive, 391
    of diet, importance of, 466
    of feed, 147
Pale, soft and exudative muscle (PSE),
        15, 21, 157, 183, 189, 252

Palm oil, 301
Palmitoleic acid, 323
Pancreas, 85, 308
Pantothenic acid, 369
Paralysis, 274
Parasites, 260
Parathyroid hormone, 364
Paratyphoid, 268
Parent
    females, 231
    males, 236
    pigs, 231
Parity number, 603
Parma, 169
    ham, 5
Particle
    size, effect upon digestibility, 292
    spread of disease by, 248
Parturition, 84, 106, 137, 447
    hormones of, 107
Parvovirus, 259, 272
Passive
    immunity, 246
    milk removal, 117
    withdrawal, 119, 140
Pasterns, 250
Pasteurella, 259, 270
Pasture grasses, 151
Pathogenic organisms, in gut, 389
Payment schedules, 30
Pedigree
    breeder, 198, 225
    breeding, 197
    breeds, 172
Pellets, 392
Pelvic
    ligaments, 106
    thrusting, 135
Pen space, 515
Penis, 135
Pentosans, 385, 387
Percentage of lean meat, 31, 33, 593
Perforated flooring, 456
Performance
    of breeder/feeder herds, 553
    of feeder herd, 558
    of sow herd, 558
    of UK pig herd, 558
    targets, 562
    test, 182, 197, 212
Perimeter fence, 263

Pesticide residues, 250
pH, 389
  of muscle, 16
Phaseolus, 315
Phases of suckling, 120, 144
Phenotype, 170
Phenotypic variation, 192
Phenylalanine, 304
Phosphorus, 364
  in slurry, 519
Physical injury, 523
  limits to feed intake, 370, 376, 378
Physiological
  control of feed intake, 376
  maturity, 37, 62
  minimum, of lipid, 65
Phytase enzymes, 388
Phytate phosphorus, 365, 388
Phytates, 315
Pietrain, 22, 177, 190
Pig
  breeder, 503
  cycle, 7
  fat, 10
  feed marketing, 469
  housing, 531
  Industry Development Authority, 213
  meat consumption, 8, 10
  meat, throughout the world, 4, 11
  mixing, 522
  Park, 149
  price, 40
  production strategies, 500
  protein, 304
Pig-keepers, in relation to welfare codes,
  163
Pigeon peas, 316
Piglet
  growth, 608
  viability, 463
  weight, 55
Pipeline feeding, 432, 528
Pituitary, 81
Placenta, 106, 107, 138
Placid behaviour, 158
Plateau, of growth rate, 76
Play, 150
Point of inflection, 54
Poison, 315
Poland China, 177
Pollution, by slurry, 519
Polysaccharide, 280

Polyunsaturated fatty acids (PUFA), 8,
  323, 325
Polyunsaturated : saturated fatty acid
  ratio, 9
Population size, 180, 198
Populations of pigs, 195
Porcine
  intestinal adenomatosis (PIA), 268
  reproductive and respiratory
    syndrome (PRRS), 275
  somatotrophin (PST), 81
  stress syndrome (PSS), 17, 23, 157,
    242, 252
Pork and bacon consumption, 11
Positive
  feedback, 375
  selection, 191
Post-pubertal female, 60
Post-slaughter changes in meat, 16
Post-weaning performance, 124, 466
Posterior pituitary gland, 85
Potassium, 360
  in slurry, 519
  requirement, 366
Potato, 290, 315, 409
  raw, 316
Potential
  expression of, 492
  for lean tissue growth, 78
  growth curve, changing the shape of,
    61
  growth, 55, 56, 58, 66, 576
  protein retention, 58
Poultry, 8
  meat consumption, 11
Pre-slaughter treatment, 209
Pre-weaning growth performance, 67
Prediction
  equations, for DE, 296
  of production response, 568
Pregnancy, 65, 84, 101
  failure, 125
  feeding, 444, 449
  gains of fat, 65, 68
  hormones, 110
  influence on growth, 59
  rate, 100
  testing, 125
  weight change, 452
Pregnant
  mare serum (PMS), 92
  sow stalls, 532

Preventative medicine, 246
Price, of feed and carcass, 440
Prices, for pig carcasses, 8
Pricing, 490
Primary joints, 27
Processing and packing plants, for meat, 251
Processors, of pig meat, 42
Product
    development, of pig genotypes, 489
    quality control, 432, 435
Production
    energy requirement for, 281
    nutrient needs for, 405
    parameters, 572
    performance, 554, 558
    trends, 2
Productivity of sows, 128
Products of conception, 105, 584
Profit, 494
    and loss production strategies, 434
    improvement, 501
Progeny test, 173, 182, 197, 211
Progesterone, 87, 90, 102
Progressive selection, 169
Prolactin, 86, 102, 113
Prolificacy, 189, 603
Propionate, 386, 391
Propionic acid, 389
Prostaglandin, 87, 89, 96, 102
Prostaglandin/Progesterone antagonism, 106
Protease inhibitor activity, 314
Proteases, 387
Protection of proteins, 314
Protein, 28, 44, 50
    accretion, 82
    anabolism, 336
    concentrates, competition for, 477
    concentrates, marketing of, 476
    concentration of diets, 401
    content of diet, prediction of, 579
    content of pig feedstuffs, 322
    content of sows, 60
    deamination, 287
    deposition in products of conception, 584
    deposition, in mammary tissue, 349
    deposition, in milk, 349
    deposition, in the uterus, 349
    digestibility, 420
    flux in the intestine, 308

growth, 76
growth, potential for, 56
in body of pig, 591
in piglets, 70
in pigs' milk, 112
limits to lean growth, 402
loss by sows, 607
mass of sows, 598
mass, 36, 58, 338
mass, at maturity, 576
requirements of growing and breeding
    pigs, 333, 348–54, 396
retention, 57, 58, 59, 73, 76, 77, 576,
    581, 584
seeds, 305
sources, 345
synthesis, 582
tissue turnover, 336, 341
value, 342, 344, 345, 398
value, responses to, 404
yield, from diet, 577
Protein feedstuffs, in compounded diets,
    393
Protein-free energy, 578
Protein : energy ratio, 395, 416
Proteinaceous secretions, 312
Proven parents, 168
PSS/PSE, in relation to DNA coding,
    242
PST, 237
Pubertal
    heat, 449
    oestrus, 603
Puberty, 84, 91
    attainment of, 64
Puberty-positive stimuli, 91
Pumping, of pig feed, 363
Pure-bred, 185, 250
    lines, 228
Pure-breeding, 186, 197
    nucleus herds, 191. 213
Pyramidal breeding structures, 216, 233

Quality
    control, 481
    of pig meat, 4
Quarantime
    for breeding stock, 488
    quarters, 262, 520
Queueing, 530

R-PST, 82, 241

Race, for moving pigs, 152
Ractopamine, 82
Radiant heat, 508
Radiation, 504
Radiators, 509
Ranking
    of animals, 145
    of candidates on test, 218
    of herds, for improved traits, 214
Rapeseed meal, 307, 315, 319
Rashers, 25
Rate of passage, 314
Ratio
    diet protein : diet energy, 132
    fat to lean, 79
Ration allowance, standardization of,
    447
Ration scale, 440
    for breeding sows, 461, 464
    for pigs, 442
Rations for pigs, 397
Ratios of nutrients, 411
Raw material
    description, for feed compounding,
    427
    purchase, 471
Recombinant technology (DNA), 82
Records, office, 553
Recovered vegetable oil (RVO), 324,
    330
Rectal
    prolapse, 252
    stricture, 252
Refractory period, 120
Regression
    method, to determine DE, 295
    of offspring on parent, 193
    relationships for carcass analysis, 215
Relative values of nutrients, 414
Relaxin, 106
Releasing hormones, 86
Rendering plants, 251
Repeat, of oestrus, 257, 258
Repetitive activities, 150
Replacement of breeding pigs, 234, 484
Repopulation, 188
    of disease-ridden pig units, 279
Reproduction, nutrient needs for, 405
Reproductive
    capacity, selection for, 224
    characteristics, 171

performance, genetic improvement of,
    209
performance, influence of lactation
    length on, 123
problems, diseases associated with,
    272
tract, 99
Residual milk, 121
Respiration, 284
Respiratory disease, general control of,
    272
Response
    to compounded feeds, 472
    to selection, 179, 207
Restraints in feed formulation, 426–9
Restricted feeding, 39
Retention of protein, 310, 311, 341, 584
Returns, to oestrus, 125
Rice bran, 300
Rickets, 365
Risk, 570
Robertson, 212
Room temperature, 249
Root crops, in compounded diets, 393
Rooting, 149, 153
Routine control, 563
Run-off, 519
Rye, 316

Saccharomyces, 390
Saddleback, 173
Saliva, 308
Salivary carbohydrase, 386
Salmonella, 251, 268
Salt requirement, 366
Saponifiable fats, 329
Saponins, 315
Saturated fatty acids, 8, 324, 329
Scandinavian Landrace breeds, 174, 175
Scarring, 146
Scoring, for sow condition, 461
Scrotum, 156
Secretory cells, 109, 111
Selection, 79
    differential, 193, 194
    for lean growth, 174
    for type, 170
    index, 208
    objectives, 210, 217
    objectives, for reproductive, meat and
        efficiency traits, 205
    of breeding stock for sale, 485

rate, 213
rates, in breeding stock for sale, 486
Selenium, 277
    requirement, 366
Self-acceleratory phase of growth, 55
Self-choice feeding, 130
Self-feed *ad libitum* hoppers, 432, 526
Self-feeder trough, 548
Selling, of breeding stock, 485
Semen, 97, 100
Seminal
    fluid, 96
    oestrogens, 96
Sensitivity analysis, 502, 570
Serotonin, 371
Serum gonadotrophin, 92
Sex, 26
    determination, 243
    effects on optimum production, 437
Sexual
    behaviour, 90
    maturity, 187, 448
Shape, 30, 41, 50
Sheep meat consumption, 8, 11
Shelter, 137, 504
Show-ring, 173
Showers, for humans, 264
Sib testing, 212
Simulation, 567
    models, 618
Single-mix, 428
Sire
    lines, 229
    selection, 193
Sires, values of, 491
Size, 48
    at maturity, 62
Skatole, 26
Skeletal damage, 249
Skeleton, 360
Skjervold, 212
Slatted
    floor, 548
    floor, part and full, 538
    house, 537
    pens for pregnant gilts, 540
Slaughter
    generation, 231
    handling pre-, peri- and post-, 20
    policy for, 246
    weight, 5, 38, 46, 62
Slaughterhouse waste, 250

Sleeping, time spent, 150
Slope-ratio assay, 321
Slurry, 388, 518
    digestion, 388
    output, 519
Small intestine, 310
SMEDI, 273
Smell, 151, 372
Smith, 185, 212
Soaps, 329
Social
    groups, 146
    rank, 147
Sodium, 360
    requirement, 366
Soft
    fat, 14, 329
    tissue, 360, 364
Soiling, 153
Solid floor, 546, 548
Solid-floor house, 535, 537
Solids, from slurry, 518
Somatomedins, 81
Somatostatin, 81
Somatotrophin (STH), 86, 112, 372
Sorghum, 315
Sow
    appetite, 383
    body fatness, 444
    productivity, rate of increase in, 3
    stalls, 147, 149
    tethers, 149
    weight change, 445
Sow : boar ratio, 254
Soya bean
    meal, 307, 319, 409
    raw, 316
Soya oil, 10, 301
Space, 517
    allowance, 147, 247, 514, 515, 518
    availability of, 150
    requirements, summary, 516
Spain, 34
Speciality lines, 233
Sperm, 253
    concentration and viability, 187
    mobility, 100
    production, 100
Sperm-rich semen, 96, 98
Spermatogenesis, 95
Spermatozoa, 90
Spices, 391

Splay-leg, 250
Splitting fat, 14
Sprains, 249
Sprays, 456
Stalls, 157, 166, 249, 529
Standing
  heat, 97, 134
  time spent, 151
Staphylococcus, 273, 276
Starch, 280
  hydrolysis, 387
  polysaccharide, 289
Starvation, 73
Statistics for UK pig herd, 558
Statutory powers, 165
Stearic acid, 14, 323
Stereotypic behaviour, 150, 158, 159
  relationship to feed level and
    restraint, 160
Sterility, 274
Stocking density, 247, 269, 381, 509
  influence on feed intake, 382, 588
Stocksmanship, in relation to welfare,
  163
Stomach, volume of, 370
Storage, for slurry, 518
Straights, 477
Stratification, 172
Straw, 300
  bedding, 547
  yards, 533, 537
Streptococcal meningitis, 275
Streptococcus, 273
Stress, 23, 157, 189, 274
  heat, 153
  reduction of, 161
Structural carbohydrates, 314
Stud books, 168
Stunning, 18
Subclinical disease, 246
Subcutaneous
  fat, 9, 49, 593
  fatty tissue, 460
  tissue loss, 71
Submissive response, 145
Substitution of breeds, 189
Succulence, 9
Suckling, 111, 138
  frequency, 113, 117
  phases of, 139
  pigs, 465
  stimulus, 111, 115, 123

unsuccessful, 140
Sugar beet pulp, 300
Sugars, 280, 372
Sulphur amino acids, 342
Sunburn, 523
Sunflower, 315
Super-nucleus, 231
Super-prolific sows, 210
Supermarkets, 21
Supplementary
  creep feed, 378, 465, 525
  heating, 504
Supplementation of diets with trace
    minerals and vitamins, 359
Supplements and pre-mixes, marketing
  of, 478
Supply schedules for replacement stock,
  492
Support services, 476
Surface, of floor, 506
Surplus fat, 68
Survivability, 256, 601
Suspended flooring, 543, 545
Sustainability, 4
Sweat, 305
Swine
  dysentery, 267
  influenza, 272
  vesicular disease, 261
Synchronization, 97
Synthesis of milk, 111, 115
Synthetic
  lines, 228
  lysine, 344

T2 toxins, 251
Tactical
  decisions, 570
  management, 563
Tag, 156
Tail, clipping of, 155
Tail-biting, 147, 155
Tail-docking, 166, 256
Taint, 26, 156
Tallow, 10, 301, 323, 331
  blends, 332
  hydrogenated, 301
Tallow : soya ratio, 332
Tamworth, 173, 177
Tannins, 315
Target
  fat, 63

nutrient specification, 406
setting, 571
Targets, 557
Taste, 9, 151, 372, 391
Tattoo, 156
Teat, 108
order, 139
sphincter, 119
Teat-seeking, 138
Technical support
for compounded feeds, 474
for vitamin and mineral marketing,
479
of improved breeding pigs, 491
Teeth clipping, 155
Temperature, 505, 517
control, 147
fluctuation and transition, 513
influence upon appetite, 381
requirements, summary, 516
Tenderness, 19
Terminal
ileum, 307, 310
sires, 229
Test
ingredient, 293
regime, 78
Testes, 85
Testicles, 156
Testicular steroid, 95
Testing regimes, 217
Tethers, for sows, 157, 166, 249, 529
Texture, 9
Thermoregulation, 361
Thin sows, 458, 599
Threats, 145
Three-times mating, 98
Threonine, 304
Thumping/barking, 269
Thwarting, 138
Thyroid hormone, 81
Thyrotrophic hormone (TH), 86
Time-based feeding scales, 221, 441
Tissue, composition of, 48
Tongs, for stunning, 18
Tooth-clipping, 166
Top-crossing sires, 20, 187, 231, 236
Townsend, 167
Toxaemia, 276
Toxic
factors, 314, 385, 409
products, 361

Toxicity, 154
Trace
elements, 357
mineral requirement, 367
Trading relationships, 473
Traits, of low heritability, 182
Transgenic pigs, 239
Transmissible gastroenteritis (TGE),
267
Transportation of pigs, 23, 35, 38, 91,
157
Trees and shrubs, 151
Treponema, 267
Trickle feed, 530
Triticale, 316
Trough
feeding, 529
length, per pig, 516
space, 147
Trucks, 262
True digestibility, 307
Truffles, 372
Tryglycerides, 324
Trypsin inhibitor, 315
Tryptophan, 26, 304
Tull, 167
Tyrosine, 304

Udder, 111
Ulcers, 252
Ultrasonic techniques, 215
Umbilical cord, 138
Unit costs of nutrients, 412
Unpalatability, 132
Unsaturated fatty acids, 323, 324
Unsaturated : saturated fatty acid ratio,
14, 332
Urea, 304
Urinary nitrogen, 283
Urine, 311, 362
as a source of energy loss, 283
Utilization of digested protein, 341
Uterine
fluids, 104, 122
horns, 101
involution, 116, 122
Uterus, 85, 99
Utilizability
of amino acids, 306, 584
of digested amino acids, 321
of ileal digested amino acids, 422
of minerals, 360

Utilizable protein, 420

Vaccination, 261
    against Aujeszky's, 274
Vaccines, general availability, 278
Vagina, 99
Valine, 304
Van Soest, 299
Variability in feed provision and sow
    condition, 523
Variable costs, 39, 558, 560, 615
Variance, 178
Variation 42, 178
    environmental, 170
    genotypic, 170
    in backfat depth, 447
    phenotypic, 170
Vegetable
    fat, 8
    oil, 323, 331
    oils in fat blends, 332
    protein, 305
Vegetarian diet, 129
Vegetation, 137
Ventilation, 256, 511, 517
    rate, 456, 509
    requirements, summary, 516
Veterinary treatments and procedures,
    551
Visual
    appraisal, 552
    assessment, 34
Vitamin
    A (retinol), 369
    B1 (thiamine), 369
    B12 (cyanocobalamin), 369
    B2 (riboflavin), 369
    B6 (pyridoxine), 369
    C (ascorbic acid), 368, 369
    D (cholecalciferol), 364, 369
    deficiencies, 249, 367
    E (alpha tocopherol), 277, 368, 369
    H (biotin), 368, 369
    K (menapthone), 369
    potency, 359
    requirements, 357, 367, 369, 630
Vocalization, 134, 140, 144
Volatile fatty acids, 280, 287, 334, 420
Voltage, for stunning, 17
Voluntary
    feed intake, 370, 383
    stalls, 534

Vomitoxins, 251
Vulva, 99
Vulva-biting, 257
Vulval discharges, 259

Wallows, 456, 508
Washing facilities, 535
Waste, 361, 516
    output of, 519
Water, 28, 50
    allowances for pigs, 363
    as an appetite stimulator, 363
    content of lean, 52
    content of pig, 361
    content of pigs diets, 394
    depletion, 361
    gains, 68
    in body of pig, 591
    in piglets, 70
    in pigs' milk, 112
    in sows, 598
    point/trough, 362
    provision, 526
    ratio with lipid, 52
    ratio with protein, 52
    requirements, 357, 361
    shortage, 362
    sprays, 508
    supply 147
    usage by pigs, 362
Water-holding capacity, of meat, 15
Water-sprays, 17
Water : dry matter ratio, 362
Water : protein ratio, 592
Weaner costs, 40, 496
Weaning, 84, 116, 122, 124
    age, 125, 166
    time of, 115
    weight, 123, 458
Weaning to conception interval, 84, 125,
    127, 448, 459, 608, 610
Weaning to oestrus interval, 454, 455,
    608
Webb, 212
Weight, 48
    and time, 52
    at birth, 449
    at sale, 495
    at slaughter, 435
    at weaning, 551, 607
    gain of piglets, potential, 606
    gains, of weaners, 71

loss during lactation, 68, 457
loss, energy value of, 68
of litter at weaning, 453
stasis, in piglets, 69
stasis, in sows, 68
Weight-based feeding scales, 441
Weighted index, 182, 203
Welfare, 162, 256, 618
  code, 165
  recommendations, 165
Well-being of animals, 130
Wet
  feed, 528
  feeding systems, 363, 417, 527
  mixing, 527
Wet-mix pumping, 363
Wetting of floors, 508
Wheat, 316, 319
  meal, 307
  offals, 319

White hybrid, 20
Wild boar, 168
Wiltshire cure, 211
Winter keep, 168
Winters, 185
Wire mesh flooring, 544
Within-population selection, 209
Wood-Gush, 130
Work heat, 286
Worm infestations, 260
Wounding, 155

Yards for sows, 529
Yield of milk, 114
Yorkshire breed, 169, 173

Zearalenone, 251
Zinc, 364
  requirement, 366
Zoo, 233